D1447869

T5-COZ-235

STATIC AND DYNAMIC ANALYSIS OF STRUCTURES
with An Emphasis on Mechanics and Computer Matrix Methods

# SOLID MECHANICS AND ITS APPLICATIONS
## Volume 6

*Series Editor:* **G.M.L. GLADWELL**
*Solid Mechanics Division, Faculty of Engineering*
*University of Waterloo*
*Waterloo, Ontario, Canada N2L 3G1*

*Aims and Scope of the Series*

The fundamental questions arising in mechanics are: *Why?*, *How?*, and *How much?*
The aim of this series is to provide lucid accounts written by authoritative researchers giving vision and insight in answering these questions on the subject of
mechanics as it relates to solids.

The scope of the series covers the entire spectrum of solid mechanics. Thus it
includes the foundation of mechanics; variational formulations; computational
mechanics; statics, kinematics and dynamics of rigid and elastic bodies; vibrations
of solids and structures; dynamical systems and chaos; the theories of elasticity,
plasticity and viscoelasticity; composite materials; rods, beams, shells and
membranes; structural control and stability; soils, rocks and geomechanics;
fracture; tribology; experimental mechanics; biomechanics and machine design.

The median level of presentation is the first year graduate student. Some texts are
monographs defining the current state of the field; others are accessible to final
year undergraduates; but essentially the emphasis is on readability and clarity.

*For a list of related mechanics titles, see final pages.*

# Static and Dynamic Analysis of Structures

with An Emphasis on
Mechanics and Computer Matrix Methods

by

JAMES F. DOYLE
*Aeronautics and Astronautics Department,*
*Purdue University, West Lafayette,*
*Indiana, U.S.A.*

KLUWER ACADEMIC PUBLISHERS
DORDRECHT / BOSTON / LONDON

Library of Congress Cataloging-in-Publication Data

```
Doyle, James F., 1951-
   Static and dynamic analysis of structures : with an emphasis on
mechanics & matrix methods / James F. Doyle.
      p.   cm. -- (Solid mechanics and its applications : v. 6)
   ISBN 0-7923-1124-8 (alk. paper)
   1. Structural analysis (Engineering)--Matrix methods--Data
processing.   I. Title.   II. Series.
   TA647.D69   1991
   624--dc20                                              91-23609
                                                          CIP
```

ISBN 0–7923–1124–8

---

Published by Kluwer Academic Publishers,
P.O. Box 17, 3300 AA Dordrecht, The Netherlands.

Kluwer Academic Publishers incorporates
the publishing programmes of
D. Reidel, Martinus Nijhoff, Dr W. Junk and MTP Press.

Sold and distributed in the U.S.A. and Canada
by Kluwer Academic Publishers,
101 Philip Drive, Norwell, MA 02061, U.S.A.

In all other countries, sold and distributed
by Kluwer Academic Publishers Group,
P.O. Box 322, 3300 AH Dordrecht, The Netherlands.

*Printed on acid-free paper*

Printed in the Netherlands

To Tom

*[signature]*

*To Linda,*
*without whom*
*this*
*and much else*
*would not be possible.*

# Contents

# Preface

This book is concerned with the static and dynamic analysis of structures. Specifically, it uses the stiffness formulated matrix methods for use on computers to tackle some of the fundamental problems facing engineers in structural mechanics. This is done by covering the *Mechanics of Structures*, its rephrasing in terms of the *Matrix Methods*, and then their *Computational* implementation, all within a cohesive setting. Although this book is designed primarily as a text for use at the upper-undergraduate and beginning graduate level, many practicing structural engineers will find it useful as a reference and self-study guide.

Several dozen books on structural mechanics and as many on matrix methods are currently available. A natural question to ask is why another text? An odd development has occurred in engineering in recent years that can serve as a backdrop to why this book was written. With the widespread availability and use of computers, today's engineers have on their desk tops an analysis capability undreamt of by previous generations. However, the ever increasing quality and range of capabilities of commercially available software packages has divided the engineering profession into two groups: a small group of specialist program writers that know the *ins and outs* of the coding, algorithms, and solution strategies; and a much larger group of practicing engineers who use the programs. It is possible for this latter group to use this enormous power without really knowing anything of its source. Therein, in my opinion, lies the potential danger — the engineer is seduced by the power, the litany of capabilities, the seeming ease of use, and forgets the old computer adage: garbage in, garbage out, or its more recent sinister form: garbage in, gospel out. We use, and we should use, commercial packages when they are available. But to make safe, efficient, and intelligent use of them we need to have some idea of their inner workings as well as the mechanics foundations on which they are built.

It would seem reasonable, therefore, that the following abilities must be considered a minimum for any structural engineer:

- Know how to idealize structures in a sensible manner, and know when (and why) a structure is beyond the capabilities of a particular computer program.

- Know how to use consistency and cross checks to validate the computer output.

Although these are the *input/output* brackets to the program, the form they take is dictated by the internal modeling. Hence, it is impossible to avoid some consideration of the inner workings of the programs. That is, to be an intelligent user of these programs requires some appreciation of the full range of assumptions and procedures on which they are based. Without doubt, an understanding of the mechanics principles is essential, but it is not sufficient, because these principles are transformed in subtle ways when converted into algorithms and code.

With the foregoing in mind, this book sets as its goal the treatment of structural dynamics starting with the basic mechanics principles and going all the way to their implementation on digital computers. I believe that only by studying this in its complete extent do the unique difficulties of computational mechanics manifest themselves. I have made an effort to ensure that this book is not just a collection of disparate topics — I sought a unity that would give meaning to the pieces as part of a coherent whole. This is achieved by covering completely the analysis of 3-D frame structures and using it as a paradigm for treating other structural systems. In the same vein, rather than discuss particular commercial packages, I have developed STADYN: a complete (but lean) program to perform each of the standard procedures used in commercial programs. Each module in this program is reasonably complete in itself, and all were written with the sole aim of clarity plus a modicum of efficiency, compactness and elegance. (I have included complete source code listings to these modules in an appendix; their electronic form can be obtained through *ikayex* SOFTWARE TOOLS, 615 ELSTON ROAD, LAFAYETTE, INDIANA 47905, USA.)

This book takes a bootstrapping approach to developing the material. That is, the elemental blocks are developed on first principles; these are then combined to model more complicated problems. It is only then that the general principles are established. For example, Chapter 2 looks at the simplest of structures, the rod. The analysis is developed fully from the governing differential equations all the way to the matrix formulation. Chapter 3 discusses beam structures and develops the analysis in nearly identical fashion. The following chapter then shows how these elementary structures can be combined to form complicated 3-D structures. The twin concepts of compatibility and equilibrium are emphasized throughout these three chapters. Chapter 5 refines the concept of equilibrium in the process of discussing structural stability. Chapter 6 introduces the energy methods, and shows how the concept of the stationary potential energy (coupled with the Ritz method) can be used to recover the results of the previous chapters in a unified and consistent fashion. The first part of the book concludes with an introduction to the computational aspects of the matrix methods in Chapter 7.

The second part of the book begins with a summary of the vibration of a simple spring-mass system. Chapters 9,10&11 are the dynamic analogs of Chapters 2,3&4; they achieve the matrix formulation for the dynamic analysis of 3-D continuous structures. Chapter 11 introduces modal analysis as a means of understanding the complicated dynamics on a structural or global level. These results are put into a unified form in Chapter 12 using the energy concepts. Again, the Ritz method emerges as a powerful technique for converting continuous systems into discrete matrix form. The computational methods for direct integration and modal analysis are developed in Chapter 13.

Appendices dealing with matrix algebra, spectral analysis, and computer source code round out the coverage of the material. Admittedly, many problems (such as those associated with thermal loading, non-linear material and non-linear geometric effects) have been left out, even though most of them are treatable by matrix methods. Consequently, an effort is made to supplement each chapter with a collection of pertinent problems that indicate extensions of the theory and the applications. These problems, combined with selected references to relevant literature, can form the basis for further study.

A book like this is impossible to complete without the help of many people, but it is equally impossible to properly acknowledge them all individually. However, I would like to single out: Graham Gladwell for his very many helpful suggestions and his understanding of what I was trying to achieve; Albert Danial, Shiv Joshi, Tim Norman, and Steve Rizzi for their constructive criticisms of the almost final version of the manuscript; and all my former students who suffered through the early drafts, their feedback was invaluable. The errors and inaccuracies remaining in this book are purely my own doing; I would deem it a kind service to be informed of them.

I used a combination of LaTeX and PostScript to typeset this manuscript; I thank all those people who made these wonderful systems available for the desktop computer.

May 1991                                                                 James F. Doyle

*"I am quite convinced that the central question is how to make computers more usable, how to make their software more comprehensible, and how to avoid the dangers imposed by the complexities of standard software in the current generation."*

C. A. R. HOARE

# Chapter 1

# Background and Scope

Structural mechanics is concerned with the behavior of solid objects (or assemblies of them) under the action of applied loads. The behavior is usually described by idealized models from which the internal forces and displacements are found. We present, in the following chapters, the stiffness formulated matrix methods as models to tackle some of the fundamental problems facing engineers in structural mechanics.

It is a happy coincidence that matrices are also a convenient means for carrying out calculations on a computer and therefore it is only natural that an intimate relation has evolved between structural analysis and computers. The two are linked and modern courses must reflect that. We will attempt to cover the *Mechanics of Structures*, its rephrasing in terms of the *Matrix Methods*, and then the *Computational* implementation, all within a cohesive setting. This chapter sets out to describe the context within which we will do this. References [11, 23, 33, 35] are very useful sources for additional background material.

## 1.1  Structural Analysis

Physical systems are usually complex and very difficult to analyze. Moreover, they often consist of a large number of components nominally acting as a single entity. (Just think of the number of separate components in an automobile!) A rational approach to the analysis of such systems starts by identifying the various components and determining their physical properties. The analyst is then in a position to construct a mathematical model to describe the behavior of each component; the goal is the simplest model consistent with the essential features of the actual component. By a process of synthesis, a model emerges which represents an idealization of the actual physical system. Notwithstanding the pitfalls, history has shown this to be an effective strategy for solving complex physical problems. This is the approach we follow.

The primary function of any structure is to support and transfer externally applied loads. It is the task of structural analysis to determine two main quantities arising as the structure performs its role: internal loads, and changes of shape (called *deforma-*

1

*tions*). It is necessary to determine the first in order to know whether the structure is capable of withstanding the applied loads. The second must be determined to assure that excessive displacements do not occur. To understand the construction of an analytical structural model, therefore, requires an understanding of the function of the particular structure as well as the requirements of the analysis. In other words, if the structure does not experience dynamic loads then a dynamic model need not be developed. Similarly, if a stability analysis is of interest then this capability must be built into the model from the beginning. It is not surprising that structural models abound; we will attempt to keep them to a minimum by stressing the underlining principles of model building. We will, however, set for ourselves the tasks of developing models to describe the static, stability, and dynamic responses of structures.

The modeling of the dynamic response of structures introduces many additional considerations probably not anticipated from a static analysis. It is therefore worth our while at this juncture to say a few words about structural dynamics. The subject of rigid–body dynamics treats the physical objects as bodies that undergo motion without any change of shape. This has many applications: the movement of machine parts; the flight of an aircraft or space vehicle; the motion of the earth and the planets. In many instances, however, the primary concern is dynamic response involving changes of shape. This is particularly so in the design of structures and structural frames as encountered in automobiles, ships, aircraft, space vehicles, offshore platforms, buildings, and bridges.

Dynamic response involving deformations is usually oscillatory in nature; the structure vibrates about a configuration of stable equilibrium. For example, suppose a building structure is in a state of static equilibrium under the gravity loads acting on it. When subjected to wind, the structure will oscillate about this position of static equilibrium. An airplane provides an example of oscillatory motion about an equilibrium configuration that involves rigid–body motion. When in flight, the whole system moves as a rigid body but is also subjected to oscillatory motion due to engine and aerodynamic loads.

The analysis of vibration response is of considerable importance in the design of structures that may be subjected to dynamic disturbances. Under certain situations vibrations may cause large displacements and severe stresses in the structure. This may happen when the frequency of the exciting force coincides with a natural frequency of the structure and *resonance* ensues. A related problem is that fluctuating stresses, even of moderate intensity, may cause material failure through fatigue and wear. Also, the transmission of vibrations to connected structures may lead to undesirable consequences: delicate instruments may malfunction or human occupants may suffer discomfort.

With the increasing use being made of lightweight, high–strength materials, structures today are more susceptible than ever before to critical vibrations. Modern buildings and bridges are lighter, more flexible, and are made of materials that provide much lower energy dissipation; all of these may contribute to more intense vibration responses.  Dynamic analysis of structures is therefore important for modern structures, and likely to become even more so.

## 1.2   Types of Structures Considered

Structures that can be satisfactorily idealized as line elements (as shown in Figure 1.1) are called *frame* or *skeletal* structures.  All the structures analyzed in later chapters are of this type.  Usually their members are assumed to be connected either by frictionless pins or by rigid joints, and are idealized as one of the following six generic types:

| Framework type | Joint type | Loads |
|----------------|------------|-------|
| Beams | Rigid/Pinned | Transverse |
| Plane trusses | Pinned | In-plane |
| Space trusses | Pinned | Any direction |
| Plane frames | Rigid/Pinned | In-plane |
| Grids | Rigid | Normal to plane |
| Space frames | Rigid/Pinned | Any direction |

The individual members of frames are modeled in terms of the following three elemental forms: Rod, Shaft, and Beam.  The difference between these is not so much their geometric shape but rather the types of loading they support and the type of resulting deformation.  In this book

- Rods support only axial loads (tensile or compressive) and the resulting deformation is only along the length.
- Shafts support only a torque acting along its length resulting in only axial twist.
- Beams are designed to carry both moments and transverse loads.

Fundamental to the description of 3-D frames is the assumption that the response of a general member is a simple superposition only of the above three actions.  As a result, these actions can be developed separately and only as a final step need they be combined.

A *truss* consists of a collection of arbitrarily oriented rod members that are interconnected at pinned joints.  They are usually loaded only at their joints and (because the joints cannot transmit bending moment) must be triangulated to avoid collapse. Plane trusses are loaded in their own plane, whereas the joints of a space truss can be loaded from any direction.

A *frame* structure, on the other hand, is one that consists of beam members which are connected rigidly or by pins at the joints.  The members of a frame can support bending (in any direction) as well as axial loads, and at the rigid joints the relative positions of the members remain unchanged after deformation.  Rigidly jointed frames

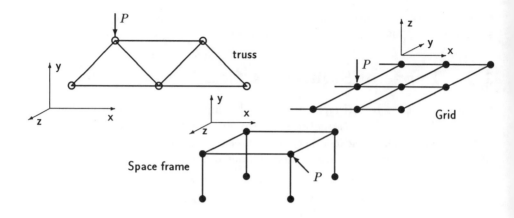

**Figure 1.1**: Types of structural frameworks.

are often loaded along their members as well as at their joints. Plane frames, like plane trusses, are loaded only in their own plane. In contrast, *grids* are always loaded normal to the plane of the structure. Space frames can be loaded in any plane. The space frame is the most complicated type of jointed framework. Each member can undergo axial deformation, torsional deformation, and flexural deformation (in two planes). Its supports may be fixed, pinned, elastic, or there may be roller supports.

We will concentrate on the space frame because all other types of jointed frameworks are special cases that can be obtained by reduction from it. Further, it is assumed throughout most of the subsequent discussions that the structures have prismatic members; that is, each member has a straight axis and uniform cross-section throughout its length. The non-prismatic case is developed as part of the exercises.

## 1.3   Mechanics of Structures

There are three important concepts in the mechanics of structures, namely; *Deformation, Equilibrium,* and *Constitutive Behavior.* One of the powerful features of the stiffness approach to structures problems is that these concepts get combined into one compact, efficient form; however, we should not lose sight of the three basic concepts, and throughout the following chapters we emphasize this by basing the derivation of the important relations directly on them.

### Deformation of Structures and Compatibility

Structures support applied loads by developing internal stresses within their members. These stresses, in turn, cause the structure to undergo deformations (or small changes in shape) and, as a consequence, points within the structure are displaced to new positions. The principle of *compatibility* is assumed to apply to all of the structures encountered in this book. That is, it is assumed that if a joint of a structure moves, then the ends of the members connected to that joint move by the same amount, consistent with the nature of the connection. For a pin-jointed truss, compatibility means that the ends of the members meeting at a joint undergo equal translation. For the rigidly jointed frame, in addition to equal translation, there must be equality of the rotations of the ends of the members meeting at the joint.

Small deflection theory assumes that the analysis of the loaded structure can be safely based upon the unloaded geometry. Fortunately, for most engineering structures the change in geometry under loading is a second order effect that can usually be ignored. Large deformations sometimes occur in flexible structures such as suspension bridges, slender arches, and some lightweight aerospace structures. In the following, the deformation description is based exclusively upon small deformation theory. There is one case, however, where we concern ourselves with the deformed geometry; this is in the analysis of structural stability where we base equilibrium on the loaded geometry.

## Equilibrium

Loads can be regarded as being either *internal* or *external*. External loads consist of applied forces and moments and the corresponding reactions. The applied loads usually have preset values, whereas the reactions assume values that will maintain the equilibrium of the structure. Internal forces are the forces generated within the structure in response to the applied loads. We will sometimes refer to them also as *member loads*. The applied loads may be concentrated forces, distributed loads, or couples.

There are two types of equilibrium encountered in this book: static and dynamic. To illustrate these two concepts, consider a single mass point under the action of a system of forces. Newton's second law gives

$$\sum \vec{F} = m\ddot{\vec{u}}$$

This can be read as saying that the resulting forces $(\sum \vec{F})$ cause the mass $(m)$ to accelerate by the amount $(\ddot{\vec{u}})$. If there are no resulting accelerations, then

$$\sum \vec{F} = 0$$

and we refer to this as static equilibrium. D'Alembert's principle rewrites Newton's law as

$$\sum \vec{F} - m\ddot{\vec{u}} = 0$$

which can be read as saying that the system is in (dynamic) equilibrium under the action of the forces $\sum \vec{F}$ and $-m\ddot{\vec{u}}$. Thus the term $-m\ddot{\vec{u}}$ is treated as a force (called the *inertia* force) and the above equation resembles the condition for static equilibrium.

Although it can be argued that static equilibrium is a special case of dynamic equilibrium, the nature of the inertia force is significantly different from the usual applied loads that the approaches and methodologies for solving both problems are very different. They are so different in fact that they must be treated as two separate subjects. When discussing a static analysis, we will draw the further distinction between *stable* and *unstable* equilibrium configurations.

## Constitutive Behavior

We shall restrict ourselves to materials that are *linearly elastic*. By elastic we mean that the stress-strain curve is the same for both loading and unloading. The restriction to linear behavior allows us to use the very important concept of superposition. Fortunately, most materials of interest in structural applications are modeled adequately by this and so we will always assume that the materials behave according to Hooke's law. For example, the one-dimensional stress-strain behavior is given by

$$\sigma = E\epsilon \qquad \text{or} \qquad F = \frac{EA}{L}u$$

The second form is that of the force-deflection relation.

There is one exception to this. The dynamic response of structures usually exhibit some type of dissipation of energy. (This is referred to as *damping*.) Since the amount of damping is usually not significant, and since (for structures) it is difficult to get precise experimental values, we will generally describe the dissipation using a linear viscoelastic model. In so doing, we are allowing the material behavior to be time dependent.

The principle of *superposition* is one of the most important concepts in structural analysis. In general terms the principle states that the effects produced by several causes can be obtained by combining the effects due the individual causes. It may be used whenever linear relations exist. This occurs whenever the following requirements are satisfied: (1) the displacements are small; (2) the material obeys Hooke's law; and (3) there is no interaction between flexural and axial effects in the members. This principle is fundamental to the stiffness method of analysis and therefore it will always be assumed that the structure being analyzed meets the stated requirements.

# 1.4   Degrees of Freedom

In an elastic structure the displacements vary continuously; such structures have an infinite number of independent displacement components. The essence of the stiffness method is the consideration of displacements and forces only at selected points, called *nodes*. Nodes are located at joints, load points, and section discontinuities of the structure. (In dynamic analyses, additional nodes are used in modeling the inertia distribution.) The analysis then reduces to determining the nodal displacements caused by the applied loading. Obtaining the responses at points other than nodes is then properly viewed as part of the post-processing operation. Thus an essential part of the description is determining the minimum number of unknowns necessary for an adequate characterization of the structural response.

The number of possible displacement components at each node is known as the *nodal degree of freedom* (DoF); the nodal degree of freedom for different framework types is shown in the following table:

| Structure type | Nodal loads | Nodal displacements | DoF |
|---|---|---|---|
| Rod | $\{P_x\}$ | $\{u\}$ | 1 |
| Plane truss | $\{P_x, P_y\}$ | $\{u, v\}$ | 2 |
| Space truss | $\{P_x, P_y, P_z\}$ | $\{u, v, w\}$ | 3 |
| Beam | $\{P_y, T_z\}$ | $\{v, \phi_z\}$ | 2 |
| Plane frame | $\{P_x, P_y, T_z\}$ | $\{u, v, \phi_z\}$ | 3 |
| Grid | $\{P_z, T_x, T_y\}$ | $\{w, \phi_x, \phi_y\}$ | 3 |
| Space frame | $\{P_x, P_y, P_z, T_x, T_y, T_z\}$ | $\{u, v, w, \phi_x, \phi_y, \phi_z\}$ | 6 |

where

$P_x, P_y, P_z$     are external force components,
$T_x, T_y, T_z$     are external moment (torque) components,
$u, v, w$         are global linear displacement components,
$\phi_x, \phi_y, \phi_z$     are global rotational displacement components.

The degree of freedom of a structure is the number of displacement components to be found during the analysis. Finding these displacements by the stiffness method involves the solution of a system of equations relating the unknown nodal displacements to the known applied loads. We will distinguish between *unconstrained* (i.e., unknown) and *constrained* (i.e., known in some way) degrees of freedom. The number of equations involved is equal to the total number of unconstrained degrees of freedom of the structural system.

The loads and displacements are conveniently stored in one dimensional arrays known as the *nodal load vector* $\{P\}$ and *nodal displacement vector* $\{u\}$, respectively. The number of entries in each vector is equal to the nodal degree of freedom of the system as shown above. When these vectors have the entries associated with the constrained degrees of freedom removed they are often referred to as *generalized* force and displacement vectors, respectively. In our description the loads will be thought of as being "applied," while the displacements are thought of as "resulting." This distinction may seem pointless since we can obviously apply a known displacement to a structure and seek the resulting force. But the distinction is actually very useful in the stiffness method because we will show that all structural problems can be phrased as

$$[K]\{u\} + [C]\{\dot{u}\} + [M]\{\ddot{u}\} = \{P\}$$

where $[K]$ is the stiffness matrix, $[M]$ the mass matrix, and $[C]$ is the damping matrix. In fact, we will even rephrase the imposed displacement problem as an applied force situation.

## 1.5   Time Varying Loads

Motions involving deformation are caused by time varying forces or dynamic disturbances. Time varying loads may, for example, be induced by rotating machinery, wind, water waves, or a blast. A dynamic disturbance may result from an earthquake during which the motion of the ground is transmitted to the supported structure.

### Periodic and Transient Loads

Rotating machines that are not fully balanced about the center of rotation will give rise to excitation forces that vary with time. For example, a motor that has an eccentric mass $m_0$ attached to it at a distance $e$ from the center of rotation experiences a centrifugal force. This in turn is transferred to the bearings which experience force components varying in time as

$$F(t) = em_0\omega^2 \sin\omega t$$

where $\omega \, r/s$ is the angular velocity of the motor. This behavior is shown in Figure 1.2. Obviously, these unbalanced forces are also transmitted to any supporting structures and therefore they too will experience the same sort of time variation.

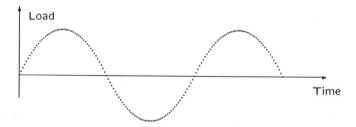

**Figure 1.2**: Periodic loading due to eccentric mass.

A dynamic load of considerable interest in the design of certain structures is that due to a blast of air striking the structure. The blast or shock wave is usually caused by the detonation of an explosive that results in the rapid release of a large amount of energy. The peak overpressure in the shock front decreases quite rapidly as the shock wave propagates outward from the center of explosion. The shock wave arriving at a structure will thus depend on both the distance from the center of the explosion and the strength of the explosive. The overpressure also rapidly decreases behind the front, and may, in fact, become negative.

A blast load can be represented by a pressure wave in which the pressure rises very rapidly and then drops off fairly rapidly as shown in Figure 1.2. Empirical equations are available for estimating the peak overpressure and its duration.

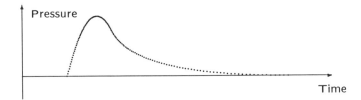

**Figure 1.3**: Pressure loading due to a blast.

A dynamic load that varies in magnitude with time, and that repeats itself at regular intervals, is called a *periodic* load. The harmonic load caused by an unbalanced rotating machine is an example of a periodic load. Loads that do not seem to repeat are called non-periodic loads. Such a load may be of a comparatively long duration, in which case the response is usually modeled as being static or quasi-static. Non-periodic loads may also be of short duration and are then called *transient* or *impulsive* loads; an explosive blast striking a building is an example.

We will investigate the response of structures to periodic and transient loads. Actually, by use of *spectral analysis* we will show that a transient signal can be represented as a collection of periodic signals of different period. (In other words, it can be represented by a spectrum of frequency components.)

## Random and Non-deterministic Loads

Ground motions resulting from earthquakes can be one of the most severe and disastrous dynamic disturbances that affect human–made structures. Earthquakes result from a fracture in the earth's crust and the consequent release of the elastic strain energy stored in the rock. This gives rise to elastic waves which propagate outward from the source fault and eventually arrive at the earth's surface. The resulting wave motion is very complex as seen in Figure 1.4.

The effect of such a motion on the supported structure is assessed by obtaining measurements of the time histories of ground accelerations by use of special measuring

instruments called *seismometers* (or seismographs), and then analyzing the structure
for the recorded motion. Such analytical studies play an important role in the design
of structures expected to undergo seismic vibrations.

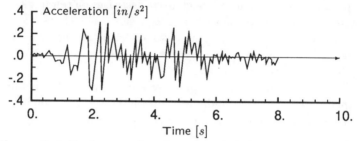

**Figure 1.4**: Surface accelerations recorded during an earthquake.

Structures subjected to wind experience aerodynamic forces which depend on the
wind velocity, the wind profile along the height of the structure, and the characteristics
of the structure. Winds close to the surface of the earth are affected by turbulence and
hence vary with time. The response of the structure is thus a dynamic phenomenon.
For the purpose of design, wind forces are often converted into equivalent static forces.
This approach, while reasonable for low rise, comparatively stiff structures, may not
be appropriate for structures that are tall, light, flexible, and possess low damping
resistance.

The design wind is usually obtained by a statistical analysis of the recorded data on
hourly mean winds. The mean wind speed, the wind profile, and the wind turbulence
together constitute the input data for a dynamic analysis for wind. It is evident
that the effect of the wind cannot be represented by a set of forces that are definite
functions of time, since the wind loads are known only in a statistical sense.

A dynamic load that varies in a highly irregular fashion with time is referred to
as a *random* load. The ground acceleration resulting from an earthquake provides an
example of a random disturbance. Loads that can be specified as definite functions
of time, irrespective of whether the time variation is regular or irregular, are called
deterministic loads, and the analysis of a structure for the effect of such loads is
deterministic analysis. Certain types of loads cannot be specified as definite functions
of time because of the inherent uncertainty in their magnitude and the form of their
variation with time. Such loads are known in a statistical sense only and are described
through certain statistical parameters. These loads are known as *non-deterministic*
loads. The analysis of a structure under these loads yields response values that
are themselves defined only in terms of statistical parameters. Such an analysis is
therefore known as non-deterministic analysis.

Wind loads are quite obviously non-deterministic. Earthquake loads are non-
deterministic also because the magnitude and frequency distribution of an acceleration
record for a possible future earthquake cannot be predicted with certainty, but can
be estimated only in a probabilistic sense. We will deal only with deterministic loads.

## 1.6   Computers and Algorithms

Before a task is written in a computer language, the programmer must design the *algorithm* on which the program is based. An algorithm is a procedure for arriving at the required results with use of the available data. It is usually designed in what is called a *top-down* fashion via their step-wise refinement. First, a general outline of the solution is established; then it is refined in several stages. Subsequently, it is coded in a particular computer language. Algorithms operate on data, both initially given and those obtained during computation, hence the make-up of these data is an integral part of formulating the algorithm.

Two widely used notations for algorithm presentation are *flowcharts* and *pseudo-code*. A flowchart is a graphical tool, whereas pseudo-code is a textual description of the algorithm. Pseudo-code is used exclusively in this book to present the algorithms because it lends itself more conveniently to comment and elaboration. It is worth keeping in mind that algorithms should be much easier to read than program codes; they serve to communicate with people whereas code serves to give orders to computers. The same basic control structures that are used in programming languages are also used in pseudo-code. The three basic ones are: the sequence (*Begin-End*), the decision (*If-Then-Else*), and the loop (*Do-EndDo*). The use of these is called *structured programming*. An additional control construct that is very useful is the jump (*GoTo*).

We present a number of algorithms throughout this book. This is done in a fashion which will suggest how they might be coded, although we do not give highly detailed algorithms — this would tend to obscure the basic flow with minutiae. However, for those readers who are interested in the coding aspects, we have also provided the complete source code to the program STADYN. This is a complete (but lean) program to perform each of the standard procedures used in commercial programs and covered in this book. Each module in this program is reasonably complete in itself, and all were written with the sole aim of clarity plus a modicum of efficiency, compactness, and elegance. The modules are written in FORTRAN because it is the language most engineers choose when writing large scale programs. It is also a highly portable language and the programs listed have been compiled and run successfully on mainframe, mini- and micro- computers alike.

We estimate the performance of the algorithms on a hypothetical benchmark machine and problem. The machine operates at 1 MFlops (one million floating point operations per second). Small, medium, and large problems refer to systems of equations with 100, 1000, and 10000 unknowns, respectively.

## 1.7   Systems of Units

Two different systems of units are in common use in engineering. One is a metric system called the International System (*SI units*), and the other is the system of

Imperial units. We use both sets of units in the examples and exercises in the following chapters.

In the SI system, the basic unit of length is the meter ($m$), of mass the kilogram ($kg$), of time the second ($s$). The unit of force is the Newton ($N$), defined as the force that will produce an acceleration of $1\,m/s^2$ on a mass of $1\,kg$. An alternative set of units for the Newton is therefore

$$N = \frac{kg \cdot m}{s^2}$$

Multiples and submultiples in the metric system go in powers of $10^3$. Thus for stress, there is the Pascal ($Pa$), the kiloPascal ($1\,kPa = 1000\,Pa$), the MegaPascal ($1\,MPa = 10^6\,Pa$), and the GigaPascal ($1\,GPa = 10^9\,Pa$). Lengths are in multiples of the meter as, for example, the millimeter ($1\,mm = 10^{-3}\,m$), and the micrometer ($1\,\mu m = 10^{-6}\,m$).

In the Imperial system, the basic unit of length is the foot ($ft$), of force the pound ($lb$), of time the second ($s$). The unit of mass (called a *slug*) is defined as the mass that will be accelerated at a rate of $1\,ft/s^2$ when acted on by a force of $1\,lb$. Hence, an alternate set of units for the *slug* is

$$slug = \frac{lb \cdot s^2}{ft}$$

Inches ($in$) are often taken as the standard unit of length. Also, multiples may go in thousands. For example, one thousand pounds is referred to as a kilopound or as a *kip*, while a thousand pounds per square inch ($1000\,psi$) is called a *ksi*.

The weight of a body is the gravitational force exerted on the mass. The value of the gravitational constant is usually taken as

$$g = 9.81\,m/s^2 = 32.17\,ft/s^2 = 386\,in/s^2$$

Thus, a mass of $1\,kg$ weighs $9.81\,N$ or $2.2\,lb$, whereas a mass of $1\,slug$ weighs about $32\,lb$ or $143\,N$. The SI units are an absolute system of units because the fundamental quantity of mass is independent of where it is measured. On the other hand, the Imperial system is referred to as a gravitational system because the fundamental unit of force (defined as the weight of a certain mass) varies with location on Earth.

The algorithms and source code presented do not have a built-in system of units and do not utilize any dimensional conversion constants. Therefore, any consistent system of units may be used for input and the corresponding calculated results will be in the same units. For example, if the Young's modulus is specified as $10,000\,ksi$, the reported values of force will be in *kips*. Since no particular system of units is preferred in this book, we generally present the results in non-dimensional form. When needed two useful conversion factors are $1\,in = 25.4\,mm$ and $1\,lb = 4.448\,N$.

Some typical material values that can be used in the examples and exercises may be taken as

| Material | Young's Modulus, $E$ | | Mass Density, $\rho$ | | Poisson's Ratio, $\nu$ |
|---|---|---|---|---|---|
| | [$msi$] | [$GPa$] | [$lb \cdot s^2/in$] | [$kg/m^3$] | |
| Steel | 30 | 210 | $0.81 \times 10^{-3}$ | 8000 | 0.30 |
| Aluminum | 10 | 70 | $0.27 \times 10^{-3}$ | 2800 | 0.33 |
| Concrete | 3.6 | 25 | $0.23 \times 10^{-3}$ | 2400 | 0.15 |
| Epoxy | 0.5 | 3.5 | $0.12 \times 10^{-3}$ | 1200 | 0.42 |

The shear modulus is calculated from $G = E/2(1 + \nu)$.

## Exercises

**1.1** Model a rocket as a simple structure with two nodes. What degrees of freedom would you give each node? Why?

**1.2** How many degrees of freedom are needed to describe the truss in Figure 1.1?

**1.3** A two-story building has relatively rigid floors. If the wind is assumed as a static load on one face only, how would you model the building? What degrees of freedom would you give each node?

**1.4** Suppose the building of the previous problem is modeled as a space frame with nodes at each corner. What is the total number of degrees of freedom? How does it compare to your previous model?

**1.5** If the grid of Figure 1.1 is stiffened by adding diagonal members, what is the increase in the number of degrees of freedom?

**1.6** A rod (length $= 100\,in$, area $= 1\,in^2$) has a force of 1000 $lb$ applied to it at one end only. What is the variation of stress along the rod? How does this compare to when the force is applied equally at both ends?

**1.7** The fundamental frequency $\omega$ for a vibrating clamped rod is

$$\omega = \frac{\pi}{L}\sqrt{\frac{E}{\rho}}$$

where $L$ is the length. Show that the same frequency (in radians per second) is obtained using either SI or Imperial units.

**1.8** The fundamental frequency for a vibrating simply supported beam is

$$\omega = \frac{\pi^2}{L^2}\sqrt{\frac{EI}{\rho A}}$$

where $L$ is the length, $A$ the area, and $I$ the second moment of area. Show that the same frequency is obtained using either SI or Imperial units.

# Chapter 2

# Rod Structures

A rod is a slender member that supports only axial loads. It is one of the simplest components and therefore a very suitable vehicle for introducing the basic concepts of the matrix analysis of structures. In this chapter, we consider only structures composed of rods arranged longitudinally; the more general arrangement is left to the chapter dealing with frames and trusses. We first review the basics of rod theory and derive the governing differential equations. These are then used to obtain the stiffness of a single rod element. The scheme for forming the structural stiffness matrix is established by using equilibrium and compatibility conditions at each node. Finally, the effects of boundary conditions are incorporated by simple row and column reduction of the matrices.

The developments in this chapter act as a blueprint for the corresponding developments in most of the other chapters; we introduce the major themes that will recur in the subsequent chapters.

## 2.1 Rod Theory

The central assumption in rod theory is that the stress at any position is only axial and is uniformly distributed on the section. For this to be true in practice, the rod must be long in comparison to its other dimensions, the cross-section must be uniform, and there must be no laterally applied loads. Therefore, in the following developments we will limit our attention to long, slender rods of constant cross-sectional area. We will use the three fundamental concepts from the mechanics of deformable bodies (Deformation, Equilibrium, Material Behavior) to establish the governing equations.

### Deformation and Strain

The axial displacement of each point is given by $u(x)$. The strain in a small segment $\Delta x$ of rod is related to the difference between the displacement of the two ends of the

segment by

$$\epsilon = \frac{(u + \Delta u) - (u)}{\Delta x} = \frac{\Delta u}{\Delta x}$$

In the limit as the segment becomes very small, we have

$$\epsilon = \lim_{\Delta x \to 0} \frac{\Delta u}{\Delta x} = \frac{du}{dx}$$

This differential relation tells us that, for example, a linear distribution of displacement along the axis results in a constant distribution of strain.

## Equilibrium

Let the distribution of the externally applied axial load per unit length be $q(x)$. The equilibrium of the small element shown in Figure 2.1 therefore leads (in the limit as $\Delta x$ becomes very small) to

$$F + \Delta F - F + q\,\Delta x = 0 \qquad \text{or} \qquad \frac{dF}{dx} = -q$$

where $F$ is the resultant axial force. This says that the distribution of internal force is constant if there is no applied loading.

Figure 2.1: Rod with infinitesimal rod element.

## Material Behavior and Stress Resultants

Since the stress is uniaxial, the one-dimensional version of Hooke's law gives

$$\sigma = E\epsilon = E\frac{du}{dx}$$

where $E$ is the Young's modulus. The axial resultants on the cross-section of area $A$ are

$$F = \int \sigma dA = EA\frac{du}{dx}, \qquad M = -\int \sigma y dA = -EA\frac{du}{dx}\int y dA = 0$$

This last relation is true since the $x$-axis is taken to go through the centroid of the section. The combination constant $EA$ is called the *axial stiffness*.

## Summary

The governing relationships for the structural quantities in the rod may now be summarized as

$$\text{Displacement}: \qquad u = u(x)$$

$$\text{Force}: \qquad F = EA\frac{du}{dx} \qquad\qquad (2.1)$$

$$\text{Loading}: \qquad q = -EA\frac{d^2u}{dx^2} \qquad\qquad (2.2)$$

It is seen from these that the displacement (via its derivatives) is the quantity that connects them together. That is, the displacement function $u(x)$ can be viewed as the fundamental unknown from which all the other quantities can be determined. The significance of this point will become apparent later when we develop the displacement method for solving structural problems.

**Example 2.1:**  Find the displacement shape for a rod with a uniformly applied load and fixed at both ends as shown in Figure 2.2.

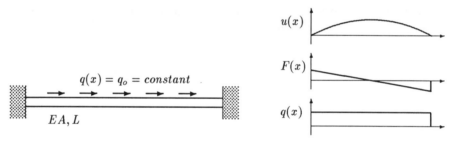

**Figure 2.2**: A fixed-fixed rod with uniformly distributed load.

Since the loading is constant, we will take this as the starting point. Thus

$$EA\frac{d^2u}{dx^2} = -q = -q_0 = \text{constant}$$

This is easily integrated to give

$$EA\frac{du}{dx} = -q_0 x + c_1$$

$$EA\,u = -\tfrac{1}{2}q_0 x^2 + c_1 x + c_2$$

where $c_1$ and $c_2$ are constants of integration. To determine these constants, it is necessary to impose some additional conditions. For the present problem, we know that the boundary conditions are

$$\text{at} \quad x = 0: \quad u = 0 = c_2$$
$$\text{at} \quad x = L: \quad u = 0 = -\tfrac{1}{2}q_0 L^2 + c_1 L + c_2$$

This gives two equations for two unknowns, allowing us to solve for the coefficients. They are

$$c_1 = \tfrac{1}{2}q_0 L, \qquad c_2 = 0$$

The displacement distribution is

$$EAu(x) = \tfrac{1}{2}q_0 x[L - x]$$

and the corresponding force distribution is

$$F(x) = \tfrac{1}{2}EAq_0[L - 2x]$$

These functions are shown plotted in Figure 2.2. Notice that while the internal (member) force goes from positive to negative, the displacement is always positive. The maximum displacement occurs at $x = L/2$ and is

$$u_{max} = \frac{q_0 L^2}{8EA}$$

The maximum force, on the other hand, occurs at $x = 0$ and $L$ and is

$$F_{max} = \tfrac{1}{2}q_0 L$$

**Example 2.2:**   Reconsider the last example, but this time using an applied loading of a single concentrated force $P$ at a distance $a$ from the first boundary.

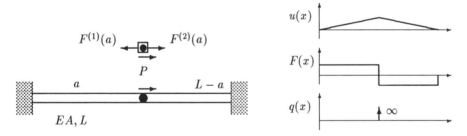

**Figure 2.3**: Fixed-fixed rod with a concentrated applied load.

It is possible to treat the rod as one complete piece of length $L$, but that will require us to write the loading $q$ as a function of $x$. However, we encounter a singularity at the point of the applied force because the loading (force per unit length) is infinite. There are ways of mathematically handling such behavior but it is conceptually simpler to divide the rod into separate sections not including the loading point. This approach is somewhat cumbersome because later we must impose additional conditions when we recombine the sections back together. Thus we can generate very many simultaneous equations. However, it will be seen that it has the significant advantage of being adaptable to computer implementation.

Since the loading is zero for points between the boundaries and the applied load, we will take the loading relation as the starting point, thus

$$0 < \quad x \quad < a$$
$$EA\frac{d^2u}{dx^2} \quad = \quad -q = 0$$
$$\frac{du}{dx} \quad = \quad c_1$$
$$u(x) \quad = \quad c_1 x + c_2$$

Similarly, for the second section

$$a < \quad x \quad < L$$
$$EA\frac{d^2u}{dx^2} \quad = \quad -q = 0$$
$$\frac{du}{dx} \quad = \quad c_1^*$$
$$u(x) \quad = \quad c_1^* x + c_2^*$$

To determine the four unknown coefficients $c_1, c_2, c_1^*, c_2^*$, we start by imposing the boundary conditions

$$\text{at} \quad x = 0: \qquad u = 0 = c_2$$
$$\text{at} \quad x = L: \qquad u = 0 = c_1^* L + c_2^*$$

This gives us two equations; hence two more conditions are required. We can obtain one of these from the compatibility requirement at $x = a$: obviously, the displacements are the same immediately to the left and right of the applied force, requiring that

$$u^{(1)}(a) = u^{(2)}(a) \qquad \text{or} \qquad c_1 a + c_2 = c_1^* a + c_2^*$$

We obtain the final condition by considering the forces at the connection; consider a small free body diagram extracted from around $x = a$, then

$$-F^{(1)}(a) + F^{(2)}(a) + P = 0 \qquad \text{or} \qquad -EA\,c_1 + EA\,c_1^* = -P$$

These four equations can now be solved to give the undetermined coefficients as

$$c_1 = \frac{P}{EA}\frac{L-a}{L}, \qquad c_2 = 0; \qquad c_1^* = \frac{-P}{EA}\frac{a}{L}, \qquad c_2^* = \frac{P}{EA}a$$

The displacement distributions are now written as

$$0 < x < a \qquad\qquad\qquad a < x < L$$
$$u(x) = \frac{PL}{EA}[1 - \frac{a}{L}][\frac{x}{L}] \qquad\qquad u(x) = \frac{PL}{EA}[1 - \frac{x}{L}][\frac{a}{L}]$$

With these in hand, we can calculate all other quantities of interest (such as strains, forces, and so on). For example, the force distributions are

$$0 < x < a \qquad\qquad\qquad a < x < L$$
$$F(x) = P[1 - \frac{a}{L}] \qquad\qquad F(x) = -P[\frac{a}{L}]$$

Notice that the force distribution is constant in each section with a jump precisely equal to the amount of the applied force..

To draw a connection between this example and the last, suppose the distributed load is lumped into a single concentrated force of $P = q_0 L$ at the center of the rod. The maximum displacement then occurs at $x = a = L/2$ and is

$$u_{max} = \frac{PL}{4EA} = \frac{q_0 L^2}{4EA}$$

The maximum force for this case occurs at $x = 0$ and $L$, and is

$$F_{max} = \tfrac{1}{2}P = \tfrac{1}{2}q_0 L$$

In comparison with the previous example, you will notice that the maximum forces are the same, but the lumping procedure has increased the maximum displacement by 100%.

**Example 2.3:**   Find the displacement shape for the rod structure with multiple sections and applied loads, shown in Figure 2.4.

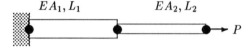

**Figure 2.4:** Rod structure with multiple sections and applied loads.

Since the loading is zero for points between the boundaries and the applied load, we will take the displacement solution as the starting point; thus for the first section

$$u(x) = c_1 x + c_2$$

Similarly, for the second section

$$u(x) = c_1^* x + c_2^*$$

Note that in this case we will use the variable $x$ to mean the local distance along the rod segment. To determine the four unknown coefficients $c_1, c_2, c_1^*, c_2^*$, we start by imposing the boundary conditions. For the first segment we have

$$\text{at}\quad x = 0:\qquad u = 0 = c_2$$

whereas, at the end of the second segment we have

$$\text{at}\quad x = L_2:\qquad F(L_2) = P = EA_2 c_1^*$$

This gives us two equations; hence two more conditions are required. We can obtain one of these from the compatibility requirement at the junction of the two segments; obviously, the displacements are the same immediately to the left and right of the junction, requiring that

$$u^{(1)}(L_1) = u^{(2)}(0)\qquad \text{or}\qquad c_1 L_1 + c_2 = c_2^*$$

We obtain the final condition by considering the forces at the connection; consider a small free body diagram extracted from around this region, then

$$-F^{(1)}(L_1) + F^{(2)}(0) = 0 \quad \text{or} \quad -EA_1\,c_1 + EA_2\,c_1^* = 0$$

These four equations can now be solved to give the undetermined coefficients as

$$c_1 = \frac{P}{EA_1}, \qquad c_2 = 0; \qquad c_1^* = \frac{P}{EA_2}, \qquad c_2^* = \frac{PL_1}{EA_1}$$

The displacement distributions are now written for the two segments as

$$1: \quad u(x) = \frac{Px}{EA_1} \qquad 2: \quad u(x) = \frac{Px}{EA_2} + \frac{PL_1}{EA_1}$$

Again, note that the variable $x$ is the local distance along the rod segment.

The approach we used in the previous examples is applicable to general rod structures made of multiple segments. That is, we divide the rod structure into sections by placing nodes (joints) where the loads are applied and where there are section discontinuities. Each section is then integrated under the condition of zero applied load. Finally, we determine the constants of integration by imposing the boundary conditions, compatibility between each section, and equilibrium at each joint.

While this is straightforward enough, it is seen that once the number of sections exceeds three or four, the simultaneous equations become unwieldy. To make the approach workable, it is necessary to marshall the equations in such a way that the treatment of each section is highly repetitious and many patterns recur. We will do this by considering a typical section under the action of typical applied loads. Once the relations for this case are established, then the relations for other sections will follow immediately as special cases. We will incorporate a significant refinement in this by arranging the equations so that compatibility between sections is automatically assured.

## 2.2   Rod Element Stiffness Matrix

Consider a rod member of length $L$ and cross-section area $A$. Denote the end points as Node 1 and Node 2, respectively, as shown in Figure 2.5. Assume that under the action of external end forces $F_1$ and $F_2$ the displacements at the nodes are $u_1$ and $u_2$, respectively. Note that the external forces and the nodal displacements point in the coordinate directions whereas the internal force is assumed 'positive on a positive face.'

A little explanation of the free body diagrams is in order here. The externally applied loads are $F_1$ and $F_2$, and these give rise to an internal (member) force distribution $F(x)$. The values of this internal force at the end points are $F(0)$ and $F(L)$, respectively. Therefore, when we consider a free body diagram of a small rod segment

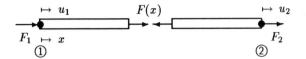

**Figure 2.5**: Sign convention for the rod element.

taken at the first end, say, then the forces acting on it are $F_1$ and $F(0)$. Equilibrium
(and the sign convention for the internal forces) requires that

$$F_1 + F(0) = 0 \quad \text{or} \quad F_1 = -F(0)$$

This procedure is set up so that we will emphasize the difference between the external
and internal forces. It is important to realize that $F_1$ and $F_2$ (which we will call
*element nodal forces*) are to be viewed as external applied forces different from the
internal force distribution $F(x)$.

If there are no loads applied in between the two nodes ($F_1$ and $F_2$ are applied at
the ends), then we may start by integrating the rod equation

$$EA\frac{d^2u}{dx^2} = -q = 0$$

We find by integrating twice that the displacement in the element is a linear function
of $x$:

$$u(x) = a_0 + a_1 x$$

where $a_0$ and $a_1$ are constants. The relation between the displacements at each node
and the coefficients is obtained by imposing

$$
\begin{aligned}
u(0) &= u_1 = a_0 \\
u(L) &= u_2 = a_0 + a_1 L
\end{aligned}
$$

This gives

$$
\begin{aligned}
a_0 &= u_1 \\
a_1 &= -(u_1 - u_2)/L
\end{aligned}
$$

allowing us to write the displacement distribution in terms of the nodal values as

$$u(x) = \left(1 - \frac{x}{L}\right)u_1 + \frac{x}{L}u_2 \equiv f_1(x)u_1 + f_2(x)u_2 \qquad (2.3)$$

The known functions $f_1(x)$ and $f_2(x)$ are called the *rod shape functions* .

As already said, if the displacements are known, then everything else about the
problem is also known. This is now modified to read that if the nodal displacements

are known then everything else is known.  For example, we can easily obtain the member force as

$$F(x) = EA\frac{du}{dx} = \frac{EA}{L}(-u_1 + u_2)$$

By equilibrium of a free body diagram near each node, the member and element nodal forces are related by

$$F_1 = -F(0), \qquad F_2 = F(L)$$

Consequently, the nodal forces are related to the nodal displacements by

$$F_1 = \frac{EA}{L}(u_1 - u_2)$$
$$F_2 = \frac{EA}{L}(u_2 - u_1)$$

We can express this in matrix notation form as

$$\left\{ \begin{matrix} F_1 \\ F_2 \end{matrix} \right\} = \frac{EA}{L} \begin{bmatrix} 1 & -1 \\ -1 & 1 \end{bmatrix} \left\{ \begin{matrix} u_1 \\ u_2 \end{matrix} \right\} \tag{2.4}$$

or symbolically,

$$\{F\} = [\ k\ ]\{u\}$$

The symmetric matrix

$$[\ k\ ] = \frac{EA}{L} \begin{bmatrix} 1 & -1 \\ -1 & 1 \end{bmatrix}$$

is called the *stiffness matrix* for the rod element, $\{F\}$ is called the *vector of element nodal forces*, and $\{u\}$ the *vector of nodal displacements*. The latter is often referred to as the *nodal degrees of freedom*.

It is noted that $[\ k\ ]$ is singular, i.e., $\det[\ k\ ] = 0$; and consequently, its inverse does not exist. This implies that given an arbitrary force vector $\{F\}$, it is not possible to find a unique solution for $\{u\}$. A singular stiffness matrix indicates that the structure is not stable. In the present case, it means that the element will translate indefinitely and boundary conditions must be specified before there is a solution.

## 2.3   Structural Stiffness Matrix

To illustrate the procedure for constructing the structural stiffness matrix, we will consider a simple rod structure composed of two members as shown in Figure 2.6. To establish the general procedure, we will assume that there is an applied axial force $P_i$ acting at each node.

The stiffness relation for each rod element is

$$\{F\}^{(12)} = [k^{(12)}]\{u\}^{(12)}, \qquad \{F\}^{(23)} = [k^{(23)}]\{u\}^{(23)}$$

where the superscript designates the element by reference to its nodes. It is necessary for us to devise a scheme for attaching these elements together subject to the requirement that the resulting assembled structure not violate any compatibility or equilibrium requirements.

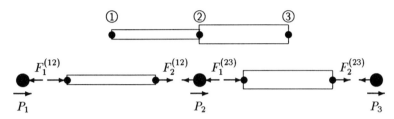

**Figure 2.6**: Simple rod structure with two members and three nodes.

## Nodal Equilibrium

Given that the structure is in a state of equilibrium, it follows that each node (or joint) must also be in a state of equilibrium. Take a free body diagram containing Node 1 as shown in the figure. Notice that we are referring to two types of nodes: one represents the joint and one the end of the element. This device is very useful because it allows us to draw the distinction between the loads applied to the structure and the element nodal loads. The applied force is denoted by $P_1$, while the nodal force acting on Member 1-2 is denoted by $F_1^{(12)}$. For equilibrium at this node, we must have

$$P_1 - F_1^{(12)} = 0 \quad \text{or} \quad P_1 = F_1^{(12)}$$

Similarly, for equilibrium at Node 2 and Node 3, it is required that

$$P_2 = F_2^{(12)} + F_2^{(23)}$$
$$P_3 = F_2^{(23)}$$

We can express the three equations above in the matrix form

$$\left\{ \begin{array}{c} P_1 \\ P_2 \\ P_3 \end{array} \right\} = \left\{ \begin{array}{c} F_1 \\ F_2 \\ 0 \end{array} \right\}^{(12)} + \left\{ \begin{array}{c} 0 \\ F_2 \\ F_3 \end{array} \right\}^{(23)}$$

This is the statement of equilibrium of each node. (It is also, consequently, a statement of the equilibrium of the structure as a whole.) There are as many equations as there are nodes, and as many vectors on the right hand side as there are members.

## Assembly

We can relate the two vectors on the right-hand side of the equilibrium relation to the nodal displacements $\{u_1, \cdots, u_3\}$ through augmented $[3 \times 3]$ element stiffness matrices as follows

$$\left\{ \begin{array}{c} \{F\} \\ 0 \end{array} \right\}^{(12)} = \left[ \begin{array}{cc} [k^{(12)}] & 0 \\ 0 & 0 \end{array} \right] \left\{ \begin{array}{c} \{u\}^{(12)} \\ u_3 \end{array} \right\}, \qquad \left\{ \begin{array}{c} 0 \\ \{F\} \end{array} \right\}^{(23)} = \left[ \begin{array}{cc} 0 & 0 \\ 0 & [k^{(23)}] \end{array} \right] \left\{ \begin{array}{c} u_1 \\ \{u\}^{(23)} \end{array} \right\}$$

Substituting these into the nodal equilibrium conditions leads us to

$$\left\{ \begin{array}{c} P_1 \\ P_2 \\ P_3 \end{array} \right\} = \left[ \begin{array}{cc} [k^{(12)}] & 0 \\ 0 & 0 \end{array} \right] \left\{ \begin{array}{c} \{u\}^{(12)} \\ u_3 \end{array} \right\} + \left[ \begin{array}{cc} 0 & 0 \\ 0 & [k^{(23)}] \end{array} \right] \left\{ \begin{array}{c} u_1 \\ \{u\}^{(23)} \end{array} \right\}$$

If we now realize that the quantity used to describe the nodal displacement of the element is precisely the same quantity used to describe the structural nodal displacement, then we can rewrite the above as

$$\left\{ \begin{array}{c} P_1 \\ P_2 \\ P_3 \end{array} \right\} = \left[ \begin{array}{ccc} k_{11}^{(12)} & k_{12}^{(12)} & 0 \\ k_{21}^{(12)} & k_{22}^{(12)} + k_{11}^{(23)} & k_{12}^{(23)} \\ 0 & k_{21}^{(23)} & k_{22}^{(23)} \end{array} \right] \left\{ \begin{array}{c} u_1 \\ u_2 \\ u_3 \end{array} \right\}$$

We can write this structural relation in shorthand as

$$\{P\} = [K]\{u\} \qquad \text{or} \qquad P_i = \sum_j K_{ij} u_j$$

Thus the stiffness property of a structure as a whole is characterized by a symmetric square matrix $[K]$ (called the *structural stiffness matrix*) which relates the external loads to the nodal displacements. Its symmetry property is inherited from the symmetry property of each element stiffness matrix.

It is worth noting that the element $K_{ij}$, which relates $P_i$ with $u_j$, is the sum of all the entries $k_{ij}$ in each element stiffness matrix that relates $P_i$ and $u_j$. Also note that while each rod element has different nodal forces, they all share common displacements at the nodal points. That is, compatibility of displacement from element to element is automatically being assured.

**Example 2.4:** A three element (four noded) rod is loaded as shown in Figure 2.7. Set up the structural stiffness matrix and the system of equations to be solved.

To construct the structural stiffness matrix, it is necessary for us to find the contributions to $[K]$ from each member. Thus, with reference to the figure, we first state each element stiffness matrix, that is

$$[k^{(12)}] = \frac{E3A}{L} \left[ \begin{array}{cc} 1 & -1 \\ -1 & 1 \end{array} \right]$$

$$[k^{(23)}] = \frac{EA}{L} \left[ \begin{array}{cc} 1 & -1 \\ -1 & 1 \end{array} \right]$$

$$[k^{(34)}] = \frac{4EA}{2L} \left[ \begin{array}{cc} 1 & -1 \\ -1 & 1 \end{array} \right]$$

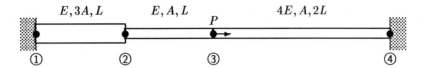

**Figure 2.7**: Four noded example.

We now augment each matrix to the size of the number of unknowns. In this case the size is $[4 \times 4]$ corresponding to the degrees of freedom $\{u_1, u_2, u_3, u_4\}$. The three augmented stiffness matrices are, respectively,

$$\frac{EA}{L} \begin{bmatrix} 3 & -3 & 0 & 0 \\ -3 & 3 & 0 & 0 \\ 0 & 0 & 0 & 0 \\ 0 & 0 & 0 & 0 \end{bmatrix}, \quad \frac{EA}{L} \begin{bmatrix} 0 & 0 & 0 & 0 \\ 0 & 1 & -1 & 0 \\ 0 & -1 & 1 & 0 \\ 0 & 0 & 0 & 0 \end{bmatrix}, \quad \frac{EA}{L} \begin{bmatrix} 0 & 0 & 0 & 0 \\ 0 & 0 & 0 & 0 \\ 0 & 0 & 2 & -2 \\ 0 & 0 & -2 & 2 \end{bmatrix}$$

We form the structural stiffness matrix by simply collecting the entries in the augmented element stiffness matrices according to their *position*. The result is

$$[K] = \frac{EA}{L} \begin{bmatrix} 3 & -3 & 0 & 0 \\ -3 & 3+1 & -1 & 0 \\ 0 & -1 & 1+2 & -2 \\ 0 & 0 & -2 & 2 \end{bmatrix} = \frac{EA}{L} \begin{bmatrix} 3 & -3 & 0 & 0 \\ -3 & 4 & -1 & 0 \\ 0 & -1 & 3 & -2 \\ 0 & 0 & -2 & 2 \end{bmatrix}$$

Note that this matrix must be symmetric as indeed it turns out to be. Also note that it is *banded*, that is, the non-zero entries are located close to the main diagonal. This is a characteristic of structural systems, and we will say more about it later when considering the computer aspects.

The system of equations for us to solve can now be established as

$$\frac{EA}{L} \begin{bmatrix} 3 & -3 & 0 & 0 \\ -3 & 4 & -1 & 0 \\ 0 & -1 & 3 & -2 \\ 0 & 0 & -2 & 2 \end{bmatrix} \begin{Bmatrix} u_1 = 0 \\ u_2 = ? \\ u_3 = ? \\ u_4 = 0 \end{Bmatrix} = \begin{Bmatrix} P_1 = ? \\ P_2 = 0 \\ P_3 = P \\ P_4 = ? \end{Bmatrix}$$

This system of equations cannot be solved directly because the right hand side contains some unknowns. That this occurred is associated with the boundary conditions and how that information is implemented. In order to proceed, it is necessary for us to rearrange these equations, and we will treat this in a general way in the next section.

It is worth noting in this last example that the knowns and unknowns in the final set of equations form mutually exclusive groups. That is, if a displacement is known at a node, then the corresponding applied load is an unknown (we refer to this as an unknown reaction). Conversely, if the applied load is known at a node, then the corresponding nodal displacement is unknown (we refer to this as a degree of freedom). Understanding this exclusivity is one of the keys to understanding the matrix approach to structural mechanics.

## 2.4  Boundary Conditions

External forces or displacements may be prescribed at the nodes of the structural system. However, when the force is given at a node, then the corresponding displacement must become an unknown quantity (and *vice versa*). In order to obtain a solution, we must rearrange and renumber the global system of equations to account for this.

Consider a case of 6 nodes (or six degrees of freedom) with the nodal values renumbered so that we can partition the arrays as

$$\{u\} = \left\{ \begin{array}{c} u_u \\ -- \\ u_k \end{array} \right\}, \qquad \{P\} = \left\{ \begin{array}{c} P_k \\ -- \\ P_u \end{array} \right\}$$

where

$$\{u_u\} \equiv \left\{ \begin{array}{c} u_1 \\ u_2 \end{array} \right\}, \qquad \{u_k\} \equiv \left\{ \begin{array}{c} u_3 \\ u_4 \\ u_5 \\ u_6 \end{array} \right\}, \qquad \{P_k\} \equiv \left\{ \begin{array}{c} P_1 \\ P_2 \end{array} \right\}, \qquad \{P_u\} \equiv \left\{ \begin{array}{c} P_3 \\ P_4 \\ P_5 \\ P_6 \end{array} \right\}$$

are the unknown and known nodal displacement and force vectors. The equations for the whole structural system are expressed in the partitioned form as

$$\begin{bmatrix} K_{uu} & K_{uk} \\ K_{ku} & K_{kk} \end{bmatrix} \left\{ \begin{array}{c} u_u \\ u_k \end{array} \right\} = \left\{ \begin{array}{c} P_k \\ P_u \end{array} \right\}$$

Multiplying this out gives

$$[K_{uu}]\{u_u\} + [K_{uk}]\{u_k\} = \{P_k\}$$
$$[K_{ku}]\{u_u\} + [K_{kk}]\{u_k\} = \{P_u\}$$

We can therefore obtain the unknown displacements from the first equation by solving

$$[K_{uu}]\{u_u\} = \{P_k\} - [K_{uk}]\{u_k\} \tag{2.5}$$

since this has only known quantities on the right hand side. The unknown external loads are obtained from the subsequent computation using the second equation above

$$\{P_u\} = [K_{ku}]\{u_u\} + [K_{kk}]\{u_k\} \tag{2.6}$$

where now, we know everything on the right hand side since $\{u_u\}$ was obtained from the previous calculation.

## Special Case

If the known nodal displacements are zero (such as for fixed boundary conditions) then the above take on the particularly simple form after substituting $\{u_k\} = 0$

$$[K_{uu}]\{u_u\} = \{P_k\}, \qquad \{P_u\} = [K_{ku}]\{u_u\} \tag{2.7}$$

For convenience, we define $[K^*] \equiv [K_{uu}]$; the system of equations for us to solve becomes simply

$$[K^*]\{u_u\} = \{P_k\}$$

where $[K^*]$ is called the *reduced structural stiffness matrix* relating the unknown displacements to the given external loads.

This method of rearranging the equations reduces the size of the system to be solved. That is, $[K^*]$ is smaller than $[K]$ because $\{u_u\}$ is smaller than $\{u\}$. What we have done is posed the problem in terms of the minimum number of unknowns; this involves the unknown displacements (degrees of freedom) but not the unknown applied forces (reactions). In the general structural principles to be developed in Chapters 6&12, we refer to $\{u_u\}$ as the *generalized coordinates* and the equations $\{u_k\} = 0$ as constraint equations.

**Example 2.5:** Complete the solution of the three element rod structure of the example in the last section.

The nodal displacements at Nodes 1 and 4 are zero, leaving the unknown nodal displacements as

$$\{u_u\} = \left\{ \begin{array}{c} u_2 \\ u_3 \end{array} \right\}$$

The corresponding known applied loads are

$$\{P_k\} = \left\{ \begin{array}{ccc} P_2 & = & 0 \\ P_3 & = & P \end{array} \right\}$$

We obtain the rearranged structural system by interchanging *both* rows and columns to get

$$\frac{EA}{L} \begin{bmatrix} 4 & -1 & -3 & 0 \\ -1 & 3 & 0 & -2 \\ -3 & 0 & 3 & 0 \\ 0 & -2 & 0 & 2 \end{bmatrix} \left\{ \begin{array}{ccc} u_2 & = & ? \\ u_3 & = & ? \\ u_1 & = & 0 \\ u_4 & = & 0 \end{array} \right\} = \left\{ \begin{array}{ccc} P_2 & = & 0 \\ P_3 & = & P \\ P_1 & = & ? \\ P_4 & = & ? \end{array} \right\}$$

The reduced stiffness matrix is just the first $[2 \times 2]$ entries giving us the reduced system of equations as

$$\frac{EA}{L} \begin{bmatrix} 4 & -1 \\ -1 & 3 \end{bmatrix} \left\{ \begin{array}{c} u_2 \\ u_3 \end{array} \right\} = \left\{ \begin{array}{c} 0 \\ P \end{array} \right\}$$

The inverse of $[K^*]$ is

$$[K^*]^{-1} = \frac{L}{11EA} \begin{bmatrix} 3 & 1 \\ 1 & 4 \end{bmatrix}$$

Thus, we can calculate the unknown displacements from

$$\{u_u\} = [K^*]^{-1}\{P_k\} = \frac{PL}{11EA} \left\{ \begin{array}{c} 1 \\ 4 \end{array} \right\}$$

Once the nodal displacements are obtained, the loads in the members can be calculated.

## Assembly with Fixed Boundaries

There is an alternative approach that allows the reduced stiffness matrix to be obtained directly. The essential aspect of the approach is to do the reductions *before* assembling into the structural matrix. The algorithm entails the following steps:

**Step 1:** Determine the size of the structural system.

**Step 2:** Determine the reduced structural system by removing zero degrees of freedom.

**Step 3:** Form the element stiffness matrix.

**Step 4:** Reduce the element stiffness in terms of the non-zero structural degrees of freedom.

**Step 5:** Add the reduced element stiffness to the reduced structural stiffness.

**Step 6:** Repeat Steps 3,4,5 for each member.

This scheme can be easily adapted to a computer. To show the approach laid out somewhat in the pedantic manner followed by the computer, we will redo the previous problem.

**Example 2.6:** Assemble the reduced stiffness matrix for the rod structure of Figure 2.7.

**Step 1:** There are 4 nodes, hence the system size is

$$size = nodes \times DoF = 4 \times 1 = 4$$

and the total degrees of freedom are

$$\{u\} = \{u_1, u_2, u_3, u_4\}$$

**Step 2:** There are fixed boundaries at Nodes 1 and 4, hence the reduced system is

$$\{u_u\} = \{u_2, u_3\}$$

**Step 3:** The element stiffness for Member 1-2 is:

$$[k^{(12)}] = \frac{3EA}{L} \left[ \begin{array}{cc} 1 & -1 \\ -1 & 1 \end{array} \right]$$

**Step 4:**   The reduced element stiffness is (after retaining rows and columns associated only with Node 2)

$$[k^{*(12)}] = \frac{EA}{L}[\ 3\ ]$$

**Step 5:**   Add to the currently unpopulated reduced structural stiffness

$$[K^*] = \frac{EA}{L}\begin{bmatrix} 3 & 0 \\ 0 & 0 \end{bmatrix}$$

**Step 6:**   For this step just repeat Steps 3,4 and 5 for each member. This will be done here in one swoop each.

**Steps 3,4,5:**

$$[k^{(23)}] = \frac{EA}{L}\begin{bmatrix} 1 & -1 \\ -1 & 1 \end{bmatrix}$$

$$[k^{*(23)}] = \frac{EA}{L}\begin{bmatrix} 1 & -1 \\ -1 & 1 \end{bmatrix}$$

$$[K^*] = \frac{EA}{L}\begin{bmatrix} 4 & -1 \\ -1 & 1 \end{bmatrix}$$

**Steps 3,4,5:**

$$[k^{(34)}] = \frac{2EA}{L}\begin{bmatrix} 1 & -1 \\ -1 & 1 \end{bmatrix}$$

$$[k^{*(34)}] = \frac{EA}{L}[\ 2\ ]$$

$$[K^*] = \frac{EA}{L}\begin{bmatrix} 4 & -1 \\ -1 & 3 \end{bmatrix}$$

This last array is the desired reduced structural stiffness matrix.

It is worth mentioning that the way this is actually programmed is not by 'scratching' rows and columns. Rather, a pointer array is maintained whose content is an identifier for each node (degree of freedom) and this identifier is a zero when the node is fixed. A refinement is to also use the array to keep track of the equation numbers; thus for the above problem the array is

$$\{IDbc\} = \{\ 0,1,2,0\ \}$$

## 2.5   Member Distributions and Reaction

The fundamental unknowns in a structures problem (as we have formulated it) are the nodal displacements. However, the quantities of engineering interest may be the member loads and structural reactions. We will obtain these quantities as a post-processing action on the displacements and thus perform it outside the main computational loop. There are advantages and disadvantages to this approach. The compelling advantage arises from the fact that an engineer does not always know in

advance the precise output desired. That is, the results of the solution are usually
investigated interactively. The post-processing arrangement lends itself to this approach, and so we will assume that the computation of member quantities needs to
be determined in this spirit.

Once we know the nodal displacements, the distribution of displacement is obtained simply from

$$u(x) = (1 - \frac{x}{L})u_1 + \frac{x}{L}u_2 \equiv f_1(x)u_1 + f_2(x)u_2$$

This shows that the distribution is linear and therefore the plot of displacement is a
straight line connecting the nodal values of $u_1$ and $u_2$ for each element.

Determining the member load distribution is only slightly more involved. Recall
that the following equation relates the member force to the global displacement

$$\{F\} = [\, k\, ]\{u\}$$

Thus the application of this to each member gives the nodal forces for each member.
The internal (member) load in the rod member is related to the nodal forces by

$$F(0) = -F_1, \qquad F(L) = +F_2$$

But we have already established that the member force in the rod element is constant.
Hence

$$F(x) = -F_1 = +F_2$$

This allows easy plotting of the distribution.

**Example 2.7:**   Plot the displacement and member force distributions for the three
element rod structure of Figure 2.7.

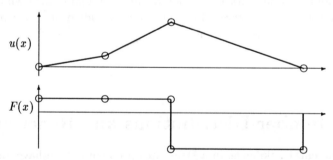

**Figure 2.8**: Displacement and force distributions.

We have already determined that the four nodal displacements are

$$u_1 = 0, \qquad u_2 = \frac{PL}{11EA}, \qquad u_3 = \frac{4PL}{11EA}, \qquad u_4 = 0$$

The displacement distribution is obtained by plotting these four points and connecting them by straight lines.

The member force distribution is obtained by first determining the nodal forces. Hence, using the stiffness relation for each member we get

$$\begin{Bmatrix} F_1 \\ F_2 \end{Bmatrix}^{(12)} = \frac{3EA}{L}\begin{bmatrix} 1 & -1 \\ -1 & 1 \end{bmatrix}\begin{Bmatrix} 0 \\ 1 \end{Bmatrix}\frac{PL}{11EA} = \begin{Bmatrix} -3 \\ 3 \end{Bmatrix}\frac{P}{11}$$

$$\begin{Bmatrix} F_1 \\ F_2 \end{Bmatrix}^{(23)} = \frac{EA}{L}\begin{bmatrix} 1 & -1 \\ -1 & 1 \end{bmatrix}\begin{Bmatrix} 1 \\ 4 \end{Bmatrix}\frac{PL}{11EA} = \begin{Bmatrix} -3 \\ 3 \end{Bmatrix}\frac{P}{11}$$

$$\begin{Bmatrix} F_1 \\ F_2 \end{Bmatrix}^{(34)} = \frac{2EA}{L}\begin{bmatrix} 1 & -1 \\ -1 & 1 \end{bmatrix}\begin{Bmatrix} 4 \\ 0 \end{Bmatrix}\frac{PL}{11EA} = \begin{Bmatrix} 8 \\ -8 \end{Bmatrix}\frac{P}{11}$$

These values of nodal force are related to the internal member forces by the previously derived equilibrium conditions. Thus for the first member

$$F(0)^{(12)} = -F_1^{(12)} = \frac{3P}{11}, \qquad F(L)^{(12)} = +F_2^{(12)} = \frac{3P}{11}$$

We know that the distribution of force is constant, and these relations show that we have a choice as to which node we use in order to determine its value. Proceeding in like manner for the other elements, we get

$$F(0)^{(23)} = -F_1^{(23)} = \frac{3P}{11}, \qquad F(L)^{(23)} = +F_2^{(23)} = \frac{3P}{11}$$

and

$$F(0)^{(34)} = -F_1^{(34)} = \frac{-8P}{11}, \qquad F(2L)^{(34)} = +F_2^{(34)} = \frac{-8P}{11}$$

Note that the discontinuity in the plot at Node 3 corresponds precisely to the amount of the applied load of $P$.

## Boundary Reactions

As pointed out before, the unknown forces (which are usually the boundary reactions) can be obtained from the calculated displacements as

$$\{P_u\} = [K_{ku}]\{u_u\} + [K_{kk}]\{u_k\}$$

When viewed as a post-processing operation, this approach requires either storing or reassembling the partial stiffnesses $[K_{ku}]$ and $[K_{kk}]$. The alternative approach favored here is to obtain the reactions when they are needed.

Let the junction between two elements be attached to a second structure as shown in Figure 2.9. The second structure is shown as being rigid, but actually the results to be established are valid even for elastic attachment. By considering the equilibrium of the common node we get

$$P_R = -F_1^{(2)} - F_2^{(1)} + P \qquad (2.8)$$

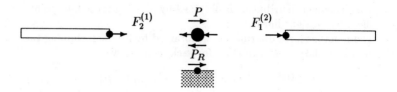

**Figure 2.9**: Sign convention for boundary reactions.

For fixed boundaries, we will generally take that the applied load $P$ is zero, but the above allows for other cases also.

**Example 2.8:**  Find the boundary reactions for the three element rod structure of Figure 2.7.

Following on from the last example, we can use the stiffness relation for each member attached to a boundary of interest to get

$$\left\{ \begin{array}{c} F_1 \\ F_2 \end{array} \right\}^{(12)} = \frac{3EA}{L} \left[ \begin{array}{cc} 1 & -1 \\ -1 & 1 \end{array} \right] \left\{ \begin{array}{c} 0 \\ 1 \end{array} \right\} \frac{PL}{11EA} = \left\{ \begin{array}{c} -3 \\ 3 \end{array} \right\} \frac{P}{11}$$

$$\left\{ \begin{array}{c} F_1 \\ F_2 \end{array} \right\}^{(34)} = \frac{2EA}{L} \left[ \begin{array}{cc} 1 & -1 \\ -1 & 1 \end{array} \right] \left\{ \begin{array}{c} 4 \\ 0 \end{array} \right\} \frac{PL}{11EA} = \left\{ \begin{array}{c} 8 \\ -8 \end{array} \right\} \frac{P}{11}$$

These values of nodal force are related to the reactions at Nodes 1 and 4 by

$$P_{R1} = -F_1^{(12)} = \frac{+3P}{11}, \qquad P_{R4} = -F_2^{(34)} = \frac{+8P}{11}$$

The plus sign in both cases indicates that the action is tending to move the boundaries from left to right. Thus, the reaction at Node 1 is tensile, while that at Node 2 is compressive.

## 2.6   Distributed Loads

We conclude this chapter by considering the effect of distributed loads on our matrix formulation. As much as possible, we will try to follow the development of Section 2.2.

We start by integrating the rod equation

$$EA\frac{d^2u}{dx^2} = -q(x)$$

We find by integrating twice that the displacement in the element is no longer a linear function of $x$ but given by:

$$u(x) = a_0 + a_1x + w(x), \qquad EA\frac{d^2w}{dx^2} \equiv -q(x)$$

where $a_0$ and $a_1$ are constants. The relation between the displacements at each node and the coefficients is obtained by imposing

$$
\begin{aligned}
u(0) &= u_1 = a_0 + w_1 \\
u(L) &= u_2 = a_0 + a_1 L + w_2
\end{aligned}
$$

where we have defined $w_1 \equiv w(0)$ and $w_2 \equiv w(L)$. This gives

$$
\begin{aligned}
a_0 &= u_1 - w_1 \\
a_1 &= -(u_1 - w_1 - u_2 + w_2)/L
\end{aligned}
$$

allowing us to write the displacement distribution in terms of the nodal values as

$$
u(x) = (1 - \frac{x}{L})u_1 + \frac{x}{L}u_2 - (1 - \frac{x}{L})w_1 - \frac{x}{L}w_2 + w(x) \tag{2.9}
$$

As already stated, if the displacements are known, then everything else about the problem is also known. We can now obtain the member forces as

$$
F(x) = EA\frac{du}{dx} = \frac{EA}{L}(-u_1 + u_2) - \frac{EA}{L}(-w_1 + w_2) + EA\frac{dw}{dx}
$$

By equilibrium of a free body diagram near each node, the member and nodal forces are related by

$$
F_1 = -F(0), \qquad F_2 = F(L)
$$

Consequently, the nodal forces are related to the nodal displacements by

$$
\left\{ \begin{array}{c} F_1 \\ F_2 \end{array} \right\} = \frac{EA}{L}\left[ \begin{array}{cc} 1 & -1 \\ -1 & 1 \end{array} \right]\left\{ \begin{array}{c} u_1 \\ u_2 \end{array} \right\} - \frac{EA}{L}\left[ \begin{array}{cc} 1 & -1 \\ -1 & 1 \end{array} \right]\left\{ \begin{array}{c} w_1 \\ w_2 \end{array} \right\} + EA\left\{ \begin{array}{c} -w_1' \\ +w_2' \end{array} \right\} \tag{2.10}
$$

The remaining step is to substitute the applied loading $q(x)$ into the expressions for $w(x)$. The results are simplified considerably if we note that

$$
\int_0^L (L - x)q\, dx = EALw_1' - EA(w_2 - w_1), \qquad \int_0^L xq\, dx = -EALw_2' + EA(w_2 - w_1)
$$

Both of these relations can be demonstrated by integrating by parts. The stiffness relation can now be expressed as

$$
\left\{ \begin{array}{c} F_1 \\ F_2 \end{array} \right\} = \frac{EA}{L}\left[ \begin{array}{cc} 1 & -1 \\ -1 & 1 \end{array} \right]\left\{ \begin{array}{c} u_1 \\ u_2 \end{array} \right\} - \int_0^L \left\{ \begin{array}{c} f_1 q \\ f_2 q \end{array} \right\} dx \tag{2.11}
$$

where the known functions $f_1(x)$ and $f_2(x)$ are the rod shape functions of Equation(2.3). From this, we see that the distributed load is replaced by concentrated loads associated with Nodes 1 and 2. Therefore, the assembled equations for the structure are given as

$$
\{P + Q\} = [K]\{u\}, \qquad \{Q\} \equiv \sum_m \int \{fq\}_m dx
$$

where $\{P\}$ are the applied concentrated loads and $\{Q\}$ are the assembled distributed loads treated as concentrated applied loads.

These results look deceptively similar to those of the point loads only case; all we have done is replaced the distributed load with concentrated loads at the nodes. Keep in mind, however, that the displacement functions are quite different in the two cases; thus for a given set of nodal displacements both would give different computed values for the member loads.

**Example 2.9:**   Use matrix methods to find the axial force and displacements in the uniformly loaded rod of Figure 2.2.

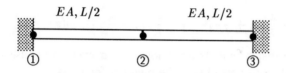

**Figure 2.10**: A uniformly loaded rod modeled using two elements.

We will model the rod with two elements (three nodes). The global degrees of freedom are then

$$\{u\} = \{u_1, u_2, u_3\}$$

The stiffness relation for element 1-2 is

$$\left\{ \begin{matrix} F_1 \\ F_2 \end{matrix} \right\}^{(12)} = \frac{EA}{L/2} \left[ \begin{matrix} 1 & -1 \\ -1 & 1 \end{matrix} \right] \left\{ \begin{matrix} u_1 \\ u_2 \end{matrix} \right\} - \int_0^{L/2} \left\{ \begin{matrix} (1 - x2/L) \\ (x2/L) \end{matrix} \right\} q_0 \, dx$$

$$= \frac{EA2}{L} \left[ \begin{matrix} 1 & -1 \\ -1 & 1 \end{matrix} \right] \left\{ \begin{matrix} u_1 \\ u_2 \end{matrix} \right\} - \frac{q_0 L}{4} \left\{ \begin{matrix} 1 \\ 1 \end{matrix} \right\}$$

Similarly, the stiffness relation for element 2-3 is

$$\left\{ \begin{matrix} F_1 \\ F_2 \end{matrix} \right\}^{(23)} = \frac{EA2}{L} \left[ \begin{matrix} 1 & -1 \\ -1 & 1 \end{matrix} \right] \left\{ \begin{matrix} u_1 \\ u_2 \end{matrix} \right\} - \frac{q_0 L}{4} \left\{ \begin{matrix} 1 \\ 1 \end{matrix} \right\}$$

The global stiffness relation is assembled to be

$$\left\{ \begin{matrix} P_1 \\ P_2 \\ P_3 \end{matrix} \right\} = \frac{EA2}{L} \left[ \begin{matrix} 1 & -1 & 0 \\ -1 & 2 & -1 \\ 0 & -1 & 1 \end{matrix} \right] \left\{ \begin{matrix} u_1 \\ u_2 \\ u_3 \end{matrix} \right\} - \frac{q_0 L}{4} \left\{ \begin{matrix} 1 \\ 2 \\ 1 \end{matrix} \right\}$$

The last vector in the above equation is the vector $\{Q\}$.

The boundary conditions impose that the degrees of freedom are zero at Nodes 1 and 3; further, there are no applied concentrated loads, hence the reduced system becomes

$$\frac{EA2}{L} [\, 2 \,]\{u_2\} = \frac{q_0 L}{4}\{2\}$$

This has the solution

$$u_2 = \frac{q_0 L^2}{8EA}$$

which is the exact value. We can now substitute this into the stiffness relation for each element to determine the nodal forces as

$$\begin{Bmatrix} F_1 \\ F_2 \end{Bmatrix}^{(12)} = \frac{EA2}{L} \begin{bmatrix} 1 & -1 \\ -1 & 1 \end{bmatrix} \begin{Bmatrix} 0 \\ 1 \end{Bmatrix} \frac{q_0 L^2}{8EA} - \frac{q_0 L}{4} \begin{Bmatrix} 1 \\ 1 \end{Bmatrix} = \frac{q_0 L}{2} \begin{Bmatrix} -1 \\ 0 \end{Bmatrix}$$

$$\begin{Bmatrix} F_1 \\ F_2 \end{Bmatrix}^{(23)} = \frac{EA2}{L} \begin{bmatrix} 1 & -1 \\ -1 & 1 \end{bmatrix} \begin{Bmatrix} 1 \\ 0 \end{Bmatrix} \frac{q_0 L^2}{8EA} - \frac{q_0 L}{4} \begin{Bmatrix} 1 \\ 1 \end{Bmatrix} = \frac{q_0 L}{2} \begin{Bmatrix} 0 \\ -1 \end{Bmatrix}$$

These values of nodal force are related to the internal member forces by the previously derived equilibrium conditions. Thus it is easy to show that

$$F(0) = \frac{q_0 L}{4}, \qquad F(L/2) = 0, \qquad F(L) = \frac{-q_0 L}{4}$$

This is the exact member force distribution.

# Problems

**2.1** A uniform rod of length $L$ is hung vertically under the action of gravity. Show that the loading per unit length is $q(x) = \rho A g$, where $\rho$ is the mass density and $g$ is the gravitational constant. Consequently, show that the displacement distribution is

$$u(x) = \frac{\rho A L}{EA} g x \left(1 - \frac{x^2}{2L}\right)$$

**2.2** Model the previous self weight problem using 3 elements. Account for the distributed mass by 'lumping' it at the nodes. Compare the distribution of displacement with the exact values.

**2.3** Show that when the cross-sectional area or the elastic modulus changes along the length of the rod that the relevant equations are

$$F(x) = EA\frac{du}{dx}, \qquad q(x) = -\frac{d}{dx}[EA\frac{du}{dx}]$$

Consider a rod of length $L$ that has a varying area of the form

$$A(x) = A_1 + (A_2 - A_1)\frac{x}{L}$$

If this is fixed at one end and a load of $P$ applied at the other, show that the displacement function is

$$u(x) = \frac{PL}{E(A_2 - A_1)} \log_n[1 + (\frac{A_2}{A_1} - 1)\frac{x}{L}]$$

**2.4** Model the previous problem using three elements of different area but each of the same length. (Assume that $A_1 = 2A_2$.) Compare the displacement distributions.

**2.5** When a rod is heated uniformly it expands by the amount $\Delta L = \alpha \Delta T L$, where $\alpha$ is the coefficient of thermal expansion, and $\Delta T$ is the change in temperature. Hooke's law for the material is modified as

$$\sigma = E(\epsilon - \alpha \Delta T)$$

Show that when temperature is taken into account, that the element stiffness relation becomes

$$\left\{ \begin{matrix} F_1 \\ F_2 \end{matrix} \right\} = \frac{EA}{L} \begin{bmatrix} 1 & -1 \\ -1 & 1 \end{bmatrix} \left\{ \begin{matrix} u_1 \\ u_2 \end{matrix} \right\} + EA\alpha\Delta T \left\{ \begin{matrix} 1 \\ -1 \end{matrix} \right\}$$

[Reference [35], pp. 65]

**2.6** What modifications (if any) are needed during the assemblage stage to account for temperature effects? If the second term on the right hand side of the previous stiffness relation is viewed as a force vector $\{q\}$, show how the global equilibrium equations are affected.

[Reference [35], pp. 131]

## Exercises

**2.1** Rework Example 2.9 but using only one element. Show that the same values of boundary reaction are obtained.

**2.2** The pendulum of a clock has a $3\,lb$ weight suspended by three parallel rods of $30\,in$ length. Two of the rods are brass $(E = 15\,msi\ diam = .10\,in)$ and the third steel $(diam = .05\,in)$. What is the force in each rod?

$$[F_s = 0.6\,lb\ F_b = 1.2\,lb]$$

**2.3** A square reinforced pier $1\,ft \times 1\,ft$ in cross-section and $4\,ft$ high is loaded axially by $150\,kip$. The concrete is stiffened with eight $1\,in^2$ steel reinforcing bars placed symmetrically about the vertical axis. What percentage of the force is supported by the steel? [33%]

**2.4** Show that the governing equations for the twisting of a shaft are

$$T(x) = GJ\frac{d\phi}{dx}, \qquad q(x) = -\frac{d}{dx}[GJ\frac{d\phi}{dx}]$$

where $T$ is the torque and $\phi$ is the angle of twist.

**2.5** Following the procedure for a rod, show that the shaft element shape functions are the same as for the rod. Consequently, show that the stiffness relation for a shaft element can be derived as

$$\left\{ \begin{matrix} T_1 \\ T_2 \end{matrix} \right\} = \frac{GJ}{L} \begin{bmatrix} 1 & -1 \\ -1 & 1 \end{bmatrix} \left\{ \begin{matrix} \phi_1 \\ \phi_2 \end{matrix} \right\}$$

Chapter 3

# Beam Structures

A *beam* is a slender structural member designed to carry transverse loads and applied couples. In response to these loads, it develops internal bending moments and shear forces. We shall refer to a beam structure as a collection of beams arranged in a collinear manner. These are sometimes referred to as continuous beams. Additional aspects of the mechanics of beams can be found in strength of materials books such as Reference [13].

This chapter develops the governing differential equations of beam theory so as to set the basis for the introduction of the beam element. This is done by emphasizing the application of the twin principles of compatibility and equilibrium at the junction of each beam connection. In this way the matrix assemblage procedure is seen as just an application of these principles. Consequently, the matrix description of these beam structures follows very closely that of the rod structures of the previous chapter. We do, however, introduce new procedures for handling elastic boundaries.

## 3.1   Beam Theory

The simplest model for describing beam behavior is that of Bernoulli-Euler. For such a beam, the depth is considered to be much less than the length and therefore shear deformation through the thickness is neglected. For similar reasons, the stress state can be treated as uniaxial. It is also assumed that the displacement of any point along the centroidal axis is vertical only and that this axis does not stretch.

### Deformation and Strain

For small deflections, it is apparent that the slope (of the centroidal axis) of the beam at an arbitrary point is related to the deflection by

$$\phi(x) = \frac{\Delta v}{\Delta x} \simeq \frac{dv}{dx}$$

37

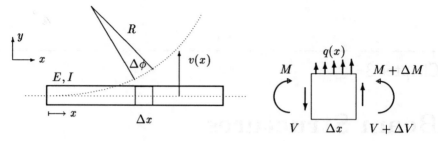

**Figure 3.1**: Beam with infinitesimal beam element.

Since the beam deforms (locally) into the arc of a circle, then the deformed length of a small line segment a distance $y$ from the axis is $(R - y)\Delta\phi$, where $R$ is the radius of curvature. The strain of the segment is therefore

$$\epsilon = \frac{(R - y)\Delta\phi - R\Delta\phi}{R\Delta\phi} = \frac{-y}{R}$$

This shows that the strain is linearly distributed on the section. From geometry, it is known that radius and curvature are related by $1/R = d^2v/dx^2$, hence we can also obtain for the strain

$$\epsilon = -y\frac{d^2v}{dx^2} = -y\frac{d\phi}{dx}$$

This gives the relation between the strain ($\epsilon$), slope ($\phi$), and curvature ($d^2v/dx^2$).

## Equilibrium

The equilibrium of the small element shown in Figure 3.1 leads to the following two equations (in the limit as $\Delta x$ becomes very small)

$$\frac{dV}{dx} + q = 0, \qquad \frac{dM}{dx} + V = 0$$

where $V$ is the resultant shear force, $M$ the resultant moment, and $q$ the distributed load per unit length. In contrast to the rod, the beam has two equilibrium equations. One consequence of this is that the variety of beam behavior will be much greater than for the rod.

## Stress Resultants

Applying Hooke's law to the axial stress state of the beam gives

$$\sigma = E\epsilon = -y\,E\frac{d\phi}{dx} = -y\,E\frac{d^2v}{dx^2}$$

Note that the stress, too, is distributed linearly on the section. Knowing the explicit form for this distribution allows us to determine the following stress resultants on the cross-section

$$F = \int \sigma \, dA = 0, \qquad M = -\int \sigma y \, dA = EI\frac{d\phi}{dx} = EI\frac{d^2v}{dx^2}$$

where $I$ is the moment of inertia (or the second moment of area) given by

$$I \equiv \int y^2 \, dA$$

For a rectangular section ($b \times h$) and a circular section (diameter $D$), this gives

$$I_{rect} = \tfrac{1}{12}bh^3, \qquad I_{circ} = \tfrac{1}{64}D^4$$

respectively. Similar expressions exist for other sections.

## Summary

All the relationships for the structural quantities may now be collected as

$$\text{Displacement} \quad : \quad v = v(x)$$

$$\text{Slope} \quad : \quad \phi = \frac{dv}{dx} \tag{3.1}$$

$$\text{Moment} \quad : \quad M = +EI\frac{d^2v}{dx^2} \tag{3.2}$$

$$\text{Shear} \quad : \quad V = -EI\frac{d^3v}{dx^3} \tag{3.3}$$

$$\text{Loading} \quad : \quad q = +EI\frac{d^4v}{dx^4} \tag{3.4}$$

It is seen from these that the deflected shape $v(x)$ can be viewed as the fundamental unknown of interest; all other quantities are obtained by differentiation. This is precisely the same conclusion already drawn from the rod analysis. It is also interesting to note that the section properties are reduced to the single combination term $EI$. This is called the *flexural stiffness*.

When solving beam problems, we may be given information at any of the five levels above, and have to carry out integrations (or differentiations) to obtain the other quantities. Integration gives rise to constants of integration which must be found from the boundary and compatibility conditions. In the general case there are four constants of integration, twice as many as for the rod. Thus twice as many conditions must be imposed at each section.

**Example 3.1:**   Find the deflected shape of the fixed-pinned beam shown in Figure 3.2. The applied load per unit length, $w_o$, is uniformly distributed.

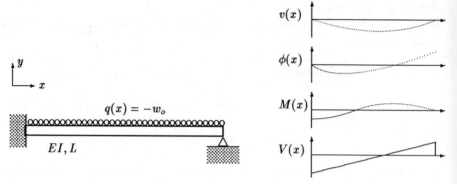

**Figure 3.2**: Uniformly loaded beam.

The loading is constant and given as $q(x) = -w(x) = -w_o$; therefore we will take this as the starting point. That is,

$$EI\frac{d^4v}{dx^4} = q = -w_o$$

Integrate to obtain

$$EI\frac{d^3v}{dx^3} = -w_o x + c_1$$

$$EI\frac{d^2v}{dx^2} = -\tfrac{1}{2}w_o x^2 + c_1 x + c_2$$

$$EI\frac{dv}{dx} = -\tfrac{1}{6}w_o x^3 + \tfrac{1}{2}c_1 x^2 + c_2 x + c_3$$

$$EIv = -\tfrac{1}{24}w_o x^4 + \tfrac{1}{6}c_1 x^3 + \tfrac{1}{2}c_2 x^2 + c_3 x + c_4$$

We will now impose the boundary conditions. At the fixed end, both the deflection and rotation (slope) are constrained to be zero. At the pinned end, only the deflection is constrained to be zero; in other words, it is free to rotate, which in turn means there is no restraining moment. The four conditions are expressed as

$$\text{at } x = 0: \quad v = 0, \quad \frac{dv}{dx} = 0$$

$$\text{at } x = L: \quad v = 0, \quad M = EI\frac{d^2v}{dx^2} = 0$$

These give, respectively,

$$0 = c_4 \qquad\qquad\qquad\qquad 0 = c_3$$
$$0 = -\tfrac{1}{24}w_o L^4 + \tfrac{1}{6}c_1 L^3 + \tfrac{1}{2}c_2 L^2 + c_3 L + c_4 \qquad 0 = -\tfrac{1}{2}w_o L^2 + c_1 L + c_2$$

After solving for the coefficients, we find the deflected shape to be

$$EIv(x) = -\tfrac{1}{48}w_o x^2[2x^2 - 5xL + 3L^2]$$

We can obtain the other quantities of the solution by differentiation. This gives the slope, moment, and shear distributions as

$$EI\phi(x) = -\tfrac{1}{48}w_o[8x^3 - 15x^2L + 6xL^2]$$
$$M(x) = -\tfrac{1}{48}w_o[24x^2 - 30xL + 6L^2]$$
$$V(x) = +\tfrac{1}{48}w_o[48x - 30L]$$

These are shown plotted in the Figure 3.2.

For comparison, the corresponding solutions for the cases of the fixed-fixed and pinned-pinned beams are

$$\text{fixed-fixed:} \qquad EIv(x) = -\tfrac{1}{48}w_ox^2[2x^2 - 4xL + 2L^2]$$
$$\text{fixed-pinned:} \qquad EIv(x) = -\tfrac{1}{48}w_ox^2[2x^2 - 5xL + 3L^2]$$
$$\text{pinned-pinned:} \qquad EIv(x) = -\tfrac{1}{48}w_ox[2x^3 - 4x^2L + 2L^3] \qquad (3.5)$$

Both the fixed-fixed and pinned-pinned cases have symmetric distributions. The maximum deflection of the fixed-pinned case lies between the maximum for these two.

**Example 3.2:**   Find the deflected shape of the fixed-fixed beam shown in Figure 3.3. The concentrated applied load is located a distance $a$ from the end.

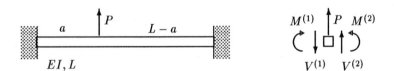

**Figure 3.3**: Fixed-Fixed beam with concentrated applied force.

Because of the change in loading due to the concentrated force, we will consider the beam to be made of two sections; compatibility and equilibrium conditions will then be imposed to match them at the connection. In the present problem, the loading on each section is zero and so it is convenient for us to start there. That is, for the first section

$$EI\frac{d^4v}{dx^4} = q(x) = 0$$
$$EIv = \tfrac{1}{6}c_1x^3 + \tfrac{1}{2}c_2x^2 + c_3x + c_4$$

We will first impose the boundary conditions of zero deflection and slope at the near end. That is,

$$\text{at } x = 0: \qquad v = 0, \qquad \frac{dv}{dx} = 0$$

We immediately obtain that $c_4 = 0$ and $c_3 = 0$, from which we can write the deflection as

$$EIv(x) = \tfrac{1}{6}c_1 x^3 + \tfrac{1}{2}c_2 x^2$$

In analyzing the second section, we will use the variable $x$ as a *local* variable, that is, it ranges over $0 \le x \le (L-a)$. With that in mind, we have

$$EI\frac{d^4 v}{dx^4} = q(x) = 0$$
$$EIv = \tfrac{1}{6}c_1^* x^3 + \tfrac{1}{2}c_2^* x^2 + c_3^* x + c_4^*$$

The boundary conditions for this section are zero deflection and slope at the end. That is

$$\text{at } x = L - a : \qquad v = 0, \qquad \frac{dv}{dx} = 0$$

Hence, we obtain that

$$c_3^* = -\tfrac{1}{2}c_1^*(L-a)^2 - c_2^*(L-a), \qquad c_4^* = \tfrac{1}{3}c_1^*(L-a)^3 + \tfrac{1}{2}c_2^*(L-a)^2$$

This allows the deflection shape to be written as

$$EIv(x) = \tfrac{1}{6}c_1^*[x^3 - 3x(L-a)^2 + 2(L-a)^3] + \tfrac{1}{2}c_2^*[x^2 - 2x(L-a) + (L-a)^2]$$

The four boundary conditions were insufficient to determine all eight of the unknown coefficients of integration. It is necessary for us to impose compatibility across the junction of the two sections. That is, at the junction

$$v^{(1)} = v^{(2)}, \qquad \frac{dv^{(1)}}{dx} = \frac{dv^{(2)}}{dx}$$

We still need more equations and to obtain them we will consider the equilibrium conditions in the vicinity of the junction. To this end, isolate a small segment near the applied load as shown in Figure 3.3, the equilibrium conditions give

$$M^{(1)} = M^{(2)}, \qquad V^{(1)} = V^{(2)} + P$$

The two compatibility and two equilibrium conditions become (when we rewrite them in terms of the unknown coefficients)

$$\tfrac{1}{6}c_1 a^3 + \tfrac{1}{2}c_2 a^2 = \tfrac{1}{3}c_1^*(a-L)^3 + \tfrac{1}{2}c_2^*(a-L)^2$$
$$\tfrac{1}{2}c_1 a^2 + c_2 a = -\tfrac{1}{2}c_1^*(a-L)^2 - c_2^*(a-L)$$
$$c_1 a + c_2 = c_2^*$$
$$-c_1 = -c_1^* + P/EI$$

These equations are now sufficient to solve for all of the coefficients, and consequently allow us to determine the deflection equation. In the special case $a = L/2$, the coefficients are

$$c_1 = -\tfrac{1}{2}P, \qquad c_2 = \tfrac{1}{4}Pa, \qquad c_1^* = \tfrac{1}{2}P, \qquad c_2^* = -\tfrac{1}{4}Pa$$

and the deflection curve is

$$EIv(x) = \frac{P}{48}[3x^2L - 4x^3] \qquad 0 \le x \le L/2$$

A similar expression can be obtained for the other half of the beam.

The important point of this last example is that when the integration of the differential equations is performed over a number of regions, then the compatibility and equilibrium conditions must be imposed at each junction. This is exactly what will happen later when we begin to model the structure as a collection of finite element beams.

## 3.2   Beam Element Stiffness Matrix

Consider a straight homogeneous beam element of length $L$ as shown in Figure 3.4. Assume that there are no external loads applied between the two ends of Node 1 and Node 2. At each node, there are two essential beam actions, namely, the bending moment $M_i$ and shear force $V_i$. The corresponding nodal degrees of freedom are the rotation $\phi_i$ (or the slope of the deflection curve at the node) and the vertical displacement $v_i$. The positive directions of nodal forces and moments and the corresponding displacements and rotations are shown in the figure.

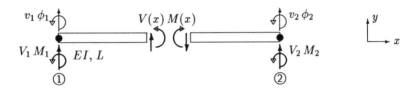

**Figure 3.4**: Sign convention for beam element.

The beam element stiffness matrix will be derived by directly integrating the governing differential equations for the beam. Since the loading between nodes is zero, we have

$$EI\frac{d^4v}{dx^4} = 0$$

giving the general solution for the deflection curve as

$$v(x) = a_0 + a_1x + a_2x^2 + a_3x^3$$

where $a_0, a_1, a_2$, and $a_3$ are constants. By using the following end conditions

$$v(0) = v_1, \qquad \frac{dv(0)}{dx} = \phi_1$$

$$v(L) = v_2, \qquad \frac{dv(L)}{dx} = \phi_2$$

we can rewrite the constants in terms of the nodal displacements $v_1$ and $v_2$, and the nodal rotations $\phi_1$ and $\phi_2$. Specifically, they are

$$a_0 = v_1$$

$$a_1 = \phi_1$$

$$a_2 = -\frac{3}{L^2}v_1 - \frac{2}{L}\phi_1 + \frac{3}{L^2}v_2 - \frac{1}{L}\phi_2$$

$$a_3 = \frac{2}{L^3}v_1 + \frac{1}{L^2}\phi_1 - \frac{2}{L^3}v_2 + \frac{1}{L^2}\phi_2$$

Substitution of these into the expression for the deflection leads us to

$$
\begin{aligned}
v(x) &= \left[1 - 3(\frac{x}{L})^2 + 2(\frac{x}{L})^3\right]v_1 + (\frac{x}{L})\left[1 - 2(\frac{x}{L}) + (\frac{x}{L})^2\right]L\phi_1 \\
&\quad + (\frac{x}{L})^2\left[3 - 2(\frac{x}{L})\right]v_2 + (\frac{x}{L})^2\left[-1 + (\frac{x}{L})\right]L\phi_2 \\
&\equiv g_1(x)v_1 + g_2(x)L\phi_1 + g_3(x)v_2 + g_4(x)L\phi_2
\end{aligned}
\tag{3.6}
$$

The functions $g_n(x)$ are called the *beam shape functions*. The complete description of the element is captured in the four nodal degrees of freedom $v_1$, $\phi_1$, $v_2$, and $\phi_2$ (since the shape functions are known explicitly). If, in any problem, these can be determined, then the solution has been obtained.

The slope, moment and shear force are obtained (in terms of the nodal degrees of freedom) simply by differentiation. For example, the moment distribution is

$$
\begin{aligned}
M(x) &= EI\frac{d^2v}{dx^2} = EI[g_1''(x)v_1 + g_2''(x)L\phi_1 + g_3''(x)v_2 + g_4''(x)L\phi_2] \\
&= \frac{EI}{L^3}[(-6L + 12x)v_1 + (-4L + 6x)L\phi_1 + (6L - 12x)v_2 + (-2L + 6x)L\phi_2]
\end{aligned}
$$

The end values are

$$M(0) = \frac{EI}{L^3}[-6Lv_1 - 4L^2\phi_1 + 6Lv_2 - 2L^2\phi_2], \quad M(L) = \frac{EI}{L^3}[6Lv_1 + 2L^2\phi_1 - 6Lv_2 + 4L^2\phi_2]$$

By considering a free body diagram of each end of the element, it is easy for us to establish the following relations among the member loads and the nodal values

$$
\begin{aligned}
V_1 &= -V(0), & M_1 &= -M(0) \\
V_2 &= +V(L), & M_2 &= +M(L)
\end{aligned}
\tag{3.7}
$$

Therefore, the nodal moments are easily related to the nodal degrees of freedom. By carrying out the indicated differentiations we can also relate the nodal forces to the nodal degrees of freedom. To summarize the above results in matrix form, we define the element nodal loads vector and the corresponding nodal displacements vector as the following column matrices

$$\{F\} \equiv \begin{Bmatrix} V_1 \\ M_1 \\ V_2 \\ M_2 \end{Bmatrix} \qquad \text{and} \qquad \{u\} \equiv \begin{Bmatrix} v_1 \\ \phi_1 \\ v_2 \\ \phi_2 \end{Bmatrix}$$

respectively. Then the nodal loads-displacement relations can be rearranged into the following form

$$\begin{Bmatrix} V_1 \\ M_1 \\ V_2 \\ M_2 \end{Bmatrix} = \{F\} = [\,k\,]\{u\} = \frac{EI}{L^3} \begin{bmatrix} 12 & 6L & -12 & 6L \\ 6L & 4L^2 & -6L & 2L^2 \\ -12 & -6L & 12 & -6L \\ 6L & 2L^2 & -6L & 4L^2 \end{bmatrix} \begin{Bmatrix} v_1 \\ \phi_1 \\ v_2 \\ \phi_2 \end{Bmatrix} \tag{3.8}$$

where $[\,k\,]$ is the *beam element stiffness matrix*. Note that this stiffness matrix is symmetric. Also note that the nodal loads satisfy the equilibrium conditions for the free body diagram of the element. This is as it should be, since the relation was obtained by integrating the differential form of equilibrium.

The beam element stiffness (unlike that for the rod) contains dimensional quantities inside the brackets. This comes about because the vector terms have mixed quantities; that is, rotation is non-dimensional but deflection has the units of length. An interesting alternative form of the above relation is

$$\begin{Bmatrix} V_1 \\ M_1/L \\ V_2 \\ M_2/L \end{Bmatrix} = \frac{EI}{L^3} \begin{bmatrix} 12 & 6 & -12 & 6 \\ 6 & 4 & -6 & 2 \\ -12 & -6 & 12 & -6 \\ 6 & 2 & -6 & 4 \end{bmatrix} \begin{Bmatrix} v_1 \\ L\phi_1 \\ v_2 \\ L\phi_2 \end{Bmatrix}$$

This gives dimension of force for all load terms and dimension of length for all degrees of freedom, and makes the terms inside the matrix dimensionless. As a result the relation is dimensionally similar to that for rods. While this form has a certain appeal, it is unsuited for our purpose — we wish to assemble many elements of (possibly) different lengths to form a structure and this task is made easier by having the vector of degrees of freedom common from element to element. In the above form, these vectors are element dependent since they contain the element length.

When the loads or displacement vectors contain mixed terms (e.g., force and moment) they are often referred to as *generalized* vectors. In Chapter 6, we will show that it is possible to establish quite general results in terms of generalized loads and displacements without having to specify them explicitly. Therefore, using mixed load vectors at this stage does not lead to any difficulty when we move onto more complicated structures.

## 3.3   Structural Stiffness Matrix

We saw in the earlier examples that when the region of integration is divided over a
number of beam segments it is necessary to enforce compatibility and equilibrium at
the junctions between the segments. We obtain the structural stiffness matrix $[\,K\,]$ by
imposing the equilibrium conditions at the nodes connecting the members together
and also by insuring compatibility of the displacements.

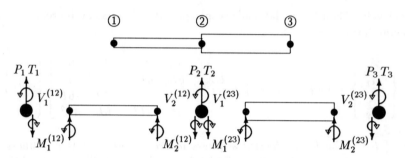

**Figure 3.5**: Beam structure with three nodes.

As an example, consider the three-noded beam structure as shown in Figure 3.5.
The external applied forces and moments are denoted by $P_1, P_2, P_3$ and $T_1, T_2, T_3$,
respectively. The superscript designates the element by its nodes.

The equilibrium equations for the three nodes are

$$
\begin{aligned}
P_1 &= V_1^{(12)} \\
T_1 &= M_1^{(12)} \\
P_2 &= V_2^{(12)} + V_1^{(23)} \\
T_2 &= M_2^{(12)} + M_1^{(23)} \\
P_3 &= V_2^{(23)} \\
T_3 &= M_2^{(23)}
\end{aligned}
$$

Note that the total number of equilibrium equations (6) is equal to the number of
nodes (3) times the number of equilibrium equations at each node (2), and that this
coincides with the total number of degrees of freedom of the system (6). We can
express the above equations in matrix form as:

$$
\begin{Bmatrix} P_1 \\ T_1 \\ P_2 \\ T_2 \\ P_3 \\ T_3 \end{Bmatrix}
=
\begin{Bmatrix} V_1 \\ M_1 \\ V_2 \\ M_2 \\ 0 \\ 0 \end{Bmatrix}^{(12)}
+
\begin{Bmatrix} 0 \\ 0 \\ V_1 \\ M_1 \\ V_2 \\ M_2 \end{Bmatrix}^{(23)}
\equiv \{F\}^{(12)} + \{F\}^{(23)}
$$

This is arranged so that there are as many column vectors on the right hand side as there are members. We now replace each element nodal load vector by use of the element stiffness relation augmented to the system size. That is,

$$\{F\}^{(12)} = \begin{bmatrix} [k^{(12)}] & 0 \\ 0 & 0 \end{bmatrix} \{u\}^{(12)}, \qquad \{F\}^{(23)} = \begin{bmatrix} 0 & 0 \\ 0 & [k^{(23)}] \end{bmatrix} \{u\}^{(23)}$$

Compatibility of displacement and rotation at the joint can be satisfied by requiring that the nodal degrees of freedom be common between the element and the structural node. That is, we write

$$\{u\}^{(12)} = \begin{Bmatrix} v_1 \\ \phi_1 \\ v_2 \\ \phi_2 \\ 0 \\ 0 \end{Bmatrix}^{(12)} = \begin{Bmatrix} v_1 \\ \phi_1 \\ v_2 \\ \phi_2 \\ v_3 \\ \phi_3 \end{Bmatrix}, \qquad \{u\}^{(23)} = \begin{Bmatrix} 0 \\ 0 \\ v_2 \\ \phi_2 \\ v_3 \\ \phi_3 \end{Bmatrix}^{(23)} = \begin{Bmatrix} v_1 \\ \phi_1 \\ v_2 \\ \phi_2 \\ v_3 \\ \phi_3 \end{Bmatrix}$$

As a result the structural equilibrium equations can be written is short form as

$$\{P\} = [K]\{u\}$$

where

$$\{P\} \equiv \begin{Bmatrix} P_1 \\ T_1 \\ P_2 \\ T_2 \\ P_3 \\ T_3 \end{Bmatrix}, \qquad \{u\} \equiv \begin{Bmatrix} v_1 \\ \phi_1 \\ v_2 \\ \phi_2 \\ v_3 \\ \phi_3 \end{Bmatrix}, \qquad [K] \equiv \begin{bmatrix} [k^{(12)}] & 0 \\ 0 & 0 \end{bmatrix} + \begin{bmatrix} 0 & 0 \\ 0 & [k^{(23)}] \end{bmatrix}$$

where $[K]$ is the *structural stiffness matrix* of the system, and is obviously the simple superposition of the respective element stiffnesses augmented to the full structural size. This is precisely the same result that occurred when we analyzed the rod structure. In fact, the pattern that is emerging when forming the structural stiffness matrix is one that is true for all linear elastic structures.

**Example 3.3:** A fixed-fixed beam, made of two sections, is loaded as shown in Figure 3.6. Obtain the response (deflection and rotation) at the load point.

We will use the quick procedure introduced in the previous chapter to assemble the reduced stiffness matrix. First divide the beam into two elements with three nodes and label them as shown in the figure. The total system degrees of freedom are

$$\{u\} = \{v_1, \phi_1 ; v_2, \phi_2 ; v_3, \phi_3\}$$

Thus the system size is $[6 \times 6]$. The boundary conditions require that

$$v_1 = \phi_1 = v_3 = \phi_3 = 0$$

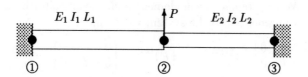

**Figure 3.6**: Two element problem.

and therefore the reduced system has only the following two free degrees of freedom

$$\{u_u\} = \{v_2, \phi_2\}$$

The corresponding $\{IDbc\}$ matrix is

$$\{IDbc\} = \{0,0;1,2;0,0\}$$

The two unknown nodal displacements and the given loads are, respectively,

$$\{u_u\} = \begin{Bmatrix} v_2 \\ \phi_2 \end{Bmatrix}, \qquad \{P_k\} = \begin{Bmatrix} P_2 \\ T_2 \end{Bmatrix} = \begin{Bmatrix} P \\ 0 \end{Bmatrix}$$

In order to form the stiffness matrix, we must first form the element stiffness matrix and then 'scratch' the rows and columns corresponding to the zero degrees of freedom. The contributing reduced element stiffness matrices for our problem are determined to be

$$[k^{*(12)}] = \frac{E_1 I_1}{L_1^3} \begin{bmatrix} 12 & -6L_1 \\ -6L_1 & 4L_1^2 \end{bmatrix}, \qquad [k^{*(23)}] = \frac{E_2 I_2}{L_2^3} \begin{bmatrix} 12 & 6L_2 \\ 6L_2 & 4L_2^2 \end{bmatrix}$$

The assembled reduced structural stiffness matrix is therefore

$$[K^*] = \frac{E_1 I_1}{L_1^3} \begin{bmatrix} 12 + 12r_1 & -6 + 6r_1 r_2 \\ -6 + 6r_1 r_2 & 4 + 4r_1 r_2^2 \end{bmatrix}, \quad r_1 \equiv (\frac{EI}{L^3})_1 / (\frac{EI}{L^3})_2, \quad r_2 \equiv (L_1)/(L_2)$$

As a special case, consider when each segment is the same, that is, $E_1 = E_2 = E$, $I_1 = I_2 = I$, $L_1 = L_2 = L$. Then

$$[K^*] = \frac{EI}{L^3} \begin{bmatrix} 24 & 0 \\ 0 & 8L^2 \end{bmatrix}$$

Note how the off-diagonal terms are zero, indicating uncoupling of the degrees of freedom. We can now obtain the inverse of this matrix by forming the reciprocal of the diagonal terms. This gives

$$[K^*]^{-1} = \frac{L}{24EI} \begin{bmatrix} L^2 & 0 \\ 0 & 3 \end{bmatrix}$$

The two unknown nodal displacements are therefore

$$\begin{Bmatrix} v_2 \\ \phi_2 \end{Bmatrix} = [K^*]^{-1} \begin{Bmatrix} P_2 \\ T_2 \end{Bmatrix} = \frac{L}{24EI} \begin{Bmatrix} L^2 P_2 \\ 3T_2 \end{Bmatrix} = \frac{PL^3}{24EI} \begin{Bmatrix} 1 \\ 0 \end{Bmatrix}$$

The rotation at Node 2 is zero showing that there is symmetry of the deflection. Both results are identical to that obtained in Example 3.2. This is a reminder that the matrix method can give the exact result.

**Example 3.4:**   A uniform beam is pinned at two points and attached to a vertical roller as shown in Figure 3.7. Set up the reduced structural stiffness matrix for this problem.

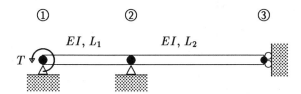

**Figure 3.7**: Three node problem.

We start by dividing the beam into two elements with three nodes, and number them as shown. The total system degrees of freedom are

$$\{u\} = \{v_1, \phi_1; v_2, \phi_2; v_3, \phi_3\}$$

giving a total system size of $[6 \times 6]$. The boundary conditions (or geometric constraints) require that

$$v_1 = 0, \qquad v_2 = 0, \qquad \phi_3 = 0$$

and therefore we can describe the reduced system in terms of

$$\{u_u\} = \{\phi_1, \phi_2, v_3\}$$

The corresponding $\{IDbc\}$ matrix is

$$\{IDbc\} = \{0, 1; 0, 2; 3, 0\}$$

The unknown displacements and known forces are

$$\{u_u\} = \begin{Bmatrix} \phi_1 \\ \phi_2 \\ v_3 \end{Bmatrix}, \qquad \{P_k\} = \begin{Bmatrix} T_1 = T \\ T_2 = 0 \\ P_2 = 0 \end{Bmatrix}$$

The contributing reduced element stiffness matrices are

$$[k^{*(12)}] = \frac{EI}{L_1^3} \begin{bmatrix} 4L_1^2 & 2L_1^2 \\ 2L_1^2 & 4L_1^2 \end{bmatrix}, \qquad [k^{*(23)}] = \frac{EI}{L_2^3} \begin{bmatrix} 4L_2^2 & -6L_2 \\ -6L_2 & 12 \end{bmatrix}$$

The reduced structural stiffness matrix is obtained as (using $L_2 = 2L_1 = 2L$)

$$[K^*] = \frac{EI}{L^3} \begin{bmatrix} 4L^2 & 2L^2 + 0 & 0 + 0 \\ 2L^2 & 4L^2 + 2L^2 & 0 - 12L/8 \\ 0 + 0 & 0 - 12L/8 & 0 + 12/8 \end{bmatrix} = \frac{EI}{L^3} \begin{bmatrix} 4L^2 & 2L^2 & 0 \\ 2L^2 & 6L^2 & -3L/2 \\ 0 & -3L/2 & 3/2 \end{bmatrix}$$

It would now be a straight-forward matter for us to solve the reduced system of equations. For future reference, the results are

$$\left\{ \begin{array}{c} \phi_1 \\ \phi_2 \\ v_3 \end{array} \right\} = \frac{TL}{28EI} \left\{ \begin{array}{c} 9 \\ -4 \\ -4L \end{array} \right\}$$

These values are used in Examples 3.6 and 3.7 to find the load distributions and reactions.

## 3.4   Equivalent Loads

In the matrix method for beams developed so far, we have restricted the applied loads to be concentrated forces and moments at the nodes. We will demonstrate that it is possible for us to develop other elements that take certain types of distributed loads into account exactly. Our main interest here, however, will be to develop a scheme that converts the distributed loads into equivalent concentrated forces and moments. In this process, the matrix solution loses its exactness; but keep in mind that any desired degree of accuracy can always be achieved by increasing the number of nodes. (This is at the expense of computing time, but has the advantage of being simple.) The question now is: Is there a best way to convert a distributed load into a set of concentrated loads? We will investigate two approximate ways of doing this.

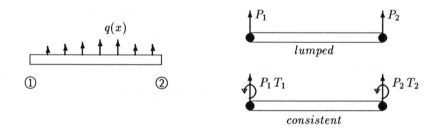

**Figure 3.8**: Replacement of arbitrary load.

### Exact Treatment of Distributed Loads

We follow the procedure used in Section 2.7 for rods; only the major steps will be reported here.

Since the loading between nodes is not zero, we have for the beam equation

$$EI\frac{d^4v}{dx^4} = q(x)$$

giving the general solution for the deflection curve as

$$v(x) = a_0 + a_1 x + a_2 x^2 + a_3 x^3 + w(x), \qquad EI\frac{d^4 w}{dx^4} \equiv q(x)$$

where $a_0, a_1, a_2$, and $a_3$ are constants. By using the following end conditions

$$v(0) = v_1, \qquad \frac{dv(0)}{dx} = \phi_1$$

$$v(L) = v_2, \qquad \frac{dv(L)}{dx} = \phi_2$$

we can rewrite the constants in terms of the nodal displacements. Substitution of these into the expression for the deflection leads us to

$$v(x) = g_1(x)(v_1 - w_1) + g_2(x)L(\phi_1 - w_1') + g_3(x)(v_2 - w_2) + g_4(x)L(\phi_2 - w_2') + w(x) \quad (3.9)$$

where the functions $g_n(x)$ are the beam shape functions of Equation(3.6) and $w_1 = w(0)$, $w_1' = dw(0)/dx$, and so on.

The moment and shear force distributions are obtained (in terms of the nodal degrees of freedom) by differentiation. After we evaluate the end values, we can arrange them in matrix form as

$$\begin{Bmatrix} V_1 \\ M_1 \\ V_2 \\ M_2 \end{Bmatrix} = \frac{EI}{L^3} \begin{bmatrix} 12 & 6L & -12 & 6L \\ 6L & 4L^2 & -6L & 2L^2 \\ -12 & -6L & 12 & -6L \\ 6L & 2L^2 & -6L & 4L^2 \end{bmatrix} \begin{Bmatrix} v_1 - w_1 \\ \phi_1 - w_1' \\ v_2 - w_2 \\ \phi_2 - w_2' \end{Bmatrix} - EI \begin{Bmatrix} w_1'' \\ -w_1''' \\ -w_2'' \\ w_2''' \end{Bmatrix}$$

The final step is to replace the function $w(x)$ by the distributed load function $q(x)$. We can show by integration by parts that the following relation holds true

$$\int_0^L g_1(x)q(x)\,dx = k_{11}w_1 + k_{12}w_1' + k_{13}w_2 + k_{14}w_2' - EI\frac{d^3 w_1}{dx^3}$$

Similar results can be obtained for the products of $q(x)$ with the other beam shape functions. They allow us to simplify the stiffness relation to

$$\begin{Bmatrix} V_1 \\ M_1 \\ V_2 \\ M_2 \end{Bmatrix} = \frac{EI}{L^3} \begin{bmatrix} 12 & 6L & -12 & 6L \\ 6L & 4L^2 & -6L & 2L^2 \\ -12 & -6L & 12 & -6L \\ 6L & 2L^2 & -6L & 4L^2 \end{bmatrix} \begin{Bmatrix} v_1 \\ \phi_1 \\ v_2 \\ \phi_2 \end{Bmatrix} - \int_0^L \begin{Bmatrix} g_1 q \\ L g_2 q \\ g_3 q \\ L g_4 q \end{Bmatrix} dx \quad (3.10)$$

We have thus succeeded in replacing the distributed load by equivalent loads associated with the nodes and given by

$$P_1 = \int_0^L q(x) g_1(x)\,dx, \qquad P_2 = \int_0^L q(x) g_3(x)\,dx$$

$$T_1 = L\int_0^L q(x) g_2(x)\,dx, \qquad T_2 = L\int_0^L q(x) g_4(x)\,dx \quad (3.11)$$

These loads are referred to as *initial stress* terms; that is, they are loads in addition to the nodal loads $\{F\}$. The initial stress loads are also called the *consistent* load representations because they involve displacement shape functions consistent with the stiffness matrix formulation. The assembled system of equations take the form

$$\{P + Q\} = [\,K\,]\{u\}$$

where $\{P\}$ is the collection of applied nodal loads and $\{Q\}$ is the assembled form of the initial stress loads. We see from this that the distributed load can be accounted for by using a set of equivalent nodal loads.

The consistent loads are a statically equivalent load system. This can be shown in general, but we demonstrate it with the following special cases. For a uniform distributed load $q_0$, for example,

$$P_1 = \tfrac{1}{2} q_0 L, \qquad\qquad P_2 = \tfrac{1}{2} q_0 L$$
$$T_1 = +\tfrac{1}{12} q_0 L^2, \qquad T_2 = -\tfrac{1}{12} q_0 L^2$$

The resultant vertical force is $P_1 + P_2 = q_0 L$ which is in agreement with the total distributed load. The resultant moment about the first node is $T_1 + T_2 + P_2 L = \tfrac{1}{2} q_0 L^2$ again in agreement with the value from the distributed load. The corresponding values for a linear distribution of load with a maximum of $q(L) = q_m$ are

$$P_1 = \tfrac{3}{20} q_m L, \qquad\qquad P_2 = \tfrac{7}{20} q_m L$$
$$T_1 = \tfrac{1}{30} q_m L^2, \qquad\qquad T_2 = -\tfrac{1}{20} q_m L^2$$

The resultant force and moment are $\tfrac{1}{2} q_m L$ and $\tfrac{1}{3} q_m L^2$, respectively. Notice that the consistent loads have moments even though the applied distributed load does not.

## Consistent Load Approximation

The exact formulation of the distributed load suggests that we can replace the distributed load with a set of equivalent nodal loads. To maintain the exact solution we must retain the initial stress term in the element stiffness relation. An approximation often resorted to is the use of the consistent load formulas but neglecting the initial stress terms. This will give values for the displacement that are exact at the nodal points, but the computed nodal loads will be in error.

## Lumped Load Approximation

For comparison, we will now develop an alternative approximate representation for a distributed load. This is worth doing because in many ways it is more intuitive than the previous developments.

Consider a beam element subjected to an arbitrary distributed load as shown in Figure 3.8. Imagine the distributed load as being made up of sandbags. Now move the sandbags to their nearest node. That is, we have lumped the total load into two statically equivalent concentrated loads at the nodes. The mathematical expressions for this are

$$P_1 = \int_0^{L/2} q(x) \, dx \,, \qquad P_2 = \int_{L/2}^{L} q(x) \, dx \qquad (3.12)$$

This is called the *lumped load* approximation and is essentially a scheme for replacing the actual distribution in terms of its average. It is evident that when $q(x)$ is highly irregular, many nodes are required to yield an accurate representation.

For a uniform distributed load $q_0$, for example, the lumped load representations are

$$P_1 = \tfrac{1}{2} q_0 L \,, \qquad P_2 = \tfrac{1}{2} q_0 L$$

That is, half of the total load is placed at each node. Whereas, if the distribution is linear with a maximum of $q(L) = q_m$, the corresponding lumped loads are

$$P_1 = \tfrac{1}{8} q_m L \,, \qquad P_2 = \tfrac{3}{8} q_m L$$

In comparison to the consistent load representation, we see that the lumped representation does not contain any end moments.

**Example 3.5:** A simply-supported beam is subjected to a uniform distributed weight of $w_0$ as shown in Figure 3.9. Solve for the maximum deflection and moment. Compare with the exact solution.

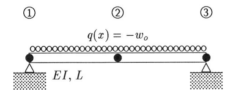

**Figure 3.9**: Simply supported beam with distributed load.

We have already quoted the exact solution in Equation( 3.5) as

$$v(x) = \frac{-w_o x}{24 EI}(x^3 - 2Lx^2 + L^3) \,, \qquad M(x) = \frac{-w_o x}{2}(x - L)$$

The maximum deflection and moment occurs as $x = L/2$ and are

$$v_{max} = \frac{-5 w_0 L^4}{384 EI} \,, \qquad M_{max} = \frac{-w_0 L^2}{8}$$

The shear force distribution is linear with a value of zero at $x = L/2$.

### Exact Matrix Solution

We will model this problem using two elements as shown in Figure 3.9. The equivalent loads obtained by using the consistent approach are

$$1-2: \qquad P_1 = -\tfrac{1}{4}w_0 L \qquad P_2 = -\tfrac{1}{4}w_0 L \qquad T_1 = -\tfrac{1}{48}w_0 L^2 \qquad T_2 = +\tfrac{1}{48}w_0 L^2$$

$$2-3: \qquad P_1 = -\tfrac{1}{4}w_0 L \qquad P_2 = -\tfrac{1}{4}w_0 L \qquad T_1 = -\tfrac{1}{48}w_0 L^2 \qquad T_2 = +\tfrac{1}{48}w_0 L^2$$

Note that the applied moments at the center cancel each other, and that the forces at the ends do not cause deflections. Therefore, the equivalent problem is that of a simply supported beam under the action of a concentrated force $-\tfrac{1}{2}w_0 L$ acting at the center and two concentrated end moments $\tfrac{1}{48}w_0 L^2$ acting in opposite directions.

Because of symmetry we actually need model only one half of the beam. In doing so, however, we must retain the $T_2$ loading for the first element even though it cancels when assembled with the second element. Imposing the constraints that $v_1 = 0$ and $\phi_2 = 0$ we get the reduced stiffness relation as

$$\left\{ \begin{array}{c} M_1 \\ V_2 \end{array} \right\} = \frac{8EI}{L^3} \left[ \begin{array}{cc} L^2 & -3L \\ -3L & 12 \end{array} \right] \left\{ \begin{array}{c} \phi_1 \\ v_2 \end{array} \right\} - \left\{ \begin{array}{c} T_1 = -w_0 L^2/48 \\ P_2 = -w_0 L/4 \end{array} \right\}$$

After assemblage, since there are no other applied loads, we get

$$\frac{8EI}{L^3} \left[ \begin{array}{cc} L^2 & -3L \\ -3L & 12 \end{array} \right] \left\{ \begin{array}{c} \phi_1 \\ v_2 \end{array} \right\} = \left\{ \begin{array}{c} w_0 L^2/48 \\ w_0 L/4 \end{array} \right\}$$

The solution gives

$$\phi_1 = \frac{-w_0 L^3}{24EI}, \qquad v_2 = \frac{-5w_0 L^4}{384EI}$$

which agrees with the exact solution. To determine the nodal loads, we need to go back to the element stiffness relation that includes the initial stress term. That is,

$$\left\{ \begin{array}{c} V_1 \\ M_1 \\ V_2 \\ M_2 \end{array} \right\} = \frac{8EI}{L^3} \left[ \begin{array}{cccc} 12 & 3L & -12 & 3L \\ 3L & L^2 & -3L & \tfrac{1}{2}L^2 \\ -12 & -3L & 12 & -3L \\ 3L & \tfrac{1}{2}L^2 & -3L & L^2 \end{array} \right] \left\{ \begin{array}{c} 0 \\ \phi_1 \\ v_2 \\ 0 \end{array} \right\} - \left\{ \begin{array}{c} -w_0 L/4 \\ -w_0 L^2/48 \\ -w_0 L/4 \\ +w_0 L^2/48 \end{array} \right\}$$

After substituting for $\phi_1$ and $v_2$ we get

$$\left\{ \begin{array}{c} V_1 \\ M_1 \\ V_2 \\ M_2 \end{array} \right\} = \frac{w_0 L}{48} \left\{ \begin{array}{c} 12 \\ -L \\ -12 \\ 7L \end{array} \right\} - \frac{w_0 L}{48} \left\{ \begin{array}{c} -12 \\ -L \\ -12 \\ L \end{array} \right\} = \frac{w_0 L}{48} \left\{ \begin{array}{c} 24 \\ 0 \\ 0 \\ 6L \end{array} \right\}$$

These results agree with the exact solution.

### Consistent Load Approximation

The assembled matrix system is identical to that of the exact matrix formulation, hence the exact displacements are obtained. The difference occurs in determining the nodal loads since the initial stress term is neglected in the element stiffness relation. Therefore, we can write from above

$$\left\{ \begin{array}{c} V_1 \\ M_1 \\ V_2 \\ M_2 \end{array} \right\} = \frac{w_0 L}{48} \left\{ \begin{array}{c} 12 \\ -L \\ -12 \\ 7L \end{array} \right\}$$

The moment at the center of the beam is in error by about 2%. A non-zero value of shear force is also found at that location.

### Lumped Load

Realizing that the length of each element is $L/2$, then the equivalent lumped loads acting on the nodes of each element are

$$1-2: \qquad P_1 = -\tfrac{1}{4}w_0 L \qquad\qquad P_2 = -\tfrac{1}{4}w_0 L$$

$$2-3: \qquad P_1 = -\tfrac{1}{4}w_0 L \qquad\qquad P_2 = -\tfrac{1}{4}w_0 L$$

Note that the loads at Nodes 1 and 3 are taken directly by the supports and do not contribute to the deflection. Therefore the equivalent problem is that of a simply supported beam under the action of a single concentrated force $-\tfrac{1}{2}w_0 L$ acting at the center. The reduced system of equations are

$$\frac{8EI}{L^3}\begin{bmatrix} L^2 & -3L \\ -3L & 12 \end{bmatrix}\left\{ \begin{array}{c} \phi_1 \\ v_2 \end{array} \right\} = \left\{ \begin{array}{c} T_1 = 0 \\ P_2 = w_0 L/4 \end{array} \right\}$$

The solution gives

$$\phi_1 = \frac{-w_0 L^3}{32EI}, \qquad v_2 = \frac{-4w_0 L^4}{384EI}$$

which are off compared to the exact solution. Actually, the deflection is off by 20%. The nodal loads are determined from the element stiffness relation

$$\left\{ \begin{array}{c} V_1 \\ M_1 \\ V_2 \\ M_2 \end{array} \right\} = \frac{8EI}{L^3}\begin{bmatrix} 12 & 3L & -12 & 3L \\ 3L & L^2 & -3L & \tfrac{1}{2}L^2 \\ -12 & -3L & 12 & -3L \\ 3L & \tfrac{1}{2}L^2 & -3L & L^2 \end{bmatrix}\left\{ \begin{array}{c} 0 \\ \phi_1 \\ v_2 \\ 0 \end{array} \right\} = \frac{w_0 L}{48}\left\{ \begin{array}{c} 12 \\ 0 \\ -12 \\ 6L \end{array} \right\}$$

The maximum moment agrees with the exact solution.

## 3.5   Elastic Supports

Sometimes, the boundary constraints (or supports) are neither "fixed" nor "free" but lie somewhere in between. We can usually treat these adequately as elastic supports with a given spring constant. We now show how these can be incorporated into the stiffness formulation.

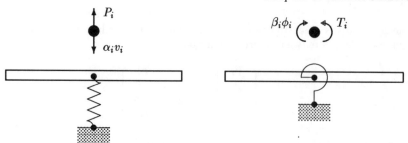

**Figure 3.10**: Spring action at nodes.

When a beam rests on a system of springs at some of the nodes, constraints are exerted on the beam system. For a linear spring and a rotational spring, for example, we have reactions of

$$P = -\alpha v, \qquad T = -\beta\phi$$

where $\alpha$ is the linear spring constant and $\beta$ is the rotational spring constant. Consider Node $i$, as shown in Figure 3.10. The resultant applied linear force is $(P_i - \alpha_i v_i)$, that is, for a positive deflection the spring exerts a restoring force. Similarly, for the torsional spring, the resultant applied moment is $(T_i - \beta_i \phi_i)$. Therefore, during the assemblage process, the equilibrium equations are modified to become

$$[K]\{u\} = \left\{ \begin{array}{c} P_1 - \alpha_1 v_1 \\ T_1 - \beta_1 \phi_1 \\ \vdots \\ P_n - \alpha_n v_n \\ T_n - \beta_n \phi_n \end{array} \right\}$$

The right hand side contains a set of unknowns (the displacements $v_i$ and rotations $\phi_i$) and so this system cannot be solved as is. We recognize, however, that the unknowns are simple functions of the system degrees of freedom; and their contribution to the load may simply be written as

$$[K_s]\{u\}$$

where we defined the spring stiffness matrix as

$$[K_s] \equiv \begin{bmatrix} \alpha_1 & 0 & \cdots & 0 & 0 \\ 0 & \beta_1 & \cdots & 0 & 0 \\ \vdots & \vdots & \ddots & \vdots & \vdots \\ 0 & 0 & \cdots & \alpha_n & 0 \\ 0 & 0 & \cdots & 0 & \beta_n \end{bmatrix}$$

It is now possible to rearrange the equilibrium equations as

$$[K]\{u\} + [K_s]\{u\} = ([K] + [K_s])\{u\} = [\bar{K}]\{u\} = \begin{Bmatrix} P_1 \\ T_1 \\ \vdots \\ P_n \\ T_n \end{Bmatrix}$$

At this stage, we have re-formed the elastic constraint problem so that it looks the same as what we have already encountered. This is not surprising if we realize that a spring is really just another elastic element that is added during assemblage. Actually, the linear spring is simply a rod element perpendicular to the beam, whereas the torsional spring is a shaft element. (Handling these sorts of elements in a 3-D arrangement is treated in the next chapter.) One further point to note is that the spring acting at Node $i$ affects only the diagonal stiffness term associated with that degree of freedom.

**Example 3.6:**   Consider the uniform beam attached to a spring at its mid-span, as shown in Figure 3.11. Determine the resulting deflection and rotation at the point of attachment for various values of spring stiffness.

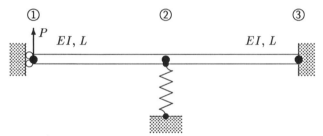

**Figure 3.11**: Beam on elastic support.

We will model this beam with two elements (three nodes). The reduced degrees of freedom are

$$\{u_u\} = \begin{Bmatrix} v_1 \\ v_2 \\ \phi_2 \end{Bmatrix}, \qquad \{P_k\} = \begin{Bmatrix} P_1 = P \\ P_2 = 0 \\ T_2 = 0 \end{Bmatrix}$$

The reduced element stiffness matrix for Elements 1-2 and 2-3 are, respectively,

$$[k^{*(12)}] = \frac{EI}{L^3} \begin{bmatrix} 12 & -12 & 6L \\ & 12 & -6L \\ sym & & 4L^2 \end{bmatrix}, \qquad [k^{*(23)}] = \frac{EI}{L^3} \begin{bmatrix} 12 & 6L \\ 6L & 4L^2 \end{bmatrix}$$

The resulting reduced structural stiffness matrix is (assuming both element lengths are the same)

$$[K^*] = \frac{EI}{L^3} \begin{bmatrix} 12 & -12 & 6L \\ & 12+12 & -6L+6L \\ sym & & 4L^2+4L^2 \end{bmatrix} = \frac{EI}{L^3} \begin{bmatrix} 12 & -12 & 6L \\ & 24 & 0 \\ sym & & 8L^2 \end{bmatrix}$$

The reduced spring constant matrix $[K_s]$ is simply the diagonal term associated with the vertical deflection at Node 2. This gives the overall system to be solved as

$$\frac{EI}{L^3} \begin{bmatrix} 12 & -12 & 6L \\ & 24 + \alpha^* & 0 \\ sym & & 8L^2 \end{bmatrix} \begin{Bmatrix} v_1 \\ v_2 \\ \phi_2 \end{Bmatrix} = \begin{Bmatrix} P \\ 0 \\ 0 \end{Bmatrix}, \qquad \alpha^* \equiv \frac{\alpha L^3}{EI}$$

The response at Node 2 is determined to be

$$v_2 = \frac{PL^3}{EI} \frac{16}{(48 + 10\alpha^*)}, \qquad \phi_2 = \frac{-PL^2}{EI} \frac{(24 + \alpha^*)}{(48 + 10\alpha^*)}$$

In the limit of a very flexible spring ($\alpha = 0$), these give, respectively,

$$v_2 = \frac{PL^3}{EI} \frac{1}{3}, \qquad \phi_2 = \frac{-PL^2}{EI} \frac{1}{2}$$

The corresponding results in the limit of a very stiff spring ($\alpha = \infty$) are, respectively,

$$v_2 = 0, \qquad \phi_2 = \frac{-PL^2}{EI} \frac{1}{10}$$

As expected, there is no vertical deflection. It is important to note, however, that there is a rotation. Indeed the very stiff spring acts as a pinned support. To achieve a fixed support, it would be necessary to also attach a very stiff torsional spring.

**Example 3.7:** Reconsider the problem of the last example, but this time place the applied load at the point of attachment of the spring.

The beam is modeled as in the last example, except that the applied load vector is different and given by

$$\{P_k\} = \begin{Bmatrix} P_1 = 0 \\ P_2 = P \\ T_2 = 0 \end{Bmatrix}$$

This gives the overall system to be solved as

$$\frac{EI}{L^3} \begin{bmatrix} 12 & -12 & 6L \\ & 24 + \alpha^* & 0 \\ sym & & 8L^2 \end{bmatrix} \begin{Bmatrix} v_1 \\ v_2 \\ \phi_2 \end{Bmatrix} = \begin{Bmatrix} 0 \\ P \\ 0 \end{Bmatrix}, \qquad \alpha^* \equiv \frac{\alpha L^3}{EI}$$

The response at Node 2 is determined to be

$$v_2 = \frac{PL^3}{EI} \frac{10}{(48 + 10\alpha^*)}, \qquad \phi_2 = \frac{-PL^2}{EI} \frac{12}{(48 + 10\alpha^*)}$$

Now in the limit of a very stiff spring ($\alpha = \infty$) these give respectively

$$v_2 = 0, \qquad \phi_2 = 0$$

As expected, there is no response at all in the beam.

Suppose the applied load is given a value calculated from

$$P = \alpha v_o = \frac{EI}{L^3} \alpha^* v_o$$

where $v_o$ is a specified displacement. Now the response at Node 2 is

$$v_2 = \frac{10\alpha^* v_o}{(48 + 10\alpha^*)}, \qquad \phi_2 = \frac{12\alpha^* v_o}{(48 + 10\alpha^*)}$$

In the limit of a very stiff spring ($\alpha = \infty$) these give, respectively,

$$v_2 = v_o, \qquad \phi_2 = \frac{1}{4} v_o$$

In other words, a deflection of a given amount, $v_o$, is achieved at Node 2.

## Imposed Displacements

Not all problems of interest are posed in terms of applied loads; it often happens that we know information in the form of imposed displacements. The exact approach to solving these problems is to rearrange the stiffness equations explicitly. For example, consider a $[3 \times 3]$ system where we wish to impose a known displacement at the second node. The rearranged system of equations are

$$\begin{bmatrix} K_{11} & 0 & K_{13} \\ 0 & 1 & 0 \\ K_{13} & 0 & K_{33} \end{bmatrix} \begin{Bmatrix} u_1 \\ u_2 \\ u_3 \end{Bmatrix} = \begin{Bmatrix} P_1 - K_{12} u_o \\ u_o \\ u_2 - K_{23} u_o \end{Bmatrix}$$

The results of the last example suggests an alternative scheme for imposing known (non-zero) displacements as a boundary condition. The required displacements can be imposed by adding the constraint equations that express the prescribed displacement conditions into the structural equilibrium equations. Assume that the displacement is to be specified at degree of freedom $i$, say $u_i = u_o$, then the constraint equation

$$k u_i = k u_o$$

is added to the equilibrium equations as was done for the spring. This means that a spring stiffness of $k$ is added to the diagonal while simultaneously a load of $P_i = k u_o$ is added to the load vector. If we insure that $k \gg \tilde{K}_{ii}$ (the assembled stiffness at $i$), then the solution of the modified equilibrium equations must now give $u_i = u_o$. We note that only the diagonal element in the stiffness matrix is affected, resulting in a numerically stable solution.

Mathematically, the procedure corresponds to an application of the *penalty method*. As regards the computer implementation, it is most conveniently achieved by actually adding a spring as an additional element.

# 3.6    Member Loads and Reactions

The main task of the stiffness method is to determine the non-zero degrees of freedom. Once these have been obtained then other quantities, such as the internal shear forces and moments, can be calculated by a subsequent computation. That is, we obtain these quantities as a post-processing action on the already calculated global degrees of freedom. Further, this is done only for selected members. It is in that spirit that the following schemes for obtaining member values and reactions are presented.

## Displacement distributions

We assume that the structural problem has already been solved and therefore we have in hand the nodal degrees of freedom $(v_i, \phi_i)$ at every node. More specifically, we know all the nodal displacements $\{v_1, \phi_1, v_2, \phi_2\}$ associated with the member of interest. The distribution of deflection and slope are then obtained by using the shape functions of Equations(3.6) separately for each element. That is,

$$
\begin{aligned}
v(x) &= g_1(x)v_1 + g_2(x)L\phi_1 + g_3(x)v_2 + g_4(x)L\phi_2 \\
\phi(x) &= g_1'(x)v_1 + g_2'(x)L\phi_1 + g_3'(x)v_2 + g_4'(x)L\phi_2
\end{aligned}
$$

where the prime refers to differentiation with respect to $x$. Because these are written for each element, the variable $x$ must be understood to be local to that element, and not the global or structural coordinate; it is zero at the first node and $L$ at the second.

Note that since $v(x)$ is a cubic function in $x$, it is not accurate to connect the nodal values with a straight line to get the distribution. This is usually done, nonetheless, just to get a sense of the distribution pattern.

## Load Distribution

Based on our particular beam element model, we have a shear force distribution that is constant throughout the length, that is,

$$
V(x) = V(0) = V(L)
$$

and a bending moment which varies linearly,

$$
M(x) = M(0) + \frac{x}{L}[-M(0) + M(L)]
$$

In both cases, the distributions are plotted simply by connecting the end values with a straight line. The task now is to obtain the end values from knowledge of the degrees of freedom.

We have already shown that the end values of the distributions of moment and shear force are related to the element nodal load values by

$$
\begin{aligned}
V(0) &= -V_1, & M(0) &= -M_1 \\
V(L) &= +V_2, & M(L) &= +M_2
\end{aligned}
$$

The nodal forces and moments of each element are, in turn, obtained from the element stiffness matrix by

$$\begin{Bmatrix} V_1 \\ M_1 \\ V_2 \\ M_2 \end{Bmatrix} = [\,k\,] \begin{Bmatrix} v_1 \\ \phi_1 \\ v_2 \\ \phi_2 \end{Bmatrix}$$

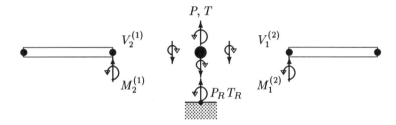

**Figure 3.12**: Sign conventions for boundary reactions.

## Reactions

Figure 3.12 shows the sign convention used to determine the boundary reactions from the nodal loads; keep in mind that for our one-dimensional problems we have at most two elements meeting at a support. Based on equilibrium, the total reactions are seen to be

$$\begin{aligned} P_R &= -V_2^{(1)} - V_1^{(2)} + P \\ T_R &= -M_2^{(1)} - M_1^{(2)} + T \end{aligned} \tag{3.13}$$

The element nodal load values are obtained from the element nodal degrees of freedom as indicated above. That is, for each member meeting at the node, we use

$$\begin{Bmatrix} V_1 \\ M_1 \\ V_2 \\ M_2 \end{Bmatrix} = [\,k\,] \begin{Bmatrix} v_1 \\ \phi_1 \\ v_2 \\ \phi_2 \end{Bmatrix}$$

and select the appropriate nodal values.

**Example 3.8:**   For the beam problem given in Figure 3.7 of Example 3.4, plot the shear force and bending moment distributions.

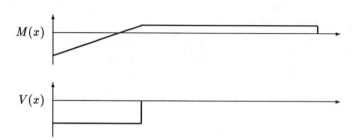

$M(x)$

$V(x)$

**Figure 3.13**: Shear force and bending moment distributions.

The total degrees of freedom were determined to be

$$\begin{Bmatrix} v_1 \\ \phi_1 \\ v_2 \\ \phi_2 \\ v_3 \\ \phi_3 \end{Bmatrix} = \frac{TL}{28EI} \begin{Bmatrix} 0 \\ 9 \\ 0 \\ -4 \\ -4L \\ 0 \end{Bmatrix}$$

The element nodal loads for the first element, Member 1-2, are determined from

$$\begin{Bmatrix} V_1 \\ M_1 \\ V_2 \\ M_2 \end{Bmatrix}^{(12)} = \frac{EI}{L^3} \begin{bmatrix} 12 & 6L & -12 & 6L \\ 6L & 4L^2 & -6L & 2L^2 \\ -12 & -6L & 12 & -6L \\ 6L & 2L^2 & -6L & 4L^2 \end{bmatrix} \begin{Bmatrix} 0 \\ 9 \\ 0 \\ -4 \end{Bmatrix} \frac{TL}{28EI} = \frac{T}{28L} \begin{Bmatrix} 30 \\ 28L \\ -30 \\ 2L \end{Bmatrix}$$

From this we can now obtain the member values as

$$V(0) = -V_1 = \frac{-30T}{28L}, \qquad M(0) = -M_1 = \frac{-28T}{28} = -T$$

$$V(L) = +V_2 = \frac{-30T}{28L}, \qquad M(L) = +M_2 = \frac{2T}{28} = \tfrac{1}{14}T$$

The distributions for this member are plotted by connecting these end values by straight lines. Working similarly for Member 2-3, we get

$$\begin{Bmatrix} V_1 \\ M_1 \\ V_2 \\ M_2 \end{Bmatrix}^{(23)} = \frac{EI}{8L^3} \begin{bmatrix} 12 & 12L & -12 & 12L \\ 12L & 16L^2 & -12L & 8L^2 \\ -12 & -12L & 12 & -12L \\ 12L & 8L^2 & -12L & 16L^2 \end{bmatrix} \begin{Bmatrix} 0 \\ -4 \\ -4L \\ 0 \end{Bmatrix} \frac{TL}{28EI} = \frac{TL}{28} \begin{Bmatrix} 0 \\ -2 \\ 0 \\ 2 \end{Bmatrix}$$

with the corresponding member values

$$V(0) = -V_1 = 0, \qquad M(0) = -M_1 = \frac{2T}{28}$$

$$V(L) = +V_2 = 0, \qquad M(L) = +M_2 = \frac{2T}{28}$$

The complete plot is shown in Figure 3.13.

Obviously there is a little redundancy in this solution since, for example, $V(0) = V(L)$ always, and only one of them need be computed. However, for the computer the extra work is negligible.

**Example 3.9:** Find the reactions at each support for the beam of the previous example.

We have already determined the element nodal loads in the last example, now we substitute these values into the relations for the reactions. At the first node, we have

$$P_{R1} = -V_1^{(12)} + P_1 = \frac{-30T}{28L}, \qquad T_{R1} = -M_1^{(12)} + T_1 = -T + T = 0$$

Notice that the moment reaction is zero, consistent with Node 1 being pinned. There are two elements meeting at the second node, hence

$$P_{R2} = -V_2^{(12)} - V_1^{(23)} + P_2 = \frac{30T}{28L}, \qquad T_{R2} = -M_2^{(12)} - M_1^{(23)} + T_2 = 0$$

Again, the zero moment reaction is in agreement with our expectation for a pinned support. Finally, for the third node

$$P_{R3} = -V_2^{(23)} + P_3 = 0, \qquad T_{R3} = -M_2^{(23)} + T_3 = \frac{-2T}{28}$$

The zero vertical force is in agreement with our expectation for vertical rollers.

## Problems

**3.1** Show that when the cross-sectional area or the elastic modulus changes along the length of the beam that the relevant equations are

$$\phi = \frac{dv}{dx}, \qquad M(x) = \frac{d^2}{dx^2}[EI\frac{d^2v}{dx^2}], \qquad V(x) = -\frac{d}{dx}[EI\frac{d^2v}{dx^2}]$$

[Reference [13], pp. 542]

**3.2** A beam is cantilevered at $x = L$ and has a vertical force applied at $x = 0$. Its moment of inertia varies as

$$I(x) = I_0[1 + r\frac{x}{L}]$$

where $r$ is a numerical factor. Show that the deflected shape is given by

$$v(x) = \frac{PL^3}{r^3EI_0}\left[(\frac{r^2x^2}{2L^2} + 2\frac{rx}{L} + 1) - (1 + \frac{rx}{L})\log(1 + \frac{rx}{L})\right] + c_1x + c_2$$

and the integration constants evaluate to

$$c_1 = \frac{PL^2}{r^2EI_0}[\log(1 + r) - (1 + r)], \qquad c_2 = \frac{PL^3}{r^3EI_0}\left[\log(1 + r)\frac{1}{2}r^2 - r - 1\right]$$

[Reference [49], pp. 180]

**3.3** Model the previous problem using two elements of different inertia but each of the same length. (Take that $r = 2$.) Compare the displacement distributions.

**3.4** An alternative way to derive an equivalent load system is to base it on the work done by the system. Show that if we define the work done by the equivalent force and moment system as

$$W^* = \tfrac{1}{2}(P_1 v_1 + T_1 \phi_1 + P_2 v_2 + T_2 \phi_2)$$

and the work done by the distributed load $q(x)$ on the beam as

$$W = \tfrac{1}{2} \int_0^L q(x)\, v(x)\, dx$$

and on replacing the deflection with its representation in terms of the beam shape functions that we recover the consistent load formulas.

[Reference [49], pp. 143]

**3.5** Show that the consistent load representation gives rise to a statically equivalent system even for a general distributed applied load $q(x)$.

**3.6** If a beam is subjected to a temperature differential $\Delta T$ between its top and bottom, then it will tend curve. Show that the curvature is

$$\frac{d^2 v}{dx^2} = \alpha \Delta T / h$$

where $\alpha$ is the coefficient of thermal expansion, and $h$ is the beam thickness.

[Reference [30], pp. 115]

**3.7** Using the result of the last exercise, show that the stiffness relation for a beam element with a temperature differential is given by

$$\left\{ \begin{array}{c} V_1 \\ M_1 \\ V_2 \\ M_2 \end{array} \right\} = \frac{EI}{L^3} \left[ \begin{array}{cccc} 12 & 6L & -12 & 6L \\ 6L & 4L^2 & -6L & 2L^2 \\ -12 & -6L & 12 & -6L \\ 6L & 2L^2 & -6L & 4L^2 \end{array} \right] \left\{ \begin{array}{c} v_1 \\ \phi_1 \\ v_2 \\ \phi_2 \end{array} \right\} + \alpha EI \Delta T \frac{1}{h} \left\{ \begin{array}{c} 0 \\ -1 \\ 0 \\ 1 \end{array} \right\}$$

[Reference [30], pp. 116]

## Exercises

**3.1** An aluminum cantilever beam of uniform width $100\,mm$ tapers from a thickness of $150\,mm$ to $75\,mm$ over its length of $4000\,mm$. If a uniformly distributed load of $0.5\,N/mm$ is applied find the tip deflection using four uniform elements.

[$12.42\,mm$]

**3.2** A cantilever beam of length $2L$ has a non-simple support at its middle. If a load $Q$ is applied at the end and a rotation $\phi$ applied at the middle support, determine the reactions at the support.

[$T = QL + (4EI/L)\phi,\ P = Q - (6EI/L^2)\phi$]

**3.3** Show that for the uniformly loaded beam of Figure 3.9 that ten elements are required before the error in maximum deflection is less that 1% when using the lumped load approximation.

**3.4** A carpenter with a power saw has a $20\,ft$ plank of uniform weight per unit length $w_o$ and two saw horses. He wishes to cut a $6\,ft$ length from the plank but in order to minimize splitting of the ends he wants to cut it at a point where the bending moment in the plank is zero. If he places one sawhorse at one end of the plank, where should he place the other? $[\approx 15.5\,ft]$

**3.5** A bookshelf is made by placing a wooden plank on two brick supports. Where should the bricks be placed so as to make the maximum bending moment as small as possible. $[\approx .21L\ .79L]$

**3.6** A $5\,m$ beam of cross-section $100\,mm \times 100\,mm$ is built in at both end. If one end slips an amount $\delta = 25\,mm$ relative to the other end, determine the maximum moment generated. $[M = 6EI\delta/L^2]$

**3.7** A cantilever beam of length $2\,m$ has a distributed load of $4800\,N/m$ over half of its length beginning at the free end. It also has a concentrated force of $3000\,N$ at the mid point and a linear spring (stiffness $200\,N/m$) attached at the tip. The beam has an $EI$ of $40\,kN/m^2$. Use two elements and the consistent load approximation to obtain the deflections at the mid point and tip. $[11.66\,mm\ \ 6.80\,mm]$

**3.8** Timber beams $18\,ft$ long, $12\,in$ deep and $4\,in$ wide, are simply supported at the ends. How far apart, center to center, should such beams be placed when supporting a floor loaded with $40\,lb/ft^2$? The beams weigh $40\,lb/ft^3$ and the maximum allowable bending moment in each beam is $6800\,lb \cdot ft$. $[3.86\,ft]$

**3.9** A beam, $20\,ft$ long, carries a uniform load of $1\,ton/ft$ run. It is simply supported at one end and at some other point. Find the position (from the free end) of the other support so that the maximum bending moment may be as small as possible. $[5.87\,ft]$

**3.10** A timber beam, $8\,in$ wide and $6\,in$ deep, is placed directly above another timber beam $6\,in$ wide and $8\,in$ deep. The beams are held apart by three solid blocks, one under each end and one under the center of the upper beam. The beams are $20\,ft$ long, and the whole rests on two supports, one under each end of the lower beam. If the upper beam is loaded with $100\,lb/ft$ run, find the deflection at the center, and the maximum bending moment in each beam. ($E = 1.5\,msi$) $[0.6\,in;\ 1800,\ 4000\,lb \cdot ft]$

Chapter 4

# Truss and Frame Analysis

A *truss* is a structure composed of rod members arranged to form one or more tri-angles. The joints are pinned (do not transmit moments) so that the members must be triangulated. A *frame*, on the other hand, is a structure that consists of arbi-trarily oriented beam members which are connected rigidly or by pins at joints. The members support bending as well as axial loads.

The essential new aspect to the study of these structures is the consideration of the element stiffness of an arbitrarily oriented member. Since differently oriented members are to be considered simultaneously, they must have common or *global* ref-erence axes. The connection between the global and the local axes is established via the 3-D rotation matrix. We will first develop the analysis for plane structures as an intermediate step to introducing the general case of space frames. References [3, 45] are excellent sources for additional details on modeling 3-D structures.

A fundamental assumption in the following developments is that the principle of superposition holds. In this way, we can assemble the general frame by combining the separate actions of the simpler cases developed in this and previous chapters.

## 4.1   Truss Analysis

The stiffness matrix, as derived in the previous chapters, is with respect to *local co-ordinates*, that is, a set of coordinates aligned with the member. Consequently, in a truss consisting of members with different orientation there are many such local coor-dinates. The essence of the direct stiffness approach is that the stiffness of the various members are referred to a common *global* coordinate system before assemblage. To that end, we must first obtain the stiffness matrix of an arbitrarily oriented element referred to the global coordinate system.

### Stiffness Matrix for a Truss Element

Consider a truss element whose longitudinal axis makes an angle $\theta$ relative to the global $x$-axis as shown in Figure 4.1. The stiffness matrix of the truss element, when

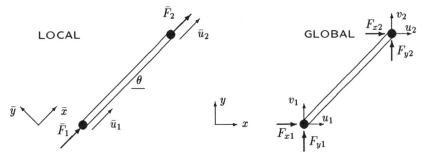

**Figure 4.1**: Load and displacement components in local and global coordinates.

referred to the local coordinates $\bar{x} - \bar{y}$, is as given for the rod. That is,

$$\left\{ \begin{array}{c} \bar{F}_1 \\ \bar{F}_2 \end{array} \right\} = \frac{EA}{L} \left[ \begin{array}{cc} 1 & -1 \\ -1 & 1 \end{array} \right] \left\{ \begin{array}{c} \bar{u}_1 \\ \bar{u}_2 \end{array} \right\}$$

The forces and displacements referred to the local coordinates are now denoted by the barred notation. We wish to give the truss member extra degrees of freedom that allow it to move in the plane. That is, we want to add the degrees of freedom $\bar{v}_1$ and $\bar{v}_2$. The corresponding stiffness relation is obtained by augmenting the above equation by adding zero forces in the $\bar{y}$-direction, and the nodal displacements $\bar{v}_1$ and $\bar{v}_2$ in the $\bar{y}$-direction at Node 1 and Node 2, respectively. The resulting element stiffness relation is

$$\left\{ \begin{array}{c} \bar{F}_1 \\ 0 \\ \bar{F}_2 \\ 0 \end{array} \right\} = \frac{EA}{L} \left[ \begin{array}{cccc} 1 & 0 & -1 & 0 \\ 0 & 0 & 0 & 0 \\ -1 & 0 & 1 & 0 \\ 0 & 0 & 0 & 0 \end{array} \right] \left\{ \begin{array}{c} \bar{u}_1 \\ \bar{v}_1 \\ \bar{u}_2 \\ \bar{v}_2 \end{array} \right\} \qquad \text{or} \qquad \{\bar{F}\} = [\,\bar{k}\,]\{\bar{u}\}$$

The axial forces $\bar{F}_1$ and $\bar{F}_2$ can be decomposed into global components in the $x$ and $y$-direction. At Node 1, for example,

$$\begin{array}{rcl} \bar{F}_1 & = & +F_{x1} \cos\theta + F_{y1} \sin\theta \\ 0 & = & -F_{x1} \sin\theta + F_{y1} \cos\theta \end{array}$$

A similar set of equations can be written at Node 2. In matrix notation, these four equations are written as

$$\{\bar{F}\} = [\,T\,]\{F\}$$

where

$$\{\bar{F}\} \equiv \left\{ \begin{array}{c} \bar{F}_1 \\ 0 \\ \bar{F}_2 \\ 0 \end{array} \right\}, \quad \{F\} \equiv \left\{ \begin{array}{c} F_{x1} \\ F_{y1} \\ F_{x2} \\ F_{y2} \end{array} \right\}, \quad [\,T\,] \equiv \left[ \begin{array}{cccc} \cos\theta & \sin\theta & 0 & 0 \\ -\sin\theta & \cos\theta & 0 & 0 \\ 0 & 0 & \cos\theta & \sin\theta \\ 0 & 0 & -\sin\theta & \cos\theta \end{array} \right]$$

The displacement components $\{\bar{u}_1, \bar{v}_1; \bar{u}_2, \bar{v}_2\}$ and $\{u_1, v_1; u_2, v_2\}$ are related in the same manner as the forces by

$$\{\bar{u}\} = [\,T\,]\{u\}$$

Substituting for the local forces and displacements into the element stiffness relation gives

$$[\,T\,]\{F\} = [\,\bar{k}\,][\,T\,]\{u\}$$

It can be easily shown that the transformation matrix $[\,T\,]$ is orthogonal, i.e., its inverse is its transpose

$$[\,T\,]^{-1} = [\,T\,]^T$$

Therefore, multiplying both sides of the stiffness relation by the transpose of $[\,T\,]$ leads to

$$\{F\} = [\,T\,]^T[\,\bar{k}\,][\,T\,]\{u\} \qquad \text{or} \qquad \{F\} = [\,k\,]\{u\}$$

The element stiffness matrix $[\,k\,]$ in the global $x - y$ coordinate system is given by

$$[\,k\,] = [\,T\,]^T[\,\bar{k}\,][\,T\,]$$

That is, the global stiffness matrix is obtained from a transformation of the stiffness matrix in local coordinates. The explicit expression for $[\,k\,]$, after multiplying through by the transformation matrices, is

$$
\begin{Bmatrix} F_{x1} \\ F_{y1} \\ F_{x2} \\ F_{y2} \end{Bmatrix}
= \frac{EA}{L}
\begin{bmatrix}
C^2 & CS & -C^2 & -CS \\
CS & S^2 & -CS & -S^2 \\
-C^2 & -CS & C^2 & CS \\
-CS & -S^2 & CS & S^2
\end{bmatrix}
\begin{Bmatrix} u_1 \\ v_1 \\ u_2 \\ v_2 \end{Bmatrix}
\tag{4.1}
$$

where $C \equiv \cos\theta$ and $S \equiv \sin\theta$. Note that while the special case of $\theta = 0$ corresponds to a horizontal rod, the resulting stiffness is not that of the rod — the presence of the $v_1, v_2$ degrees of freedom means that the member can move transversely. To recover the one-dimensional rod behavior, it is necessary to suppress these degrees of freedom by imposing some transverse constraints.

This element stiffness relation allows a simple expression for the axial force to be determined from

$$\bar{F} = F_{x2}\cos\theta + F_{y2}\sin\theta$$

By substituting for the force components we get

$$\bar{F} = \frac{EA}{L}\left[(u_2 - u_1)\cos\theta + (v_2 - v_1)\sin\theta\right] \tag{4.2}$$

This is useful as a quick means to obtain the axial force once the global degrees of freedom are known.

## Structural Stiffness Matrix for a Truss

We will consider a simple truss composed of three members, as shown in Figure 4.2, in order to illustrate the procedure for constructing the structural stiffness matrix. The procedure itself is essentially the same as already used for the rod and beam structures. That is, we will develop it by imposing compatibility and equilibrium at each of the joints.

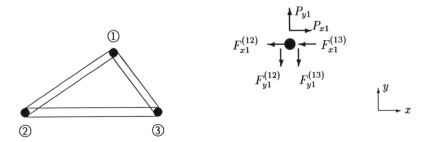

**Figure 4.2**: Simple truss with three members.

Compatibility of displacements at the joints is insured by simply using the same displacement variables for the joint as for the element nodes. Then automatically the displacements across a joint are equal. To impose equilibrium consider, for example, the free body diagram of Node 1 shown in Figure 4.2. Two members meet at the node hence there are two sets of nodal loads. These must be in equilibrium with the applied loads, hence we obtain from the equilibrium equations that

$$P_{x1} = F_{x1}^{(12)} + F_{x2}^{(31)}$$
$$P_{y1} = F_{y1}^{(12)} + F_{y2}^{(31)}$$

This free body diagram of the node is typical of the other nodes also because all force components are written in terms of the global reference frame. To amplify on this point, the specific orientation of a member does not enter these free body diagrams; that information is included in the stiffness matrix. Therefore, we can write similar equilibrium equations at Node 2 and Node 3. The six equations in all can be expressed in the form

$$\begin{Bmatrix} P_{x1} \\ P_{y1} \\ P_{x2} \\ P_{y2} \\ P_{x3} \\ P_{y3} \end{Bmatrix} = \begin{Bmatrix} F_{x1} \\ F_{y1} \\ F_{x2} \\ F_{y2} \\ 0 \\ 0 \end{Bmatrix}^{(12)} + \begin{Bmatrix} 0 \\ 0 \\ F_{x1} \\ F_{y1} \\ F_{x2} \\ F_{y2} \end{Bmatrix}^{(23)} + \begin{Bmatrix} F_{x2} \\ F_{y2} \\ 0 \\ 0 \\ F_{x1} \\ F_{y1} \end{Bmatrix}^{(31)}$$

There is a force vector on the right hand side for each member in the structure. These vectors are related to the nodal displacements through the [4 × 4] element stiffness matrices $[k^{(12)}]$, $[k^{(23)}]$ and $[k^{(31)}]$. By augmenting these to a system size of $[6 \times 6]$ corresponding to the degrees of freedom $\{u_1, \cdots, v_3\}$, we can obtain, for example,

$$\left\{ \begin{matrix} F \\ 0 \end{matrix} \right\}^{(12)} = \left[ \begin{matrix} [k^{(12)}] & 0 \\ 0 & 0 \end{matrix} \right] \{u\}, \qquad \left\{ \begin{matrix} 0 \\ F \end{matrix} \right\}^{(23)} = \left[ \begin{matrix} 0 & 0 \\ 0 & [k^{(23)}] \end{matrix} \right] \{u\}$$

Combining all the stiffnesses together leads to the structural equilibrium equations

$$\{P\} = [\,K\,]\{u\} \qquad \text{or} \qquad P_i = \sum_j K_{ij} u_j = \sum_j \sum_m k_{ij}^{(m)} u_j$$

The stiffness property of the truss as a whole is characterized by the square symmetric matrix $[\,K\,]$ which relates the external loads to the nodal displacements. The entry $K_{ij}$ which relates $P_i$ with $u_j$, is the sum of all the entries in the element stiffness matrices that relate $P_i$ and $u_j$.

## Connectivities

In order to use the general form of the element stiffness matrix, we must be consistent in our designation of the orientation. For example, a member that is upright can be said to have an orientation of either $\theta = 90°$ or $\theta = -90°$ depending on which end is considered the 'first' node. We will state this information in the form of a *connectivity*. That is, we will state (as a property of the member) the two nodes that it connects. Note that the actual numbering sequence of the nodes is not important, only which of the two is to be considered the first node.

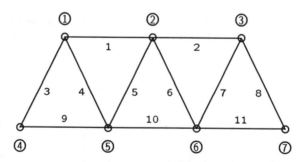

**Figure 4.3**: Seven noded truss.

Consider the numbering of the frame shown in Figure 4.3. The numbering of the members is quite arbitrary but, as will be seen later, the numbering of the nodes can be very important to the computational efficiency. The connectivities can be specified in a variety of ways; the following is one possible set:

| member | connectivity | member | connectivity |
|--------|--------------|--------|--------------|
| 1 :    | $1 - 2$      | 7 :    | $3 - 6$      |
| 2 :    | $2 - 3$      | 8 :    | $3 - 7$      |
| 3 :    | $1 - 4$      | 9 :    | $4 - 5$      |
| 4 :    | $1 - 5$      | 10 :   | $5 - 6$      |
| 5 :    | $2 - 5$      | 11 :   | $6 - 7$      |
| 6 :    | $2 - 6$      |        |              |

The *connectivity matrix* is a scheme for indicating which nodes are connected together. The general term $C_{ij}$ indicates that Node $i$ is connected to Node $j$. It is formed by making a square array of node numbers. Then for each node in turn an entry is made for each node that is attached to it. For the numbering and connectivities above it gives

$$[C_{ij}] = \begin{bmatrix} & 1 & 2 & 3 & 4 & 5 & 6 & 7 \\ 1 & \# & + & & + & + & & \\ 2 & + & \# & + & & + & + & \\ 3 & & + & \# & & & + & + \\ 4 & + & & & \# & + & & \\ 5 & + & + & & + & \# & + & \\ 6 & & + & + & & + & \# & + \\ 7 & & & + & & & + & \# \end{bmatrix}$$

This gives a view of how populated the system matrix will be. There are many zeros (blanks) but they are not grouped in any particular arrangement. (This is called a *sparse* array.) It is important to realize that if the nodes are renumbered then the sparsity changes. In fact, it is possible to make it more banded and less sparse as we shall see in Chapter 7.

As seen from the above, the connectivity matrix has some of the features of the stiffness matrix such as symmetry. Since it shows what nodes are connected, it can help in deciding how to renumber the nodes.

**Example 4.1:**    A three-bar truss is loaded as shown in Figure 4.4.  Find the displacements at the load application point. Each member has the same $EA$.

The power of the matrix methods is that similar solution procedures can be used irrespective of the particular structural shape. Thus we can use the approach introduced for the rod and beam structures. Labelling the nodes as shown gives the total degrees of freedom as

$$\{u\} = \{u_1, v_1; u_2, v_2; u_3, v_3; u_4, v_4\}$$

The displacements at Nodes 2, 3, and 4 are zero leaving the unknown nodal displacements and the corresponding known loads as

$$\{u_u\} = \left\{ \begin{array}{c} u_1 \\ v_1 \end{array} \right\}, \qquad \{P_k\} = \left\{ \begin{array}{ccc} P_{x1} & = & 1000 \\ P_{y1} & = & -2000 \end{array} \right\}$$

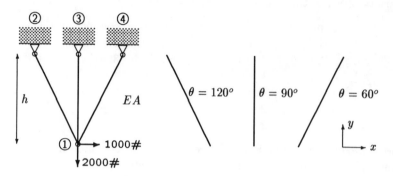

**Figure 4.4**: Four noded truss.

These correspond to a *IDbc* matrix of

$$\{IDbc\} = \{1,2;0,0;0,0;0,0\}$$

since only the first two degrees of freedom are non-zero.

To construct the reduced stiffness matrix $[K^*]$, it is necessary for us to find the contributions from each member. We will state the connectivities of the three members as, 1-2, 1-3, 1-4, respectively. In each case, then, the reduced element stiffness is of the form

$$[k^*] = \frac{EA}{L} \left[ \begin{array}{cc} C^2 & CS \\ CS & S^2 \end{array} \right]$$

With reference to the figure, the orientation of the three members are 120°, 90°, and 60°, respectively. Substituting these values of orientation gives

$$\begin{array}{rcl}
[k^{*(12)}] & = & \dfrac{\sqrt{3}EA}{2h} \left[ \begin{array}{rr} 0.25 & -0.433 \\ -0.433 & 0.45 \end{array} \right] \\[3mm]
[k^{*(13)}] & = & \dfrac{EA}{h} \left[ \begin{array}{cc} 0 & 0 \\ 0 & 1 \end{array} \right] \\[3mm]
[k^{*(14)}] & = & \dfrac{EA}{h} \dfrac{\sqrt{3}}{2} \left[ \begin{array}{cc} 0.25 & 0.433 \\ 0.433 & 0.75 \end{array} \right]
\end{array}$$

The reduced structural stiffness matrix is formed by collecting the elements in the reduced element stiffness matrices according to their position *vis-a-vis* the global degrees of freedom. In this case each matrix shares the same global degrees of freedom, hence the result is

$$[K^*] = \frac{\sqrt{3}EA}{2h} \left[ \begin{array}{cc} 0.5 & 0 \\ 0 & 2.655 \end{array} \right]$$

Inverting $[K^*]$ gives

$$[K^*]^{-1} = \frac{2h}{\sqrt{3}EA} \left[ \begin{array}{cc} 2 & 0 \\ 0 & 0.3767 \end{array} \right]$$

Thus the unknown displacements can be calculated from

$$\{u\} = [K^*]^{-1}\{P_k\} = \frac{2h}{\sqrt{3}EA} \left\{ \begin{array}{c} 2000 \\ -735.4 \end{array} \right\}$$

Once the nodal displacements are obtained, the loads in the truss members can be calculated.

**Example 4.2:**   Determine the deflections of the equilateral truss shown in Figure 4.5. One joint is on a horizontal roller.

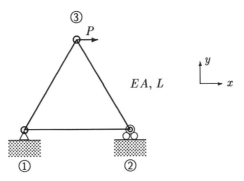

**Figure 4.5**: Simple equilateral truss.

Labelling the nodes as shown gives the total degrees of freedom as

$$\{u\} = \{u_1\,,v_1\,;u_2\,,v_2\,;u_3\,,v_3\}$$

Both displacements at Node 1 are zero, while the vertical displacement at Node 2 is zero. This gives the unknown nodal displacements and the corresponding known loads as

$$\{u_u\} = \left\{ \begin{array}{c} u_2 \\ u_3 \\ v_3 \end{array} \right\}, \qquad \{P_k\} = \left\{ \begin{array}{c} P_{x2} = 0 \\ P_{x3} = P \\ P_{y1} = 0 \end{array} \right\}$$

These correspond to a $IDbc$ matrix of

$$\{IDbc\} = \{0\,,0\,;1\,,0\,;2\,,3\}$$

We will state the connectivities of the three members as, 1-2, 2-3, 3-1, respectively. For Member 1-2, the orientation is $0°$, giving the reduced stiffness matrix as

$$[k^{(*12)}] = \frac{EA}{L}[\,1\;]$$

For Member 2-3, the orientation is $120°$, giving the reduced stiffness matrix as

$$[k^{(*23)}] = \frac{EA}{4L} \left[ \begin{array}{ccc} 1 & -1 & \sqrt{3} \\ -1 & 1 & -\sqrt{3} \\ \sqrt{3} & -\sqrt{3} & 3 \end{array} \right]$$

The associated degrees of freedom are $\{u_2, u_3, v_3\}$. Finally, for Member 3-1 the orientation is $-120°$, giving the reduced stiffness matrix as

$$[k^{(*31)}] = \frac{EA}{4L} \begin{bmatrix} 1 & \sqrt{3} \\ \sqrt{3} & 3 \end{bmatrix}$$

for the associated degrees of freedom $\{u_3, v_3\}$. The reduced structural stiffness matrix is formed by collecting the elements in the reduced element stiffness matrices according to their position *vis-a-vis* the global degrees of freedom. The result is

$$[K^*] = \frac{EA}{4L} \begin{bmatrix} 5 & -1 & \sqrt{3} \\ -1 & 2 & 0 \\ \sqrt{3} & 0 & 6 \end{bmatrix}$$

The system of equations to be solved is

$$\frac{EA}{4L} \begin{bmatrix} 5 & -1 & \sqrt{3} \\ -1 & 2 & 0 \\ \sqrt{3} & 0 & 6 \end{bmatrix} \begin{Bmatrix} u_2 \\ u_3 \\ v_3 \end{Bmatrix} = \begin{Bmatrix} 0 \\ P \\ 0 \end{Bmatrix}$$

Thus the unknown displacements are found to be

$$\begin{Bmatrix} u_2 \\ u_3 \\ v_3 \end{Bmatrix} = \frac{PL}{12EA} \begin{Bmatrix} 6 \\ 27 \\ -\sqrt{3} \end{Bmatrix}$$

Once the nodal displacements are obtained, the loads in the truss members can be calculated.

## 4.2   Plane Frame Analysis

A frame is a collection of beam members; because they have arbitrary orientations, they must also support axial loading. Thus the first step in deriving the frame stiffness is to add this axial behavior to the beam. We will assume that there is no interaction between the axial and flexural loadings, that is, the member acts as a beam for transverse loads and as a rod for axial loads. In the next chapter, we take up more fully the question of the interaction of axial and flexural behaviors.

### Stiffness Matrix for a Frame Element

Consider a beam subjected to both axial and bending loads. Assume that the flexural deformation and axial deformation are uncoupled; that is, the force-displacement relation can be written separately for each action. For the axial deformation, we have

$$\begin{Bmatrix} \bar{F}_1 \\ \bar{F}_2 \end{Bmatrix} = \frac{EA}{L} \begin{bmatrix} 1 & -1 \\ -1 & 1 \end{bmatrix} \begin{Bmatrix} \bar{u}_1 \\ \bar{u}_2 \end{Bmatrix}$$

and for bending deformation

$$\left\{ \begin{array}{c} \bar{V}_1 \\ \bar{M}_1 \\ \bar{V}_2 \\ \bar{M}_2 \end{array} \right\} = \frac{EI}{L^3} \left[ \begin{array}{cccc} 12 & 6L & -12 & 6L \\ 6L & 4L^2 & -6L & 2L^2 \\ -12 & -6L & 12 & -6L \\ 6L & 2L^2 & -6L & 4L^2 \end{array} \right] \left\{ \begin{array}{c} \bar{v}_1 \\ \bar{\phi}_1 \\ \bar{v}_2 \\ \bar{\phi}_2 \end{array} \right\}$$

The element stiffness relation for the beam element with axial loading can be expressed as

$$\{\bar{F}\} = [\ \bar{k}\ ]\{\bar{u}\}$$

where the following augmented matrices have been introduced

$$\{\bar{F}\} \equiv \left\{ \begin{array}{c} \bar{F}_1 \\ \bar{V}_1 \\ \bar{M}_1 \\ \bar{F}_2 \\ \bar{V}_2 \\ \bar{M}_2 \end{array} \right\}, \qquad \{u\} \equiv \left\{ \begin{array}{c} \bar{u}_1 \\ \bar{v}_1 \\ \bar{\phi}_1 \\ \bar{u}_2 \\ \bar{v}_2 \\ \bar{\phi}_2 \end{array} \right\}$$

and

$$[\ \bar{k}\ ] \equiv \frac{EA}{L} \left[ \begin{array}{cccccc} 1 & 0 & 0 & -1 & 0 & 0 \\ 0 & 0 & 0 & 0 & 0 & 0 \\ 0 & 0 & 0 & 0 & 0 & 0 \\ -1 & 0 & 0 & 1 & 0 & 0 \\ 0 & 0 & 0 & 0 & 0 & 0 \\ 0 & 0 & 0 & 0 & 0 & 0 \end{array} \right] + \frac{EI}{L^3} \left[ \begin{array}{cccccc} 0 & 0 & 0 & 0 & 0 & 0 \\ 0 & 12 & 6L & 0 & -12 & 6L \\ 0 & 6L & 4L^2 & 0 & -6L & 2L^2 \\ 0 & 0 & 0 & 0 & 0 & 0 \\ 0 & -12 & -6L & 0 & 12 & -6L \\ 0 & 6L & 2L^2 & 0 & -6L & 4L^2 \end{array} \right]$$

The beam under combined loading is seen to be represented by a $[6 \times 6]$ stiffness matrix. We will generally leave it in separated form like this so as to emphasize that it arose as a simple superposition of the axial and flexural actions.

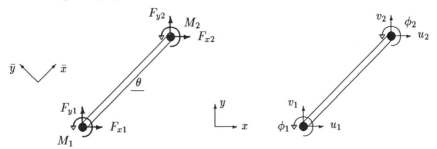

**Figure 4.6**: Global coordinate system for a frame element.

Before we analyze a general frame, it is necessary for us to know the stiffness matrix of an arbitrarily oriented member. To this end, consider the local coordinate system $\bar{x} - \bar{y}$ and the global coordinate system $x - y$ of the member shown in the Figure 4.6. The nodal forces referred to these two systems are denoted by

$$\{\bar{F}\} = \begin{Bmatrix} \bar{F}_1 \\ \bar{V}_1 \\ \bar{M}_1 \\ \bar{F}_2 \\ \bar{V}_2 \\ \bar{M}_2 \end{Bmatrix} \quad \text{and} \quad \{F\} = \begin{Bmatrix} F_{x1} \\ F_{y1} \\ M_1 \\ F_{x2} \\ F_{y2} \\ M_2 \end{Bmatrix}$$

respectively. The relation between $\{\bar{F}\}$ and $\{F\}$ is obtained by using the transformation for vector components. Thus, for example, at the first node

$$\begin{aligned} \bar{F}_1 &= \cos\theta F_{x1} + \sin\theta F_{y1} \\ \bar{V}_1 &= -\sin\theta F_{x1} + \cos\theta F_{y1} \\ \bar{M}_1 &= M_1 \end{aligned}$$

There are similar expressions at the second node. We can write all these equations in the symbolic form

$$\{\bar{F}\} = [\,T\,]\{F\}$$

where

$$[\,T\,] = \begin{Bmatrix} \cos\theta & \sin\theta & 0 & 0 & 0 & 0 \\ -\sin\theta & \cos\theta & 0 & 0 & 0 & 0 \\ 0 & 0 & 1 & 0 & 0 & 0 \\ 0 & 0 & 0 & \cos\theta & \sin\theta & 0 \\ 0 & 0 & 0 & -\sin\theta & \cos\theta & 0 \\ 0 & 0 & 0 & 0 & 0 & 1 \end{Bmatrix}$$

It is easy to confirm that this matrix is orthogonal. Proceeding in like manner, we obtain for the nodal displacement vectors $\{\bar{u}\}$ and $\{u\}$

$$\{\bar{u}\} \equiv \begin{Bmatrix} \bar{u}_1 \\ \bar{v}_1 \\ \phi_1 \\ \bar{u}_2 \\ \bar{v}_2 \\ \phi_2 \end{Bmatrix}, \quad \{u\} \equiv \begin{Bmatrix} u_1 \\ v_1 \\ \phi_1 \\ u_2 \\ v_2 \\ \phi_2 \end{Bmatrix}, \quad \{\bar{u}\} = [\,T\,]\{u\}$$

Substitution for the barred displacements and forces into the element stiffness relation gives

$$[\,T\,]\{F\} = [\,\bar{k}\,][\,T\,]\{u\}$$

from which the following is obtained

$$\{F\} = [\,T\,]^T[\,\bar{k}\,][\,T\,]\{u\} = [\,k\,]\{u\}$$

since $[\,T\,]$ is orthogonal. The explicit expression for the frame element stiffness relation is given as

$$
\left\{
\begin{array}{c}
F_{x1} \\
F_{y1} \\
M_1 \\
F_{x2} \\
F_{y2} \\
M_2
\end{array}
\right\}
=
\frac{EA}{L}
\left[
\begin{array}{cccccc}
C^2 & & & & \text{sym} & \\
CS & S^2 & & & & \\
0 & 0 & 0 & & & \\
-C^2 & -CS & 0 & C^2 & & \\
-CS & -S^2 & 0 & CS & S^2 & \\
0 & 0 & 0 & 0 & 0 & 0
\end{array}
\right]
\tag{4.3}
$$

$$
+\frac{EI}{L^3}
\left[
\begin{array}{cccccc}
12S^2 & & & & \text{sym} & \\
-12CS & 12C^2 & & & & \\
-6LS & 6LC & 4L^2 & & & \\
-12S^2 & 12CS & 6LS & 12S^2 & & \\
12CS & -12C^2 & -6LC & -12CS & 12C^2 & \\
-6LS & 6LC & 2L^2 & 6LS & -6LC & 4L^2
\end{array}
\right]
\left\{
\begin{array}{c}
u_1 \\
v_1 \\
\phi_1 \\
u_2 \\
v_2 \\
\phi_2
\end{array}
\right\}
$$

where, as before, the abbreviations $C \equiv \cos\theta$, $S \equiv \sin\theta$ are used. Note that both of these matrices reduce to the respective matrices when $\theta = 0$. But also note that the first matrix is the augmented global stiffness for the truss element. Therefore, in a sense, the truss behavior is embedded in the frame.

This relationship shows that even for the arbitrarily oriented frame member that the axial and flexural actions are uncoupled. This decomposition of the frame stiffness is due to our initial assumption.

## Structural Stiffness Matrix for Frames

The structural stiffness matrix for the frame is assembled as for the truss, rod and beam structures. At each node there are three applied nodal loads $P_x$, $P_y$ and $T$ (the applied moment or torque), and three equations of equilibrium. This eventually leads to

$$\{P\} = [\,K\,]\{u\}$$

The vector $\{u\}$ contains the three nodal degrees of freedom, $u, v, \phi$, at each node and $\{P\}$ contains the applied nodal loads at each node.

**Example 4.3:**   A right-angled two-member frame is loaded as shown in Figure 4.7. One end is fixed and the other end is on horizontal rollers. Each member has the same material and sectional properties. Find the deflections.

Label the nodes as shown, then the total degrees of freedom are

$$\{u\} = \{u_1, v_1, \phi_1\,; u_2, v_2, \phi_2\,; u_3, v_3, \phi_3\}$$

The boundary conditions at the first and third nodes require that

$$u_1 = v_1 = \phi_1 = 0, \qquad v_3 = 0$$

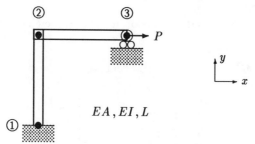

**Figure 4.7**: Two member frame.

giving the reduced degrees of freedom and known loads as

$$\{u_u\} = \{u_2, v_2, \phi_2, u_3, \phi_3\}, \qquad \{P_k\} = \{0, 0, 0, P, 0\}$$

The corresponding $IDbc$ matrix

$$\{IDbc\} = \{0, 0, 0; 1, 2, 3; 4, 0, 5\}$$

We will take the connectivities for the members as 1-2 and 2-3. The reduced element stiffnesses are obtained by substituting $\theta = 90^o$ and $\theta = 0^o$ for Members 1-2 and 2-3, respectively. This gives for Member 1-2

$$[k^*] = \frac{EA}{L} \begin{bmatrix} 0 & 0 & 0 \\ 0 & 1 & 0 \\ 0 & 0 & 0 \end{bmatrix} + \frac{EI}{L^3} \begin{bmatrix} 12 & 0 & 6L \\ 0 & 0 & 0 \\ 6L & 0 & 4L^2 \end{bmatrix}$$

and for Member 2-3

$$[k^*] = \frac{EA}{L} \begin{bmatrix} 1 & 0 & 0 & -1 & 0 \\ 0 & 0 & 0 & 0 & 0 \\ 0 & 0 & 0 & 0 & 0 \\ -1 & 0 & 0 & 1 & 0 \\ 0 & 0 & 0 & 0 & 0 \end{bmatrix} + \frac{EI}{L^3} \begin{bmatrix} 0 & 0 & 0 & 0 & 0 \\ 0 & 12 & 6L & 0 & 6L \\ 0 & 6L & 4L^2 & 0 & 2L^2 \\ 0 & 0 & 0 & 0 & 0 \\ 0 & 6L & 2L^2 & 0 & 4L^2 \end{bmatrix}$$

The reduced structural stiffness matrix is obtained as

$$[K^*] = \frac{EA}{L} \begin{bmatrix} 1 & 0 & 0 & -1 & 0 \\ 0 & 1 & 0 & 0 & 0 \\ 0 & 0 & 0 & 0 & 0 \\ -1 & 0 & 0 & 1 & 0 \\ 0 & 0 & 0 & 0 & 0 \end{bmatrix} + \frac{EI}{L^3} \begin{bmatrix} 12 & 0 & 6L & 0 & 0 \\ 0 & 12 & 6L & 0 & 6L \\ 6L & 6L & 8L^2 & 0 & 2L^2 \\ 0 & 0 & 0 & 0 & 0 \\ 0 & 6L & 2L^2 & 0 & 4L^2 \end{bmatrix}$$

The unknown nodal displacements can now be solved for and are

$$u_2 = \frac{P}{D}[7\tfrac{EA}{L} + 12\tfrac{EI}{L^3}]\frac{L^3}{EI}$$

$$v_2 = \frac{P}{D}[18\tfrac{EI}{L^3}]\frac{L^3}{EI}$$

$$L\phi_2 = \frac{-6P}{D}[\frac{EA}{L} + 3\frac{EI}{L^3}]\frac{L^3}{EI}$$

$$u_3 = \frac{P}{D}[7\frac{EA}{L} + 30\frac{EI}{L^3}]\frac{L^3}{EI}$$

$$L\phi_3 = \frac{3P}{D}[\frac{EA}{L} - 6\frac{EI}{L^3}]\frac{L^3}{EI}$$

$$D = 12[4\frac{EA}{L} + 3\frac{EI}{L^3}]$$

This form emphasizes that the two stiffness terms interact in a complicated fashion
to give the final deflections. It also shows that the sense of the rotation at Node 3
depends on the relative values of the axial and flexural stiffnesses.

**Example 4.4:**   A two-member frame is loaded as shown in Figure 4.8. One end
is fixed while the end with the applied load is fixed to a horizontal rollers. Both
members have the same material and section properties, and the joint connecting
them is at 90°.
    Find the deflections.

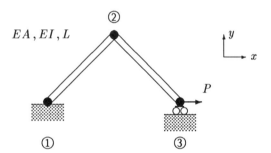

**Figure 4.8**: Two member frame.

Number the nodes as shown. The total degrees of freedom are

$$\{u\} = \{u_1, v_1, \phi_1; u_2, v_2, \phi_2; u_3, v_3, \phi_3\}$$

The boundary conditions require that

$$u_1 = v_1 = \phi_1 = 0, \qquad v_3 = \phi_3 = 0$$

The unknown nodal displacements and known applied loads are easily identified as,
respectively,

$$\{u_u\} = \begin{Bmatrix} u_2 \\ v_2 \\ \phi_2 \\ u_3 \end{Bmatrix} \quad \text{and} \quad \{P_k\} = \begin{Bmatrix} 0 \\ 0 \\ 0 \\ P \end{Bmatrix}$$

and the corresponding $IDbc$ matrix is

$$\{IDbc\} = \{0,0,0;1,2,3;4,0,0\}$$

We will take the connectivities for the two members as $1-2$ and $2-3$, respectively. For member 1-2 we have $\theta = 45^\circ$, hence the reduced stiffness is

$$[k^{*(12)}] = \frac{EA}{2L}\begin{bmatrix} 1 & 1 & 0 \\ 1 & 1 & 0 \\ 0 & 0 & 0 \end{bmatrix} + \frac{EI}{L^3}\begin{bmatrix} 6 & -6 & 3L\sqrt{2} \\ -6 & 6 & -3L\sqrt{2} \\ 3L\sqrt{2} & -3L\sqrt{2} & 4L^2 \end{bmatrix}$$

For member 2-3 we have $\theta = -45^\circ$, hence its reduced stiffness is

$$[k^{*(23)}] = \frac{EA}{2L}\begin{bmatrix} 1 & -1 & 0 & -1 \\ -1 & 1 & 0 & 1 \\ 0 & 0 & 0 & 0 \\ -1 & 1 & 0 & 1 \end{bmatrix} + \frac{EI}{L^3}\begin{bmatrix} 6 & 6 & 3L\sqrt{2} & -6 \\ 6 & 6 & 3L\sqrt{2} & -6 \\ 3L\sqrt{2} & 3L\sqrt{2} & 4L^2 & -3L\sqrt{2} \\ -6 & -6 & -3L\sqrt{2} & 6 \end{bmatrix}$$

The reduced structural system of equations is

$$\left[\frac{EA}{2L}\begin{bmatrix} 2 & 0 & 0 & -1 \\ 0 & 2 & 0 & 1 \\ 0 & 0 & 0 & 0 \\ -1 & 1 & 0 & 1 \end{bmatrix} + \frac{EI}{L^3}\begin{bmatrix} 12 & 0 & 6L\sqrt{2} & -6 \\ 0 & 12 & 0 & -6 \\ 6L\sqrt{2} & 0 & 8L^2 & -3L\sqrt{2} \\ -6 & -6 & -3L\sqrt{2} & 6 \end{bmatrix}\right]\begin{Bmatrix} u_2 \\ v_2 \\ \phi_2 \\ u_3 \end{Bmatrix} = \begin{Bmatrix} 0 \\ 0 \\ 0 \\ P \end{Bmatrix}$$

From the first and third equation get that

$$2u_2 = u_3$$

Using this in the first equation shows that

$$\phi_2 = 0$$

The unknown nodal displacements are

$$\begin{Bmatrix} u_2 \\ v_2 \\ u_3 \end{Bmatrix} = \frac{PL}{4EA}\begin{Bmatrix} 2 \\ 1 \\ 4 \end{Bmatrix} + \frac{PL^3}{48EI}\begin{Bmatrix} 2 \\ -1 \\ 4 \end{Bmatrix}$$

Note that if the axial flexibility is small enough, this result predicts that the center can actually move upwards.

## 4.3    Space Frames

All real structures, of course, are three-dimensional, but when the structural action is predominantly planar then the analysis using a two-dimensional model (such as a plane frame or a grid) is usually adequate. There are many situations, however, when the analysis should be conducted using a three-dimensional model. The purpose of this section is to show how this general case is handled.

## Element Stiffness Matrix in Local Axes

The displacement of each node of a space frame is described by three translational and three rotational components of displacement, giving six degrees of freedom at each unrestrained node.  Corresponding to these degrees of freedom are six nodal loads. The notations we will use for the displacement and force vectors at each node are, respectively,

$$
\begin{Bmatrix} u \\ v \\ w \\ \phi_x \\ \phi_y \\ \phi_z \end{Bmatrix}, \qquad
\begin{Bmatrix} F_x \\ F_y \\ F_z \\ M_x \\ M_y \\ M_z \end{Bmatrix}
$$

For each member, the forces are related to the displacements by the following partitioned matrices

$$
\{\bar{F}\} = \begin{bmatrix}
\bar{k}_{11} & \bar{k}_{12} & \bar{k}_{13} & \bar{k}_{14} \\
\bar{k}_{21} & \bar{k}_{22} & \bar{k}_{23} & \bar{k}_{24} \\
\bar{k}_{31} & \bar{k}_{32} & \bar{k}_{33} & \bar{k}_{34} \\
\bar{k}_{41} & \bar{k}_{42} & \bar{k}_{43} & \bar{k}_{44}
\end{bmatrix} \{\bar{u}\} = [\,\bar{k}\,]\{\bar{u}\}
$$

At this point it should be noted that every stiffness coefficient associated with a space frame element has already been encountered when dealing with the simpler structures of the previous sections. Hence there is little difficulty in showing that the above matrices take the following form when applied to a space frame member

$$
\begin{bmatrix} \bar{k}_{11} & \bar{k}_{12} \\ \bar{k}_{21} & \bar{k}_{22} \end{bmatrix} = \frac{+1}{L}
\begin{bmatrix}
EA & 0 & 0 & 0 & 0 & 0 \\
0 & 12EI_z/L^2 & 0 & 0 & 0 & 6EI_z/L \\
0 & 0 & 12EI_y/L^2 & 0 & -6EI_y/L & 0 \\
0 & 0 & 0 & GI_x & 0 & 0 \\
0 & 0 & -6EI_y/L & 0 & 4EI_y & 0 \\
0 & 6EI_z/L & 0 & 0 & 0 & 4EI_z
\end{bmatrix}
$$

$$
\begin{bmatrix} \bar{k}_{13} & \bar{k}_{14} \\ \bar{k}_{23} & \bar{k}_{24} \end{bmatrix} = \frac{-1}{L}
\begin{bmatrix}
EA & 0 & 0 & 0 & 0 & 0 \\
0 & 12EI_z/L^2 & 0 & 0 & 0 & -6EI_z/L \\
0 & 0 & 12EI_y/L^2 & 0 & 6EI_y/L & 0 \\
0 & 0 & 0 & GI_x & 0 & 0 \\
0 & 0 & -6EI_y/L & 0 & -2EI_y & 0 \\
0 & 6EI_z/L & 0 & 0 & 0 & -2EI_z
\end{bmatrix}
$$

$$
\begin{bmatrix} \bar{k}_{31} & \bar{k}_{32} \\ \bar{k}_{41} & \bar{k}_{42} \end{bmatrix} = \frac{-1}{L}
\begin{bmatrix}
EA & 0 & 0 & 0 & 0 & 0 \\
0 & 12EI_z/L^2 & 0 & 0 & 0 & 6EI_z/L \\
0 & 0 & 12EI_y/L^2 & 0 & -6EI_y/L & 0 \\
0 & 0 & 0 & GI_x & 0 & 0 \\
0 & 0 & 6EI_y/L & 0 & -2EI_y & 0 \\
0 & -6EI_z/L & 0 & 0 & 0 & -2EI_z
\end{bmatrix}
$$

$$
\begin{bmatrix} \bar{k}_{33} & \bar{k}_{34} \\ \bar{k}_{43} & \bar{k}_{44} \end{bmatrix} = \frac{+1}{L} \begin{bmatrix} EA & 0 & 0 & 0 & 0 & 0 \\ 0 & 12EI_z/L^2 & 0 & 0 & 0 & -6EI_z/L \\ 0 & 0 & 12EI_y/L^2 & 0 & 6EI_y/L & 0 \\ 0 & 0 & 0 & GI_x & 0 & 0 \\ 0 & 0 & 6EI_y/L & 0 & 4EI_y & 0 \\ 0 & -6EI_z/L & 0 & 0 & 0 & 4EI_z \end{bmatrix}
$$

$$(4.4)$$

For regularity of notation, the torsional stiffness is written as $GI_x = GJ$. The special cases of plane frame and grid, for example, can be obtained by setting the appropriate degrees of freedom to zero. This will be done in a later section.

A point to note is the regularity in the repetition of terms. For example, the first quadrant of $[\,\bar{k}\,]$ is almost the same as the fourth except for the signs of the off-diagonal terms.

## Assemblage in Global Axes

The transformation of the components of a vector $\{v\}$ from the local to the global axes is given by

$$\{\bar{v}\} = [\,R\,]\{v\}$$

We will discuss the specific form of $[\,R\,]$ in the next section. The same matrix will transform the vectors of nodal forces and displacements. To see this, note that the element nodal displacement vector is composed of four separate vectors, namely,

$$\{u\} = \Big\{\{u_1,\, v_1,\, w_1\};\ \{\phi_{x1},\, \phi_{y1},\, \phi_{z1}\};\ \{u_2,\, v_2,\, w_2\};\ \{\phi_{x2},\, \phi_{y2},\, \phi_{z2}\}\Big\}$$

Each of these are separately transformed by the $[3 \times 3]$ rotation matrix $[\,R\,]$. Hence the complete transformation is

$$\{\bar{F}\} = [\,T\,]\{F\}, \qquad \{\bar{u}\} = [\,T\,]\{u\}$$

where

$$[\,T\,] \equiv \begin{bmatrix} R & 0 & 0 & 0 \\ 0 & R & 0 & 0 \\ 0 & 0 & R & 0 \\ 0 & 0 & 0 & R \end{bmatrix}$$

is a $[12 \times 12]$ matrix. Substituting for the barred vectors into the element stiffness relation allows us to obtain the global stiffness as

$$[\,k\,] = [\,T\,]^T[\,\bar{k}\,][\,T\,]$$

This is formally the same relation as obtained for the truss and plane frame. In practice, however, we take advantage of the special nature of $[\,T\,]$ to reduce this further to

$$[k_{11}] = [\,R\,]^T[\bar{k}_{11}][\,R\,], \qquad [k_{12}] = [\,R\,]^T[\bar{k}_{12}][\,R\,] \qquad \text{or} \qquad [k_{ij}] = [\,R\,]^T[\bar{k}_{ij}][\,R\,]$$

This is a transform of the $[3 \times 3]$ partial stiffnesses.

The assembly process follows that of the other structures already encountered. Formally, each member stiffness is rotated to the global coordinate system and then augmented to the system size. The structural stiffness matrix is then

$$[K] = \sum_m [\,T\,]_m^T [\,\bar{k}\,]_m [\,T\,]_m$$

where the summation is over each member. In practice, there is no need to augment the member stiffness since we assemble the reduced global stiffness directly. It is important to realize that the transformation occurs before the assembly.

## 4.4   Determining the Rotation Matrix

The rotation matrix required to transform one cartesian coordinate system to another sharing a common origin is

$$[R] = \begin{bmatrix} l_x & m_x & n_x \\ l_y & m_y & n_y \\ l_z & m_z & n_z \end{bmatrix}$$

where $l_x, m_x$ and $n_x$ are the cosines of the angles that the $\bar{x}$ axis makes with the global $x, y, z$ axes, respectively, as shown in Figure 4.9. The other terms give the orientations of the $\bar{y}$ and $\bar{z}$ axes, respectively. Since the member axis of a frame or truss coincides with $\bar{x}$, then the direction cosines of the first row can also be written as

$$l_x = (x_j - x_i)/L_{ij} , \qquad m_x = (y_j - y_i)/L_{ij} , \qquad n_x = (z_j - z_i)/L_{ij} \qquad (4.5)$$

where $(x_i, y_i, z_i)$ and $(x_j, y_j, z_j)$ are the coordinates of the first and second nodes, respectively, and $L_{ij}$ is the length of the member. The problem here is to find the remaining elements of $[\,R\,]$, as $l_x, m_x$, and $n_x$ define only the orientation of the member $\bar{x}$ axis.

In order to determine the rotation matrix we will assume that the member arrived at its current position by successive rotations of axes. Consider the member to be initially oriented along the $x$ axis. The first rotation is through an angle $\alpha$ about the $z$ axis. The second is a rotation through an angle $\beta$ about the $\bar{y}$ axis. (This sequence leaves the member $\bar{y}$-axis always oriented so as to lie in the global $x - y$ plane.) The resulting rotation matrix is therefore

$$[R] = [R_\beta][R_\alpha] = \begin{bmatrix} \cos\beta & 0 & \sin\beta \\ 0 & 1 & 0 \\ -\sin\beta & 0 & \cos\beta \end{bmatrix} \begin{bmatrix} \cos\alpha & \sin\alpha & 0 \\ -\sin\alpha & \cos\alpha & 0 \\ 0 & 0 & 1 \end{bmatrix}$$

Multiplying these matrices together, we get

$$[R] = \begin{bmatrix} \cos\beta\cos\alpha & \cos\beta\sin\alpha & \sin\beta \\ -\sin\alpha & \cos\alpha & 0 \\ -\sin\beta\cos\alpha & -\sin\beta\sin\alpha & \cos\beta \end{bmatrix}$$

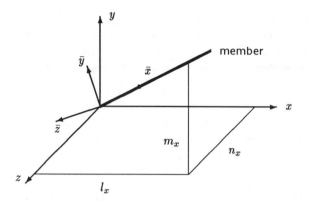

**Figure 4.9**: Direction cosines for general member.

Equating the first row to the direction cosines of the member gives

$$l_x = \cos \beta \cos \alpha, \qquad m_x = \cos \beta \sin \alpha, \qquad n_x = \sin \beta$$

Therefore, the functions $\cos \alpha$, $\sin \alpha$, $\cos \beta$ and $\sin \beta$ may be expressed in terms of the direction cosines of the member by

$$\cos \alpha = \frac{l_x}{D}, \qquad \sin \alpha = \frac{m_x}{D}, \qquad \cos \beta = D, \qquad \sin \beta = n_x, \qquad D \equiv \sqrt{l_x^2 + m_x^2}$$

Substitution of these expressions into the above gives

$$[\,R\,] = \begin{bmatrix} l_x & m_x & n_x \\ -m_x/D & l_x/D & 0 \\ -l_x n_x/D & -m_x n_x/D & D \end{bmatrix}$$

This is the rotation matrix $[\,R\,]$ for a space truss member and is also valid for space frames when the member has a symmetric cross-section.

## Non-symmetric Sections

A space frame member may have its principal axes in general directions. There are various ways in which the orientation of these cases can be defined and the one chosen involves specifying the orientation by means of an angle of rotation about the member $\bar{x}$ axis. In order to visualize clearly how such an angle is measured, consider three successive rotations from the structure axes to the member axes. The first two rotations through the angles $\alpha$ and $\beta$ are exactly the same as before. The third transformation consists of a final rotation through the angle $\gamma$ about the member $\bar{x}$ axis, resulting in the $\bar{y}$ and $\bar{z}$ axes coinciding with the principal axes of the cross section.

The rotation of axes through the angle $\gamma$ about the member axis requires the introduction of a rotation matrix $[R_\gamma]$ given by

$$[R_\gamma] = \begin{bmatrix} 1 & 0 & 0 \\ 0 & \cos\gamma & \sin\gamma \\ 0 & -\sin\gamma & \cos\gamma \end{bmatrix}$$

Multiplication of the three successive rotations gives $[R] = [R_\gamma][R_\beta][R_\alpha]$ and when multiplied out in terms of the direction cosines this becomes

$$[R] = \begin{bmatrix} l_x & m_x & n_x \\ \dfrac{-m_x\cos\gamma - l_x n_x \sin\gamma}{D} & \dfrac{l_x\cos\gamma - m_x n_x \sin\gamma}{D} & D\sin\gamma \\ \dfrac{m_x\sin\gamma - l_x n_x \cos\gamma}{D} & \dfrac{-l_x\sin\gamma - m_x n_x \cos\gamma}{D} & D\cos\gamma \end{bmatrix} \qquad (4.6)$$

This rotation matrix is expressed in terms of the direction cosines of the member (which are readily computed from the coordinates of the joints, Equation(4.5)) and the angle $\gamma$, which must be given as part of the description of the structure itself. Note that if this angle is equal to zero, the matrix $[R]$ reduces to the form given previously for a space truss member.

## Special Case

When the member axes are specified in the manner just described, there is no ambiguity about their orientations except in the special case of a member oriented along the global z-axis. There is no unique rotation to get to that orientation, i.e., $\alpha = 0°, \beta = 90°$ or $\alpha = 90°, \beta = 90°$. To overcome this difficulty, the additional specification will be made that the member $\bar{y}$-axis is always taken to be along the global y-axis for these cases. That is $\alpha = 0°, \beta = 90°$.

The complete set of direction cosines is therefore

$$[R_z] = \begin{bmatrix} 1 & 0 & 0 \\ 0 & \cos\gamma & \sin\gamma \\ 0 & -\sin\gamma & \cos\gamma \end{bmatrix} \begin{bmatrix} 0 & 0 & n_x \\ 0 & 1 & 0 \\ -n_x & 0 & 0 \end{bmatrix} = \begin{bmatrix} 0 & 0 & n_x \\ -n_x\sin\gamma & \cos\gamma & 0 \\ -n_x\cos\gamma & -\sin\gamma & 0 \end{bmatrix}$$

This is for the general case of a member of non-symmetric section. All that is necessary is to substitute for the direction cosine $n_x$ its appropriate value, which is either 1 or -1.

**Example 4.5:**   A truss structure has joints at the following coordinates.

| node | $x$ | $y$ | $z$ |
|---|---|---|---|
| 1 : | 0 | 0 | 0 |
| 2 : | 100 | 0 | 0 |
| 3 : | 100 | -200 | 0 |
| 4 : | 100 | -200 | 100 |

Determine the rotation matrices for members with connectivities $1 - 4$ and $3 - 4$.
The length of Member 1-4 is

$$L = \sqrt{(100 - 0)^2 + (-200 - 0)^2 + (-100 - 0)^2} = 100\sqrt{6}$$

The direction cosines of the member are

$$
\begin{aligned}
l_x &= (100 - 0)/(100\sqrt{6}) = 1/\sqrt{6} \\
m_x &= (-200 - 0)/(100\sqrt{6}) = -2/\sqrt{6} \\
n_x &= (-100 - 0)/(100\sqrt{6}) = -1/\sqrt{6}
\end{aligned}
$$

This gives $D = \sqrt{5/6}$. The structure is specified as being a truss, hence the angle $\gamma$ is zero, therefore the rotation matrix can be obtained by substituting into Equation 4.6. The result is

$$[\,R\,] = \begin{bmatrix} .4082 & -.8165 & -.4082 \\ .8944 & .4472 & .0000 \\ .1826 & -.3652 & .3727 \end{bmatrix}$$

The length of Member 3-4 is

$$L = \sqrt{(100 - 100)^2 + (-200 + 200)^2 + (-100 - 0)^2} = 100$$

The direction cosines of the member are

$$l_x = (100 - 100)/100 = 0, \quad m_x = (-200 + 200)/100 = 0, \quad n_x = (-100 - 0)/100 = -1$$

This is the special case with $n_x = -1$, hence the rotation matrix is

$$[\,R\,] = \begin{bmatrix} 0 & 0 & -1 \\ 0 & 1 & 0 \\ 1 & 0 & 0 \end{bmatrix}$$

## 4.5   Special Considerations

Having developed the analysis for the general case of a space frame, we now mention some special considerations that are of use when dealing with practical problems.

### Global Reductions

There are six degrees of freedom at each node in the space frame. Many problems, however, do not need this many; for example, the plane frame only requires three, while the plane truss uses two. Obviously to analyze a 2-D structure as a 3-D frame is a waste of computer resources.

The key to understanding the reduction of the general case is the idea of imposing constraints. We saw in the case of fixed boundary conditions that we specify the degree of freedom as zero, and consequently 'scratch' the associated rows and columns in the stiffness relation. In essence we do the same here; we specify constraints on the degrees of freedom as if they were boundary conditions. This process is illustrated next for a grid structure.

A close companion of the 2-D frame is the *grid*. The grid is essentially a plane frame but with the loads applied laterally to its plane. Consequently, members must also support axial twisting. That is, the member action is a combination of that of a beam and a shaft. To recover this behavior from the space frame, we assume the grid lies in the $x - y$ plane and impose the following constraints

$$u = v = 0, \qquad \phi_z = 0$$

at each node. The non-zero degrees of freedom are the out of plane displacement $w$, and the two in plane rotations $\phi_x$, $\phi_y$. The member nodal forces and degrees of freedom are

$$\{F\} = \left\{ \begin{array}{c} F_{z1} \\ M_{x1} \\ M_{y1} \\ F_{z2} \\ M_{x2} \\ M_{y2} \end{array} \right\}, \qquad \{u\} = \left\{ \begin{array}{c} w_1 \\ \phi_{x1} \\ \phi_{y1} \\ w_2 \\ \phi_{x2} \\ \phi_{y2} \end{array} \right\}$$

The stiffness matrix of the grid member in local coordinates is

$$[\bar{k}] = \frac{GJ}{L} \begin{bmatrix} 0 & 0 & 0 & 0 & 0 & 0 \\ 0 & 1 & 0 & 0 & -1 & 0 \\ 0 & 0 & 0 & 0 & 0 & 0 \\ 0 & 0 & 0 & 0 & 0 & 0 \\ 0 & -1 & 0 & 0 & 1 & 0 \\ 0 & 0 & 0 & 0 & 0 & 0 \end{bmatrix} + \frac{EI}{L^3} \begin{bmatrix} 12 & 0 & -6L & -12 & 0 & -6L \\ 0 & 0 & 0 & 0 & 0 & 0 \\ -6L & 0 & 4L^2 & 6L & 0 & 2L^2 \\ -12 & 0 & 6L & 12 & 0 & 6L \\ 0 & 0 & 0 & 0 & 0 & 0 \\ -6L & 0 & 2L^2 & 6L & 0 & 4L^2 \end{bmatrix}$$

where $GJ$ is the torsional rigidity of the member and $EI$ is the flexural rigidity for bending out of the plane. The rotation matrix

$$[R] = \begin{bmatrix} l_x & m_x & 0 \\ -m_x & l_x & 0 \\ 0 & 0 & 1 \end{bmatrix}$$

is used to obtain the stiffness matrix for an arbitrarily oriented grid member.

Similar global reductions can be done to recover the cases already considered as well some new ones such as the shaft in torsion.

## Boundary Conditions and Reactions

The boundary reactions mimic the loading of the general node; that is there are six components, in all. It is possible to find these reactions by summing the contributions from each member acting at the support. This is the procedure used in the simple rod and beam structures.

For large scale problems, however, it is much simpler to find the boundary reactions by adding a 'boundary element' to the structure. The basic idea is to connect the structure to a rigid support using a small frame element. The nodal loads of this element are the required boundary reactions. This is feasible since many elements are already being used to model the structure.

Actually this idea of the boundary element adds a lot of flexibility to the matrix analysis method. We already saw how it allowed the specification of displacements as boundary conditions, we next show how it can also be used to implement oblique supports.

We have formulated the stiffness approach in terms of a global coordinate system. Therefore the allowable constraints must also be in terms of the global coordinates. Consider the case of a frame with *oblique supports*, that is, the frame is attached to rollers on an inclined surface. The boundary condition is that the displacement normal to the surface is zero. This is a constraint condition written as

$$u_{normal} = 0 = -u\sin\theta + v\cos\theta$$

where $\theta$ is the slope of the incline. Thus neither of the global degrees of freedom is zero. It is possible, of course, to reformulate the stiffness relation to allow for the incorporation of such constraint conditions. A simpler approach, however, is to use another boundary element. That is, we replace the actual support by a relatively stiff member having its longitudinal axis in the direction normal to the inclined surface. If this member is pinned at both ends, then all the motion will be perpendicular to it, thus simulating the effect of an inclined roller.

Variations on this idea can be used to simulate other types of boundary conditions.

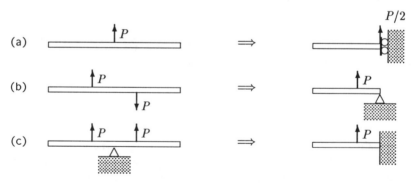

**Figure 4.10**: Equivalent boundary conditions.

## Use of Symmetry and Anti-symmetry

In a practical problem, some of the displacements may be obvious simply by inspection. For example, there is no $w$ displacement in a plane frame. In other cases we can

infer this information from the symmetry (or anti-symmetry) of the geometry and
loading conditions. We can then implement these constraints by use of equivalent
boundary conditions and thereby simplify the problem.

Figure 4.10 shows a few examples for a beam, but the idea is quite general. The
first structure shown has a loading symmetric with respect to the center of the beam.
Thus, at the center, $\phi = 0$ but $v \neq 0$. An equivalent boundary support is as shown
on the right hand side. It should be noted that only half of the load acts on the
equivalent model. In fact, the rigidity of members lying in a plane of symmetry must
also be halved in order to divide the structure into two equal parts. The second case
corresponds to an anti-symmetric loading condition achieved with either a couple or
moment. This corresponds to a pinned support condition. The symmetric version of
the problem corresponds to a fixed support.

The use of symmetry and anti-symmetry does not involve any approximation and
therefore when the opportunity arises, advantage should be taken of it. It is worth
keeping in mind that this can be done as long as the structure is symmetric, because
any unsymmetrical loading can be decomposed into the sum of a symmetric and
anti-symmetric load.

**Example 4.6:**   A two-member frame is loaded symmetrically as shown in Figure 4.11. The joint forms an angle of 90° and the loads are applied at pinned rollers.
Find the deflections.

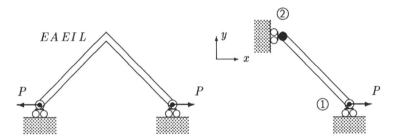

**Figure 4.11**: Two member frame.

Due to the symmetry, only half of the structure need be analyzed. This is done
by modeling it as shown where the vertical displacement at the apex is left free while
the rotational and horizontal degrees of freedom are suppressed. Number the nodes
as shown, then the total degrees of freedom are

$$\{u\} = \{u_1, v_1, \phi_1 ; u_2, v_2, \phi_2\}$$

The boundary conditions require that

$$v_1 = 0, \qquad u_2 = 0, \qquad \phi_2 = 0$$

The unknown nodal displacements and known applied loads are easily identified as, respectively,

$$\{u_u\} = \left\{ \begin{array}{c} u_1 \\ \phi_1 \\ v_2 \end{array} \right\} \quad \text{and} \quad \{P_k\} = \left\{ \begin{array}{c} P \\ 0 \\ 0 \end{array} \right\}$$

and the corresponding $IDbc$ matrix is

$$\{IDbc\} = \{1,0,2;0,3,0\}$$

Since there is only one element the reduced structural stiffness matrix is obtained directly as (the connectivity is 1-2, and $\theta = 135°$)

$$[K^*] = \frac{EA}{2L} \begin{bmatrix} 1 & 0 & 1 \\ 0 & 0 & 0 \\ 1 & 0 & 1 \end{bmatrix} + \frac{EI}{\sqrt{2}L^3} \begin{bmatrix} 6\sqrt{2} & -6L & -6\sqrt{2} \\ -6L & 4\sqrt{2}L^2 & 6L \\ -6\sqrt{2} & 6L & 6\sqrt{2} \end{bmatrix}$$

The unknown nodal displacements are

$$\left\{ \begin{array}{c} u_1 \\ \phi_1 \\ v_2 \end{array} \right\} = [K^*]^{-1} \left\{ \begin{array}{c} P \\ 0 \\ 0 \end{array} \right\} = \frac{P}{2EA} \left\{ \begin{array}{c} L \\ 0 \\ L \end{array} \right\} + \frac{PL^2}{12EI} \left\{ \begin{array}{c} 2L \\ 3\sqrt{2} \\ -2L \end{array} \right\}$$

An interesting aspect to this solution is that if $EA/L = 3EI/L^3$, then there is no vertical deflection.

## Inextensible Structures

Many frame structures are more flexible in bending than in axial deformation, and, consequently, the axial deformation can often be neglected. The structure is then referred to as being *inextensible*. Sometimes, the complexity of a structures problem may be reduced by making use of this approximation.

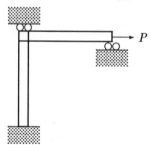

**Figure 4.12:** Remodeling of Frame 4.7 due to inextensibility.

The inextensible approximation is usually justified when the contribution to the overall stiffness of the bending rigidity $EI/L^3$ is much smaller than that axial rigidity $EA/L$. In other words, if

$$\frac{EI}{L^3} \ll \frac{EA}{L}$$

it is usual to assume that the structure is inextensible. For example, reconsider the frame problem in Figure 4.7. In the limit as the axial stiffness is very large then we have

$$u_2 = \frac{7PL^3}{48EI}$$

$$v_2 = 0$$

$$L\phi_2 = \frac{-PL^3}{8EI}$$

$$u_3 = \frac{7PL^3}{48EI} = u_2$$

$$L\phi_3 = \frac{PL^3}{16EI}$$

In that case, $v_2 \approx 0$ and the horizontal displacements at Nodes 2 and 3 are the same.

The inextensibility assumption is implemented in the form of a set of new boundary constraints. Figure 4.12 shows the modeling of the above problem. The vertical deflection at Node 2 (being assumed small) is forced to be zero by attaching the node to a horizontal roller by pins. The inextensibility of the horizontal member is implemented by the constraint relation

$$u_3 = u_2$$

It should be noted that when axial deformation is neglected, the axial force cannot be calculated from the nodal displacement solution. Instead, equilibrium must be used.

**Example 4.7:**  Simplify the modeling of a three story building by assuming it to have inextensible vertical walls and very stiff floors.

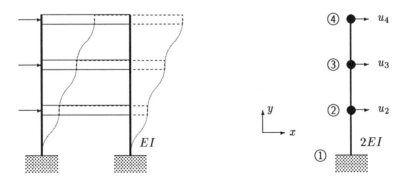

**Figure 4.13**: Equivalent modeling of a building.

There are three degrees of freedom $\{u, v, \phi\}$ at each joint. The inextensibility of the vertical members results in

$$v = 0$$

at each joint. The inextensibility of the floor, however, gives the constraint

$$u_{left} = u_{right}$$

Finally, because the floor is also flexurally very stiff then

$$\phi_{left} = \phi_{right}$$

Effectively, the floors of the building just move horizontally as shown in Figure 4.13; there is no rotation and no vertical movement. (This is similar to a shearing action and so the model is sometimes called a *shear building* model.) The equivalent model is therefore as shown in Figure 4.13. Each floor is given just a horizontal degree of freedom. Note that the effective flexural rigidity is $2EI$. Each wall is modeled as a beam with zero rotation at the ends. Thus for each section

$$\left\{ \begin{array}{c} V_1 \\ V_2 \end{array} \right\} = \frac{EI}{L^3} \left[ \begin{array}{cc} 12 & -12 \\ -12 & 12 \end{array} \right] \left\{ \begin{array}{c} u_1 \\ u_2 \end{array} \right\} = \frac{12EI}{L^3} \left[ \begin{array}{cc} 1 & -1 \\ -1 & 1 \end{array} \right] \left\{ \begin{array}{c} u_1 \\ u_2 \end{array} \right\}$$

This resembles the rod stiffness relation, but actually the displacements are transverse to the axis. This element is called a *shear element.*

In this way we reduced the system size from 18 down to 3. This is a very useful approximation in the dynamic analysis of buildings.

## 4.6   Substructuring

We conclude this chapter with an introduction to the approach used to analyze complex structures. The division of a structure into components for separate analysis is called *substructuring.* The main advantage of the approach is the replacement of a large problem by many smaller ones each of which may be more manageable for a given set of resources. Also, design changes on a component do not affect the work already done on the other components. As compared with a single pass analysis, the computational cost is usually much greater, the coding is usually much more complicated and needs to be customized to specific problems. When the structure is made of repetitive substructures, this approach is very efficient. The basic idea, as will be shown, is essentially that of a *super-element* — the substructure is used in the same way as an individual finite element with internal degrees of freedom that are eliminated prior to the element assemblage process.

Consider the case of a structure divided into two. Now look at one of those substructures. We can partition the arrays as

$$\{u\} = \left\{ \begin{array}{c} u_i \\ - \\ u_c \end{array} \right\}, \qquad \{P\} = \left\{ \begin{array}{c} P_i \\ - \\ F_c \end{array} \right\}$$

where the subscript $i$ refers to interior nodes, and $c$ refers to common or connection nodes. The equations for the substructural system are expressed in the partitioned form as

$$\left[ \begin{array}{cc} K_{ii} & K_{ic} \\ K_{ci} & K_{cc} \end{array} \right] \left\{ \begin{array}{c} u_i \\ u_c \end{array} \right\} = \left\{ \begin{array}{c} P_i \\ F_c \end{array} \right\}$$

For concreteness, we will assume that we have already reduced the zero degrees of freedom from the system. Multiplying this out gives

$$[K_{ii}]\{u_i\} + [K_{ic}]\{u_c\} = \{P_i\}$$
$$[K_{ci}]\{u_i\} + [K_{cc}]\{u_c\} = \{F_c\}$$

We will condense the interior degrees of freedom from this. From the first equation get

$$\{u_i\} = [K_{ii}]^{-1}[\{P_i\} - [K_{ic}]\{u_c\}] \tag{4.7}$$

Substitute this into the second equation to give

$$[[K_{cc}] - [K_{ci}][K_{ii}]^{-1}[K_{ic}]]\{u_c\} = \{F_c\} - [K_{ci}][K_{ii}]^{-1}\{P_i\}$$

Introducing the new arrays

$$[K_{ss}] \equiv [[K_{cc}] - [K_{ci}][K_{ii}]^{-1}[K_{ic}]], \qquad \{Q_c\} \equiv [K_{ci}][K_{ii}]^{-1}\{P_i\}$$

then we can express the above in the form of a stiffness relation

$$\{F_c\} = [K_{ss}]\{u_c\} + \{Q_c\}$$

This is the stiffness relation for the substructure considered as a super-element. We call it an element because it is written in terms of only the connection nodes. We know from previous chapters that if all the substructures are so formulated as elements, then they can be assembled in the usual manner and solved. Keep in mind that the system size at that stage is only the sum of all the connection nodes.

It is of interest to look a little closer at the terms in the above. For example, the quantity $[K_{ii}]^{-1}\{P_i\}$ can be interpreted as the displacement field of the substructure when all the connection nodes are fixed. That is

$$[K_{ii}]\{\tilde{u}\} = \{P_i\}$$

The load term $\{Q_c\} = [K_{ci}]\{\tilde{u}\}$ contains the reactions at the fixed connection points.

**Example 4.8:**   Analyze the truss structure shown in Figure 4.14 by substructuring. Each member has the same length and section properties.

The original structure will be divided at Node 2; thus there is only one connection node and two connection degrees of freedom. Both substructures are similar and therefore have the same reduced structural stiffnesses. These are formed in the usual way and can easily be shown to be

$$[K^*] = \frac{EA}{4L} \begin{bmatrix} 5 & -\sqrt{3} & -4 & 0 \\ -\sqrt{3} & 3 & 0 & 0 \\ -4 & 0 & 5 & \sqrt{3} \\ 0 & 0 & \sqrt{3} & 3 \end{bmatrix}$$

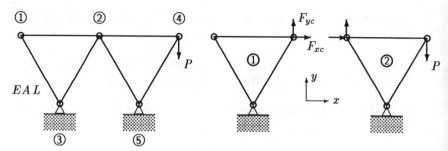

**Figure 4.14**: Substructural modeling.

The [6 × 6] stiffness matrix has been reduced by scratching the rows and columns associated with the degrees of freedom at Node 3 or Node 5.

For Substructure 1, the partitioned arrays are

$$[K_{ii}] = \frac{EA}{4L} \begin{bmatrix} 5 & -\sqrt{3} \\ -\sqrt{3} & 3 \end{bmatrix} \qquad [K_{ic}] = \frac{EA}{4L} \begin{bmatrix} -4 & 0 \\ 0 & 0 \end{bmatrix}$$

$$[K_{ci}] = \frac{EA}{4L} \begin{bmatrix} -4 & 0 \\ 0 & 0 \end{bmatrix} \qquad [K_{cc}] = \frac{EA}{4L} \begin{bmatrix} 5 & \sqrt{3} \\ \sqrt{3} & 3 \end{bmatrix}$$

The corresponding load vectors are

$$\{P_i\} = \begin{Bmatrix} 0 \\ 0 \end{Bmatrix}, \qquad \{F_c\} = \begin{Bmatrix} F_{xc} \\ F_{yc} \end{Bmatrix}$$

The inverse matrix is

$$[K_{ii}]^{-1} = \frac{L}{3EA} \begin{bmatrix} 3 & \sqrt{3} \\ \sqrt{3} & 5 \end{bmatrix}$$

This gives the structural stiffness of the substructure as

$$[K_{ss}] = \frac{EA}{4L} \begin{bmatrix} 5 & \sqrt{3} \\ \sqrt{3} & 3 \end{bmatrix} - \frac{EA}{4L} \begin{bmatrix} 4 & 0 \\ 0 & 0 \end{bmatrix}$$

The stiffness relation for the substructure is (the load term $\{Q_c\}$ is zero)

$$\begin{Bmatrix} F_{xc} \\ F_{yc} \end{Bmatrix} = \frac{EA}{4L} \begin{bmatrix} 1 & \sqrt{3} \\ \sqrt{3} & 3 \end{bmatrix} \begin{Bmatrix} u_c \\ v_c \end{Bmatrix}$$

For Substructure 2, the partitioned arrays are

$$[K_{ii}] = \frac{EA}{4L} \begin{bmatrix} 5 & \sqrt{3} \\ \sqrt{3} & 3 \end{bmatrix} \qquad [K_{ic}] = \frac{EA}{4L} \begin{bmatrix} -4 & 0 \\ 0 & 0 \end{bmatrix}$$

$$[K_{ci}] = \frac{EA}{4L} \begin{bmatrix} -4 & 0 \\ 0 & 0 \end{bmatrix} \qquad [K_{cc}] = \frac{EA}{4L} \begin{bmatrix} 5 & -\sqrt{3} \\ -\sqrt{3} & 3 \end{bmatrix}$$

with the corresponding load vectors

$$\{P_i\} = \begin{Bmatrix} P_x \\ P_y \end{Bmatrix}, \qquad \{F_c\} = \begin{Bmatrix} F_{xc} \\ F_{yc} \end{Bmatrix}$$

The inverse matrix is

$$[K_{ii}]^{-1} = \frac{L}{3EA} \begin{bmatrix} 3 & -\sqrt{3} \\ -\sqrt{3} & 5 \end{bmatrix}$$

This gives the structural stiffness of the substructure as

$$[K_{ss}] = \frac{EA}{4L} \begin{bmatrix} 5 & -\sqrt{3} \\ -\sqrt{3} & 3 \end{bmatrix} - \frac{EA}{4L} \begin{bmatrix} 4 & 0 \\ 0 & 0 \end{bmatrix}$$

The loading term is non-zero and determined as

$$\{Q_c\} = \frac{EA}{4L} \begin{bmatrix} -4 & 0 \\ 0 & 0 \end{bmatrix} \frac{L}{3EA} \begin{bmatrix} 3 & -\sqrt{3} \\ -\sqrt{3} & 5 \end{bmatrix} \begin{Bmatrix} P_x \\ P_y \end{Bmatrix} = \frac{1}{12} \begin{bmatrix} -12 & 4\sqrt{3} \\ 0 & 0 \end{bmatrix} \begin{Bmatrix} P_x \\ P_y \end{Bmatrix}$$

The stiffness relation for the substructure is therefore

$$\begin{Bmatrix} F_{xc} \\ F_{yc} \end{Bmatrix} = \frac{EA}{4L} \begin{bmatrix} 1 & -\sqrt{3} \\ -\sqrt{3} & 3 \end{bmatrix} \begin{Bmatrix} u_c \\ v_c \end{Bmatrix} + \begin{Bmatrix} -P_x + P_y/\sqrt{3} \\ 0 \end{Bmatrix}$$

The assembled system now becomes

$$\begin{Bmatrix} 0 \\ 0 \end{Bmatrix} = \frac{EA}{4L} \begin{bmatrix} 2 & 0 \\ 0 & 6 \end{bmatrix} \begin{Bmatrix} u_c \\ v_c \end{Bmatrix} + \begin{Bmatrix} P_x - \dfrac{1}{\sqrt{3}} P_y \\ 0 \end{Bmatrix}$$

This gives the solution

$$u_c = \frac{2L}{EA}\left(P_x - \frac{1}{\sqrt{3}} P_y\right), \qquad v_c = 0$$

Once the nodal displacements are obtained, the loads in the members can be calculated.

## Problems

**4.1** Maxwell's *reciprocal theorem* states that for an elastic structure the deflection produced at some point $i$ by a unit load at point $j$ is equal to the deflection at $j$ due to a unit load at $i$. Show that this statement is true for systems described by the stiffness relation

$$\{P\} = [\, K\, ]\{u\}$$

[Reference [45], pp. 43]

**4.2** Show that a frame attached rigidly to rollers on an incline can be modeled using the idea of two boundary elements separated a distance and connected by a cross-bar.

[Reference [45], pp. 375]

**4.3** In a truss problem, it is possible to implement a relative displacement constraint (say $u_1 = u_2$) by connecting the nodes by a very stiff element. Which entries in the structural stiffness matrix are affected? What are the dangers in implementing such a scheme?

[Reference [12], pp. 159]

## Exercises

**4.1** Give an example that proves it is possible that the diagonal term $k_{ii}$ of the element stiffness matrix can be zero. Show, by a simple example, that it cannot be less than zero.

**4.2** A simple rigid body motion is described by

$$v_2 = v_1 + \phi_1 L, \qquad \phi_1 = \phi_2$$

Show that for the plane frame, the forces required to produce such a motion are zero.

**4.3** For the truss of Figure 4.3, renumber the nodes in an attempt to reduce the bandwidth. State the connectivities and assemble the connectivity matrix.

**4.4** Solve the problem of Figure 4.4, but choose the connectivities as 2-1, 3-1, 4-1.

**4.5** A simply supported beam is modeled by two elements of equal length. A uniform lateral load acts along the left half. How can this problem be analyzed as the sum of a symmetric and an anti-symmetric loading? Does this approach have any advantage over a single analysis?

**4.6** Solve the problem of Figure 4.4, but make use of symmetry and anti-symmetry.

**4.7** For the truss problem of Figure 4.14, solve it making use of symmetry and anti-symmetry.

**4.8** Show that the rotation matrix obtained in Equation(4.6) is orthogonal.

**4.9** A steel portal frame ABCD, fixed at A and D, is 8 $ft$ high and 12 $ft$ wide. It has a distributed load of 500 $lb/ft$ along BC and a 3000 $lb$ horizontal force applied at B. Determine the deflections.     $[\{u_B\} = \{0.092\,in, -0.001\,in, -.0014\,r\}]$

**4.10** An aluminum frame is fixed at A and C with AB vertical of length 2 $m$ and BC horizontal of length 1 $m$. A vertical load of 6000 $N$ is applied at B and a horizontal load of 6000 $N$ applied midway along AB. If the cross-section of both members is 24 $mm$ × 24 $mm$, find the deflected shape and plot the shear force and bending moment diagrams.     $[\{u_{mid}\} = \{162\,mm, -0.04\,mm, .033\,r\}]$

Chapter 5

# Structural Stability

In the previous chapters we investigated the equilibrium of various structural systems. Now we wish to consider a very important property of systems in equilibrium, namely *stability*. Basically, we are interested in what happens to the structure when it is disturbed slightly from its equilibrium position: does it tend to return to its equilibrium position, or does it tend to depart even further? We term the former case *stable equilibrium* and the latter *unstable equilibrium*. A load carrying structure in a state of unstable equilibrium is unreliable and hazardous — a small disturbance can cause catastrophic changes. Reference [20] gives a good background setting for the study of structural stability while Reference [41] is an excellent compendium of examples and solutions. Mention should also be made of Reference [51] which places stability in a much broader context.

We will reconsider our previous developments of equilibrium and compatibility, but for the structure in a slightly disturbed (deformed) state. This is an essential step for stability analysis even when the deformation is small. We will see that the loss of structural stiffness is the key to understanding stability.

## 5.1   Elastic Stability

Consider the simple pinned structure shown in Figure 5.1. The initially vertical bar is assisted in remaining vertical by the action of the horizontal spring; the spring is unstretched when the bar is vertical. An equilibrium analysis in the undeformed state gives the axial force in the bar as $P$ and in the spring as $Q$. Consideration of compatibility shows that the resulting displacements are related to the forces through the following stiffness relations

$$[\,k\,]u = Q\,, \qquad [\,k_b\,]v = -P \tag{5.1}$$

where $k_b$ is the axial stiffness of the bar. (Identical results are obtained if the matrix methods of Chapter 4 are used.) For the purpose of the later discussion, let the bar be very stiff, from which we conclude that the displacement is only horizontal, i.e., $v = 0$. Keep in mind that all displacements are considered to be very small.

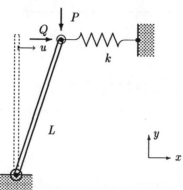

**Figure 5.1**: Disturbed equilibrium state of a pinned bar.

We will now look at equilibrium again, but this time based on the deformed configuration. Consider the situation when the bar has already displaced by an amount $u$ as shown in Figure 5.1. Since it is in equilibrium, summing the moments about the base gives

$$-QL - Pu + kuL = 0$$

This can be put in the form of a stiffness relation

$$[k - P/L]u = Q \tag{5.2}$$

In general, we can solve this for the unknown displacement $u$ if both loads $P$ and $Q$ are specified. Contrast this stiffness relation with the one obtained in Equation(5.1); using the deformed position as the reference state for application of the equilibrium equations has coupled both applied loads in the stiffness relation. The consequences of this, as we shall see, are what dictates the stability behavior of a structure. For example, in the present case a critical situation arises when the applied load $P$ approaches a value near $kL$ since then an inordinately large displacement is indicated. Indeed, even if $Q$ is almost zero very large deflections are calculated. Obviously, the structure is experiencing some critical behavior; in fact, it is becoming unstable. The critical condition is when $P = kL$ and this value of load is called the *critical* load $P_{cr}$. The critical load is the borderline (or transition) between the structure being stable and unstable, and is the primary concern in stability analysis. Additional insight can be gained by inquiring as to what force ($Q$) is necessary to a achieve a given displacement $u$. At or near the critical load, it takes virtually no load $Q$ to move the structure any amount. We call the situation *neutral equilibrium*.

Some important points can be drawn from this example. First, the question of stability arose only after we used the deformed configuration to establish the equilibrium conditions. Second, the critical behavior occurred when the stiffness tended to zero. That is, instability is associated with a loss of stiffness; there is no longer a unique displacement solution for a given set of loads. Third, while we call the term in square brackets in Equation(5.2) a stiffness, it is unlike any of the stiffnesses encountered in the previous chapters because it includes one of the applied loads. Actually, this is the *total stiffness* and is comprised of an elastic stiffness $k$ and a *geometric stiffness* $P/L$. The latter is due to the geometry change of the equilibrium configuration, hence its name and why it is load dependent. Finally, the critical value of $P$ does not depend on the value of $Q$. But it is important for us to realize that $Q$ is necessary to actually cause instability. That is why our concept of stability involves imagining the consequences as the structure is displaced slightly, in other words, we will assume there is always an *upsetting force* present.

The concept of stability we will develop is rooted in the linear, small deflection analysis of structures. After instability has occurred, it may well be that the structure achieves a new equilibrium position corresponding to the large deflection position. What we are concerned with is occurrence of a neutral equilibrium state under small deformation conditions.

## 5.2   Stability of Truss Structures

The following formulation takes into account the fact that the shape of the structure changes during deformation. That is, we look at equilibrium in the deformed state even though the deformations will be considered small. To make the analysis consistent with that of Chapter 4 we will assume the truss members still carry only axial loads irrespective of their new orientation.

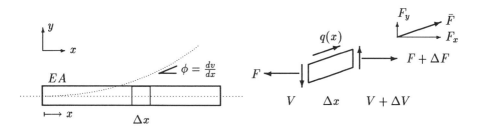

**Figure 5.2**: Deflection of small segment of truss member.

Consider a typical member shown in Figure 5.2; for convenience the reference axis is taken along the member and the deflection is assumed only in the plane. This figure is similar to that of Figure 2.1 except that the infinitesimal element is shown slightly displaced in the $y$ direction and slightly rotated about the $z$ axis. The axial force $\bar{F}$ acts along the member, but since we want a set of equations written with respect to the global axes, we write the components of the force as

$$F_x = \bar{F} \cos \phi \approx \bar{F}, \qquad F_y = \bar{F} \sin \phi \approx \bar{F} \phi \approx \bar{F} \frac{dv}{dx}$$

This last relation shows that the member has a shear force that depends on the amount of rotation. Thus the equilibrium relations must also take this shear into account. To make the relations look similar to those of the previous chapters, we introduce the notations

$$F \equiv F_x, \qquad V \equiv F_y$$

Summing the forces on the element gives (in the limit of $\Delta x$ becoming very small)

$$\frac{dF}{dx} = -q, \qquad \frac{dV}{dx} = -q \frac{dv}{dx}$$

A slightly more convenient form of the second equation is obtained by summing the moments, this results in

$$V = \bar{F} \frac{dv}{dx}$$

These equilibrium equations are actually coupled (through $\bar{F}$) and this makes them very difficult to use. In fact, they are a system of nonlinear equations. Since our whole development is for small displacements, we take advantage of this and solve the equations approximately in two steps. First we solve them with respect to the original configuration and compute an estimate of the axial force. Call this force the *initial force* $\bar{F}_o$ and for reasons to become apparent later call this analysis the *pre-buckle analysis*. We then re-solve the equations, this time using $\bar{F}_o$ as a parameter in the shear expression. This effectively uncouples the above equations.

All the relationships for the structural quantities may now be collected as

$$\begin{aligned} \text{Axial Displacement}: && u &= u(x) \\ \text{Axial Force}: && F &= +EA\frac{du}{dx} && (5.3) \\ \text{Loading}: && q &= -EA\frac{d^2u}{dx^2} && (5.4) \\ \text{Transverse Displacement}: && v &= v(x) \\ \text{Shear Force}: && V &= +\bar{F}_o\frac{dv}{dx} && (5.5) \end{aligned}$$

Note that the initial force term $\bar{F}_o$ plays a role similar to that of the axial stiffness $EA$. As we have seen so many times before, the displacements ($u$ and $v$) can be viewed as the fundamental unknowns of interest. In the special case when the loading is zero, we have for these displacements

$$u(x) = c_1 x + c_2, \qquad v(x) = b_1 x + b_2$$

Since the displacements are linear, we conclude that the axial force and shear force are constant along the section.

The manner in which we have posed the above problem makes it a continuation of the methods already developed in the previous chapters. That is, the solution to a particular problem is obtained by determining the constants of integration for each section by imposing equilibrium and compatibility at the junctions. We give an example to demonstrate the approach.

**Example 5.1:**   Determine the displacement distributions in the horizontal member of the two member truss shown in Figure 5.3. Use equilibrium based on the deformed configuration.

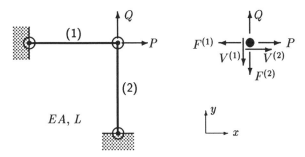

**Figure 5.3:** Two member truss.

We first solve the equilibrium equations in the undisturbed configuration so as to determine the initial loads. This analysis gives that

$$\bar{F}_o^{(1)} = P, \qquad \bar{F}_o^{(2)} = Q$$

Normally, this stage of the solution can be quite extensive, requiring a full analysis of the static problem. We will see this in some of the later examples.

For the horizontal member, the first boundary is pinned, hence the deflections are

$$u(x) = c_1 x, \qquad v(x) = b_1 x$$

Choosing a local $x$ for the vertical member, we have

$$u(x) = c_1^* x, \qquad v(x) = b_1^* x$$

Keep in mind that these are referred to the member. For compatibility of displacements at the joint, we must have

$$u^{(1)} = -v^{(2)}, \qquad v^{(1)} = u^{(2)}$$

This gives the relation among the coefficients as

$$c_1 = -b_1^*, \qquad b_1 = c_1^*$$

Applying equilibrium at the joint gives, with reference to Figure 5.3,

$$F^{(1)} - V^{(2)} = P, \qquad F^{(2)} + V^{(1)} = Q$$

After substituting for the displacements we get

$$EA\,c_1 - \bar{F}_o^{(2)} b_1^* = P, \qquad EA\,b_1 + \bar{F}_o^{(1)} c_1^* = Q$$

These equations are now sufficient to determine all of the coefficients. The deflection distributions in the horizontal member are then found to be

$$u(x) = \frac{x}{[EA + \bar{F}_o^{(2)}]}P, \qquad v(x) = \frac{x}{[EA + \bar{F}_o^{(1)}]}Q$$

It is apparent from these relations that as $P$ and $Q$ are made larger, the system is made stiffer; this is precisely how a cable acts.

**Example 5.2:** Determine the sensitivity of the joint displacements of the last example to small variations in the applied loading.

We will first rewrite the above results in the form of stiffness relations for the joint $(x = L)$

$$[\frac{EA}{L} + \frac{\bar{F}_o^{(2)}}{L}]u_L = P, \qquad [\frac{EA}{L} + \frac{\bar{F}_o^{(1)}}{L}]v_L = Q$$

It is important to remember that the initial loads $\bar{F}_o^{(1)}$ and $\bar{F}_o^{(2)}$ are obtained from the undeformed configuration and therefore they will always have the relation

$$\bar{F}_o^{(1)} = P, \qquad \bar{F}_o^{(2)} = Q$$

irrespective of the actual displacements $u_L$ and $v_L$. That is, we view the pre-buckle analysis as establishing the stiffness properties of the structure which can then be treated as constant for the subsequent analysis.

It should be pointed out that the contribution to the stiffness from the initial load is quite small in the above example. For example, for the contributions to be equal would require the axial stress to be of magnitude $E$. This is much too large since the yield stress of a material is usually of the order $E/100$.

## Truss Stability as an Eigenvalue Problem

An interesting aspect of the stiffnesses of the last example occurs when the initial loads are compressive. Then we observe a decrease in stiffness. Indeed, when either $P = -EA$ or $Q = -EA$, the structure experiences a complete loss of stiffness. In that state, it takes virtually no additional force to place the joint at any position. That is, the structure is in neutral equilibrium and at the borderline of instability.

To investigate the connection between loss of stiffness and stability we will consider the following example.

**Example 5.3:** Determine the relation between the displacements at the apex of the truss shown in Figure 5.4 and the applied loads. Use equilibrium based on the deformed configuration.

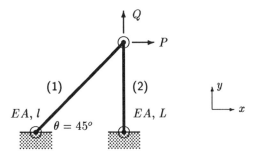

**Figure 5.4:** Two member truss.

Proceeding as in the last example, we first solve the equilibrium equations in the undisturbed configuration to determine the initial loads. This analysis gives that

$$\bar{F}_o^{(1)} = P\sqrt{2}, \qquad \bar{F}_o^{(2)} = -P + Q$$

We get the same form of displacement functions of

$$u(x) = c_1 x, \qquad v(x) = b_1 x$$

for the first member and

$$u(x) = c_1^* x, \qquad v(x) = b_1^* x$$

for the second member. For compatibility of displacements at the joint, we must have

$$u^{(1)}\cos\theta - v^{(1)}\sin\theta = -v^{(2)}, \qquad u^{(1)}\sin\theta + v^{(1)}\cos\theta = u^{(2)}$$

Applying equilibrium at the joint gives

$$F^{(1)}\cos\theta - V^{(1)}\sin\theta - V^{(2)} = P, \qquad F^{(1)}\sin\theta + V^{(1)}\cos\theta + F^{(2)} = Q$$

After substituting for the displacements we get the coefficients from

$$\frac{1}{\sqrt{2}}\begin{bmatrix} EA + \bar{F}_o^{(2)}l/L & -\bar{F}_o^{(1)} - \bar{F}_o^{(2)}l/L \\ EA + EAl/L & \bar{F}_o^{(1)} + EAl/L \end{bmatrix}\begin{Bmatrix} c_1 \\ b_1 \end{Bmatrix} = \begin{Bmatrix} P \\ Q \end{Bmatrix}$$

where we have taken $\theta = 45°$ and $l = \sqrt{2}L$. Since we actually want the stiffness relation, we can avoid having to solve this system by writing the coefficients in terms of the displacements. That is,

$$c_1 l = u_L \cos\theta + v_L \sin\theta, \qquad b_1 l = -u_L \sin\theta + v_L \cos\theta$$

Substituting these we get the stiffness relation as

$$\frac{1}{\sqrt{2}}\begin{bmatrix} EA + \bar{F}_o^{(1)} + 2\bar{F}_o^{(2)}l/L & EA - \bar{F}_o^{(2)} \\ EA - \bar{F}_o^{(2)} & EA + \bar{F}_o^{(1)} + 2EAl/L \end{bmatrix}\begin{Bmatrix} u_L \\ v_L \end{Bmatrix} = \begin{Bmatrix} P \\ Q \end{Bmatrix}$$

While the stiffness matrix has a linear dependence on the initial loads, the resultant effect these loads have on the displacements is not clear.

We can solve for the displacements only if the determinant associated with this system of equations is non-zero. Thus the question arises: Is it possible for the determinant of this stiffness matrix to be zero? As a particular case, suppose that $P = 0$, then the initial loads are $\bar{F}_o^{(1)} = 0$ and $\bar{F}_o^{(2)} = Q$. The determinant of the stiffness matrix is

$$\det = \frac{1}{\sqrt{2}}\left[(EA + 2Ql/L)(EA + 2EAl/L) - (EA)(EA)\right]$$

It is apparent that $Q$ can be chosen such that the determinant is zero. This value is given by

$$Q = \frac{-EA}{1 + 2\sqrt{2}}$$

Note that $Q$ must act so as to put one of the members in compression. Even in the general case, we can find values of $P$ and $Q$ so that the determinant is zero. In contrast to the earlier example where individual stiffness terms went to zero, we see that for a general system the indication of neutral equilibrium is that the determinant of the stiffness matrix goes to zero.

To see the consequences of this, recall that we always set up a system of simultaneous relations as

$$[K]\{u\} = \{P\}$$

where $\{P\}$ represents the known applied loads. (Generally, we will have already incorporated the known displacement boundary conditions.) A condition used implicitly in solving all the simultaneous equations of the last few chapters is that $\det[K] \neq 0$. As shown above, however, it is possible to choose initial loads to make $\det[K] = 0$. At precisely those values there is not a unique connection between the displacements and the boundary conditions. This means that two or more of the equations are linearly dependent and therefore the solution is not unique. In other words, there is a second solution $\{u^*\}$ that also satisfies the equations

$$[K]\{u^*\} = \{B\}$$

Let the difference of the two solutions be $\{\phi\} = \{u\} - \{u^*\}$; it therefore satisfies the homogeneous equations

$$[K]\{\phi\} = \{0\}$$

In general, such an homogeneous equation admits only the trivial solution $\{\phi\} = 0$, but when $\det[K]$ is zero there are other solutions. These are the ones we are interested in.

From the above it is clear that we need only consider the homogeneous problem in order to determine the critical points at which the structure attains neutral stability.

That is, we consider the problem

$$[K(\lambda)]\{\phi\} = \{0\}$$

where $\lambda$ is some parameter (in our case associated with the critical load), special values of which cause the determinant of $[K]$ to be zero. This is known as an *eigenvalue problem*. We obtain these special values of $\lambda$ (called *eigenvalues*) by setting the determinant to zero. Corresponding to each eigenvalue, we can find a solution for $\{\phi\}$; this is called an *eigenvector* or *mode shape*. For all truss problems, the critical loads appear in the equations in a linear fashion and therefore we will deal with a simpler version of the eigenvalue problem in the form

$$[[K_1] - \lambda[K_2]]\{c\} = \{0\}$$

Since there are standard solution procedures available for this equation, we will even approximate the frame analysis so as to cast it in this form.

## 5.3   Matrix Formulation for Truss Stability

With reference to Figures 5.2 and 4.1, since the axial and transverse displacements are linear functions of position (when the loading $q(x)$ is zero), we can write

$$u(x) = (1 - \frac{x}{L})u_1 + (\frac{x}{L})u_2$$
$$v(x) = (1 - \frac{x}{L})v_1 + (\frac{x}{L})v_2$$

where $u_i$ and $v_i$ are degrees of freedom at the ends of a truss member of length $L$. Realizing that $F_1 = -F(0)$ and $V_1 = -V(0)$, then the nodal forces corresponding to the degrees of freedom are

$$F_1 = \frac{EA}{L}(u_1 - u_2), \qquad V_1 = \frac{\bar{F}_o}{L}(v_1 - v_2)$$

Similar expressions can be determined for $F_2$ and $V_2$. This is put into matrix form as

$$\begin{Bmatrix} F_1 \\ V_1 \\ F_2 \\ V_2 \end{Bmatrix} = \frac{EA}{L} \begin{bmatrix} 1 & 0 & -1 & 0 \\ 0 & 0 & 0 & 0 \\ -1 & 0 & 1 & 0 \\ 0 & 0 & 0 & 0 \end{bmatrix} \begin{Bmatrix} u_1 \\ v_1 \\ u_2 \\ v_2 \end{Bmatrix} + \frac{\bar{F}_o}{L} \begin{bmatrix} 0 & 0 & 0 & 0 \\ 0 & 1 & 0 & -1 \\ 0 & 0 & 0 & 0 \\ 0 & -1 & 0 & 1 \end{bmatrix} \begin{Bmatrix} u_1 \\ v_1 \\ u_2 \\ v_2 \end{Bmatrix}$$

The first term is recognized as the augmented element stiffness for a rod and is referred to as the elastic stiffness. The second term is associated with the change in geometry of the structure and is referred to as the geometric matrix. For future reference, we will write this as

$$\{F\} = \left[[\bar{k}_E] + \bar{F}_o[\bar{k}_G]\right]\{u\}$$

We are interested in assembling structures of arbitrarily oriented members, hence following the direct stiffness approach developed in the other chapters we form the element stiffness matrix in global coordinates by transformation as

$$[\,k\,] = [\,T\,]^T\Big[[\bar{k}_E] + \bar{F}_o[\bar{k}_G]\Big]\,[\,T\,] = [\,T\,]^T[\bar{k}_E][\,T\,] + \bar{F}_o[\,T\,]^T[\bar{k}_G][\,T\,]$$

In obtaining this relation we have considered that the transformation is applied only to the vector displacement $\{u\}$ and vector load $\{F\}$, and that the axial force $\bar{F}_o$ is left intact. We do this because the axial force is a member quantity and therefore independent of our choice of global coordinates. The expression for $[\,k\,]$ is a general one and if the appropriate form for $[\,T\,]$ is used it can be valid even for the 3-D truss. As a special case, we get the total stiffness matrix for the plane truss in global coordinates as

$$[\,k\,] = \frac{EA}{L}\begin{bmatrix} C^2 & CS & -C^2 & -CS \\ CS & S^2 & -CS & -S^2 \\ -C^2 & -CS & C^2 & CS \\ -CS & -S^2 & CS & S^2 \end{bmatrix} + \frac{\bar{F}_o}{L}\begin{bmatrix} S^2 & -SC & -S^2 & SC \\ -SC & C^2 & SC & -C^2 \\ -S^2 & SC & S^2 & -SC \\ SC & -C^2 & -SC & C^2 \end{bmatrix}$$

$$(5.6)$$

where, as before, the abbreviations $C \equiv \cos\theta$, $S \equiv \sin\theta$ are used.

Assembling the global stiffness matrix is now a matter of adding all the element stiffnesses. The only significant point is that during assemblage there is an initial force $\bar{F}_o$ for each member. Thus the assemblage occurs in two stages. First, we assemble only the elastic stiffness and use it to find the initial force in each member by solving

$$[K_E]\{u\} = \{P\}$$

where $[K_E]$ is the assembled elastic global stiffness. This corresponds to the pre-buckle analysis. In the analysis of stability problems, we do not know the loading $\{P\}$; we do, however, know the distribution of loads, therefore, let these be normalized as $P\{p\}$. That is, all loads are assumed proportional to a single parameter $P$. Thus the pre-buckle problem is solved using $\{p\}$ as the load vector. From this, the axial force in each member is obtained. Let the axial forces obtained in this way be called $\bar{f}_o$, then the axial forces due to $\{P\}$ are

$$\bar{F}_o = P\bar{f}_o$$

In this manner the eigenvalue problem associated with truss stability can be established as

$$\Big[[K_E] + P[\bar{f}_o K_G]\Big]\{u\} = \{0\}$$

where $[\bar{f}_o K_G]$ is the assembled global geometric stiffness. The notation $[\bar{f}_o K_G]$ is a reminder that each element may have a different initial axial load. Indeed, some members may even have zero axial forces. These points will be made clearer by the following example.

**Example 5.4:**  Find the critical load for the truss shown in Figure 5.5. All joints are assumed pinned and both members have the same properties.

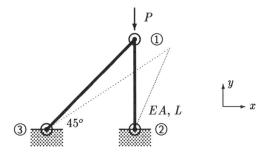

**Figure 5.5**: Two member truss.

We first solve the pre-buckle problem in order to get the initial axial force in each member. Since the only non-zero degrees of freedom are $\{u_1, v_1\}$, the reduced element elastic stiffnesses are

$$[k^{*(12)}] = \frac{EA}{L} \begin{bmatrix} 0 & 0 \\ 0 & 1 \end{bmatrix}, \qquad [k^{*(13)}] = \frac{EA}{2\sqrt{2}L} \begin{bmatrix} 1 & 1 \\ 1 & 1 \end{bmatrix}$$

The applied load is

$$\{P\} = \{P_1, P_2\} = P\{0, -1\}$$

where we have ratioed the applied loads to $P$. The structural system of equations are therefore

$$\frac{EA}{L2\sqrt{2}} \begin{bmatrix} 1 & 1 \\ 1 & 1 + 2\sqrt{2} \end{bmatrix} \begin{Bmatrix} u_1 \\ v_1 \end{Bmatrix} = \begin{Bmatrix} 0 \\ -1 \end{Bmatrix}$$

Note that the scaling factor $P$ does not appear. Solving this system gives the displacements at Node 1 as

$$u_1 = \frac{L}{EA}, \qquad v_1 = -\frac{L}{EA}$$

The axial loads in each member can now be obtained from the stiffness relation or by using the general formula

$$\bar{F}^{(ij)} = \left(\frac{EA}{L}\right)_{ij} [\cos\theta_{ij}(u_j - u_i) + \sin\theta_{ij}(v_j - v_i)]$$

and gives

$$\bar{f}_o^{(12)} = \left(\frac{EA}{L}\right)[-0(u_2 - u_1) - 1(v_2 - v_1)] = -1$$

$$\bar{f}_o^{(13)} = \left(\frac{EA}{\sqrt{2}L}\right)[-\frac{1}{\sqrt{2}}(u_3 - u_1) - \frac{1}{\sqrt{2}}(v_3 - v_1)] = 0$$

Both of these values could have been obtained by inspection, but it is instructive to go through the general procedure.

The total element stiffnesses can now be established as

$$[k^{*(12)}] = \frac{EA}{L}\begin{bmatrix} 0 & 0 \\ 0 & 1 \end{bmatrix} - \frac{P}{L}\begin{bmatrix} 1 & 0 \\ 0 & 0 \end{bmatrix}, \qquad [k^{*(13)}] = \frac{EA}{2\sqrt{2}L}\begin{bmatrix} 1 & 1 \\ 1 & 1 \end{bmatrix}$$

giving the assembled total stiffness matrix as

$$[K^*] = \frac{EA}{2\sqrt{2}L}\begin{bmatrix} 1 - P2\sqrt{2}/EA & 1 \\ 1 & 1 + 2\sqrt{2} \end{bmatrix}$$

For the stability problem, we want to know what value of $P$ causes buckling. This is obtained by finding the value of $P$ that makes the determinant of the total stiffness matrix zero. That is,

$$\det = 1 - \frac{P}{EA}2\sqrt{2} + 2\sqrt{2} - \frac{P}{EA}8 - 1 = 0$$

or

$$P_{cr} = \frac{EA}{(1 + 2\sqrt{2})}$$

Note that only one value of the buckling load is obtained even though there are two degrees of freedom in the system. Actually the second critical load is infinite.

The value of critical load we obtained is very large, and in a real structure it is more than likely that either plastic yielding or a column type buckling would have occurred. However, there are two circumstances when this type of buckling could be significant. First, if the members forming the triangulated system form very shallow angles the structure becomes susceptible to this type of failure. Second, as the structure becomes more complex, made of a greater number of cells of about the same size, then the load to cause buckling decreases.

**Example 5.5:** Find the critical load for the truss shown in Figure 5.5 but apply the force $P$ horizontally in the $x$ direction.

We first solve the pre-buckle problem in order to get the initial axial force in each member. It is straightforward to show that

$$\bar{f}_o^{(12)} = -P, \qquad \bar{f}_o^{(13)} = \sqrt{2}P$$

In this case both members have an axial component of force.

The total element stiffness is almost identical to the last example except for the addition of the geometric stiffness of Member 1-2.

$$[k^{*(12)}] = \frac{EA}{L}\begin{bmatrix} 0 & 0 \\ 0 & 1 \end{bmatrix} - \frac{P}{L}\begin{bmatrix} 1 & 0 \\ 0 & 0 \end{bmatrix}, \qquad [k^{*(13)}] = \frac{EA}{2\sqrt{2}L}\begin{bmatrix} 1 & 1 \\ 1 & 1 \end{bmatrix} + \frac{P}{2L}\begin{bmatrix} 1 & -1 \\ -1 & 1 \end{bmatrix}$$

giving the assembled stiffness matrix as

$$[K^*] = \frac{EA}{2\sqrt{2}L}\begin{bmatrix} 1 - \sqrt{2}\lambda & 1 - \sqrt{2}\lambda \\ 1 - \sqrt{2}\lambda & 1 + 2\sqrt{2} + \sqrt{2}\lambda \end{bmatrix}$$

where $\lambda \equiv P/EA$. We set the determinant of the total stiffness matrix to zero to get the characteristic equation to determine the buckling loads. That is,

$$(1 - \sqrt{2}\lambda)(1 + \lambda) = 0$$

This gives the two critical values of

$$P_{cr} = -EA/\sqrt{2} \quad \text{or} \quad EA$$

The negative sign in the first value means that the horizontal force to cause buckling acts in the negative $x$ direction. Note that in this problem two finite values of the buckling load are obtained.

## 5.4  Beams with Axial Forces

Our development here is similar to that of Chapter 3 except that we look at equilibrium of the beam in a slightly displaced position and we add the effect of an axial load. The axial force may be either tensile or compressive, but compression is of more interest in this section because of its effect on stability.

We consider equilibrium in two steps. The first step corresponds to Figure 3.1 where we establish the forces and moments acting on the beam segment; the beam is still in its initial configuration. This we call the *pre-buckle analysis*. Then we displace the beam segment slightly leaving the loads intact as shown in Figure 5.6. This will give us the measure of the sensitivity of the equilibrium condition to changes in configuration. Since the change in configuration is only slight, we can retain all the usual assumptions of the elementary beam theory.

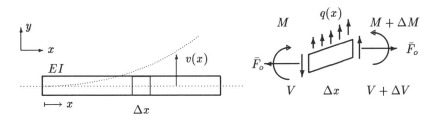

**Figure 5.6**: Equilibrium of small beam segment slightly deformed.

It is apparent that the axial force causes a bending action and therefore will enter the beam equilibrium equations. Thus, balance of force and moment on the small segment of Figure 5.6 gives (for small slopes and deflections), respectively,

$$\frac{dV}{dx} = -q, \qquad \frac{dM}{dx} + V - \frac{dv}{dx}\bar{F}_o = 0$$

The governing differential equation for the deflection shape is given by combining these and using the moment-deflection relation to get

$$EI\frac{d^4v}{dx^4} - \bar{F}_o\frac{d^2v}{dx^2} = q \tag{5.7}$$

We shall refer to this equation as the *coupled beam equation*. When the initial axial force $\bar{F}_o$ is compressive only, then it is referred to as the *beam-column equation*.

All the relationships for the structural quantities may now be collected as

Displacement :   $v = v(x)$

$$\text{Slope :} \qquad \phi = \frac{dv}{dx} \tag{5.8}$$

$$\text{Moment :} \qquad M = +EI\frac{d^2v}{dx^2} \tag{5.9}$$

$$\text{Shear :} \qquad V = -EI\frac{d^3v}{dx^3} + \bar{F}_o\frac{dv}{dx} \tag{5.10}$$

$$\text{Loading :} \qquad q = +EI\frac{d^4v}{dx^4} - \bar{F}_o\frac{d^2v}{dx^2} \tag{5.11}$$

The only difference in comparison with the beam equations of Chapter 3 is the addition of the $\bar{F}_o$ related terms in the expressions for the loading and shear. Thus with $\bar{F}_o$ treated as a parameter, it is seen from these that the transverse displacement $v(x)$ (via its derivatives) can be viewed as the fundamental unknown of interest.

When solving the coupled beam equations, information may be given at any of the five levels above, thus requiring integrations (or differentiations) to obtain the other quantities. Integration gives rise to constants of integration which must be found from the boundary and compatibility conditions. In the general case, there are four constants of integration (since the highest derivative is four).

We reiterate that for the purpose of integrating these governing equations, $\bar{F}_o$ is considered known from the pre-buckle analysis. As an example, to integrate the loading relation, we will assume that the material properties $EI$, the axial force $\bar{F}_o$, and the loading $q$, are all constant. Integrating twice gives

$$EI\frac{d^2v}{dx^2} - \bar{F}_o v = c_1^* x + c_2^* + \tfrac{1}{2}q_o x^2$$

This is an inhomogeneous second order differential equation. The complete solution comprises a solution to the homogeneous problem plus a particular solution. The character of the solution differs depending on whether $\bar{F}_o$ is tensile or compressive. The general solutions are

$$v(x) = c_1\cos kx + c_2\sin kx + c_3 x + c_4 + \frac{q_o}{2EIk^2}x^2 \qquad k^2 \equiv \frac{-\bar{F}_o}{EI} \tag{5.12}$$

for compressive loading and

$$v(x) = c_1 \cosh kx + c_2 \sinh kx + c_3 x + c_4 + \frac{q_o}{2EIk^2}x^2 \qquad k^2 \equiv \frac{\bar{F}_o}{EI} \qquad (5.13)$$

for tensile loading. The constants of integration $c_1, c_2, c_3, c_4$ are evaluated by imposing particular boundary conditions. The shear distribution is given by the simple
expression

$$V(x) = \pm k^2 EI\, c_3 - q_o x$$

with the $\pm$ corresponding to tensile and compressive axial loading, respectively. When
the distributed loading is zero, the shear is constant and simply related to the constant
$c_3$.

**Example 5.6:**    Consider the cantilevered beam shown in Figure 5.7 with the
combination of axial and transverse forces. Determine the deflection shape.

**Figure 5.7**: Cantilevered beam with end loads.

Let us first solve this problem in the manner of the previous chapters (this will
constitute the pre-buckle analysis). The axial behavior gives a displacement and
force distribution of

$$u(x) = -\frac{Px}{EA}, \qquad F(x) = -P$$

That is, the axial analysis simply gives that the axial force is the same everywhere
along the beam and is $P$ compressive.

Now consider the flexural behavior. Since the loading is zero, the general deflected shape is

$$v(x) = c_1 x^3 + c_2 x^2 + c_3 x + c_4$$

Imposing the boundary conditions of zero displacements and slope at one end, zero
moment and an applied force of $Q$ at the other, allows the constants of integration
to be determined. Specifically, at $x = 0$, we have

$$v = 0 \quad \Rightarrow \quad 0 = c_4$$

$$\frac{dv}{dx} = 0 \quad \Rightarrow \quad 0 = c_3$$

At $x = L$, we have

$$M = EI\frac{d^2 v}{dx^2} = 0 \quad \Rightarrow \quad 0 = EI[2c_2 + 6c_1 x]$$

$$V = -EI\frac{d^3 v}{dx^3} = Q \quad \Rightarrow \quad Q = -EIc_1$$

Solving for the coefficients gives the deflection shape as

$$v(x) = \frac{Q}{6EI}[3xL^2 - x^3]$$

In essence, we have solved for the transverse displacements independent of the axial behavior.

We will now solve the coupled beam equations. The applied loading is zero and the axial force is compressive, therefore, the general deflected shape is

$$v(x) = c_1 \cos kx + c_2 \sin kx + c_3 x + c_4$$

We impose the following boundary conditions to determine the coefficients. At $x = 0$, we have

$$v = 0 \quad \Rightarrow \quad 0 = c_1 + c_4$$
$$\frac{dv}{dx} = 0 \quad \Rightarrow \quad 0 = c_2 k + c_3$$

At $x = L$, we have

$$M = EI\frac{d^2v}{dx^2} = 0 \quad \Rightarrow \quad 0 = EI[-c_1 k^2 \cos kL - c_2 k^2 \sin kL]$$
$$V = -EI\frac{d^3v}{dx^3} - P\frac{dv}{dx} = Q \quad \Rightarrow \quad Q = -EIk^2 c_3$$

Solving directly gives (noting that $P = k^2 EI$)

$$c_1 = -\frac{Q}{k^3 EI}\frac{\sin kL}{\cos kL}, \qquad c_2 = \frac{Q}{k^3 EI}, \qquad c_3 = -\frac{Q}{k^2 EI}, \qquad c_4 = \frac{Q}{k^3 EI}\frac{\sin kL}{\cos kL}$$

As a result, the deflected shape is determined to be

$$v(x) = \frac{Q}{k^3 EI}\frac{1}{\cos kL}[-\sin kL \cos kx + \cos kL \sin kx - kx \cos kL + \sin kL]$$

A couple of points to note about this solution. First, if $Q$ is removed then the beam returns to its initially straight position. That is, it behaves linearly elastic; a doubling of $Q$ results in a doubling of deflection. Second, the axial load appears in the solution in a complicated fashion; the deflection is not linear with respect to this load. A final point is that we should recover the uncoupled case (straight beam) by appropriately letting $P \to 0$. We now look at varying $P$.

**Example 5.7:** Determine the effect of the initial axial load on the stiffness relation for the cantilever beam.

Since we are eventually interested in stiffness type relations, consider the above solution for a particular point. For convenience, choose the end point $x = L$, then

$$v(L) = \frac{Q}{k^3 EI}\left[\frac{\sin kL - kL \cos kL}{\cos kL}\right]$$

Rearrange this as a stiffness relation in the form

$$\frac{EI}{L^3}\left[\frac{k^3 L^3 \cos kL}{\sin kL - kL \cos kL}\right] v(L) = Q$$

Figure 5.8 shows a plot of the stiffness as a function of the initial axial load. The axial load is seen to have a profound effect on the stiffness. Indeed, there are multiple points when the stiffness is zero and other points where it is infinite (both positive and negative).

To show that these results are consistent with the straight beam, choose $P$ small so that the parameter $\xi \equiv kL$ is small also, then

$$v(L) \approx \frac{Q}{k^3 EI}\left[\frac{\xi - \xi^3/6 \cdots - \xi(1 - \xi^2/2 \cdots)}{1 - \xi^2/6 \cdots}\right] = \frac{Q}{k^3 EI}\left[\frac{\xi^3/3 \cdots}{1 - \xi^2/6 \cdots}\right] = \frac{QL^3}{3EI}$$

This indeed is the uncoupled value obtained above.

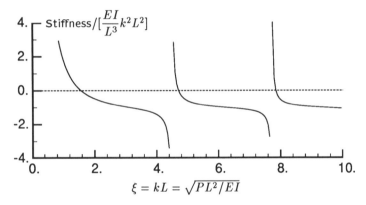

**Figure 5.8**: Stiffness as a function of axial load.

The zero crossings in this example correspond to critical loads because at these points the structure has complete loss of stiffness and is in a state of neutral equilibrium. That is, even for a nominal $Q$ the deflections are indicated to be very large. (An alternative view point is that any specified position can be obtained with very little $Q$.) These critical points occur at

$$\cos kL = 0 \quad\text{or}\quad kL = \tfrac{1}{2}\pi, \tfrac{3}{2}\pi, \tfrac{5}{2}\pi, \cdots, \tfrac{1}{2}(n+1)\pi$$

On substituting for $k$ in terms of the force, this gives

$$P = (n+1)^2 \pi^2 \frac{EI}{4L^2}$$

There are many critical loads. The minimum load to cause buckling is at $n = 0$, hence

$$P_{cr} = EI\left(\frac{\pi}{2L}\right)^2$$

When we analyzed the truss, we saw that the critical loads were of the order $EA$, but in the present case we see that

$$P_{cr} = EA\frac{I}{A}(\frac{\pi}{2L})^2 \approx EA(\frac{h}{L})^2$$

The ratio $(h/L)$ is called the *slenderness ratio* and relates the thickness $h$ of the beam $h$ to its length $L$. When the length of a beam is 10 times it thickness, we see that the critical load has already decreased to 1% of $EA$.

**Example 5.8:**  Determine the deflected shapes corresponding to the critical loads for the beam loaded as in Figure 5.7.

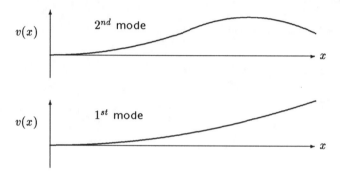

**Figure 5.9**: Deflected shapes for the first and second buckling modes.

The deflection at any load is obtained by substituting for $kL$ into the expressions for $v(x)$ as

$$v(x) = \frac{Q}{k^3EI}\frac{1}{\cos kL}\left[-\sin kL\cos kx + \cos kL\sin kx - kx\cos kL + \sin kL\right]$$

However, if we do this for the critical values $kL = \frac{1}{2}(n+1)\pi$ we see that the cosine term goes to zero and the deflection will become infinite. To overcome this, we will normalize the deflection with respect to the tip deflection

$$v(L) = \frac{Q}{k^3EI}\left[\frac{\sin kL - kL\cos kL}{\cos kL}\right]$$

The expression for the deflection now becomes

$$v(x) = \frac{v(L)}{[\sin kL - kL\cos kL]}\left[-\sin kL\cos kx + \cos kL\sin kx - kx\cos kL + \sin kL\right]$$

This expression is valid for any value of $kL$, in particular, if we substitute the critical values we obtain the deflection as

$$v(x) = v(L)[1 - \cos kx], \qquad kL = \frac{1}{2}(n+1)\pi$$

These are shown plotted in Figure 5.9 for the first two modes. It is important to realize that the absolute position of any point on the beam is unknown because the structure is in neutral equilibrium; however, the relative position of points (mode shapes) take on very definite forms.

# 5.5   Beam Buckling

In a buckling analysis we seek only the critical values of load and possibly the corresponding deflection shapes. We saw in the case of the truss, that we can get this information from an eigenvalue analysis. This section shows that buckling of beams can also be treated as eigenvalue problem.

To see this, recall that there are four constants of integration which are to be determined from the boundary conditions. Thus we always set up a system of simultaneous relations as

$$[A]\{c\} = \{B\}$$

where $\{c\} = \{c_1, c_2, c_3, c_4\}$ and $\{B\}$ are the specific form of boundary conditions. To solve for $c_1$, say, we could use Cramer's rule but for this to be valid we must have $\det[A] \neq 0$. Note, however, that it is possible for $\det[A]$ to be zero and at precisely those values Cramer's rule breaks down and there is not a unique connection between the coefficients and the boundary conditions. This means that one or more of the equations are linearly dependent and therefore the solution is not unique. In other words, there is a second solution $\{c^*\}$ that also satisfies the equations

$$[A]\{c^*\} = \{B\}$$

Let the difference of the two solutions be $\{\phi\} = \{c\} - \{c^*\}$; it satisfies the homogeneous equations

$$[A]\{\phi\} = \{0\}$$

In general, this admits only the trivial solution, but when $\det[A]$ is zero there are other solutions.

From the above it is clear that we need only consider the homogeneous problem in order to determine the critical loads. That is, we consider the problem

$$[A(\lambda)]\{\phi\} = \{0\}$$

where $\lambda$ is some parameter (in our case the critical load), special values of which cause the determinant of $[A]$ to be zero. This is known as an *eigenvalue problem*; actually, since $[A]$ contains transcendental functions it is usually referred to as a transcendental eigenvalue problem. We obtain the special values of $\lambda$ (called *eigenvalues*) by setting the determinant to zero. Corresponding to each eigenvalue, we can find a solution for $\{\phi\}$, this is called an *eigenvector*.

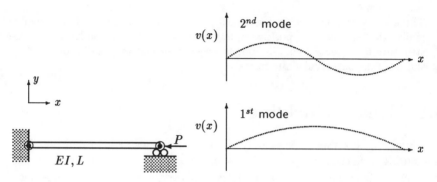

**Figure 5.10**: Pinned-pinned beam.

**Example 5.9:**   Determine the buckling loads for the pinned-pinned beam shown in Figure 5.10.

Except for $P$ the problem does not state any other applied loads, therefore, by inspection get that $\bar{F}_o = -P$. The boundary conditions are

$$\text{at } x = 0: \qquad v = 0 \quad \Rightarrow \quad 0 = c_1 + c_4$$

$$M = EI\frac{d^2 v}{dx^2} = 0 \quad \Rightarrow \quad 0 = -c_1 k^2$$

$$\text{at } x = L: \qquad v = 0 \quad \Rightarrow \quad 0 = c_1 \cos kL + c_2 \sin kL + c_3 L + c_4$$

$$M = EI\frac{d^2 v}{dx^2} = 0 \quad \Rightarrow \quad 0 = -c_1 k^2 \cos kL - c_2 k^2 \sin kL$$

These four equations can be put into the matrix form as

$$
\begin{bmatrix}
1 & 0 & 0 & 1 \\
-k^2 & 0 & 0 & 0 \\
\cos kL & \sin kL & L & 1 \\
-k^2 \cos kL & -k^2 \sin kL & 0 & 0
\end{bmatrix}
\begin{Bmatrix}
c_1 \\ c_2 \\ c_3 \\ c_4
\end{Bmatrix} = 0
$$

We now inquire if the determinant of this matrix can be zero. On multiplying out get

$$\det = k^4 L \sin kL = 0$$

There are many values of $k$ when this equation is satisfied. The obvious one of $k = 0$ corresponds to the trivial case of zero axial load. The other possibilities are

$$kL = \pi, \, 2\pi, \, 3\pi, \, \cdots, \, n\pi$$

On substituting for $k$ in terms of the force, this gives

$$P_{cr} = n^2 \pi^2 \frac{EI}{L^2}$$

There are many critical loads, and corresponding to each there is a different deflected shape. To determine these, let us reconsider the relation among the coefficients. At the special values of $kL = n\pi$, the matrix for determining $\{c\}$ reduces to

$$\begin{bmatrix} 1 & 0 & 0 & 1 \\ -k^2 & 0 & 0 & 0 \\ 1 & 0 & L & 1 \\ -k^2 & 0 & 0 & 0 \end{bmatrix} \begin{Bmatrix} c_1 \\ c_2 \\ c_3 \\ c_4 \end{Bmatrix} = 0$$

From this we get

$$c_1 = 0, \qquad c_3 = 0, \qquad c_4 = 0$$

But we cannot determine $c_2$. In other words, the equations are satisfied for any value of $c_2$. Hence the mode shape is

$$v(x) = c_2 \sin kx, \qquad k = n\pi/L$$

The first two mode shapes are shown in Figure 5.10. Note that since $c_2$ is unknown we have determined the *shape* of the deflection but not the actual deflection.

**Example 5.10:**  It is desired to investigate the effect of an elastic support on the buckling load of a beam. This is of interest because the spring represents (in a simple way) the effect of a structural attachment.

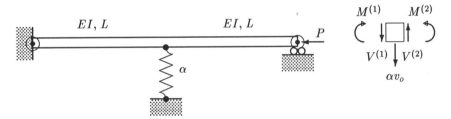

**Figure 5.11**: Pinned-pinned beam on elastic support.

We will assume that the spring does not exert any horizontal restraint on the beam, and therefore conclude that the axial force is the same over the complete length and given by $\bar{F}_o = -P$. The spring, however, acts as a concentrated force (of magnitude $\alpha v_o$) hence the beam must be divided into two regions of integration.

For the first section we have at $x = 0$

$$v = 0 \quad \Rightarrow \quad 0 = c_1 + c_4$$
$$M = EI\frac{d^2v}{dx^2} = 0 \quad \Rightarrow \quad 0 = -c_1 k^2$$

From these we get that the deflected shape is

$$v(x) = c_2 \sin kx + c_3 x$$

For this section $k^2 = P/EI$.

For the second section, we have at $x = L$ (we use $x$ to mean the local distance for the section)

$$v = 0 \quad \Rightarrow \quad 0 = c_1^* \cos kL + c_2^* \sin kL + c_3^* L + c_4^*$$

$$M = EI\frac{d^2v}{dx^2} = 0 \quad \Rightarrow \quad 0 = -k^2 c_1^* \cos kL - k^2 c_2^* \sin kL$$

where $k^2 = P/EI$ is the same as for the first section. From these we get for the deflected shape

$$v(x) = c_1^* \cos kx + c_2^* \sin kx + c_3^*[x - L], \qquad c_1^* \cos kL + c_2^* \sin kL = 0$$

We do not substitute for $c_2^*$ in terms of $c_1^*$ just in case $\sin kL$ is zero.

The compatibility and equilibrium equations must be satisfied at the junction of the two beam segments. Thus continuity of deflection and slope across the two sections gives

$$v^{(1)} = v^{(2)} \quad \Rightarrow \quad c_2 \sin kL + c_3 L = c_1^* - c_3^* L$$

$$\frac{dv^{(1)}}{dx} = \frac{dv^{(2)}}{dx} \quad \Rightarrow \quad c_2 k \cos kL + c_3 = c_2^* k + c_3^*$$

The final equations to be imposed are equilibrium at the junction of the two beam sections and the spring. With reference to the free body diagram in Figure 5.11, we have

$$M^{(1)} = M^{(2)} \quad \Rightarrow \quad EI[-k^2 c_2 \sin kL] = EI[-k^2 c_1^*]$$

$$V^{(1)} = V^{(2)} - \alpha v_o \quad \Rightarrow \quad -EI[k^2 c_3] = -EI[k^2 c_2^*] - \alpha[c_1^* - c_3^* L]$$

It does not matter which set of coefficients we use for $v_o$. From the first and third of these equations we get

$$c_1^* = c_2 \sin kL, \qquad c_3^* = -c_3$$

The remaining equations for both beam sections can be put into a reduced matrix form as

$$\begin{bmatrix} k\cos kL & 2 & -k \\ -\alpha^* \sin kL & 2k^2 - \alpha^* L & 0 \\ \sin kL \cos kL & 0 & \sin kL \end{bmatrix} \begin{Bmatrix} c_2 \\ c_3 \\ c_2^* \end{Bmatrix} = 0$$

where $\alpha^* \equiv \alpha/EI$. The determinant of this is

$$\det = 4k^3 \cos kL \sin kL + 2\alpha^*[\sin kL - kL \cos kL] \sin kL = 0$$

This equation cannot be solved explicitly for $k$. There are multiple modes but computing them requires solving a transcendental equation.

The complete results for an arbitrary spring are shown in the Figure 5.12. These were computed numerically. We see that increasing the spring stiffness causes the buckling loads to increase. That is, the structure becomes less sensitive to buckling.

**Example 5.11:** For the previous example, consider the special limits of a very flexible spring, and a very stiff spring, respectively.

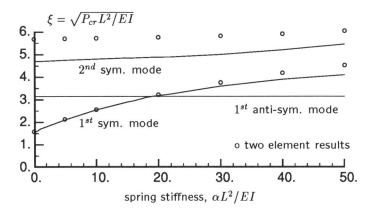

**Figure 5.12**: Effect of spring stiffness on the first few critical loads.

Note that if $\alpha^* = 0$ (i.e., no spring) then a zero determinant can be obtained from

$$\sin kL \cos kL = 0 \qquad \text{or} \qquad kL = \pi/2,\, \pi,\, 3\pi/2,\, 2\pi, \cdots$$

This gives the buckling loads as

$$P_{cr} = \frac{n^2\pi^2 EI}{(2L)^2}$$

which is the result already obtained for the pinned-pinned beam.

On the other hand, if the spring is very stiff (i.e., $\alpha^* \to \infty$) then the determinant equation becomes

$$[\sin kL - kL \cos kL] \sin kL = 0$$

One set of solutions are obtained from

$$\sin kL = 0 \qquad \text{or} \qquad kL = \pi,\, 2\pi, \cdots$$

This corresponds to a pinned-pinned beam of length $L$. These are the anti-symmetric modes and they are unaffected by the spring (since there is no displacement at the attachment point).

The remaining solutions must be obtained using some numerical or other approximate means. We present here a simple graphical scheme. First introduce the variable $\xi \equiv kL$ and rearrange the determinant equation as

$$\tan \xi = \xi$$

Now make a plot of the functions $f_1(\xi) \equiv \tan \xi$ and $f_2(\xi) \equiv \xi$ for different values of $\xi$ as shown in Figure 5.13. The intersections of these plots are at the roots of the characteristic equation. The first few roots are

$$kL = \xi = 4.493\,,\, 7.725\,,\, 10.904$$

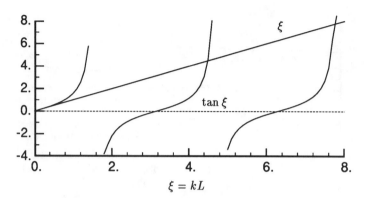

**Figure 5.13**: Graphical scheme for solving transcendental equation $\tan \xi = \xi$.

For future reference, Table 5.1 gives results for some other boundary conditions. The critical load is given by

$$P_{cr} = \alpha \pi^2 (kL)^2 \frac{EI}{L^2}$$

The numerical parameter $\alpha$ is given in the table.

| BC | mode=1 | 2 | 3 | Characteristic Equation |
|---|---|---|---|---|
| cantilever | 0.25 | 2.25 | 6.25 | $\cos kL = 0$ |
| pinned-pinned | 1.00 | 4.00 | 9.00 | $\sin kL = 0$ |
| fixed-pinned | 2.05 | 6.05 | 12.05 | $\tan kL - kL = 0$ |
| fixed-fixed | 4.00 | 8.18 | 16.00 | $2(\cos kL - 1) + kL \sin kL = 0$ |

Table 5.1: Critical load factor for the first three buckling modes.

## 5.6   Matrix Analysis of Stability of Beams

We will use the governing differential equations to derive a matrix formulation for beam buckling. The resulting stiffness matrix, however, is not in a form suitable for eigenanalysis, so we will proceed to present an approximate linearized version of the stiffness.

### Exact Total Stiffness

The basic idea in establishing the stiffness relation is the same as used in Chapter 3: having integrated the governing differential equations, we rewrite the constants of integration in terms of the nodal degrees of freedom. Using the displacement function, we are then in a position to determine the member moment and shear distributions. These, in turn, can be related to the nodal loads.

With reference to Figures 5.6 and 3.4, we assume that the loading is zero over the element length and that the axial force is constant and compressive. The general deflected shape is

$$v(x) = c_1 \cos kx + c_2 \sin kx + c_3 x + c_4 , \qquad k^2 = -\bar{F}_o / EI$$

We write the coefficients $c_n$ in terms of the nodal degrees of freedom. For example, at $x = 0$

$$v(0) \equiv v_1 = c_1 + c_4 , \qquad \frac{dv(0)}{dx} \equiv \phi_1 = kc_2 + c_3$$

Consequently, the deflected shape is

$$v(x) = c_1[\cos kx - 1] + c_2[\sin kx - kx] + v_1 + \phi_1 x$$

Similarly at the other node, we write

$$
\begin{aligned}
v(L) \equiv v_2 &= c_1[\cos kL - 1] + c_2[\sin kL - kL] + v_1 + \phi_1 L \\
\frac{dv(L)}{dx} \equiv \phi_2 &= -kc_1 \sin kL + kc_2 \cos kL + \phi_1
\end{aligned}
$$

The system of equations to determine the remaining coefficients can now be arranged as

$$
\begin{bmatrix} (1-C) & (\xi - S) \\ \xi S & \xi(1-C) \end{bmatrix}
\begin{Bmatrix} c_1 \\ c_2 \end{Bmatrix} =
\begin{Bmatrix} v_1 + \phi_1 L - v_2 \\ \phi_1 L - \phi_2 L \end{Bmatrix}
$$

where we have used the notations $\xi \equiv kL$, and $C \equiv \cos \xi$, $S \equiv \sin \xi$. We will use Cramer's rule to solve this. First we get the determinant and rearrange it as

$$\Delta = \xi[2 - 2C - \xi S]$$

The solution for $c_1$ is now given as

$$c_1 = \det \begin{vmatrix} v_1 + \phi_1 L - v_2 & (\xi - S) \\ \phi_1 L - \phi_2 L & \xi(1-C) \end{vmatrix} / \Delta$$

Multiplying this out gives

$$c_1 = [v_1 \xi(1 - C) + \phi_1 L(S - \xi C) - v_2 \xi(1 - C) + \phi_2 L(\xi - S)]/\Delta$$

We can obtain the other coefficient in a similar manner

$$c_2 = [-v_1 \xi S + \phi_1 L(1 - C - \xi S) + v_2 \xi S + \phi_2 L(C - 1)]/\Delta$$

At this stage we can rewrite the deflection function just in terms of the nodal degrees of freedom. From this, the moment and shear force distributions can be obtained. For example, the moment distribution is

$$M(x) = EI\frac{d^2v}{dx^2} = EI[-k^2c_1 \cos kx - k^2c_2 \sin kx]$$

The moment at $x = 0$ is

$$M_1 = -M(0) = EI\frac{-\xi^2}{L^2}[v_1\xi(1 - C) + \phi_1 L(S - \xi C) - v_2\xi(1 - C) + \phi_2 L(\xi - S)]/\Delta$$

The shear force distribution is

$$V(x) = -EI\frac{d^3v}{dx^3} + \bar{F}_o\frac{dv}{dx} = -EI[k^2c_3] = -EIk^2[\phi_1 - kc_2]$$

giving a value at $x = 0$ of

$$V_1 = -V(0) = EI\frac{\xi^3}{L^3}[v_1\xi S - \phi_1 L(1 - C) - v_2\xi S - \phi_2 L(C - 1)]/\Delta$$

Similar expressions can be obtained for the nodal loads $M_2$ and $V_2$, from which the stiffness relation can be established. The complete stiffness matrix is

$$[\,k\,] = \frac{EI}{L^3}\begin{bmatrix} \xi^2 S & \xi L(1 - C) & -\xi^2 S & \xi L(1 - C) \\ \xi L(1 - C) & -L^2(\xi C - S) & -\xi L(1 - C) & L^2(\xi - S) \\ -\xi^2 S & -\xi L(1 - C) & \xi^2 S & -\xi L(1 - C) \\ \xi L(1 - C) & L^2(\xi - S) & -\xi L(1 - C) & L^2(S - \xi C) \end{bmatrix}\frac{\xi^2}{\Delta} \quad (5.14)$$

This is symmetric and exhibits many of the same repetitions as the beam element stiffness.

**Example 5.12:** Solve for the tip deflection of the cantilever problem in Figure 5.7 using the exact stiffness matrix.

The procedure we follow is the same as for other matrix methods. Number the nodes with 1 at the fixed end and 2 at the other, then, the unknown degrees of freedom are

$$\{u_u\} = \{v_2, \phi_2\}$$

There is only one element, hence the reduced structural stiffness matrix is just the fourth quadrant of the element stiffness. The total stiffness relation is therefore

$$\frac{EI}{L^3}\begin{bmatrix} \xi^2 S & -\xi L(1 - C) \\ -\xi L(1 - C) & L^2(S - \xi C) \end{bmatrix}\frac{\xi^2}{\Delta}\begin{Bmatrix} v_2 \\ \phi_2 \end{Bmatrix} = \begin{Bmatrix} Q \\ 0 \end{Bmatrix}$$

The second of these equations gives

$$\phi_2 = \frac{\xi}{L}\frac{(1 - C)}{(S - \xi C)}v_2$$

Using this in the first equation and canceling the determinant term gives

$$v_2 = \frac{QL^3}{EI}\frac{(S - \xi C)}{\xi^3 C}$$

This is the same as previously obtained.

## Approximate Geometric Stiffness

The stiffness matrix as developed is exact; however, the unknown critical loads (which are related to $\bar{F}_o = -\xi^2 EI/L^2$) are deeply imbedded in its transcendental form. What we are now interested in is a simplification of this matrix to allow easier determination of the critical loads. More precisely, we want to linearize this matrix with respect to $\bar{F}_o$. The price is that the stiffness relation is no longer exact.

The axial load appears in the term $\xi \equiv kL$ and in the trigonometric terms $C \equiv \cos\xi$, $S \equiv \sin\xi$. We will make a Taylor series expansion of these with respect to $\xi$. This will give a polynomial, but we will retain only enough terms so that the stiffness, ultimately, is linear in the loading $\bar{F}_o$. Note that this expansion on small $\xi$ is equivalent to requiring a small element length $L$.

We will start with the determinant. Using the series expansion for the sine and cosine terms, we get

$$
\begin{aligned}
\Delta &= \xi[2 - 2C - \xi S] \\
&\approx \xi[2 - 2(1 - \xi^2/2 + \xi^4/24 - \xi^6/720 + \cdots) - \xi(\xi - \xi^3/6 + \xi^5/120 - \cdots)] \\
&\approx \xi^5[1 - \xi^2/15 + \cdots]/12
\end{aligned}
$$

It is important to realize that the series has been truncated precisely as indicated above because that is the first non-trivial point at which the determinant can exhibit a zero. We will need the reciprocal of the determinant and this is approximated as

$$
\frac{1}{\Delta} \approx \frac{12}{\xi^5}(1 + \xi^2/15 + \cdots)
$$

We now do the expansion on the stiffness terms out to the same order. For example

$$
k_{11} \approx \frac{EI}{L^3}[\xi^4(\xi - \xi^3/6 + \cdots)]\frac{12}{\xi^5}(1 + \xi^2/15 + \cdots) = \frac{EI}{L^3}12[1 - \xi^2/10 + \cdots]
$$

Replacing $\xi$ in terms of the axial load, we finally get

$$
k_{11} = \frac{EI}{L^3}[12] + \frac{\bar{F}_o}{L}[\frac{12}{10}]
$$

The first term is recognized as the $k_{11}$ of the beam element stiffness; the second term is very similar to that of truss buckling. In like manner, we can expand for all the stiffness terms to finally get the complete approximate element stiffness matrix as

$$
[\,k\,] = \frac{EI}{L^3}
\begin{bmatrix}
12 & 6L & -12 & 6L \\
6L & 4L^2 & -6L & 2L^2 \\
-12 & -6L & 12 & -6L \\
6L & 2L^2 & -6L & 4L^2
\end{bmatrix}
+ \frac{\bar{F}_o}{30L}
\begin{bmatrix}
36 & 3L & -36 & 3L \\
3L & 4L^2 & -3L & -L^2 \\
-36 & -3L & 36 & -3L \\
3L & -L^2 & -3L & 4L^2
\end{bmatrix}
\tag{5.15}
$$

This is symmetric, and the axial load appears in a linear fashion. We can make this stiffness matrix resemble the results for the truss by writing

$$[\,k\,] = [\,k_E\,] + \bar{F}_o[\,k_G\,]$$

where $[\,k_E\,]$ is the element elastic stiffness matrix, and $\bar{F}_o[\,k_G\,]$ is the element geometric stiffness. Note that $\bar{F}_o$ may be positive or negative, the former means that the structure gets stiffer with tensile loading.

If we compare the diagonal terms of the approximate total stiffness matrix with the exact values, we see they go through a zero only once as $\bar{F}_o$ is made more compressive. Therefore, at most, only four buckling loads are obtained for this element. Contrast this with the infinite number obtainable from the exact stiffness matrix. However, we can improve the approximate solution simply by using many elements for a given member length.

## Assemblage

Assembling the global stiffness matrix in stability problems is done in exactly the same way as in the other chapters. The only significant difference is that during assemblage there is an initial force $\bar{F}_o$ for each member. These axial forces are not known; indeed, these are precisely what we are trying to determine in a buckling analysis.

Following the procedure used for the truss, we assume there is a load or a group of loads $\{P\}$ applied and they are normalized as $P\{p\}$. We solve the pre-buckle problem so that the initial axial force can be obtained as

$$\bar{F}_o = P\bar{f}_o$$

for each member. In this manner the assembled matrices of the eigenvalue problem can be established as

$$\left[[K_E] + P[\bar{f}_o K_G]\right]\{u\} = 0$$

where $[\bar{f}_o K_G]$ is the assembled global geometric stiffness. That is, we have extracted a common $P$ from each member, and now the problem resembles an eigenvalue problem. The notation $[\bar{f}_o K_G]$ is a reminder that each element may have a different axial load.

**Example 5.13:**  Obtain an approximation for the buckling load for the clamped-clamped beam shown in Figure 5.14. Use two elements.

The first step is to solve the pre-buckled equations in order to determine the distribution of initial axial load. In this problem, $\bar{F}_o$ is the same throughout the beam and simply equal to the applied axial load as $\bar{F}_o = -P$. Therefore, we can go directly to the eigenvalue problem.

Number the nodes as shown, then, the unknown degrees of freedom are

$$\{u_u\} = \{v_2, \phi_2\}$$

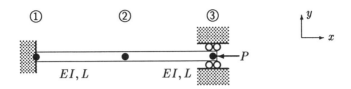

**Figure 5.14**: Clamped-clamped beam modeled with two elements.

The reduced element stiffness matrices are (using the fact that $\bar{F}_o = -P$ for both elements)

$$[k^{*(12)}] = \frac{EI}{L^3}\begin{bmatrix} 12 & -6L \\ -6L & 4L^2 \end{bmatrix} - \frac{P}{30L}\begin{bmatrix} 36 & -3L \\ -3L & 4L^2 \end{bmatrix}$$

$$[k^{*(23)}] = \frac{EI}{L^3}\begin{bmatrix} 12 & 6L \\ 6L & 4L^2 \end{bmatrix} - \frac{P}{30L}\begin{bmatrix} 36 & 3L \\ 3L & 4L^2 \end{bmatrix}$$

The assembled stiffness matrix is

$$[K^*] = \frac{EI}{L^3}\begin{bmatrix} 24 & 0 \\ 0 & 8L^2 \end{bmatrix} - \frac{P}{30L}\begin{bmatrix} 72 & 0 \\ 0 & 8L^2 \end{bmatrix} = \frac{EI}{L^3}\begin{bmatrix} 24 - \lambda 72/30 & 0 \\ 0 & 8L^2(1 - \lambda/30) \end{bmatrix}$$

where $\lambda \equiv PL^2/EI$. The determinantal equation, after rearranging, becomes

$$(24 - \lambda 72/30)(1 - \lambda/30)8L^2 = 0$$

In this case, there are two values of the critical load. Solving these gives the critical loads as

$$P_{cr} = 24\frac{EI}{L^2}\frac{30}{72} = \frac{10EI}{L^2}, \qquad P_{cr} = \frac{EI}{L^2}30 = \frac{30EI}{L^2}$$

Compare these with the exact solutions (keeping in mind that the beam is of length $2L$)

$$P_{cr} = \frac{4\pi^2 EI}{4L^2} = \frac{9.87EI}{L^2}, \qquad P_{cr} = \frac{16\pi^2 EI}{4L^2} = \frac{39.48EI}{L^2}$$

The difference in the lowest critical load is less than 1%. The second corresponds to the antisymmetric mode.

It must be remembered that even though there is no distributed load, the matrix solution is an approximate one. Hence, to obtain a better solution it is necessary to use more elements. Recall that for a given length of beam, this makes $L$ of the element smaller and therefore extends the range of validity of the approximation. Further, the number of eigenvalues obtained is directly proportional to the number of elements used. Again, this is equivalent to extending the range of the Taylor series expansion. A rule of thumb is that if N eigenvalues are computed, then approximately the first N/2 of them are reasonably accurate. Table 5.2 indicates the convergence of the computed critical loads as a function of the number of elements used; more or less the rule of thumb seems to be substantiated.

| Elements | DoF | mode= 1 | 2 | 3 | 4 | 5 | 6 |
|----------|-----|---------|-------|-------|-------|-------|-------|
| 2 | 2 | 4.05 | 12.16 | - | - | - | - |
| 4 | 6 | 4.03 | 8.39 | 16.21 | 30.44 | 52.17 | 80.05 |
| 8 | 14 | 4.00 | 8.20 | 16.12 | 24.57 | 37.20 | 50.61 |
| 16 | 30 | 4.00 | 8.18 | 16.00 | 24.21 | 36.09 | 48.40 |
| Exact | | 4.00 | 8.18 | 16.00 | 24.20 | 36.00 | 48.20 |

Table 5.2: Convergence of the critical loads for a clamped-clamped beam.

**Example 5.14:**   Consider the pinned-pinned beam with an elastic support as shown in Figure 5.15. Use two elements to model the beam and determine the first few critical loads.

Number the nodes as shown, then the total degrees of freedom are

$$\{u\} = \{v_1, \phi_1; v_2, \phi_2; v_3, \phi_3\}$$

The pinned boundary conditions require that

$$v_1 = 0, \qquad v_3 = 0$$

giving the reduced system as

$$\{u_u\} = \{\phi_1, v_2, \phi_2, \phi_3\}$$

The corresponding numbered equations are

$$\{IDbc\} = \{0, 1; 2, 3; 0, 4\}$$

A simple static analysis shows that the axial force is the same in each element (and of compressive value $P$), hence we can proceed directly to assembling the elastic and geometric stiffnesses. The reduced element stiffnesses are

$$[k^{*(12)}] = \frac{EI}{L^3} \begin{bmatrix} 4L^2 & -6L & 2L^2 \\ -6L & 12 & -6L \\ 2L^2 & -6L & 4L^2 \end{bmatrix} - \frac{P}{30L} \begin{bmatrix} 4L^2 & -3L & -L^2 \\ -3L & 36 & -3L \\ -L^2 & -3L & 4L^2 \end{bmatrix}$$

$$[k^{*(23)}] = \frac{EI}{L^3} \begin{bmatrix} 12 & 6L & 6L \\ 6L & 4L^2 & 2L^2 \\ 6L & 2L^2 & 4L^2 \end{bmatrix} - \frac{P}{30L} \begin{bmatrix} 36 & 3L & L^2 \\ 3L & 4L^2 & -L^2 \\ 3L & L^2 & 4L^2 \end{bmatrix}$$

The reduced structural stiffness matrix is

$$[K^*] = \frac{EI}{L^3} \begin{bmatrix} 4L^2 & -6L & 2L^2 & 0 \\ -6L & 24+\alpha^* & 0 & 6L \\ 2L^2 & 0 & 8L^2 & 2L^2 \\ 0 & 6L & 2L^2 & 4L^2 \end{bmatrix} - \frac{P}{30L} \begin{bmatrix} 4L^2 & -3L & -L^2 & 0 \\ -3L & 72 & 0 & 3L \\ -L^2 & 0 & 8L^2 & -L^2 \\ 0 & 3L & -L^2 & 4L^2 \end{bmatrix}$$

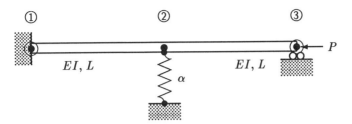

**Figure 5.15**: Beam with an elastic support.

where $\alpha^* \equiv \alpha L^3/EI$. Note that the elastic stiffness $[K_E]$ changes by the presence of the spring, but $[K_G]$ remains the same as in the previous example.

Generally, we solve for the critical loads by expanding the determinant. However, the equation to be solved is then quartic in the load, so we will take a different tack. Because of the geometric symmetry of the problem, we know that the results will contain separate symmetric and anti-symmetric modes. We will simplify the problem by considering each separately.

For the symmetric modes, we have that

$$\phi_2 = 0, \qquad \phi_1 = -\phi_3$$

From this we can reduce the eigenvalue problem to

$$\left[ \begin{bmatrix} 4L^2 & -6L \\ -6L & 12+\alpha^* \end{bmatrix} - \frac{\lambda}{30} \begin{bmatrix} 4L^2 & -3L \\ -3L & 72 \end{bmatrix} \right] \begin{Bmatrix} \phi_1 \\ v_2 \end{Bmatrix} = 0$$

where $\lambda \equiv PL^2/EI$. The determinant of this system gives the quadratic equation

$$135\lambda^2 - \lambda[156 + 2\alpha^*]30 + [12 + 2\alpha^*]900 = 0$$

This is solved to give

$$45\lambda = 78 + \alpha^* \pm \sqrt{4464 - 114\alpha^* + \alpha^{*2}}$$

It is instructive to consider the limiting cases of a very flexible spring and a very stiff spring. In the former case, we have

$$P_{cr} = \frac{EI}{L^2}[2.49 + 0.41\alpha^*], \qquad P_{cr} = \frac{EI}{L^2}[32.18 - 0.03\alpha^*]$$

The second critical value is hardly affected by the spring. The lowest critical load increases linearly by the presence of the spring. In the case of a stiff spring, we have

$$P_{cr} = \frac{EI}{L^2}[12.7], \qquad P_{cr} = \frac{EI}{L^2}[0.45\alpha^*]$$

The first critical value is not affected by the spring, whereas, the second critical load increases to infinity. Indeed, if the spring stiffness is large enough this load may exceed that of the second mode. What this means is that buckling is being prevented from occurring in that mode. The full results are shown in Figure 5.12 where they are compared to the exact results.

For the anti-symmetric modes we have that

$$v_2 = 0 , \qquad \phi_1 = \phi_3$$

From this we can reduce the eigenvalue problem to

$$\left[ \begin{bmatrix} 4 & 2 \\ 2 & 4 \end{bmatrix} - \frac{\lambda}{30} \begin{bmatrix} 4 & -1 \\ -1 & 4 \end{bmatrix} \right] \begin{Bmatrix} \phi_1 \\ \phi_2 \end{Bmatrix} = 0$$

where $\lambda \equiv PL^2/EI$. The spring does not enter these equations since the displacement at the attachment is zero. The determinant of this system is

$$\lambda^2 - \lambda 72 + 720 = 0$$

This is solved to give

$$\lambda = 12 \quad \text{or} \quad 60$$

The critical loads are therefore

$$P_{cr} = \frac{EI}{L^2} [12] , \qquad P_{cr} = \frac{EI}{L^2} [60]$$

What these results show is that for a flexible spring the first antisymmetric mode lies between the first and second symmetric modes. However, as the spring stiffness increases, it is possible for the anti-symmetric mode to become the lowest buckling load.

# 5.7   Stability of Space Frames

The approach to establishing the geometric stiffness relation for the space frame is essentially the same as used in Chapter 4 for establishing the elastic stiffness. That is, we consider the total stiffness of each load system to be independent of each other. Therefore, it is simply a matter of augmenting the geometric stiffnesses to the full degrees of freedom of the frame and superposing them. We can achieve this even for the coupled-beam loading by treating the axial loads as parameters obtained from a pre-buckle analysis.

## Assemblage of the Geometric Stiffness

The displacement of each node of a space frame is described by three translational and three rotational components of displacement, giving six degrees of freedom at each unrestrained node. More specifically, the displacement and force vectors at each node are, respectively,

$$\begin{Bmatrix} u \\ v \\ w \\ \phi_x \\ \phi_y \\ \phi_z \end{Bmatrix} , \quad \begin{Bmatrix} F_x \\ F_y \\ F_z \\ M_x \\ M_y \\ M_z \end{Bmatrix}$$

These forces and displacements are related to each other through the following $[12 \times 12]$ partitioned matrix

$$\{\bar{F}\} = \begin{bmatrix} \bar{k}_{11} & \bar{k}_{12} & \bar{k}_{13} & \bar{k}_{14} \\ \bar{k}_{21} & \bar{k}_{22} & \bar{k}_{23} & \bar{k}_{24} \\ \bar{k}_{31} & \bar{k}_{32} & \bar{k}_{33} & \bar{k}_{34} \\ \bar{k}_{41} & \bar{k}_{42} & \bar{k}_{43} & \bar{k}_{44} \end{bmatrix} \{\bar{u}\}$$

The only mode of buckling we have considered is due to the axial force – bending moment interaction, hence we need to superpose the geometric stiffness for bending about the $y$ and $z$ axes. We neglect torsional buckling. Such buckling would occur if the torsional rigidity of the section is very low (as for a bar of thin-walled open section, say). There is little difficulty in showing that the geometric portion of the stiffness matrix takes the following form when augmented to the full size

$$\begin{bmatrix} \bar{k}_{11} & \bar{k}_{12} \\ \bar{k}_{21} & \bar{k}_{22} \end{bmatrix}_G = \frac{\bar{F}_o}{30L} \begin{bmatrix} 0 & 0 & 0 & 0 & 0 & 0 \\ 0 & 36 & 0 & 0 & 0 & 3L \\ 0 & 0 & 36 & 0 & -3L & 0 \\ 0 & 0 & 0 & 0 & 0 & 0 \\ 0 & 0 & -3L & 0 & 4L^2 & 0 \\ 0 & 3L & 0 & 0 & 0 & 4L^2 \end{bmatrix}$$

The remaining three quadrants can be filled easily since they contain the same repetitions as in the elastic stiffness.

The assemblage process is similar to that used for trusses but is repeated here for completeness. First, the stiffness matrix for the arbitrarily oriented member is determined by rotation of the member stiffnesses

$$[k] = [T]^T \left[ [\bar{k}_E] + \bar{F}_o[\bar{k}_G] \right][T] = [T]^T[\bar{k}_E][T] + \bar{F}_o[T]^T[\bar{k}_G][T]$$

In obtaining this relation, we have considered that the transformation is applied only to the vector displacement $\{u\}$ and vector load $\{F\}$, and that the initial axial force $\bar{F}_o$ is left intact. We do this because the axial force is a member quantity and therefore independent of our choice of global coordinates.

Assembling the global stiffness matrix is now a matter of adding all the element stiffnesses. The only significant point is that during assemblage there is an initial force $\bar{F}_o$ for each member. Thus the assemblage occurs in two stages. First we assemble only the elastic stiffness and use it to find the axial force in each member by solving

$$[K_E]\{u\} = \{p\}$$

where $[K_E]$ is the assembled elastic global stiffness and $\{p\}$ is the normalized load distribution related to the actual loading by $\{P\} = P\{p\}$. The initial problem is then solved to give the axial force in each member (represented by $\bar{f}_o$). The axial forces due to $\{P\}$ are therefore

$$\bar{F}_o = P\bar{f}_o$$

In this manner the eigenvalue problem can be set up as

$$\left[[K_E] + P[\bar{f}_o K_G]\right]\{u\} = \{0\}$$

where $[\bar{f}_o K_G]$ is the assembled global geometric stiffness. This notation emphasizes that the initial axial force needs to be known before the geometric stiffness can be assembled.

## Plane Frames

The total global stiffness matrix for an arbitrary frame element in two-dimensions reduces to

$$[k] = \frac{EA}{L}\begin{bmatrix} C^2 & & & & & \\ CS & S^2 & & sym & & \\ 0 & 0 & 0 & & & \\ -C^2 & -CS & 0 & C^2 & & \\ -CS & -S^2 & 0 & CS & S^2 & \\ 0 & 0 & 0 & 0 & 0 & 0 \end{bmatrix} \tag{5.16}$$

$$+ \frac{EI}{L^3}\begin{bmatrix} 12S^2 & & & & & \\ -12CS & 12C^2 & & sym & & \\ -6LS & 6LC & 4L^2 & & & \\ -12S^2 & 12CS & 6LS & 12S^2 & & \\ 12CS & -12C^2 & -6LC & -12CS & 12C^2 & \\ -6LS & 6LC & 2L^2 & 6LS & -6LC & 4L^2 \end{bmatrix}$$

$$+ \frac{\bar{F}_o}{30L}\begin{bmatrix} 36S^2 & & & & & \\ -36CS & 36C^2 & & sym & & \\ -3LS & 3LC & 4L^2 & & & \\ -36S^2 & 36CS & 3LS & 36S^2 & & \\ 36CS & -36C^2 & -3LC & -36CS & 36C^2 & \\ -3LS & 3LC & -L^2 & 3LS & -3LC & 4L^2 \end{bmatrix}$$

where the abbreviations $C \equiv \cos\theta$, $S \equiv \sin\theta$ are used. The first two matrices are recognized as the elastic stiffness for a plane frame member.

**Example 5.15:**  Find the buckling loads for the plane frame shown in Figure 5.16. Each member has the same material and section properties. The load of $\sqrt{2}P$ is applied at an angle of $45°$ to the horizontal.

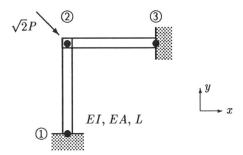

**Figure 5.16**: Frame with fixed-fixed supports.

We will use two elements to model the problem. Numbering the nodes as shown, the total degrees of freedom are

$$\{u\} = \{u_1, v_1, \phi_1; u_2, v_2, \phi_2; u_3, v_3, \phi_3\}$$

The fixed boundary conditions require that

$$u_1 = v_1 = \phi_1 = 0 \qquad \text{and} \qquad u_3 = v_3 = \phi_3 = 0$$

giving the reduced system as

$$\{u_u\} = \{u_2, v_2, \phi_2\}, \qquad \{P\} = P\{1, -1, 0\}$$

First, we solve the problem neglecting the geometric stiffness. The reduced element stiffness matrices for both members are for the non-zero degree of freedoms $\{u_2, v_2, \phi_2\}$. For Member 1-2 with connectivity 1 to 2, the orientation is $\theta = 90°$, giving

$$[k^{(*12)}] = \frac{EA}{L} \begin{bmatrix} 0 & 0 & 0 \\ 0 & 1 & 0 \\ 0 & 0 & 0 \end{bmatrix} + \frac{EI}{L^3} \begin{bmatrix} 12 & 0 & 6L \\ 0 & 0 & 0 \\ 6L & 0 & 4L^2 \end{bmatrix}$$

For Member 2-3 with connectivity 2 to 3, the orientation is $\theta = 0°$, giving

$$[k^{(*23)}] = \frac{EA}{L} \begin{bmatrix} 1 & 0 & 0 \\ 0 & 0 & 0 \\ 0 & 0 & 0 \end{bmatrix} + \frac{EI}{L^3} \begin{bmatrix} 0 & 0 & 0 \\ 0 & 12 & 6L \\ 0 & 6L & 4L^2 \end{bmatrix}$$

The reduced structural stiffness relation is therefore obtained as

$$\left[ \frac{EA}{L} \begin{bmatrix} 1 & 0 & 0 \\ 0 & 1 & 0 \\ 0 & 0 & 0 \end{bmatrix} + \frac{EI}{L^3} \begin{bmatrix} 12 & 0 & 6L \\ 0 & 12 & 6L \\ 6L & 6L & 8L^2 \end{bmatrix} \right] \begin{Bmatrix} u_2 \\ v_2 \\ \phi_2 \end{Bmatrix} = P \begin{Bmatrix} 1 \\ -1 \\ 0 \end{Bmatrix}$$

Solving this gives

$$u_2 = -v_2 = P\frac{L}{(EA + 12EI/L^2)}, \qquad \phi_2 = 0$$

This gives the initial axial loads as

$$\bar{F}_o^{(12)} = P\frac{-EA}{(EA + 12EI/L^2)} = P\bar{f}_o^{(12)}, \quad \bar{F}_o^{(23)} = P\frac{-EA}{(EA + 12EI/L^2)} = P\bar{f}_o^{(23)}$$

In this case, both axial loads are the same and compressive. Indeed the pre-buckle problem is symmetric.

It is now possible to form the geometric stiffnesses. These are

$$[k_G^{*(12)}] = \frac{P\bar{f}_o^{(12)}}{30L} \begin{bmatrix} 36 & 0 & 3L \\ 0 & 0 & 0 \\ 3L & 0 & 4L^2 \end{bmatrix}, \quad [k^{*(23)}] = \frac{P\bar{f}_o^{(23)}}{30L} \begin{bmatrix} 0 & 0 & 0 \\ 0 & 36 & 3L \\ 0 & 3L & 4L^2 \end{bmatrix}$$

The reduced structural stiffness matrix can therefore be assembled. Introducing the notations

$$\alpha \equiv \frac{EA}{L}, \quad \beta \equiv \frac{EI}{L^3}, \quad \gamma \equiv \frac{-\bar{f}_o^{(12)}}{30L} = \frac{-\bar{f}_o^{(23)}}{30L} = \frac{\alpha}{(\alpha + 12\beta)L}$$

we get the eigenvalue problem as

$$\left[ \begin{bmatrix} \alpha + 12\beta & 0 & \beta 6L \\ 0 & \alpha + 12\beta & \beta 6L \\ \beta 6L & \beta 6L & \beta 8L^2 \end{bmatrix} - P\gamma \begin{bmatrix} 36 & 0 & 3L \\ 0 & 36 & 3L \\ 3L & 3L & 8L^2 \end{bmatrix} \right] \begin{Bmatrix} u_2 \\ v_2 \\ \phi_2 \end{Bmatrix} = 0$$

The characteristic equation is

$$(\alpha + 12\beta - 36P\gamma)[135P^2\gamma^2 - P\gamma 4(\alpha + 39\beta) + 4\beta(\alpha + 3\beta)] = 0$$

One of the roots is

$$P = \frac{\alpha + 12\beta}{36\gamma} = \frac{(\alpha + 12\beta)^2 L}{36\alpha}$$

Substituting this into the equations of the eigenvalue problem, we get

$$\phi_2 = 0 \quad \text{and} \quad u_2 = -v_2$$

This is the symmetric mode. In the inextensible case ($\alpha$ very large) we get

$$P_{cr} = 0.833\alpha L$$

which shows that the frame will not buckle in this mode.

The other two critical loads occur at

$$P = \frac{2}{135\gamma}\left[(\alpha + 39\beta) \pm \sqrt{\alpha^2 - 57\alpha\beta + 1116\beta^2}\right]$$

Again, suppose that the axial stiffness is very large, then

$$P_{cr} = 30\,L\beta, \quad 0.777\,L\alpha$$

One of these critical loads approaches infinity, indicating the impossibility of buckling. The remaining buckling mode corresponds to

$$u_2 = v_2 = 0, \quad \phi_2 \neq 0$$

This deformation is the same as if the structure is pinned at Node 2.

This example clearly shows that even though the geometry and loading is symmetric, that the anti-symmetric buckling loads could be the dominant ones. Also, the use of the inextensibility assumption can profoundly influence the prediction of the buckling modes. Thus in the above problem, of the three possible modes, only one was predicted to occur.

## Problems

**5.1** In reference to Figure 5.5, let the inclined member be at an angle of $\theta$. Show that the dependence of the buckling load on $\theta$ is given by

$$P_{cr} = \frac{EA \sin \theta \cos^2 \theta}{(1 + \sin^3 \theta)}$$

From this it is clear that as the inclined member becomes more vertical that the buckling load decreases dramatically.

[Reference [41], pp. 145]

**5.2** In reference to the previous exercise, at what angle $\theta$ is the critical load 1% of $EA$?

**5.3** Show that for a cantilevered beam loaded by an axial force and concentrated moment applied at the free end

$$v(x) = \frac{M_o L^2}{EI} \left[ \frac{1 - \cos kx}{k^2 \cos kL} \right]$$

[Reference [20], pp. 162]

**5.4** Recover the uncoupled solution from the previous exercise, and show that it is in agreement with the straight-beam solution.

**5.5** If the cantilever beam of Figure 5.7 has a tensile axial force applied, show that

$$v(L) = \frac{Q}{k^3 EI} \left[ \frac{\sinh kL - kL \cosh kL}{\cosh kL} \right]$$

Show that in the limit as $P$ becomes very large that the deflection goes to zero.

**5.6** Show that the characteristic equation associated with the buckling of a clamped-clamped beam is
$$2 - 2 \cos kL - kL \sin kL = 0$$

Show that one set of solutions is given by

$$P_{cr} = 4n^2 \pi^2 \frac{EI}{L^2}, \qquad v(x) = \left( \cos \frac{2n\pi}{L} x - 1 \right)$$

[Reference [41], pp. 54]

**5.7**  Show that the lowest critical load of the second set of solutions of the previous exercise is given by

$$P_{cr} = 8.18\pi^2\frac{EI}{L^2}$$

[Reference [41], pp. 55]

**5.8**  Consider a cantilever beam with a rigid 'handle' of length $a$ welded to the free end. The handle is oriented along the length of the beam with its tip closest to the fixed end of the beam. Show that if a force is applied to the handle and pointing toward the fixed end of the beam that the characteristic equation for the buckling load is

$$kL \tan kL = L/a$$

[Reference [51], pp. 56]

**5.9**  If the force is reversed in the previous exercise so that the beam is in tension, show that buckling occurs when

$$kL \tanh kL = L/a$$

**5.10**  In reference to the spring supported beam of Figure 5.11, an interesting special case arises when the spring has the special values $\alpha^* = 2k^2/L$. Show that the critical loads and the corresponding mode shapes are given by

$$P_{cr} = n^2\pi^2\frac{EI}{L^2}, \qquad \alpha = \frac{2\pi^2 EI}{L^3}, \qquad v(x) = c_2\left[\sin\frac{n\pi}{L}x \pm n\pi\frac{x}{2L}\right]$$

**5.11**  Consider a pinned-pinned beam of length $2L$ with a second axial load applied at its middle. Show that the displacement shape in each beam segment can be taken as

$$v^{(1)}(x) = c_2 \sin kx + c_3 x$$
$$v^{(2)}(x) = c_1^* \cos k^*x + c_2^* \sin k^*x + c_3^*[x - L], \quad c_1^* \cos k^*L + c_2^* \sin k^*L = 0$$

where, for the first section we have $k^2 = (P_2 + P_3)/EI$, and for the second section $k^{*2} = P_3/EI$. Show that the eigenvalue system is

$$\begin{bmatrix} \sin kL & L & -1 & 0 & L \\ k\cos kL & 1 & 0 & -k^* & 1 \\ EIk^2\sin kL & 0 & -EIk^{*2} & 0 & 0 \\ 0 & -EIk^3 & 0 & 0 & EIk^{*3} \\ 0 & 0 & \cos k^*L & \sin k^*L & -1 \end{bmatrix} \begin{Bmatrix} c_2 \\ c_3 \\ c_1^* \\ c_2^* \\ c_3^* \end{Bmatrix} = 0$$

and consequently that the determinant is

$$\begin{aligned} det = {} & (k^{*2} - k^2)(k^{*3} - k^3)\sin kL \sin k^*L - L^2(k^{*3} + k^3)\sin kL \cos k^*L\, k^*L \\ & -kLk^{*2}(k^{*3} + k^3)\sin k^*L \cos kL \end{aligned}$$

**5.12** Explore some of the special cases of the last exercise (such as $k^* = k$ or $P_2 = 0$) and show that the reduce to the known results. An interesting situation arises when $P_2 = -P_3$.

**5.13** A long homogeneous timber of rectangular cross-section $[a \times b]$ floats in water with its top face horizontal. Prove that a necessary and sufficient condition for the timber to be in stable equilibrium is $s^2 - s + b^2/6a^2 > 0$, where $s$ is the specific gravity of the wood.

[Reference [27], pp. 38]

## Exercises

**5.1** Plot the shear force and bending moment diagrams for a beam loaded as in Figure 5.7.

**5.2** Compare the $k_{11}$ term from the approximate beam stiffness with its exact values. At what value of $kL$ do they differ by 5%?

**5.3** A simple beam carries a concentrated load at the center. If the beam is subjected to a tensile force equal to three times the buckling load, by what percentage is the deflection reduced? [74.26%]

**5.4** A steel tube of length $4\,ft$ has a $1\,in$ outer diameter and a thickness of $0.036\,in$. Find the load which may be applied axially so as not to exceed one-third of the Euler buckling load. [544 $lb$]

**5.5** A steel beam of square cross-section is to be used as a cantilever column to support a weight of $3000\,lb$. The length of the column is to be $5\,ft$. What should the cross-sectional dimension be so that the load which would cause buckling is $12000\,lb$.

**5.6** Find the buckling loads of the aluminum truss of Figure 4.4 if only a horizontal load is applied. Let $h = 10\,m$, $A = 100\,mm^2$ and $\theta = 45°$. [15.38 $MN$]

**5.7** A steel truss similar to Figure 4.5 has a length of $5\,m$ and cross-sectional area of $250\,mm^2$. Use one element per member to obtain the buckled mode shapes. What would happen if more elements were use? [30.34 $MN$]

**5.8** An aluminum frame similar to Figure 5.16 but with cross-section $1\,in \times 1\,in$ and length $100\,in$, is modeled with four elements. What are the critical loads and mode shapes? [$P_{cr} = 1725\,lb$ 3332 $lb$]

**5.9** A steel portal frame (two vertical members of length $L$ with a cross member of length $2L$) has a horizontal load applied at the joint. Use four elements to determine the buckling loads. The cross-section is $1\,in \times 1\,in$ and length $10\,ft$. [$P_{cr} = 2310\,lb$ 5084 $lb$]

**5.10** Add a diagonal member (same properties) to the portal frame of the previous problem and determine the critical loads. What has happened? [$P_{cr} = 4960\,lb$]

# Chapter 6

# General Structural Principles I

The previous chapters developed the analysis of particular structural systems by the consistent use of the twin concepts of compatibility and equilibrium. In each case, we derived a set of governing differential equations, integrated them, and then determined the constants of integration by satisfying the appropriate boundary and compatibility conditions. This approach will work in every case, but as a practical tool it suffers from a number of drawbacks. Primary among these is that the solution can become very cumbersome when more than one integration region is involved. It may also happen that the governing differential equations cannot be integrated in a closed form. When approximations are used the differential equations are not in a suitable form for direct manipulation. For example, we obtained the approximate geometric matrix in Chapter 5 by first obtaining its exact form, and then using a Taylor series expansion to obtain the linearized version. But if the cross-sectional area varies, for example, this approach is not feasible at all; it would be much more convenient if we could go directly to the approximate solution.

The purpose of this chapter is to introduce alternative methods for establishing equilibrium and compatibility requirements for structural behavior. Since the equilibrium conditions involve forces, and the compatibility conditions involve displacements, then a process through which both conditions are satisfied involves a quantity dependent on both force and displacement. This quantity is called *work*, and we will use it to show that, when a structure is in equilibrium, certain structural quantities achieve a stationary value. This powerful idea will then form the basis for our approximate analysis of structures. References [20, 27] are excellent introductions to these energy approaches.

## 6.1  Work and Strain Energy

When loads are applied to a structure, they cause internal stressing and straining of the members. During the stage of applying the loads, work is done. In a general sense, therefore, we see a connection between these loads and the internal response; this section establishes the connection.

We will restrict the following analysis to structures for which the strains and displacements are small and there is no dissipation of energy during loading. Although the material behavior is elastic, it is not necessarily linear. A typical force-deflection or stress-strain curve is shown in Figure 6.1. The elasticity requirement is that both the loading and unloading paths coincide.

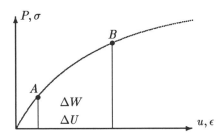

**Figure 6.1**: Typical elastic stress-strain behavior.

## Work

The work done by a force at a point is the scalar product of the force and the displacement at the point. For example, in terms of our global coordinate system,

$$dW = \vec{F} \cdot d\vec{u} = F_x du + F_y dv + F_z dw$$

where the arrow indicates a vector. The force is understood to be constant during the infinitesimal displacement $d\vec{u}$. When the force moves along a path from State $A$ to State $B$, the work done is

$$\Delta W = \int_A^B dW = \int_A^B \vec{F} \cdot d\vec{u} = \int_A^B (F_x du + F_y dv + F_z dw)$$

The loading curve of Figure 6.1 can be interpreted as a sequence of possible equilibrium states as the load (or deflection) is changed. Thus, A and B are two possible equilibrium states with different load conditions. We use the $\Delta$ symbol to signify that an *increment* of work is performed in moving from one state to the other — the change in configuration may be small or large. In the special case when the initial configuration is the unstressed, unstrained, virgin state we will simply use $W$ without the $\Delta$ for the work done in reaching a certain state.

The systems we are interested in have multiple forces and moments, so we will generalize the above expression for work to

$$\Delta W = \sum_n \left[ \int_A^B P_n du_n \right] = \int_A^B \{P\}^T \{du\} \tag{6.1}$$

It is understood that $P$ and $u$ are components of generalized forces and displacements, respectively. Thus the individual contribution to the work could refer to terms such as $P_3 du_3$ or $T_9 d\phi_9$.

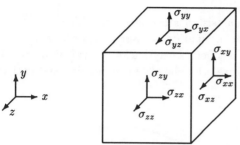

**Figure 6.2**: Stressed infinitesimal element.

## Strain Energy

If external forces $\{P\}$ act on the structure, stresses are set up inside the body. Consider a typical small volume of dimensions $\Delta x$, $\Delta y$, $\Delta z$, as shown in Figure 6.2. This volume is under the action of the following system of stresses

$$\{\sigma\}^T = \{\sigma_{xx}\,,\,\sigma_{yy}\,,\,\sigma_{zz}\,,\,\tau_{xy}\,,\,\tau_{yz}\,,\,\tau_{xz}\}$$

The small volume is in a deformed state and has a set of strains

$$\{\epsilon\}^T = \{\epsilon_{xx}\,,\,\epsilon_{yy}\,,\,\epsilon_{zz}\,,\,\gamma_{xy}\,,\,\gamma_{yz}\,,\,\gamma_{xz}\}$$

Note that there are only six independent components of stress and strain, since the shear components are related through relations such as

$$\tau_{xy} = \tau_{yx}\,,\qquad \gamma_{xy} = \gamma_{yx}$$

and so on. The strains are related to the displacements by

$$\epsilon_{xx} = \frac{\partial u}{\partial x}\,,\qquad \epsilon_{yy} = \frac{\partial v}{\partial y}\,,\qquad \epsilon_{zz} = \frac{\partial w}{\partial z}$$

$$\gamma_{xy} = \frac{\partial u}{\partial y} + \frac{\partial v}{\partial x}\,,\quad \gamma_{yz} = \frac{\partial v}{\partial z} + \frac{\partial w}{\partial y}\,,\quad \gamma_{zx} = \frac{\partial w}{\partial x} + \frac{\partial u}{\partial z} \qquad (6.2)$$

When we have a need to summarize these, we will write

$$\{\epsilon\} = [\,\mathcal{D}\,]\{\mathcal{U}\}$$

where $[\,\mathcal{D}\,]$ is the matrix of partial differential operators and $\{\mathcal{U}\} = \{u, v, w\}$.

A line that was originally of length $\Delta x$ in the undeformed state has a new length $\Delta x(1 + \epsilon_{xx})$ in the deformed state. Now consider an additional infinitesimal deformation added to the body; a typical change in length is $\Delta x(d\epsilon_{xx})$. The work done by the $\sigma_{xx}$ component of stress during this change is

$$dW = (F)du = (\sigma_{xx}\Delta y\Delta z)\Delta x d\epsilon_{xx} = \sigma_{xx}d\epsilon_{xx}\Delta \forall$$

where the symbol $\forall$ represents the volume. Hence the work increment for the whole body is obtained by allowing the typical volume to become infinitesimal and then integrating over the total volume. That is

$$\Delta W = \int_A^B \int_\forall \sigma_{xx} d\epsilon_{xx} d\forall$$

For the complete system of stresses we have the generalization

$$\Delta W = \int_A^B \int_\forall [\sigma_{xx} d\epsilon_{xx} + \sigma_{yy} d\epsilon_{yy} + \tau_{xy} d\gamma_{xy} + \cdots] d\forall = \int_A^B \int_\forall \{\sigma\}^T \{d\epsilon\} d\forall \qquad (6.3)$$

Because of perfect elasticity, the work expended can be regained if the loads are gradually decreased. That is, the work is stored in the elastically distorted body in the form of energy known as *strain energy* or *elastic energy*. We will use the definition

$$\Delta U = \int_A^B \int_\forall \{\sigma\}^T \{d\epsilon\} d\forall \qquad (6.4)$$

to mean the increment of strain energy. Although related, work and energy are distinctly different concepts; we can say that forces perform work, but the system possesses energy. Therefore, when work is performed on a system, a change of energy occurs.

The formal equivalence of the two measures of work, Equation 6.1 and Equation 6.3, can be shown; we will take it as self evident. The *principle of conservation of energy* can now be stated as

$$\Delta U - \Delta W_e = 0 \qquad \text{or} \qquad \int_A^B \int_\forall \{\sigma\}^T \{d\epsilon\} d\forall - \int_A^B \{P\}^T \{du\} = 0 \qquad (6.5)$$

where $\Delta W_e$ is the *external* work done by the forces. This says that for elastic systems, the external work done is equal to the strain energy stored. It is to be kept in mind that the changes in going from State A to State B could be large.

## 6.2   Linear Elastic Structures

Consider an elastic system with generalized applied loads $\{P\} = \{P_1, \ldots, P_N\}$. (Note that these can be moments as well as forces.) Denote the corresponding generalized degrees of freedom (displacements and rotations) by $\{u\} = \{u_1, \ldots, u_N\}$. It is of interest now to establish some properties of the general relation between these loads and responses.

### Strain Energy

Let the material obey Hooke's law in the form

$$\epsilon_{xx} = \frac{1}{E}[(1+\nu)\sigma_{xx} - \nu(\sigma_{xx} + \sigma_{yy} + \sigma_{zz})]$$

$$\epsilon_{yy} = \frac{1}{E}[(1+\nu)\sigma_{yy} - \nu(\sigma_{xx} + \sigma_{yy} + \sigma_{zz})]$$

$$\epsilon_{zz} = \frac{1}{E}[(1+\nu)\sigma_{zz} - \nu(\sigma_{xx} + \sigma_{yy} + \sigma_{zz})]$$

$$\gamma_{xy} = \frac{1}{G}\tau_{xy}, \quad \gamma_{yz} = \frac{1}{G}\tau_{yx}, \quad \gamma_{zx} = \frac{1}{G}\tau_{zx} \tag{6.6}$$

where $E$ is the Young's modulus, $\nu$ the Poisson's ratio, and $G = E/2(1+\nu)$ the shear modulus. The reciprocal of these for the normal components are

$$\sigma_{xx} = 2G[\epsilon_{xx} + \frac{\nu}{1+\nu}(\epsilon_{xx} + \epsilon_{yy} + \epsilon_{zz})]$$

$$\sigma_{yy} = 2G[\epsilon_{yy} + \frac{\nu}{1+\nu}(\epsilon_{xx} + \epsilon_{yy} + \epsilon_{zz})]$$

$$\sigma_{zz} = 2G[\epsilon_{zz} + \frac{\nu}{1+\nu}(\epsilon_{xx} + \epsilon_{yy} + \epsilon_{zz})]$$

These stress-strain relations can be summarized in the matrix forms

$$\{\epsilon\} = [\,\mathcal{C}\,]\{\sigma\}, \qquad \{\sigma\} = [\,\mathcal{E}\,]\{\epsilon\}, \qquad [\,\mathcal{C}\,] = [\,\mathcal{E}\,]^{-1}$$

with the obvious interpretation of the square matrices.

The general expression for the strain energy increment is given in Equation(6.4); we will now consider the initial state to be the zero strain state and since the stress-strain relation is linear the strain energy becomes

$$U = \tfrac{1}{2}\int_V [\sigma_{xx}\epsilon_{xx} + \sigma_{yy}\epsilon_{yy} + \tau_{xy}\gamma_{xy} + \cdots]d\forall = \tfrac{1}{2}\int_V \{\sigma\}^T\{\epsilon\}d\forall$$

Using Hooke's law this can be put in the alternate forms

$$U = \tfrac{1}{2}\int_V \{\epsilon\}^T[\,\mathcal{E}\,]\{\epsilon\}d\forall = \tfrac{1}{2}\int_V \{\sigma\}^T[\,\mathcal{C}\,][\,\sigma\,]d\forall \tag{6.7}$$

The work done by the external forces is

$$W_e = \tfrac{1}{2}\{P\}\{u\} = \tfrac{1}{2}\sum_i P_i u_i$$

The above relations will now be particularized to the structural systems of interest by writing the distributions of stress and strain in terms of resultants. We have already developed these structures in the previous chapters, so the assumptions and restrictions as stated there are still assumed to apply here.

For the rod member, there is only an axial stress present and it is uniformly distributed on the cross-section. Let $F$ be the resultant force; then $\sigma = F/A = E\epsilon$ and

$$\text{axial:} \qquad U = \tfrac{1}{2}\int_0^L \frac{F^2}{EA}dx = \tfrac{1}{2}\int_0^L EA\left(\frac{du}{dx}\right)^2 dx$$

For the beam member in bending, there is only an axial stress but it is distributed linearly on the cross-section in such a way that there is no resultant axial force. Let $M$ be the resultant moment; then $\sigma = -My/I = E\epsilon$ and

$$\text{bending:} \qquad U = \frac{1}{2}\int_0^L \frac{M^2}{EI}\,dx = \frac{1}{2}\int_0^L EI\left(\frac{d^2v}{dx^2}\right)^2 dx$$

The shear forces in a beam can also do some work. Let the shear stress be assumed to be uniformly distributed on the cross-section, and the resultant shear force be $V$, then $\tau = V/A = G\gamma$ and

$$\text{shear:} \qquad U = \frac{1}{2}\int_0^L \frac{V^2}{GA}\,dx = \frac{1}{2}\int_0^L GA\left(\frac{dv}{dx}\right)^2 dx$$

For a circular shaft in torsion, the shear stress is linearly distributed on the radius giving $\tau = Tr/J = G\gamma$ and

$$\text{torsion:} \qquad U = \frac{1}{2}\int_0^L \frac{T^2}{GJ}\,dx = \frac{1}{2}\int_0^L GJ\left(\frac{d\phi}{dx}\right)^2 dx$$

where $\phi$ is the twist per unit length, $T$ is the resultant torque, $G$ the shear modulus, and $J$ is the polar moment of inertia.

There are, of course, other types of structures, and an energy expression can be set up for these also. However, they will all have a similar form. For instance, for the cases considered above, there are the resultant loads

$$F, \quad M, \quad V, \quad T$$

the corresponding deformations

$$\frac{du}{dx}, \quad \frac{d^2v}{dx^2}, \quad \frac{dv}{dx}, \quad \frac{d\phi}{dx}$$

and the associated stiffnesses

$$EA, \quad EI, \quad GA, \quad GJ$$

In each case, the energy expression is of the form

$$energy = \frac{1}{2}\int_0^L \frac{[load]^2}{(stiffness)}\,dx = \frac{1}{2}\int_0^L (stiffness)[deformation]^2\,dx$$

Note that even the general expression, Equation(6.7), follows this form.

## Flexibility and Stiffness Matrices

The displacement at point $i$ is produced by load $P_i$ and loads acting at all the other points. Thus for linear elastic structures, we can assume

$$
\begin{aligned}
u_1 &= C_{11}P_1 + C_{12}P_2 + \cdots + C_{1N}P_N \\
u_2 &= C_{21}P_1 + C_{22}P_2 + \cdots + C_{2N}P_N \\
&\vdots \qquad \vdots \\
u_N &= C_{N1}P_1 + C_{N2}P_2 + \cdots + C_{NN}P_N
\end{aligned}
$$

These equations can be expressed as

$$
\{u\} = [\,C\,]\{P\} \qquad \text{or} \qquad u_i = \sum_{j=1}^{N} C_{ij}P_j \qquad (i = 1 \text{ to } N)
$$

The quantities $C_{ij}$ $(i, j = 1, N)$ are called *compliance coefficients, influence coefficients* or the *flexibility matrix* when viewed as an $[N \times N]$ matrix. Consider the special case with $P_1 = 1$, $P_2 = P_3 = \cdots = P_N = 0$; then the corresponding displacements are

$$
u_1 = C_{11}, \qquad u_2 = C_{21}, \qquad \cdots, \qquad u_N = C_{N1}
$$

Thus, $C_{11}$ is the displacement at Point 1 caused by application of a unit load (in the $P_1$ direction) at Point 1 (with all other loads vanishing); $C_{21}$ is the displacement at Point 2 caused by the application of a unit load at Point 1. In general, $C_{ij}$ can be interpreted as the displacement at Point $i$ due to a unit load applied at Point $j$.

If the flexibility matrix $[\,C\,]$ is not singular, i.e., $\det[C_{ij}] \neq 0$, then its inverse $[\,C\,]^{-1}$ exists. Let

$$
[\,K\,] \equiv [\,C\,]^{-1}
$$

then we obtain

$$
\{P\} = [\,K\,]\{u\} \qquad \text{or} \qquad P_i = \sum_{j=1}^{N} K_{ij}u_j \qquad (i = 1, N)
$$

The matrix $[\,K\,]$ is called the *stiffness matrix*. This is the same stiffness matrix that has already been encountered in the earlier chapters.

## Reciprocal Theorem and Symmetry

One of the basic assumptions about an elastic system is that the load sequence will not affect the final deformation; consequently, the total strain energy stored in the structure will remain the same for different load sequences. However, it should be pointed out that the work done by each individual force is affected by the load sequence, although the total work done by all forces is independent of the load sequence.

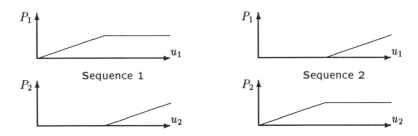

**Figure 6.3**: Work done by separate load sequences.

As an example, consider an elastic body subjected to two forces $P_1$ and $P_2$. For load sequence 1, the force $P_1$ is applied before applying $P_2$; then the work done by $P_1$ can be represented by the area under the load-deflection curve in Figure 6.3. During the application of $P_2$ the force $P_1$ is kept constant but travels an additional distance of $C_{12}P_2$ produced at Point 1 by $P_2$. The total work done by the two forces is

$$W_e^{(1)} = [\tfrac{1}{2}C_{11}P_1^2 + (C_{12}P_2)P_1] + [\tfrac{1}{2}C_{22}P_2^2]$$

For load sequence 2 on the other hand, $P_2$ is applied first and then followed by $P_1$. These are also shown in Figure 6.3. The work done by both forces during the entire loading is

$$W_e^{(2)} = [\tfrac{1}{2}C_{22}P_2^2 + (C_{21}P_1)P_2] + [\tfrac{1}{2}C_{11}P_1^2]$$

which is different from that given by load sequence 1. The total work done by $P_1$ and $P_2$ with load sequence 1 and with load sequence 2 should be the same, however. That is,

$$\tfrac{1}{2}C_{11}P_1^2 + C_{12}P_1P_2 + \tfrac{1}{2}C_{22}P_2^2 = \tfrac{1}{2}C_{11}P_1^2 + C_{21}P_1P_2 + \tfrac{1}{2}C_{22}P_2^2$$

This can be true only if

$$C_{12} = C_{21}$$

In general, it can be shown that for the total work done to be independent of the path sequence then

$$C_{ij} = C_{ji}$$

That is, the flexibility matrix $[\,C\,]$ is symmetric (and by inference, so also is the stiffness matrix). This property is referred to as *Maxwell's reciprocal theorem* and states that, in a linear elastic structure, the displacement at point $i$ in direction $n_1$ due to a unit force at point $j$ in direction $n_2$, is equal to the displacement at point $j$ in direction $n_2$ due to a unit force at point $i$ in direction $n_1$. This is a special case of some more general principles obeyed by linear elastic structures as described by the fundamental matrices $[\,C\,]$ and $[\,K\,]$. We look at two such principles next.

## Castigliano's Principles

Using the relations involving the stiffness and flexibility matrices allows the work and energy expressions to be written alternatively as

$$U = \tfrac{1}{2}\sum_i\left\{\sum_k K_{ik}u_k\right\}u_i = \tfrac{1}{2}\sum_i\sum_k u_i K_{ik}u_k = \tfrac{1}{2}\{u\}^T[\,K\,]\{u\}$$

$$U = \tfrac{1}{2}\sum_i P_i\left\{\sum_k C_{ik}P_k\right\} = \tfrac{1}{2}\sum_i\sum_k P_i C_{ik}P_k = \tfrac{1}{2}\{P\}^T[\,C\,]\{P\}$$

Thus, the work and energy are quadratic in the forces and displacements. These expressions also show that the strain energy can be considered either as a function of displacements, or of the loads. Use will be made of this in the following.

Consider the partial derivative of the strain energy with respect to the displacements. For example,

$$\begin{aligned}\frac{\partial U}{\partial u_1} &= \tfrac{1}{2}\sum_i\sum_k\left\{\frac{\partial u_i}{\partial u_1}K_{ik}u_k + u_i K_{ik}\frac{\partial u_k}{\partial u_1}\right\}\\ &= \tfrac{1}{2}\sum_k K_{1k}u_k + \tfrac{1}{2}\sum_i K_{i1}u_i = \sum_k K_{1k}u_k\\ &= P_1\end{aligned}$$

where use has been made of the symmetry property of $K_{ij}$. In general, it is possible to write

$$P_i = \frac{\partial U}{\partial u_i} \tag{6.8}$$

This is *Castigliano's first theorem* and is essentially an alternative statement of equilibrium. That is, it determines those forces that are consistent with a given compatible displacement field. In a similar manner, consider the derivatives of the strain energy with respect to the forces. That is,

$$\begin{aligned}\frac{\partial U}{\partial P_j} &= \tfrac{1}{2}\sum_i\sum_k\left\{\frac{\partial P_i}{\partial P_j}C_{ik}P_k + P_i C_{ik}\frac{\partial P_k}{\partial P_j}\right\}\\ &= \tfrac{1}{2}\sum_k C_{jk}P_k + \tfrac{1}{2}\sum_i C_{ij}P_i = \sum_k C_{jk}P_k\\ &= u_j\end{aligned}$$

where, again, use is made of the symmetry of $C_{ij}$. In general, it is possible to write

$$u_i = \frac{\partial U}{\partial P_i} \tag{6.9}$$

This relationship is usually called *Castigliano's second theorem* and is essentially an alternative statement of compatibility. That is, it determines those displacements that are consistent with a given equilibrated force system

Castigliano's principles are our first indication that elastic structures obey certain minimum principles. For example, consider a point on the structure where the applied load is zero. Now imagine that this point is displaced in a compatible fashion by the application of a force. As this is done, the strain energy of the structure changes (and, of course, we expend work in moving the point), but it is clear that as we traverse back through the original equilibrium point the force exerted is zero and we have

$$P_o = \frac{\partial U}{\partial u_o} = 0$$

That is, the strain energy achieves a minimum. We will develop this concept more fully in the next section.

It is also possible to show by further differentiation that

$$\frac{\partial^2 U}{\partial u_1 \partial u_1} = \sum_j K_{1j} \frac{\partial u_j}{\partial u_1} = K_{11}, \qquad \frac{\partial^2 U}{\partial u_1 \partial u_2} = \sum_j K_{1j} \frac{\partial u_j}{\partial u_2} = K_{12}$$

Thus, in general, the elements of the stiffness matrix are related to the second derivatives of the strain energy as

$$K_{ij} = \frac{\partial^2 U}{\partial u_i \partial u_j}$$

This indicates that if the strain energy can be established in terms of nodal displacements (degrees of freedom) associated with a compatible displacement field, then the stiffness matrix can be obtained simply by differentiation. Indeed, this will be demonstrated later in this chapter.

**Example 6.1:**   Determine the total strain energy for the cantilever beam that is loaded as shown in Figure 6.4. Then determine the vertical deflection and rotation at the tip of the beam.

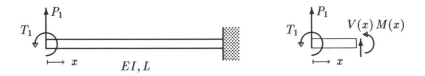

**Figure 6.4**: Cantilevered beam.

For beam structures, the total strain energy is

$$U = \frac{1}{2EI} \int_0^L [M(x)]^2 \, dx + \frac{1}{2GA} \int_0^L [V(x)]^2 \, dx$$

To evaluate this, we need to know the shear force and bending moment distributions. For the present case these distributions are easily established from the free body diagram as

$$V(x) = -P_1, \qquad M(x) = -T_1 + P_1 x$$

This force system is in equilibrium since we obtained it from use of a free body diagram. The strain energy is therefore

$$U = \frac{1}{2EI} \int_0^L [-T_1 + P_1 x]^2 \, dx + \frac{1}{2GA} \int_0^L [-P_1]^2 \, dx$$

giving, after integration,

$$U = \frac{1}{2EI}[T_1^2 L - T_1 P_1 L^2 + \tfrac{1}{3} P_1^2 L^3] + \frac{1}{2GA}[P_1^2 L]$$

Comparing the various contributions of $P_1$, it is seen that when the length of the beam is large then the shear contribution is small. More specifically, the shear effect can be neglected when

$$\frac{GA}{L} \gg \frac{EI}{L^3}$$

Indeed, this assumption is inherent in the Bernoulli-Euler beam model. In most of the work in this book, the shear effect is assumed to be small and usually neglected.

The deflections and rotations are obtained by use of Castigliano's second principle as

$$v_1 = \frac{\partial U}{\partial P_1}, \qquad \phi_1 = \frac{\partial U}{\partial T_1}$$

That is, the deflections are determined by differentiating the energy expression. Alternatively, it is usually more convenient to do the differentiating before the integration over the length. This will be illustrated here giving

$$v_1 = \frac{\partial U}{\partial P_1} = \frac{1}{2EI} \int_0^L 2[-T_1 + P_1 x] x \, dx = \frac{1}{EI}\left[ -\tfrac{1}{2} T_1 L^2 + \tfrac{1}{3} P_1 L^3 \right]$$

$$\phi_1 = \frac{\partial U}{\partial T_1} = \frac{1}{2EI} \int_0^L 2[+T_1 - P_1 x] \, dx = \frac{1}{EI}\left[ +T_1 L - \tfrac{1}{2} P_1 L^2 \right]$$

This can be rearranged as

$$\left\{ \begin{array}{c} v_1 \\ \phi_1 \end{array} \right\} = \frac{L}{6EI} \left[ \begin{array}{cc} 2L^2 & -3L \\ -3L & -6 \end{array} \right] \left\{ \begin{array}{c} P_1 \\ T_1 \end{array} \right\}$$

which is a compliance or flexibility relationship between the loads and displacements at the tip of the cantilever. Note that it is symmetric, and since the determinant is non-zero, it can be inverted to give a stiffness relation.

**Example 6.2:**  Determine the deflection at a point a distance $a$ from the tip of a cantilevered beam. Neglect the strain energy due to shear deformation.

A force is not present at the location where we desire to know the deflection, consequently we cannot use Castigliano's principle directly. A way around this is to introduce a fictitious force at the point of interest (as shown in Figure 6.5) and then use Castigliano's principle. Subsequently putting the force to zero will give the solution. This is sometimes referred to as the *dummy load method*.

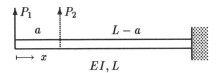

**Figure 6.5**: Cantilevered beam with dummy load $P_2$.

The moment distribution is piece-wise linear, hence the integral for the strain energy can be divided up as

$$U = \frac{1}{2EI} \int M^2 dx = \frac{1}{2EI} \int_0^a [P_1 x]^2 dx + \frac{1}{2EI} \int_a^L [P_1 x + P_2(x-a)]^2 dx$$

This gives, after expanding and integrating,

$$U = \frac{1}{6EI} \left[ P_1^2 L^3 + P_1 P_2 (L-a)^2 (2L+a) + P_2^2 (L-a)^3 \right]$$

Again, this force system is in equilibrium because we obtained the moment distribution from application of the free body diagram. Note that if $a = L$ then only $P_1$ contributes to the energy; forces at fixed supports do not contribute to the strain energy. On the other hand, if $a = 0$ then the energy is

$$U = \frac{L^3}{6EI} \left[ P_1^2 + 2P_1 P_2 + P_2^2 \right] = \frac{L^3}{6EI} [P_1 + P_2]^2$$

It is apparent from this that the total strain energy of a combined loading system is not the sum of the energy of the individual components.

We now use the general expression for the strain energy to obtain the deflection at 2 as

$$
\begin{aligned}
v_2 &= \frac{\partial U}{\partial P_2} = 0 + \frac{1}{2EI} \int_a^L 2[P_1 x + P_2(x-a)](x-a) dx \\
&= \frac{1}{EI} \left\{ P_1 \left( \frac{1}{3}L^3 - \frac{1}{2}aL^2 + \frac{1}{6}a^3 \right) + P_2 \frac{1}{3}(L-a)^3 \right\}
\end{aligned}
$$

But $P_2$ is actually zero, so the true deflection is

$$v_2 = \frac{1}{EI} P_1 \left( \frac{1}{3}L^3 - \frac{1}{2}aL^2 + \frac{1}{6}a^3 \right) = \frac{1}{6EI} P_1 (L-a)^2 (2L+a)$$

If $P_2$ is retained then by differentiation of $U$ with respect to both $P_1$ and $P_2$, we get

$$
\begin{Bmatrix} v_1 \\ v_2 \end{Bmatrix} = \frac{1}{6EI} \begin{bmatrix} 2L^3 & (L-a)^2(2L+a) \\ (L-a)^2(2L+a) & 2(L-a)^3 \end{bmatrix} \begin{Bmatrix} P_1 \\ P_2 \end{Bmatrix}
$$

Now the relationship between deflection and force is quite clear; we have introduced the concept of a flexibility at a point even if there is no concentrated force present.

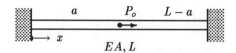

**Figure 6.6**: Clamped-clamped rod.

An important point to note is that if we are interested in the structural response of $N$ points, say, it is necessary to introduce $N \times N$ entries in the flexibility matrix.

**Example 6.3:**   Determine the load $P_o$ necessary to give an axial displacement of amount $u_o$ at the center of the clamped-clamped rod shown in Figure 6.6.

To use Castigliano's first principle, we must start with a compatible displacement field. From our developments in Chapter 2, we know that the displacement distribution is linear in regions where the applied loading is zero. We can therefore take the displacements in the rod to be

$$u(x) \;=\; \frac{x}{a} u_o \qquad\qquad 0 \le x \le a$$

$$u(x) \;=\; \left(\frac{L-x}{L-a}\right) u_o \qquad\qquad a \le x \le L$$

The displacement distribution is piece-wise linear, hence the integral for the strain energy can be divided up as

$$U = \tfrac{1}{2} \int EA(\frac{du}{dx})^2 dx = \tfrac{1}{2} \int_0^a EA(\frac{u_o}{a})^2 dx + \tfrac{1}{2} \int_a^L EA(\frac{-u_o}{L-a})^2 dx$$

This gives, after expanding and integrating,

$$U = EAu_o^2(\frac{L}{a(L-a)})$$

The force necessary to cause the deflection $u_o$ is now obtained from Castigliano's first theorem:

$$P_o = \frac{\partial U}{\partial u_o} = 2EAu_o(\frac{L}{a(L-a)})$$

This result is in agreement with what would be obtained using two rod elements as done in Chapter 2.

# 6.3   Virtual Work

This section is devoted to the concepts of virtual work and energy (as distinguished from the real work and energy of the two previous sections). This concept is quite versatile because the displacements involved may be due to influences other than the applied loads. This subtle idea provides additional techniques for establishing equilibrium and compatibility conditions; and will lead to some very powerful minimum theorems.

## Virtual Work for a Particle

Consider a particle in a state of equilibrium under the action of a set of forces as shown in Figure 6.7. The particle is attached to a linear spring with a spring constant $k$, and for generality, has acting on it a body force $F^b$, as well. The particle has achieved a static equilibrium position under the spring extension $u$. The equilibrium conditions are

$$\sum F_{ix} + F_x^b - ku = 0, \qquad \sum F_{iy} + F_y^b = 0, \qquad \sum F_{iz} + F_z^b = 0$$

where $F_{ix}$ is the component of the $i^{th}$ force in the $x-$direction and similarly for the $y-$ and $z-$directions.

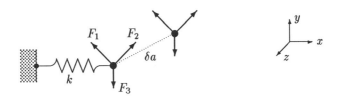

**Figure 6.7**: Virtual displacement.

Now imagine that the particle is displaced a small amount $\delta a$ and the forces move with it, but with no change in the magnitude or the direction of any of the forces. It is important to reiterate that we are not speaking of the actual displacement of the particle caused by the action of the applied forces; we are speaking of an imagined displacement in which the forces are imagined to behave as stated. For this reason, the displacement $\delta a$ is referred to as a *virtual displacement* to distinguish it from the actual displacement. Throughout this chapter, the symbol $\delta$ before any quantity will indicate that it is a virtual or imaginary quantity.

The work done on the particle during the virtual displacement is the sum of all the components of force in the direction of $\delta a$, times $\delta a$. This is called the *virtual work*. Let the virtual displacement have components $\delta u, \delta v, \delta w$ in the coordinate directions; we have for the virtual work

$$\delta W = \left(\sum F_{ix} + F_x^b - ku\right)\delta u + \left(\sum F_{iy} + F_y^b\right)\delta v + \left(\sum F_{iz} + F_y^b\right)\delta w$$

By virtue of the fact that the particle is already in equilibrium (that is, the terms in parenthesis are zero) we conclude that the virtual work is zero,

$$\delta W = 0 \qquad\qquad (6.10)$$

Thus, if a particle is in equilibrium under the action of a set of forces, the total virtual work done by the forces during a virtual displacement is zero. Note that the converse statement is not true; just because the virtual work is zero we cannot conclude that the body is in equilibrium. (This is easily demonstrated by a particle acted on by a horizontal force but given a virtual displacement in the vertical direction.) It is correct to say that if the virtual work is zero *for every possible* virtual displacement then the body is in equilibrium, but this would be an impractical approach to testing for equilibrium. We can simplify matters considerably by introducing the idea of *independent* virtual displacements; an arbitrary displacement of a point can always be written as the sum of three independent displacements. This leads to the much more useful statement that if the virtual work is zero *for every independent* virtual displacement then the body is in equilibrium. With this seemingly simple result we have managed to interpret Equation(6.10) as a condition that defines equilibrium rather than a result of equilibrium. Furthermore, some unknown forces that would appear in a free body diagram (reactions, for example) will not appear in the virtual work statement if they do no work.

This simple example identifies some important features for the general statement of virtual work. First note that there are three states of the particle:

(A) an original state where there are no loads and the spring is unstrained,

(B) a final equilibrium state after the loads are applied and the spring has been stretched an amount $u$, and

(C) an imaginary state where a virtual displacement $\delta a$ is present.

We emphasize that the true displacement of the particle is the difference in displacement between States A and B, and that the virtual displacement is the difference between States B and C. Equation(6.10) can be viewed as either a consequence of equilibrium or a definition of equilibrium. In the former case, the virtual displacements $\delta u$, etc., need not actually be 'displacements' — they could be velocities or pressure or just a purely mathematical function. In the latter case of defining equilibrium, we must restrict the statement to independent virtual displacements — essentially we refer to the generalized displacements or the degrees of freedom of the body. A final point to note is that the condition that the forces do not change during the virtual displacement also applies to the internal spring force $ku$, even though it is clearly dependent on the real displacement.

## Virtual Work for a General Deformable Body

Consider a general structure. It may be acted on by body forces that have units of force per unit volume, and surface forces with units of force per unit area. Distributed line loads and concentrated loads can be considered as special cases of surface loads. The structure is geometrically constrained so that no rigid body motions can occur. By and large, we wish to consider one of the general structures analyzed in Chapter 4.

As steps in developing the principle of virtual work, we first identify three states

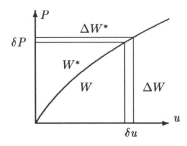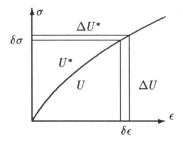

**Figure 6.8**: Work and strain energy.

of the body. In State A, constraints are in place, but the body is unloaded. To keep things simple, we consider the case where the constraints all impose zero displacements and so the body is undeformed and unstressed. State B is the true state of loading and deformation of the body; that is, the true loads are applied to the body defined in State A and it is allowed to deform to its true state of deformation and achieve a state of static equilibrium. Finally, State C is the virtual state of the body obtained by assigning virtual displacements. It is in this state that the principle of virtual work provides us with a suitable statement of equilibrium.

We take as given that the geometric constraints imposed in State A can never be violated by the true displacements in State B or the virtual displacements in State C. A deformable body has an infinite number of material points and hence has an infinite number of degrees of freedom. We therefore must assume that all real and virtual displacements are continuous (differentiable) functions of the coordinates $x, y$, and $z$. Functions that have these properties are called *kinematically admissible functions*.

In the following development we will concentrate on a single component of force $P$ and a single component of displacement $u$; later we will generalize the results to a system of arbitrary size.

With reference to Figure 6.8, the curved path can be viewed as the sequence of possible equilibrium states as the load is increased. (Although we deal only with linear elastic systems, it is more insightful to consider the non-linear elastic case at this stage.) The states $A$ and $B$ can therefore be any two points along this curve. Suppose $B$ is relatively close to $A$ then a Taylor series expansion about $A$ gives

$$\Delta W \approx \delta W + \tfrac{1}{2}\delta^2 W + \cdots = P\delta u + \tfrac{1}{2}\delta P\delta u + \cdots$$

where we used the trapezoidal rule to compute the area $\Delta W$. A similar expansion for the strain energy gives

$$\Delta U \approx \delta U + \tfrac{1}{2}\delta^2 U + \cdots = \int_V [\sigma\delta\epsilon + \tfrac{1}{2}\delta\sigma\delta\epsilon + \cdots]\,d\forall$$

The virtual strains are related to the virtual displacements by

$$\delta\epsilon_{xx} = \delta\frac{\partial u}{\partial x} = \frac{\partial}{\partial x}(\delta u)$$

and so on. Consider only the contributions of the first terms in the work and energy, then

$$\Delta W \approx \delta W = P\delta u, \qquad \Delta U \approx \delta U = \int_V [\sigma \delta \epsilon]\, dV$$

This is equivalent to saying that

$$\delta P = 0, \qquad \delta \epsilon = 0$$

during the deformation. This is not physically realizable since the load-deformation curves show a monotonic increase. For this reason, the small displacements $\delta u$ are called *virtual displacements*, and the corresponding small strains $\delta \epsilon$ *virtual strains*.

For a deformable body the principle of virtual work may be stated as follows;

*A deformable solid body is in equilibrium if and only if the total virtual work is zero for every independent kinematically admissible virtual displacement.*

The mathematical statement of this principle is

$$\delta W = \delta W_e - \delta U = 0 \qquad \text{or} \qquad P\delta u - \int_V [\sigma \delta \epsilon]\, dV = 0$$

This says that for every virtual displacement, a necessary and sufficient condition for equilibrium of an elastic body is that the virtual work done by the external loads be identically equal to the virtual elastic strain energy stored. The principle of virtual work is as valid for establishing equilibrium as Newton's laws.

It is important to realize the difference between this principle and that of the conservation of energy since both resemble each other closely. In essence, the latter is concerned with total energies whereas the former is concerned with small changes or what are called *variations*. As a result, virtual strain energy and work are calculated on the assumption that the forces remain unchanged during the variation of the state of strain.

The virtual work of a deformable solid body is divided into two parts as follows:

$$\delta W = \delta W_e + \delta W_i = (\delta W^s + \delta W^b) + \delta W_i = (\delta W^s + \delta W^b) - \delta U$$

where $\delta W_e$ is the virtual work of external forces and can be broken further into a contribution $\delta W^s$ of the surface forces and a contribution $\delta W^b$ of the body forces. These terms are given explicitly as

$$\delta W^s = \int_S (f_x^s \delta u + f_y^s \delta v + f_z^s \delta w)dS = \int_S \{\delta \mathcal{U}\}^T \{f^s\}dS \qquad (6.11)$$

$$\delta W^b = \int_V (f_x^b \delta u + f_y^b \delta v + f_z^b \delta w)dV = \int_V \{\delta \mathcal{U}\}^T \{f^b\}dV \qquad (6.12)$$

where the integrations are carried out over the surface and volume, respectively, of the body, and the vector $\{\mathcal{U}\}$ contains the three displacement components $u$, $v$, $w$.

Concentrated and distributed line forces on the surface can be taken as special forms of these equations.

The other term $\delta W_i$ is the virtual work of the internal forces, in this case the work due to the elastic straining of the body. That is, the internal virtual work is given by

$$\delta U = \int_V [\sigma_{xx}\delta\epsilon_{xx} + \sigma_{yy}\delta\epsilon_{yy} + \cdots] dV = \int_V \{\sigma\}^T \{\delta\epsilon\} dV$$

In summary, the statement for virtual work can be put in matrix form as

$$\begin{aligned}
\delta W &= \delta W^s + \delta W^b - \delta U \\
&= \int_S \{\delta U\}^T \{f^s\} dS + \int_V \{\delta U\}^T \{f^b\} dV - \int_V \{\delta\epsilon\}^T \{\sigma\} dV = 0 \qquad (6.13)
\end{aligned}$$

## Complementary Work and Energy

Looking at Figure 6.8, we see that corresponding to the small areas $\Delta W$ and $\Delta U$, there are areas $\Delta W^*$ and $\Delta U^*$. When these areas are integrated over the full path we see that $(W, W^*)$ and $(U, U^*)$ are complementary on the rectangles $Pu$ and $\sigma\epsilon$, respectively. Although these complementary areas are equal for linear elastic structures, it is still useful to distinguish the two.

Working through similar developments, we get (for the one-dimensional case)

$$\delta W^* = \delta W_e^* - \delta U^* = 0 \qquad \text{or} \qquad u\delta P - \int_V [\epsilon\delta\sigma] dV = 0$$

where we call $\delta W^*$ complementary work and $\delta U^*$ complementary strain energy corresponding to the virtual force $\delta P$ and virtual stress $\delta\sigma$, respectively. Virtual forces are referred to as *statically admissible forces* because they must satisfy the conditions of static equilibrium. We can now state the *principle of complementary virtual work*:

*For an elastic structure in equilibrium, a necessary and sufficient condition for compatibility of the deformation is that the complementary virtual work be zero for any virtual force.*

It is worth recalling that the principle of virtual work establishes the equilibrium conditions when the deformation is already compatible, the complementary virtual work principle establishes the conditions for compatibility when the loads are already equilibrated. This duality can also be seen in Castigliano's principles. To show this, apply the virtual work principles to a system with a collection of applied loads

$$\delta W = \sum P_i \delta u_i - \delta U = 0, \qquad \delta W^* = \sum u_i \delta P_i - \delta U^* = 0$$

Consider the strain energy to be dependent only on the deformation as $U = U(u)$, and the complementary strain energy to dependent only on the loads $U^* = U^*(P)$. Then the variations are given by

$$\delta U = \sum \frac{\partial U}{\partial u_i} \delta u_i, \qquad \delta U^* = \sum \frac{\partial U^*}{\partial P_i} \delta P_i$$

If we now substitute these into the virtual work relations, and realize that the variations $\delta u_i$ and $\delta P_i$ are arbitrary, then we conclude that

$$P_i = \frac{\partial U}{\partial u_i}, \qquad u_i = \frac{\partial U^*}{\partial P_i}$$

For linear elastic structures we have that $U = U^*$; the only difference is that we write the energy as a function of displacement or load.

The matrix methods we have developed in the previous chapters are called 'displacement formulated' because we always take the displacements as the fundamental set of unknowns. In so doing we always enforce compatibility and use the stiffness relation as the statement of equilibrium. Even when the displacements are approximate (as in beam buckling, for example) compatibility is assured. Hence, in the later sections we will not make use of complementary work since it assumes equilibrated load systems.

**Example 6.4:** Use the principle of virtual work to determine the forces in members $AB$ and $AC$ of the truss shown in Figure 6.9.

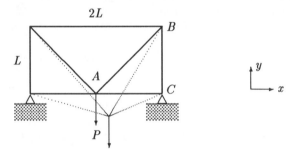

**Figure 6.9**: Truss with virtual displacements $\delta u$ and $\delta v$.

Consider a virtual displacement $\delta v$ at the point of application of the load. Since this is a virtual displacement, we will specify that all other nodes remain in the same position. Consequently, the new lengths of members $AB$ and $AC$ are (assuming small deflections)

$$AB \approx \sqrt{2}L + \delta v/\sqrt{2}, \qquad AC \approx L$$

The virtual strains in the two members are

$$\epsilon_{AB} = \frac{\delta v}{2L}, \qquad \epsilon_{AC} = 0$$

The structure is symmetric, so the virtual work becomes

$$P\delta v = \int_V \sigma \delta \epsilon \, dV = \int_L F \delta \epsilon \, dx = 2F_{AB} \frac{\delta v}{2L} \sqrt{2}L \qquad \text{or} \qquad F_{AB} = \sqrt{2}P$$

The virtual displacements need not resemble what would actually occur with the actual applied loading. So let us take a virtual displacement $\delta u$ at the load point.

Consequently, the new lengths of members $AB$ and $AC$ are (again assuming small deflections)

$$AB \approx \sqrt{2}L - \delta u/\sqrt{2}, \qquad AC \approx L - \delta u$$

The virtual strains in the two members are

$$\epsilon_{AB} = -\frac{\delta u}{2L}, \qquad \epsilon_{AC} = -\frac{\delta u}{L}$$

The virtual work done by $P$ in the virtual displacement $\delta u$ is zero, hence we have for all four members

$$0 = 2F_{AB}\frac{\delta u}{2L}\sqrt{2}L + 2F_{AC}\frac{\delta u}{L}L \qquad \text{or} \qquad F_{AC} = F_{AB}/\sqrt{2}$$

We have already obtained a value for $F_{AB}$, hence we conclude that $F_{AC} = P$.

# 6.4 Stationary Potential Energy

This section introduces a concept that is of fundamental significance in the mechanics of structures. To put this in perspective, recall from the early chapters that we presented equilibrium and compatibility as the twin fundamental mechanics principles. In the previous section, however, we saw that these concepts could be replaced by the equivalent ones of virtual work and virtual complementary work

$$\delta W = \delta W_e - \delta U = 0, \qquad \delta W^* = \delta W_e^* - \delta U^* = 0$$

The new concept introduced here is to see these relations (and consequently, equilibrium and compatibility) as the achievement of a minimization of certain quantities in the structure. While this very powerful idea can be used to recover all of the previous results, it turns out that its most fruitful application is in the approximate analysis of structures.

## Virtual Potential Energy

A system is *conservative* if the work done in moving the system around a closed path is zero. That is, we can write the work in the form of an exact differential as

$$\Delta W = \int_A^B F\,du = \int_A^B -dV = -(V(B) - V(A)) \qquad \text{or} \qquad V(B) = V(A) - \Delta W$$

$V$ is also called a potential function. The negative sign in the definition of $V$ is arbitrary but choosing it so gives us the interpretation of $V$ as the capacity (or potential) to do work. For example, consider a particle resting on a table (State A); it is raised against gravity to a height $h$ (State B). The work done is $\Delta W = -mgh$. The potential at B is

$$V(B) = V(A) - \Delta W = V(A) - (-mgh) = V(A) + mgh$$

If State A is considered to be a zero state, then we see that we can recover work of amount $mgh$.

Assume that we can obtain the external forces from a potential function; specifically, let it be a function of the displacements such that

$$P_i = -\frac{\partial V}{\partial u_i} \qquad \text{or} \qquad \{P\} = -\{\frac{\partial V}{\partial u}\}$$

We will take as the potential for the external forces $V = -\{P\}^T\{u\}$. The external work term now becomes

$$\delta W_e = \{P\}^T\{\delta u\} = -\{\frac{\partial V}{\partial u}\}^T\{\delta u\} = -\delta V$$

The principle of virtual work can be rewritten as

$$\delta U + \delta V = 0 \qquad \text{or} \qquad \delta\Pi \equiv \delta[U + V] = 0 \qquad (6.14)$$

The term inside the brackets is called the *total potential energy*. This relation is called the *principle of stationary potential energy*. We may now restate the virtual work theorem as:

*For an elastic body to be in equilibrium, the first order variation in the total potential energy must vanish for every independent admissible virtual displacement.*

Another way of stating this is that among all the displacement states of a conservative system that satisfy compatibility and the boundary constraints, those that also satisfy equilibrium make the potential energy stationary. In comparison to the conservation of energy theorem, this is much richer, because instead of one equation it leads to as many equations are there are degrees of freedom.

To elaborate on this idea, we are saying that a structures problem can be represented by a function $\Pi$ of the variables $a_i$, (the $a_i$ are called *state variables* or degrees of freedom). The conditions for equilibrium and compatibility of the structure are recovered by determining the set of values $a_i$ for which the function $\Pi(a_1, \ldots, a_N)$ has a stationary value. That is,

$$\delta\Pi = 0 \qquad \text{or} \qquad \frac{\partial\Pi}{\partial a_i} = 0 \qquad i = 1, \ldots, n$$

It is important to realize that the single equation $\delta\Pi = 0$ is actually a system of $N$ equation if there are $N$ degrees of freedom in the system. Note that the second derivatives of $\Pi$ with respect to the $a_i$ decide whether the function corresponds to a maximum, minimum, or saddle point. In the analysis of a linear structural system, the displacements are the state variables, and $\Pi$ is the total potential of the system.

**Example 6.5:**  Consider a simple spring of stiffness $K$ subjected to an applied load $P$. Establish the total potential energy of the system and then derive the conditions for equilibrium.

Let $u$ be the resulting displacement at the point of application of the load. We have for the strain energy and potential of the force, respectively,

$$U = \tfrac{1}{2}Ku^2, \qquad V = -Pu$$

The total potential energy of the system is, therefore,

$$\Pi = \tfrac{1}{2}Ku^2 - Pu$$

These terms are shown plotted in Figure 6.10 for different values of displacement $u$. It is apparent that $\Pi$ can achieve a minimum; the principle indicates that this minimum occurs at the equilibrium position.

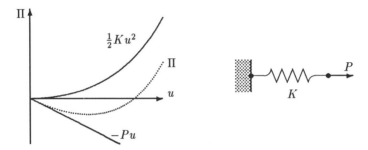

**Figure 6.10**: Potential energy for a single degree of freedom system.

Thus if we identify $u$ as the degree of freedom, then a stationary potential energy requires that

$$\frac{\partial \Pi}{\partial u} = 0 \qquad \Rightarrow \qquad Ku - P = 0$$

We recognize this as indeed the equilibrium equation for a spring

$$Ku = P$$

Note that the second derivative of the potential gives

$$\frac{\partial^2 \Pi}{\partial u^2} = K$$

This is positive thus confirming that the stationary value is actually a minimum in this case.

The significant point of this example is that the minimizing of a potential can be equivalent to deriving the system equations of equilibrium.

**Example 6.6:**   Use the principle of stationary potential energy to determine the equilibrium equations for a system of springs and applied loads.

The strain energy and potential energy of the applied forces are, respectively,

$$U = \tfrac{1}{2}\{u\}^T[\,K\,]\{u\}, \qquad V = -\{u\}^T\{P\}$$

The total potential energy is

$$\Pi = \tfrac{1}{2}\{u\}^T[\,K\,]\{u\} - \{u\}^T\{P\} = \tfrac{1}{2}\sum_m\sum_n K_{mn}u_m u_n - \sum_n u_n P_n$$

Invoking the stationarity of $\Pi$ with respect to the degrees of freedom $u_n$ requires that

$$\frac{\partial \Pi}{\partial u_n} = 0 \qquad \Rightarrow \qquad \sum_m K_{mn}u_m - P_n = 0$$

This gives a system of equations which, on rearranging, are

$$[\,K\,]\{u\} = \{P\}$$

We recognize this as the structural stiffness relation.

An interesting secondary point of this example is that we can write the total strain energy as

$$U = \tfrac{1}{2}\{u\}^T\Big(\sum_m[K^{(m)}]\Big)\{u\} = \tfrac{1}{2}\sum_m\{u\}^T[K^{(m)}]\{u\} = \tfrac{1}{2}\sum_m U^{(m)}$$

where $[K^{(m)}]$ is the augmented element stiffness. This relation shows that an alternative view of the assemblage process is one of summing the strain energies of the individual components. In the previous chapters we achieved assemblage by imposing equilibrium at the connections of the components, here equilibrium is taken care of by our stationarity principle. We reiterate that this is possible only because we have imposed compatibility between the components (by use of a set of common nodal degrees of freedom $\{u\}$).

## Boundary Conditions

When we apply the principle of stationary potential energy, we need to identify two classes of boundary conditions, called *essential* and *natural* boundary conditions. The essential boundary conditions are also called *geometric* boundary conditions because they correspond to prescribed displacements and rotations. The natural boundary conditions are sometimes called the *force* boundary conditions because they correspond to prescribed boundary forces and moments.

To see the different roles played by the boundary conditions, we will reconsider the simple problem of a rod with body forces and an applied load. We will show that by invoking the stationarity condition on $\Pi$, that the governing differential equation of the problem and the corresponding natural boundary condition can be derived.

**Example 6.7:** The total potential energy and essential boundary conditions for the rod shown in Figure 6.11 are

$$\Pi = \int_0^L \tfrac{1}{2}EA\left(\frac{du}{dx}\right)^2 dx - \int_0^L uf^b dx - u_L P\,, \qquad u(0) = u_0 = 0$$

**Figure 6.11**: Rod with axial load.

where $f^b$ is the body force per unit length of the rod. Note that the highest derivative in the potential is one.

The stationarity condition gives

$$\frac{\partial \Pi}{\partial a} = 0 \quad \Rightarrow \quad \int_0^L EA \frac{du}{dx} \frac{\partial}{\partial a}\left(\frac{du}{dx}\right) dx - \int_0^L \frac{\partial u}{\partial a} f^b dx - \frac{\partial u_L}{\partial a} P = 0$$

Interchanging the partial differentiations, i.e.,

$$\frac{\partial}{\partial a}\left(\frac{du}{dx}\right) = \frac{d}{dx}\left(\frac{\partial u}{\partial a}\right)$$

assuming that $EA$ is constant, and using integration by parts, yields

$$-\int_0^L \left[EA\frac{d^2u}{dx^2} + f_b\right]\frac{\partial u}{\partial a} dx + \left[EA\frac{du}{dx}|_{x=L} - P\right]\frac{\partial u_L}{\partial a} - \left[EA\frac{du}{dx}|_{x=0}\right]\frac{\partial u_0}{\partial a} = 0$$

To obtain the governing differential equation and natural boundary conditions, we use the argument that $\partial u/\partial a$ is arbitrary at all points interior to the region, and therefore the integrand must be zero, giving

$$EA\frac{\partial^2 u}{\partial x^2} + f^b = 0$$

It is also required that the other terms be zero, separately. Thus we must have

$$\left[EA\frac{\partial u}{\partial x}|_{x=L} - P\right] = 0 \quad \text{or} \quad \frac{\partial u_L}{\partial a} = 0$$

$$-\left[EA\frac{\partial u}{\partial x}|_{x=0}\right] = 0 \quad \text{or} \quad \frac{\partial u_0}{\partial a} = 0$$

When an essential boundary condition is specified then its variation (with respect to the state variable $a$) is zero, hence we see that this is the case in the first boundary condition but not in the second, hence the boundary condition at $x = L$ is

$$EA\frac{\partial u}{\partial x}|_{x=L} = P$$

This is a natural boundary condition. We notice that this relation could also be derived by considering a small free body diagram of the vicinity of the applied load.

Thus the natural boundary conditions are connected with the equilibrium of the body.

**Example 6.8:**  Use the stationarity of the total potential energy functional to recover the differential equation governing the static buckling of the beam shown in Figure 6.12.

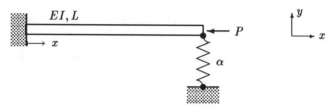

**Figure 6.12**: Beam with buckling load.

There are three contributions to the strain energy, giving the total potential energy as

$$\Pi = \int_0^L \tfrac{1}{2} EI \left(\frac{d^2 v}{dx^2}\right)^2 dx + \bar{F}_o \int_0^L \tfrac{1}{2}\left(\frac{dv}{dx}\right)^2 dx + \tfrac{1}{2}\alpha v_L^2$$

Note that the axial force $\bar{F}_o = -P$ does no work, but instead appears in the form of a strain energy. The essential boundary conditions at $x = 0$ are

$$v = 0, \qquad \frac{dv}{dx} = 0$$

Again, we will invoke the stationarity of $\Pi$ to derive the governing differential equations and also the natural boundary conditions at $x = L$.

The stationarity condition yields

$$\frac{\partial \Pi}{\partial a} = 0 \quad \Longrightarrow \quad \int_0^L EIv''\frac{\partial v''}{\partial a}dx - P\int_0^L v'\frac{\partial v'}{\partial a}dx + \alpha v_L \frac{\partial v_L}{\partial a} = 0$$

where we use the notation $v' \equiv dv/dx$ and so on. Using the fact that

$$\frac{\partial v''}{\partial a} = \frac{d}{dx}\left(\frac{\partial v'}{\partial a}\right)$$

and $EI$ is constant, then integrating by parts, we obtain

$$\int_0^L EIv''\frac{\partial v''}{\partial a}dx = EIv''\frac{\partial v'}{\partial a}\Big|_0^L - EI\int_0^L v'''\frac{\partial v'}{\partial a}dx$$

If we continue to use integration by parts, we eventually obtain

$$\int_0^L [EIv'''' + Pv'']\frac{\partial v}{\partial a}dx + [EIv''\frac{\partial v'}{\partial a}]_L - [EIv''\frac{\partial v'}{\partial a}]_0$$

$$-[(EIv'''' + Pv'' + \alpha v_L)\frac{\partial v}{\partial a}]_L + [(EIv''' + Pv')\frac{\partial v}{\partial a}]_0 \;=\; 0$$

Since the variations on $v$ and $v'$ must satisfy the essential boundary conditions, we have that at $x = 0$

$$\frac{\partial v}{\partial a} = 0, \qquad \frac{\partial v'}{\partial a} = 0$$

It follows that the third and fifth terms are therefore zero. The variations on $v$ and $v'$ are arbitrary at all other points (including $x = L$), hence to satisfy $\delta \Pi = 0$ we conclude that the following equations must be satisfied:

$$EIv'''' + Pv'' = 0, \qquad EIv''|_{x=L} = 0, \qquad [EIv''' + Pv' - \alpha v]|_{x=L} = 0$$

These give the governing differential equation and the two natural boundary conditions. We should note that these boundary conditions correspond to moment and shear equilibrium at $x = L$, respectively.

## Weak Form of Problems

A number of observations can now be made about the use of stationary potential energy. First, if the functional contains derivatives up to order $m$, then there must be continuity of displacement derivative up to $m - 1$, and the order of the highest derivative that is present in the governing differential equation is then $2m$. For example, in the beam problem above, $m = 2$ because the highest derivative in the functional is of order 2, and hence there must be continuity of $v$ and $dv/dx$. The reason for obtaining a derivative of order $2m = 4$ in the governing differential equation is that integration by parts is employed $m = 2$ times. A second observation is that through the stationarity condition we obtained the governing differential equations *and* the proper boundary conditions. Hence, the effect of the natural boundary conditions are implicitly contained in the expression for the potential $\Pi$. (Note that the essential boundary conditions must be stated separately.)

The examples of this section show that we have two alternative ways of stating our problem. The first is by a set of differential equations plus a set of associated boundary conditions; this is known as the *strong* form or *classical* form of the problem. The alternate way is by extremizing a functional; this is known as the *weak* form or *variational* form of the problem. They are both equivalent (as shown by the examples) but lend themselves to approximation in different ways. What we wish to pursue in the following sections is approximation arising from the weak form; specifically, we will approximate the functional itself and use the variational principle to obviate consideration of the natural boundary conditions.

# 6.5   Ritz Approximate Analysis

The value of the principle of stationary potential energy is in solving complicated problems where exact answers (and even precise governing differential equations) are difficult or impossible to obtain. For example, beams with discontinuous distributed loadings, multiple supports, or variable cross-sections are difficult to solve by the differential equation method. We obtain very effective approximate solutions by approximating the functional $\Pi$ itself. The reason for this effectiveness lies in the way some boundary conditions (namely, the natural boundary conditions) can be approximated when using this approach. Our formal way of doing this will be through the Ritz approximation.

In general, a continuously distributed deformable body consists of an infinity of material points and therefore has infinitely many degrees of freedom. The Ritz method is an approximate procedure by which continuous systems are reduced to systems with finite degrees of freedom. The fundamental characteristic of the method is that we operate on the functional corresponding to the problem, and that we choose the functional to be the total potential energy $\Pi$. The basic step is to assume a solution for the displacements in the form

$$u(x, y, z) = \sum_{i=1}^{N} a_i g_i(x, y, z)$$

where the $g_i$ are linearly independent *trial functions*, and the $a_i$ are multipliers to be determined in the solution. The trial functions satisfy the essential (geometric) boundary conditions but not necessarily the natural boundary conditions. Hence, in the Ritz solution, there is error in the satisfaction of the differential equations of equilibrium and the natural boundary conditions but we attempt to minimize this error. By substituting the trial functions into $\Pi$ we can generate $N$ simultaneous equations for the parameters $a_i$ using the stationarity condition of $\Pi$,

$$\delta \Pi = 0 \qquad \text{or} \qquad \frac{\partial \Pi}{\partial a_i} = 0 \qquad i = 1, 2, \ldots, N$$

An important consideration is the selection of the trial functions $g_i$. Selecting efficient admissible functions may not be easy; fortunately, many problems closely resemble other problems that have been solved before, and the literature is full of examples that can serve as a guide. It must also be kept in mind that these functions need only satisfy the essential boundary conditions and not (necessarily) the natural boundary conditions. For practical analyses, this is a significant point and largely accounts for the effectiveness of the displacement-based finite element analysis procedure.

Therefore, for convergence in a Ritz analysis, the trial function need only satisfy the essential boundary conditions. Actually, assuming a given number of trial functions, it can be expected that in most cases the solution will be more accurate if these functions also satisfy the natural boundary conditions. However, it can be very difficult to find such trial functions and it is generally more effective to use a larger number of functions that only satisfy the essential boundary conditions. We demonstrate the use of the Ritz method in the following examples.

**Example 6.9:**  Use the Ritz method to formulate the equations from which can be obtained an approximate buckling load for the beam of Figure 6.12.

The functional and associated boundary conditions governing the problem were already given as

$$\Pi = \frac{1}{2} \int_0^L EI \left(\frac{d^2v}{dx^2}\right)^2 dx - \frac{1}{2} \int_0^L P \left(\frac{dv}{dx}\right)^2 dx + \frac{1}{2}\alpha v_L^2, \qquad v(0) = 0, \qquad \frac{dv(0)}{dx} = 0$$

Assume that the Ritz functions are given by

$$v(x) = a_0 + a_1 x + a_2 x^2 + a_3 x^3$$

In order for these to satisfy the essential boundary conditions, we must have that $a_0 = 0$ and $a_1 = 0$. It remains now to determine the other two coefficients. Substituting for $v(x)$ into the potential function gives

$$\Pi = \frac{1}{2} \int_0^L EI \left(2a_2 + 6a_3 x\right)^2 dx - \frac{1}{2} \int_0^L P \left(2a_2 x + 3a_3 x^2\right)^2 dx + \frac{1}{2}\alpha(a_2 L^2 + a_3 L^3)^2$$

Invoking the stationarity of $\Pi$ with respect to $a_2$ and $a_3$ gives, respectively,

$$\frac{\partial \Pi}{\partial a_2} = \int_0^L EI \left(2a_2 + 6a_3 x\right) 2 dx - \int_0^L P \left(2a_2 x + 3a_3 x^2\right) 2x\, dx + \alpha(a_2 L^2 + a_3 L^3)L^2$$

$$\frac{\partial \Pi}{\partial a_3} = \int_0^L EI \left(2a_2 + 6a_3 x\right) 6 dx - \int_0^L P \left(2a_2 x + 3a_3 x^2\right) 3x^2 dx + \alpha(a_2 L^2 + a_3 L^3)L^3$$

From these, we obtain the system of equations

$$\left[ EI \begin{bmatrix} 4L & 6L^2 \\ 6L^2 & 12L^3 \end{bmatrix} + \alpha L^4 \begin{bmatrix} 1 & L \\ L & L^2 \end{bmatrix} - \frac{PL^3}{30} \begin{bmatrix} 40 & 45L \\ 45L & 54L^2 \end{bmatrix} \right] \begin{Bmatrix} a_2 \\ a_3 \end{Bmatrix} = \begin{Bmatrix} 0 \\ 0 \end{Bmatrix}$$

Not surprisingly, we end up with an eigenvalue problem. The solution of this eigenproblem gives two values of $P$ for which $v(x)$ is nonzero. The smaller value of $P$ represents an approximation to the lowest buckling load of the structure.

Suppose we take as the displacement shape the following reduced function

$$v(x) = a_0 + a_1 x + a_2 x^2$$

Again, in order for these functions to satisfy the essential boundary conditions, we must have that $a_0 = 0$ and $a_1 = 0$ leaving only $a_2$ to be determined. Following the above developments we get

$$\left[ EI[4L] + \alpha L^4 [1] - \frac{PL^3}{30}[40] \right] a_2 = 0$$

This gives only one buckling load with a value of

$$P_{cr} = \frac{EI}{L^2} \frac{3}{4}(4 + \alpha L^2)$$

Notice that the reduced eigenvalue problem corresponds to considering only the first entry in the $[2 \times 2]$ matrices. That is, as we increase the expansion of $v(x)$, then each additional term adds a row and column to the matrices but otherwise the existing matrices are unaffected.

**Example 6.10:**   Consider a non-uniform rod fixed at one end and subjected to an axial concentrated force at the other end, as shown in Figure 6.13. The variation of axial stiffness is $EA(x) = EA_o(1 + x/L)^2$. Obtain a Ritz approximate solution, and compare with the exact solution.

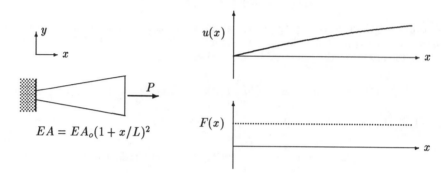

$$EA = EA_o(1 + x/L)^2$$

**Figure 6.13**: Rod with variable cross-section.

The boundary conditions for this problem are:

$$\text{essential:} \quad u|_{x=0} = 0 \qquad \text{natural:} \quad EA\frac{du}{dx}\Big|_{x=L} = P$$

The exact solution is easily calculated using the methods of Chapter 2, and gives

$$u(x) = \int_0^x \epsilon(x)dx = \int_0^x \frac{P}{EA_o(1 + x/L)^2}dx = \frac{PL}{EA_o}\frac{x/L}{(1 + x/L)}$$

The exact force distribution in the bar is

$$F(x) = EA\frac{du}{dx} = P$$

This last result is in agreement with what would be obtained from the equilibrium of simple free body diagrams. Both the displacement distribution and force distribution are shown plotted in Figure 6.13. We will use these results to evaluate the quality of the Ritz approximate solutions. Specifically, we wish to investigate the use of different trial functions.

The total potential energy of the structure is

$$\Pi = \tfrac{1}{2}\int_0^L EA\left(\frac{du}{dx}\right)^2 dx - Pu_L$$

We will calculate the displacement and force distributions using the following assumed form for the displacement:

$$u(x) = a_0 + a_1 x + a_2 x^2$$

This must satisfy the essential boundary condition, hence $a_0 = 0$. Note that the remaining polynomial does not necessarily satisfy the natural boundary condition. Substituting the assumed displacements into the total potential energy expression, we obtain

$$\Pi = \frac{1}{2} \int_0^L EA_o(1 + x/L)^2(a_1 + 2a_2x)^2 dx - P(a_1L + a_2L^2)$$

Invoking the stationarity of $\Pi$ with respect to the coefficients $a_n$, we obtain the following equations for $a_1$ and $a_2$ after differentiating

$$\frac{\partial \Pi}{\partial a_1} = \int_0^L EA_o(1 + x/L)^2(a_1 + 2a_2x)dx - PL = 0$$

$$\frac{\partial \Pi}{\partial a_2} = \int_0^L EA_o(1 + x/L)^2(a_1 + 2a_2x)2x\, dx - PL^2 = 0$$

Performing the required integrations gives

$$\frac{EA_o}{30} \begin{bmatrix} 70L & 85L^2 \\ 85L^2 & 124L^3 \end{bmatrix} \left\{ \begin{array}{c} a_1 \\ a_2 \end{array} \right\} = \left\{ \begin{array}{c} PL \\ PL^2 \end{array} \right\}$$

Note that this is symmetric. Solving this system gives for the two coefficients

$$a_1 = \frac{78}{97} \frac{P}{EA_o}, \qquad a_2 = \frac{-30}{97} \frac{P}{EA_oL}$$

This Ritz analysis, therefore, yields the approximate solution

$$u(x) = \frac{78P}{97EA_o}[x - \frac{10}{26L}x^2]$$

and the force distribution is

$$F(x) = EA\frac{du}{dx} = \frac{78P}{97}[1 - \frac{10}{13L}x](1 + x/L)^2$$

These results are shown in Table 6.1 as the 2-term columns. The most striking fact of these results is the accuracy of the displacements and yet the axial force is not constant and equal to $P$. This reiterates the fact that the Ritz approach only approximates equilibrium.

An interesting result is obtained if we use only a linear expansion for the displacements. In this circumstance, after imposing the essential boundary condition, we have the one term expansion

$$u(x) = a_1x$$

Substituting this into the potential energy expression and minimizing, we obtain

$$\frac{\partial \Pi}{\partial a_1} = \int_0^L EA_o(1 + x/L)^2(a_1)dx - PL = 0$$

| x/L | exact | displ 1 term | displ 2 term | displ bi-linear | exact | force 1 term | force 2 term | force bi-linear |
|-----|-------|--------|--------|-----------|-------|--------|--------|-----------|
| 0.0 | 0.0 | 0.0 | 0.0 | 0.0 | 1.0 | 0.428 | 0.804 | 0.6316 |
| 0.5 | 0.3333 | .2143 | .3247 | .3158 | 1.0 | 0.96 | 1.113 | 1.421 |
| 0.5 | 0.3333 | .2143 | .3247 | .3158 | 1.0 | 0.96 | 1.113 | 0.729 |
| 1.0 | 0.5 | .4285 | .4948 | .4779 | 1.0 | 1.714 | 0.740 | 1.297 |

Table 6.1: Displacement and force results for the non-uniform rod.

Performing the required integration gives

$$\frac{EA_o}{30}[70L]a_1 = PL$$

We recognize this as precisely the first term in the above [2 × 2] matrix form. That is, as we increase the expansion of $u(x)$, then each additional term adds a row and column to the matrices but otherwise the existing matrices are unaffected.

Solving for $a_1$ gives

$$a_1 = \frac{3}{7}\frac{P}{EA_o}$$

The approximate solution for the displacement and force are, respectively,

$$u(x) = \frac{3P}{7EA_o}[x], \qquad F(x) = \frac{3P}{7}[1](1 + x/L)^2$$

These results are also shown in Table 6.1 as the 1-term columns. Note that this force distribution does not satisfy the differential equation of equilibrium.

**Example 6.11:**  As a second Ritz solution to the rod problem of Figure 6.13, assume that the displacements are given in a piece-wise linear form as

$$u(x) = \frac{2x}{L}u_2 \qquad\qquad 0 \le x \le L/2$$

$$u(x) = \left(\frac{2L - 2x}{L}\right)u_2 + \left(\frac{2x - L}{L}\right)u_3 \qquad L/2 \le x \le L$$

where $u_2$ and $u_3$ are the displacements at points mid-way and the end of the rod. These displacements satisfy the essential boundary condition at $x = 0$, and also the continuity of displacement condition at $x = L/2$. There is no continuity of the first derivative $du/dx$ at $x = L/2$, but that is permissible since the highest derivative *in the potential is du/dx*.

Using these trial functions in the potential energy gives

$$\Pi = \tfrac{1}{2}\int_0^{L/2} EA_o(1 + x/L)^2 \left(\frac{2u_2}{L}\right)^2 dx + \tfrac{1}{2}\int_{L/2}^{L} EA_o(1 + x/L)^2 \left(-\frac{2u_2}{L} + \frac{2u_3}{L}\right)^2 dx - Pu_3$$

In this case, the displacements $u_2$ and $u_3$ are the state variables (degrees of freedom). Invoking that $\Pi$ is stationary with respect to these, we obtain

$$\frac{EA_o}{6L}\begin{bmatrix} 56 & -37 \\ -37 & 37 \end{bmatrix}\begin{Bmatrix} u_2 \\ u_3 \end{Bmatrix} = \begin{Bmatrix} 0 \\ P \end{Bmatrix}$$

Hence, we now have

$$u_2 = \frac{P}{EA_o}\frac{222}{703}, \qquad u_3 = \frac{P}{EA_o}\frac{336}{703}$$

The displacement distributions are linear. The forces vary with $x$ and are given by

$$F(x) = P\frac{444}{703}(1 + x/L)^2 \qquad 0 \le x \le L/2$$

$$F(x) = P\frac{228}{703}(1 + x/L)^2 \qquad L/2 \le x \le L$$

These results are shown in Table 6.1 as the bi-linear columns. Note again that the displacements are quite accurate, but the forces are significantly off. Indeed, there is not even equilibrium at the joint.

This last example is very important in demonstrating the relationship between a Ritz analysis and a finite element analysis. Indeed, a Ritz analysis can be regarded as a finite element analysis and *vice versa*, and we make use of that in the next section to derive some elements.

At this stage, it is worthwhile to summarize some of the characteristics of the Ritz method. The most important ones are:

- Usually, the accuracy of the assessed displacement is increased with an increase in the number of trial functions.

- While fairly accurate expressions for the displacements are obtained, the corresponding forces may differ significantly from the exact values.

- Equilibrium is satisfied in an average sense through minimization of the total potential energy. Therefore, forces (computed on the basis of the displacements) do not, in general, satisfy the equilibrium equations.

- The approximate system is stiffer than the actual system and therefore buckling loads and vibration resonances are overestimated.

Since the Ritz solution is approximate, sometimes it can be confusing to know what check conditions the solution should satisfy. It often helps to realize that the approximate solution is actually the exact solution to some other problem. For example, in the bi-linear approximation of the rod, we could think of the solution as replacing the original problem with a piece-wise constant rod, each section of length $L/2$ and of stiffnesses

$$\frac{19}{6}EA_o, \qquad \frac{37}{6}EA_o$$

respectively. This problem now resembles a typical one from Chapter 2 and thus we expect conditions such as equilibrium at the joints to be satisfied. Indeed, when viewed in this manner, we find that the axial force is constant and given by $F(x) = P$. Thus a safe interpretation of the Ritz approximation is that we replace the original system with a new physical system comprised of simpler elements; the effective properties of these elements are obtained by use of the principle of stationary potential energy.

## 6.6    The Finite Element Method

One disadvantage of the conventional Ritz analysis is that the trial functions are defined over the whole region. This causes a particular difficulty in the selection of appropriate functions; in order to solve accurately for large stress gradients, say, we may need many functions. However, these functions are defined over the regions in which the stresses vary rather slowly and where not many functions are required. Another difficulty arises when the total region is made up of subregions with different kinds of strain distributions. As an example, consider a building modeled by plates for the floors and beams for the vertical frame. In this situation, the trial functions used for one region (e.g., the floor) are not appropriate for the other region (e.g., the frame), and special displacement continuity conditions and boundary relations must be introduced. We conclude that the conventional Ritz analysis is, in general, not particularly computer-oriented.

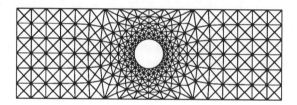

**Figure 6.14**: Continuous domain discretized as finite elements.

We can view the finite element method as an application of the Ritz method, where instead of the trial functions spanning the complete domain, the individual functions span only subdomains (the finite elements) of the complete region. Figure 6.14 shows an example of a bar with a hole modeled as a collection of many triangular regions. The use of relatively many functions in regions of high strain gradients is made possible simply by using many elements as shown around the hole in the figure. The combination of domains with different kinds of strain distributions may be achieved by using different kinds of elements to idealize the domains.

In order that a finite element solution be a Ritz analysis, it must satisfy the essential boundary conditions. However, in the selection of the displacement functions, no special attention need be given to the natural boundary conditions, because these conditions are imposed with the load vector and are satisfied approximately in the Ritz solution. The accuracy with which these natural boundary conditions are satisfied depends on the specific trial functions employed, and on the number of elements used to model the problem. This idea was demonstrated in the last example where it was shown that improvement in the approximate solution could be obtained either by increasing the number of functions, or by increasing the number of sub-domains.

## Element Stiffnesses

In order to develop the element stiffness relation we view the element itself as a structure with essential boundary conditions at the nodal points (these are the degrees of freedom). An application of the stationarity principle then gives the approximate equilibrium conditions that the nodal loads satisfy.

Let the distributed displacement fields $\{\mathcal{U}(x, y, z)\}$ be represented by an expression of the form

$$u(x, y, z) = \lfloor g(x, y, z) \rfloor \{u\}$$

with similar expressions for the other displacements $v(x, y, z)$ and $w(x, y, z)$. Here $\lfloor g(x, y, z) \rfloor$ is a set of known admissible functions of the coordinates, and $\{u\}$ is a set of constants. Let us insert this into the expression for the potential energy for a general elastic body

$$
\begin{aligned}
\Pi &= \tfrac{1}{2} \int \{\epsilon\}^T \{\sigma\} d\mathbb{V} - \int \{\mathcal{U}\}^T \{f^s\} dA - \int \{\mathcal{U}\}^T \{f^b\} d\mathbb{V} \\
&= \tfrac{1}{2} \int \{u\}^T [\, B\,]^T [\, \mathcal{E} \,][\, B\,]\{u\} d\mathbb{V} - \int \{u\}^T [\, g\,]^T \{f^s\} dA - \int \{u\}^T [\, g\,]^T \{f^b\} d\mathbb{V} \\
&= \tfrac{1}{2} \{u\}^T [\, k\,]\{u\} - \{u\}^T \{F\}
\end{aligned}
$$

where $[\, B\,] \equiv [\, \mathcal{D}\,][\, g\,]$ and

$$[\, k\,] \equiv \int_V [\, B\,]^T [\, \mathcal{E}\,][\, B\,] d\mathbb{V}, \qquad \{F\} \equiv \int_A [\, g\,]^T \{f^s\} dA + \int_V [\, g\,]^T \{f^b\} d\mathbb{V}$$

Since, for a given problem, everything within these integrals is either a constant or a known function of the coordinates, the indicated integrations can be carried out explicitly. The only unknowns are the constants $\{u\}$, and from the principle of stationary potential energy, the best choice for these make the potential energy a minimum. This minimum may be found by setting

$$\frac{\partial \Pi}{\partial u_i} = \frac{\partial}{\partial u_i} \left( \tfrac{1}{2} \{u\}^T [\, k\,]\{u\} - \{u\}^T \{F\} \right) = 0, \qquad i = 1, 2, 3, \ldots, N$$

Performing the indicated differentiation leads to

$$[\, k\,]\{u\} = \{F\}$$

This is exactly the same set of equations we derived in the early chapters using the direct approach. What is different is that we can talk about the accuracy of a given choice of functions and the convergence of the solutions when more admissible functions are added. Generally, adding a term will reduce the value of the potential energy, thus improving the answer. At worst, an admissible but otherwise inappropriate term (for example, an antisymmetrical function in a solution which is clearly symmetrical) will have little or no effect because the $u_i$ for that term will prove to be very small or even zero.

We are reminded that in the Ritz method, we are replacing an infinite degree of freedom system by one which has finite degrees of freedom. This tends to overestimate the stiffness of the system so stress, strain, and deflection err on the side of a stiffer structure.

## General Rod Stiffness

We will begin by recovering the element stiffness matrix for the simple rod used in the earlier chapters. Recall that the displacements for the rod segment can be written in terms of the nodal values as

$$u(x) = (1 - \frac{x}{L})u_1 + (\frac{x}{L})u_2 \equiv f_1(x)u_1 + f_2(x)u_2$$

This displacement function satisfies the essential boundary conditions in that it gives the displacements at both nodes. We therefore identify $f_1(x)$ and $f_2(x)$ as the Ritz functions, and $u_1$ and $u_2$ as the state variables. The normal strain in the element is

$$\epsilon = \frac{du}{dx} = \frac{u_2 - u_1}{L}$$

giving the strain energy stored in the element as

$$U = \frac{1}{2} \int_0^L EA(\frac{du}{dx})^2 \, dx = \frac{EA}{2L}(u_2 - u_1)^2$$

The potential of the nodal forces is

$$V = -F_1 u_1 - F_2 u_2$$

The total potential energy is therefore

$$\Pi = \frac{EA}{2L}(u_2 - u_1)^2 - F_1 u_1 - F_2 u_2$$

To obtain the element stiffness relation (which is essentially a statement of equilibrium) we minimize this with respect to the parameters $u_1$ and $u_2$. That is

$$\frac{\partial \Pi}{\partial u_1} = \frac{EA}{L}(u_1 - u_2) - F_1 = 0$$

$$\frac{\partial \Pi}{\partial u_2} = \frac{EA}{L}(u_2 - u_1) - F_2 = 0$$

This can be expressed in matrix notation as

$$\begin{Bmatrix} F_1 \\ F_2 \end{Bmatrix} = \frac{EA}{L} \begin{bmatrix} 1 & -1 \\ -1 & 1 \end{bmatrix} \begin{Bmatrix} u_1 \\ u_2 \end{Bmatrix}$$

which is the stiffness matrix for the rod element with respect to the local coordinates.

We will now use the general expression to derive the general form of the rod stiffness. From the shape functions we get

$$[g(x)] = \lfloor f_1(x), f_2(x) \rfloor$$

From this we get

$$[B] = [D][g(x)] = [\frac{d}{dx}]\lfloor f_1(x), f_2(x) \rfloor = \lfloor f_1'(x), f_2'(x) \rfloor$$

The element stiffnesses are therefore given by

$$k_{ij} = \int_L EA f_i'(x) f_j'(x)\, dx$$

The corresponding element nodal forces are

$$\{F\} = \int_S \lfloor f_1, f_2 \rfloor^T \{f^s\} dS = \int_L \begin{Bmatrix} f_1 q \\ f_2 q \end{Bmatrix} dx$$

Note that if $q(x)$ is replaced by concentrated forces at the end and a distributed load along the length, then we recover the usual nodal forces, but in addition, we get the rule for treating distributed loads. This example shows the natural manner in which the approximate stiffness can be obtained for complicated problems.

## General Beam Stiffness

The procedure for determining the element stiffness matrix for beams proceeds as for the rod; to make it a little more interesting we will simultaneously derive the geometric stiffness matrix. Again, the deflection is taken as the basic unknown and we recall that it can be represented in terms of the nodal values as

$$\begin{aligned}
v(x) &= \left[ 1 - 3(\frac{x}{L})^2 + 2(\frac{x}{L})^3 \right] v_1 + (\frac{x}{L}) \left[ 1 - 2(\frac{x}{L}) + (\frac{x}{L})^2 \right] L\phi_1 \\
&\quad + (\frac{x}{L})^2 \left[ 3 - 2(\frac{x}{L}) \right] v_2 + (\frac{x}{L})^2 \left[ -1 + (\frac{x}{L}) \right] L\phi_2 \\
&= g_1(x)v_1 + g_2(x)L\phi_1 + g_3(x)v_2 + g_4(x)L\phi_2
\end{aligned}$$

This satisfies the essential boundary conditions. The total strain energy stored in the beam element is (neglecting shear)

$$U_B = \frac{1}{2} \int_0^L EI[\frac{d^2v}{dx^2}]^2 dx = \frac{1}{2} \int_0^L EI[g_1''(x)v_1 + g_2''(x)L\phi_1 + g_3''(x)v_2 + g_4''(x)L\phi_2]^2 dx$$

The strain energy due to the axial force is

$$U_G = \frac{1}{2} \int_0^L \bar{F}_o[\frac{dv}{dx}]^2 dx = \frac{1}{2} \int_0^L \bar{F}_o[g_1'(x)v_1 + g_2'(x)L\phi_1 + g_3'(x)v_2 + g_4'(x)L\phi_2]^2 dx$$

The potential of the nodal forces is

$$V = -V_1 v_1 - M_1 \phi_1 - V_2 v_2 - M_2 \phi_2$$

The total potential for the problem in therefore

$$\Pi = U_B + U_G + V$$

The entities in the total stiffness matrix can be obtained by extremizing the total potential to give

$$k_{ij} = \frac{\partial^2 \Pi}{\partial u_i \partial u_j} = \int_0^L EI g_i''(x) g_j''(x)\, dx + \int_0^L \bar{F}_o g_i'(x) g_j'(x)\, dx$$

Carrying out these integrations under the condition of constant $EI$ and constant $\bar{F}_o$ gives the beam element stiffness matrix and the beam geometric stiffness matrix as already obtained. In the present form they are applicable to non-uniform sections. Note that it is the symmetry of the terms $g_i''(x) g_j''(x)$ and $g_i'(x) g_j'(x)$ that insures the symmetry of the stiffness matrices.

## 6.7   Stability Reconsidered

We saw in the last few sections that the stationary potential energy principle could recover the equilibrium conditions for a system. We also saw in Chapter 5 that stability is a property of equilibrium. We conclude this chapter by looking at the connection between stability and the potential energy.

Consider a structure to be in a state of equilibrium. Now give it a small variation in displacement $\delta u_i$. The change in total potential energy can be expressed as

$$\Delta \Pi \approx \delta \Pi + \tfrac{1}{2} \delta^2 \Pi + \cdots = \sum_i \frac{\partial \Pi}{\partial u_i} \delta u_i + \frac{1}{2} \sum_i \sum_j \frac{\partial^2 \Pi}{\partial u_i \partial u_j} \delta u_i \delta u_j + \cdots$$

We have already shown that the condition $\delta \Pi = 0$ governs equilibrium; a study of the second order terms $\delta^2 \Pi$ therefore governs the nature of the equilibrium, that is,

$$\delta^2 \Pi > 0 \qquad \text{:stable equilibrium}$$
$$\delta^2 \Pi = 0 \qquad \text{:neutral equilibrium}$$
$$\delta^2 \Pi < 0 \qquad \text{:unstable equilibrium}$$

Actually, if $\delta^2 \Pi = 0$ then we must check the higher order terms also. We conclude that for stable equilibrium the potential energy is a minimum.

To clarify the meaning of the second variation, let us reconsider the example of Figure 5.1. The total potential for the problem is

$$\Pi = \tfrac{1}{2} k u^2 - P v - Q u$$

Since we assume the vertical bar to be rigid, then there is a geometric constraint that relates the displacements $u$ and $v$. Specifically, we have

$$v = L - \sqrt{L^2 - u^2} \approx \frac{u^2}{2L}$$

This last approximation is consistent with our assumption that all deflections are small. Consequently, the potential for the problem becomes

$$\Pi = \tfrac{1}{2}ku^2 - \tfrac{1}{2}\left(\frac{P}{L}\right)u^2 - Qu$$

Note that while $P$ originally appeared in the work term $Pv$, it now appears in a strain energy like term. The different orders of variation of the potential are

$$\delta\Pi = \frac{\partial\Pi}{\partial u}\delta u = [ku - Pu/L - Q]\delta u$$

$$\tfrac{1}{2}\delta^2\Pi = \tfrac{1}{2}\frac{\partial^2\Pi}{\partial u^2}\delta u^2 = \tfrac{1}{2}[k - P/L]\delta u^2$$

$$\tfrac{1}{6}\delta^3\Pi = 0$$

The first equation gives the equilibrium condition; when the system is in equilibrium this term is zero, therefore it is the second equation that determines the sign of $\Delta\Pi$. Depending on the value of $P$ this second term can be either positive or negative. What is most important to note is that the relation does not contain the other applied load $Q$. In fact, the potential of the applied loads is always linear in displacement, hence they will never contribute to the second variation of the potential and thus affect stability. (However, they do affect the initial load term $\bar{F}_o$.)

Note that the second variation in potential energy is related to the total stiffness as

$$K_{ij} = \sum_i\sum_j\frac{\partial^2\Pi}{\partial a_i\partial a_j}$$

Hence the conditions for stability of equilibrium can be applied to the total stiffness matrix; that is, we inspect whether this matrix is positive definite, positive semi-definite, or negative definite. A necessary and sufficient condition that $\delta^2\Pi$ be positive definite is that the determinant of the matrix $[K]$ and all its principal minors be positive. Usually, it is unnecessary to examine a sequence of minors since the determinant itself vanishes before any of its minors. Thus our criterion

$$\det[K] = 0$$

for the onset of instability is also justified by energy considerations.

## Problems

**6.1** A uniform rod is held between two rigid walls. Determine the strain energy and complementary strain energy stored in the rod as it is heated.

[Reference [10], pp. 87]

**6.2** Show that the functional

$$\Pi = \int_0^{10} [\tfrac{1}{2}(\frac{du}{dx})^2 - 100u]\, dx$$

can be used to recover the following differential equation

$$\frac{d^2u}{dx^2} + 100 = 0 \qquad 0 \le x \le 10$$

subject to the boundary conditions $u(0) = u(10) = 0$.

[Reference [39], pp. 118]

**6.3** Obtain a Ritz solution to the previous exercise using the trial function $f(x) = x(10 - x)$. Compare this approximate solution with the exact solution.

**6.4** In general, it is not possible to obtain a functional for problems whose governing differential equation contains odd-power derivatives. A special exception is the following case

$$\frac{d^2u}{dx^2} + a\frac{du}{dx} + bu = 0$$

where $a$ and $b$ are constants. Show that the functional

$$\Pi = \int [\tfrac{1}{2}e^{ax}(\frac{du}{dx})^2 - bu]\, dx$$

can be used to recover the differential equation.

**6.5** The total potential of a certain system is

$$\Pi = \tfrac{1}{5}x^5 - \tfrac{1}{4}ax^4 - \tfrac{2}{3}ax^3 + a^2$$

where $a$ is a parameter and $x$ is the generalized coordinate. Determine all of the equilibrium configurations and indicate which ones are stable and unstable.

[Reference [33], pp. 356]

**6.6** Suppose we wish to derive a "higher-order" rod element and to that end we take the deflection function in the cubic form

$$u(x) = a_0 + a_1 x + a_2 x^2 + a_3 x^3$$

Introducing the nodal degree of freedoms $u_1 = u(0)$, $\phi_1 = du(0)/dx$, $u_2 = u(L)$, $\phi_2 = du(L)/dx$, show that the displacement

can be represented in terms of the nodal values as

$$u(x) = g_1(x)u_1 + g_2(x)L\phi_1 + g_3(x)u_2 + g_4(x)L\phi_2$$

where the functions $g_n(x)$ are identical to those for the beam shape functions of Equation(3.6). Show that the higher-order rod element stiffness matrix is

$$[\,k\,] = \frac{EA}{30L} \begin{bmatrix} 36 & 3L & -36 & 3L \\ 3L & 4L^2 & -3L & -L^2 \\ -36 & -3L & 36 & -3L \\ 3L & -L^2 & -3L & 4L^2 \end{bmatrix}$$

Note that this is almost identical to the geometric stiffness matrix for the beam; Why?

[Reference [49], pp. 207]

**6.7** Suppose that only the axial forces $F_1$, $F_2$ are taken as the nodal loads in the previous exercise, show that the elementary rod stiffness relation is recovered. Propose some nodal loads that do not give the trivial result.

**6.8** If in the derivation of the beam geometric stiffness matrix, we use the rod shape functions instead of the beam shape functions, i.e.,

$$v(x) = (1 - \frac{x}{L})v_1 + (\frac{x}{L})v_2 = f_1(x)v_1 + f_2(x)v_2$$

Show that the derived inconsistent geometric stiffness matrix is given by

$$[\,k_G\,] = \frac{\bar{F}_o}{L} \begin{bmatrix} 1 & 0 & -1 & 0 \\ 0 & 0 & 0 & 0 \\ -1 & 0 & 1 & 0 \\ 0 & 0 & 0 & 0 \end{bmatrix}$$

Where might such an element be more useful than its consistent counterpart?

[Reference [46], pp. 319]

**6.9** With reference to Figure 6.6, use the Ritz method to find an approximate solution using

$$u(x) = a_0 + a_1 x + a_2 x^2$$

[Reference [10], pp. 13]

**6.10** Determine a constraint element by adding

$$\tfrac{1}{2}k[\alpha_i u_i + \alpha_j u_j - \alpha_o]^2$$

to the potential energy term. Generalize the results for more than two constraints.

[Reference [10], pp. 75]

# Exercises

**6.1** A particle in the $(x, y)$ plane moves in the force field $F_x = -ky$, $F_y = kx$ ($k$ is a constant). Prove that when the particle describes any closed path in the counterclockwise sense, the work performed on the particle is $2kA$ where $A$ is the area enclosed by the path.

**6.2** With reference to Figure 4.14, verify Maxwell's reciprocal theorem for a horizontal load at Node 1 and a vertical load at Node 4.

**6.3** Use Castigliano's second principle to find the reactions for the rod in Figure 6.6.

The procedure is to replace the fixity constraints by a system of unknown forces and then use the principle to determine the displacements at the end points. By setting these displacements to zero a sufficient number of equations are obtained to determine the reactions.

**6.4** Consider a cantilever beam, fixed at the end $x = 0$ and subjected to a given displacement $v_L = c$ at the other. Show that the following is a set of admissible displacements and obtain the corresponding Ritz solution.

$$v(x) = \frac{cx^2}{L^2} + \sum a_n[1 - \cos(2n\pi x/L)]$$

**6.5** Consider a cantilever beam, fixed at the end $x = 0$ and subjected to a concentrated lateral applied force at the other. Using the Ritz method, show that the following displacements

$$v(x) = a_0 + a_1 x + a_2 x^2 + a_3 x^3$$

leads to the exact solution. Show that the addition of extra terms have zero contributions.

**6.6** A uniformly loaded beam is simply supported at both ends. Find the deflection and bending moment at the center using the Ritz method. First use a quadratic function in $x$ and then use a sine function in $x$. Compare the results with the exact solution and say why the second solution is better than the first.

**6.7** Suppose the rotation distribution in a circular shaft can be written in terms of the nodal rotations as

$$\phi(x) = (1 - \frac{x}{L})\phi_1 + (\frac{x}{L})\phi_2 \equiv f_1(x)\phi_1 + f_2(x)\phi_2$$

Use minimum potential energy to derive the torsion element.

**6.8** For a uniform column with clamped ends, assume $v = ax^2(L - x)^2$ and determine the critical load. $[P_{cr} = 42EI/L^2]$

**6.9** For a uniform column with clamped at one end and free at the other, assume $v = ax^2$ and determine the critical load. $[P_{cr} = 2.5EI/L^2]$

**6.10** For a uniform column with clamped at one end and free at the other, assume the mode shape is the static deflection shape due to a point load at the tip and determine the critical load. $[P_{cr} = 42EI/17L^2]$

Chapter 7

# Computer Methods I

The use of the computer is essential for modern structural analysis; therefore, it is important that the engineer have some understanding of how the computer is actually used to accomplish this analysis. This chapter introduces some of the basic computer methods and algorithms used in structural analysis; these include schemes for solving simultaneous equations, solving eigenvalue problems, and for efficient data storage. In addition, it attempts to describe the computer environment in which the analysis takes place. References [4, 32, 45] consider many of the general aspects of using computers for structural analysis.

It is difficult to describe computer algorithms without describing the complete programming context. Therefore, most of the discussion will refer to the algorithms implemented in the program STADYN; the source code is listed in Appendix B. The availability of this program also gives the opportunity to discuss some of the design aspects of a large structural analysis program.

## 7.1  Computers and Data Storage

A clear understanding of how software utilizes the computer is helpful to understanding some of the design constraints imposed on a structural analysis program. In general, all computers have four main components: a central processing unit (CPU), internal random access memory (RAM), external storage in the form of disk or tape drives, and the operating system (OS). The CPU executes the instructions of the program and generally its use is transparent to the user. That is, its use is handled by the programming language and the operating system. Generally, the users of a structural analysis program need not concern themselves with the particulars of the CPU or the OS.

Available computer memory (both internal and external) has direct implications for the user because it limits the size of problems that can be solved; we therefore discuss its role in more detail. RAM (random access memory) is the area where an active program and its data is stored in a computer. We will show that the amount of

RAM available affects the program in two ways; first, in the accuracy of the numbers computed, and second in the size of problems that can be solved. Much of the art in programming revolves around minimizing the number of computations and the memory requirements.

## Data Representation

To get a clearer idea of how RAM is used to store data, consider a floating point number 123456.78 represented in the form .12345678E6. We call the portion .12345678 the *mantissa* and the portion E6 the *exponent.* Numbers such as these are represented in a computer in terms of *bits*; a bit is the smallest unit of memory and has a value of either 0 or 1. According to the IEEE standard [32], the standard real variable (float) is stored as a 32 bit (4 byte) word. In FORTRAN, this is referred to as *single precision.* According to the standard, 23 bits are reserved for the mantissa resulting in approximately $23 \log_{10}(2) = 6.9$ decimal digit accuracy. Eight bits are reserved for the exponent and this gives a dynamic range of about $2^{255} = 10^{\pm 38}$. The remaining bit is the sign bit.

It is apparent that because of the manner in which numbers are stored in the computer, only a finite set of real numbers can be represented. This gives rise to side effects that would not be present with exact arithmetic. One such effect is that if a computation obtains a number outside the dynamic range, for example, $10^{+40}$ or $10^{-40}$ then an *overflow* or *underflow* error occurs, respectively. As an illustration of how easy it is to cause an overflow, consider computing the determinant of the stiffness matrix. If the matrix is diagonal the determinant is the product of all the diagonal terms. Since each of these terms is of order $10^6$ in imperial units (power 11 in metric units), then a system size of only 4 to 8 is sufficient to cause overflow. Obviously computations like this must be done with some circumspection; usually this means scaling the numbers before multiplication. Another consequence of the finite representation of numbers in the computer is that every operation (addition, multiplication, and so on) results in loss of accuracy due to *round-off error.* It must be kept in mind that these errors are quite separate from errors due to modeling. To help alleviate both of these problems, programs may make use of higher precision representation of numbers. For example, the *double precision* model represents floating point numbers as 64 bit (8 byte) words. This results in approximately 16 decimal digit accuracy and a dynamic range of $10^{\pm 308}$. However, each number requires double the amount of computer storage.

It appears that all these difficulties may be overcome by making more RAM available. Sometimes the RAM limit is due to the operating system; this is the case with the DOS limit of 640kB on the PC microcomputer. More often the limit is due to the cost of memory. Thus, for example, workstations running Unix often have from 4 to 8 MB of RAM even though the operating system can handle much more. Some programs (and operating systems as in the case of Unix) can give the effect of having large memory by using the disk drive as a *virtual memory source*. However, the speed with which data can be accessed is the critical performance factor of an external storage device; these devices are always much slower than internal memory, and thus their use is usually restricted. For example, this approach is adequate for static problems but can be very slow for dynamic problems because of the very frequent writing and reading of the disk.

## Matrix Storage

A matrix is *banded* if all nonzero coefficients cluster about the diagonal. The stiffness, mass, and geometric matrices encountered in structural analyses are often of this type. Banded storage is a simple way to exploit matrix sparsity because zeros outside the band need not be stored or processed.

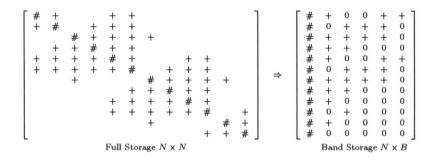

A simple storage format for a banded symmetric matrix is shown above. The $N$ principal diagonal elements are stored in the first column, the $(N-1)$ elements of the next diagonal are stored in the first $(N-1)$ positions of the second column with zero in the last position, and so on for the diagonals which have non-zero elements. The minimum number of columns needed is called the *semi-bandwidth*, $B$. It is six in the above case. The total bandwidth includes coefficients to the left of the diagonal and is therefore $2B - 1$. The entire information content of a symmetric matrix resides in the $[N \times B]$ matrix. In practice, the bandwidth $B$ may be significantly smaller, by a factor of ten or more, than the matrix order $N$, so there is merit in storing just $NB$ coefficients rather than all $N^2$ matrix coefficients. Also, as will be seen in the next section, the equation-solving expense decreases by a factor of about $B^2/N^2$ if we operate on only $NB$ coefficients instead of on all coefficients in the upper triangle. Because of the fewer operations, the solutions are more accurate.

Other schemes for efficient data storage are available and a clear discussion of them can be found in Reference [46].

## 7.2   Structural Analysis Programs

In the use of modern computer workstations, it is popular to refer to the *computing environment*; what is meant is that there is a collection of tools available to the analyst. These include tools for data preparation, graphic display of results, and for various numerical analyses. Thus a commercial structural analysis program in all likelihood is not a single executable program but rather a series of programs with separate functions.

A functional schematic of a structural analysis program is shown in Figure 7.1. The three functions can be conceptualized as separate, stand-alone programs that communicate with each other through the files stored to disk (this is represented by the dashed boxes). It is noted that these functions may be integrated seamlessly to the extent that, to the user, it appears as if there were only one 'program'.

The labelling, numbering, listing of coordinates, and so on, associated with the description of a structures problem, is stored in the *structure datafile*. Preparing this file is a time-consuming and arduous task; it is also likely to lead to numerous data errors if done manually. Consequently, an essential component of the *pre-processor* is an automatic data generator. This should allow, with minimal input from the user, the generation of a validated structure data file. For continuum type problems, an automatic mesh generator should also be included.

The *processor* is the most important unit of the program. We will consider this in more detail later, but essentially it takes as input (from the disk) the structure datafile, forms the stiffness matrix and load vector, solves for the unknown displacements, and stores the results back to disk.

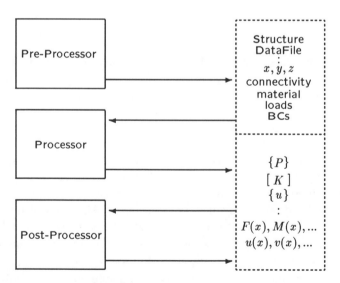

**Figure 7.1**: Functional schematic of a structural analysis program.

In many large scale problems, the volume of possible output can be gigantic. The ability to interpret the results quickly and use them effectively is of critical importance. The *post-processor* manages and presents the results of the computations. Typical tasks are the plotting of deformed shapes, or the contours of constant values of stress. Sometimes error estimates are provided with recommendations for mesh refinement. The post-processor may also have self-adaptive features incorporated with appropriate feed-back to the pre-processor and processor units.

## Design Considerations

In any design project compromises must be made between competing goals; in this respect, the design of a large scale computer program is no exception. Ideally, we would like a program that is blindingly fast, can have access to as much memory as needed, utilizes every hardware enhancement available, and gives output in just the right form. This is not to be achieved (yet) in either the mainframe or microcomputing world. The following specific guidelines were used to coordinate the design decisions when writing STADYN:

- STADYN is to be a special purpose program for the static and dynamic analysis of 3-D frame structures.
- Dynamic problems are emphasized over static ones.
- The essential data must be kept in-core during the computationally intensive stages.
- A disk drive may be used as a virtual memory source, but only for secondary and seldom used variables.
- STADYN is to be developed as a processing engine; therefore it should have hooks to pre- and post-processors.
- The program must run reasonably well on a variety of platforms from micro to mainframe computers.
- All modules must be written in a high level language, in a style that is portable and easily maintained. The modules are to be re-usable for other special purpose programs.

These guidelines, to a greater or lesser extent, shaped the choice of the algorithms to be described later. The resulting design has proven to be robust and quite suitable for microcomputers.

It is emphasized that all programs (intentionally or not) are developed with particular guidelines in mind and this can profoundly affect their suitability for certain applications. For example, a program designed to handle very large static problems may make copious use of a disk drive as a secondary memory source; this program would probably perform very poorly if applied to a dynamic problem. On a more global level, some general purpose programs are designed to make available to the user the complete library of elements and routines; these programs demand vast computer resources and usually are restricted to mainframe environments. On the other hand, specialized programs can target specific applications (structural analysis of buildings, say) and thus provide performance, problem capacity, and accuracy exceeding that of the general purpose programs. Invariably, these programs can also run in a smaller computing environment.

The point is that no single program is the 'best' for all situations, and the user must be aware of the inherent limitations in each program and not attempt to use it inappropriately.

## Anatomy of the Processor

The arrangement of modules in the processor is shown in Figure 7.2; the dashed boxes refer to storage space such as on the disk drive. This is a simple structure, but if we need more functions (e.g., for solving stability or dynamic problems), it is easy to attach these modules within this framework. Each module is comprised of a series of subroutines and functions. Communication between subroutines occurs in three possible forms. The first is through their formal argument lists. For the major modules, this is generally reserved for apportioning the memory into the local array variables. The second is by way of COMMON blocks; this is usually reserved for file handles and other global variables. Finally, information is also transferred by first writing the data to disk and then re-reading it as needed. This may seem inefficient, but as long as it is outside any major do-loops the overhead is minimal. The great advantage of this approach is that it adds flexibility in organizing the arrangements of the modules. In STADYN, this technique is used for such variables as the stiffness matrix and the material properties.

The MAIN module acts as the coordinator for all the other routines that carry out the specific parts of the analysis. It keeps track of the opening of files, as well as the allocation of memory. STADYN attempts to maximize the in-core RAM available by using some dynamic memory management techniques, and by shifting most of the secondary variables onto the disk drive. For dynamic memory management, a large reserved array is dimensioned initially, and in subsequent calls to the various subroutines, it is divided up into the appropriate sizes as needed. As a result, the same storage is re-used many times, thus maximizing its utilization. This is important since single precision usually does not have sufficient accuracy for solving large systems of equations, and therefore the programmer must resort to making all crucial variables double precision.

The description of the structure and its material properties is kept in a separate file (here called the *structure datafile*) to be read by the program. It describes the position of the nodes, the connectivities, the material properties, the boundary conditions, the applied loads, plus any global constraints. The integrity of this file is essential, therefore, as it is read in by module DataIn, numerous types of checks are performed on it so as to confirm its validity. The structural data is converted to degrees of

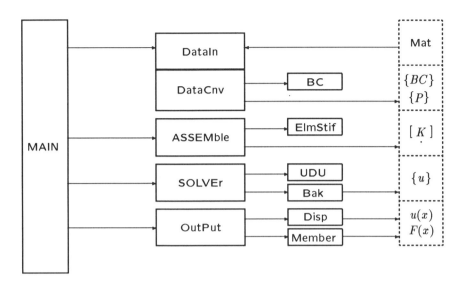

**Figure 7.2**: Anatomy of the processor.

freedom, and the boundary conditions imposed, in module `DataCnv`. The global system of equations must be rearranged and renumbered to account for this. It is at this stage that `STADYN` does any special global reductions such as restricting the frame to be 2-D only. The book-keeping for the reduced degrees of freedom is taken care of in the array $\{IDbc\}$. This is a vector array whose elements identify each node (in sequence) and the six degrees of freedom at each node. The constrained degrees of freedom contain a 0 while the non-zero ones contain the equation number associated with the global system of equations.

During the input of the structure data file, every element is scanned and the largest nodal number difference (for nodes with non-zero degrees of freedom) is taken as the bandwidth. Thus on exit from the input routine, the storage requirements are known even though the stiffness matrix has not yet been assembled. The module `ASSEMble` forms the stiffness matrix. For each element, the local element stiffness is established and then transformed to global coordinates. This is entered into the structural stiffness matrix by associating the appropriate nodal numbers. Those nodes which have a zero degree of freedom are not entered at all. After assemblage, the stiffness matrix is stored to disk in compact form ready for later re-use.

The solution of the system of equations is accomplished in module `SOLVEr`. This first decomposes the stiffness matrix and then uses forward and back substitution to obtain the solution. The obtained nodal degrees of freedom are stored to disk for use by other modules. The routines in module `OutPut` convert the nodal degrees of freedom into other informational forms (such as member loads). Members can be interrogated individually or collectively.

There are other modules listed in Appendix B that are not shown in Figure 7.2; these are associated with solving other specific problems. For example, module `EIGEN` solves eigenvalue problems associated with stability and vibration, module `TRANSient` solves dynamic wave propagation problems. As modules, they have the same relationship to `MAIN` as does `SOLVEr`, for example.

# 7.3    Node Renumbering

Imagine a case where a node must be added to a given structure data file. It would be convenient if the new node could be given the next available number, but unfortunately this would probably destroy the compact bandedness of the arrays. If small bandwidth is to be preserved, then the entire structure must be renumbered.

The bandwidth is directly dependent on the numbering of the nodes. A small bandwidth is usually achieved by placing consecutive node numbers across the shorter dimension of a structure. For large problems, this is difficult to do by inspection, so various schemes have been developed to renumber the nodes automatically. Two of these schemes are described here. Keep in mind that renumbering schemes rarely attempt to find the absolute minimum bandwidth, nor do they even guarantee improvement; but the goal of reducing the bandwidth is usually achieved, especially if the original numbering was produced automatically by patching multiple grids together. Both schemes to be presented ignore the number of degrees of freedom per node; that is, they minimize the *nodal bandwidth* by considering only node numbers.

### Wave Front Method

The simplest method of renumbering is based on the idea that connected nodes are usually near each other. Hence, the wavefront scheme simply numbers the nodes based on the distance of each node from some reference point.

The algorithm is as follows:

**Step 1:** Pick a reference point in space, typically away from the structure.

**Step 2:** Calculate the distance of each node from the reference point.

**Step 3:** Order the nodes according to their proximity to the reference point.

**Step 4:** Assign the new node numbers based on this ordering.

**Step 5:** Compute the new bandwidth.

**Step 6:** Repeat from Step 1 until a satisfactory minimum bandwidth is achieved.

This is a very effective scheme when the mesh is relatively uniform as is the case in 2-D and 3-D continuum problems.

## Cuthill-McKee Algorithm

The Cuthill-McKee method [12] is a significantly more sophisticated renumbering scheme than the wave front method. It explicitly considers the connectivities at each node and is therefore good for both uniform and highly non-uniform meshes.

The algorithm is as follows:

**Step 1:** Scan all the nodes and order them according to their degree. The *degree* of a node is the number of other nodes connected to it.

**Step 2:** Pick a starting node, usually one of low degree, and number it 1. Node 1 is said to be in *Level 1*.

**Step 3:** Nodes connected to Node 1 are said to be in Level 2. These are then numbered 2, 3, and so on, in order of increasing degree.

**Step 4:** Un-numbered nodes connected to nodes in Level 2 are then numbered, again in order of increasing degree.

**Step 5:** Subsequent levels are treated similarly.

**Step 6:** Compute the new bandwidth.

**Step 7:** Repeat from Step 1 starting with a different node until a satisfactory minimum bandwidth is achieved.

A variation on this algorithm, called the *reverse* Cuthill-McKee method, proceeds as above but then reverses the numbers. That is, the last numbered node becomes Node 1, the second last becomes Node 2, and so on. This does not affect the bandwidth, but it does affect the distribution of zeros inside the band.

**Example 7.1:** Use the Cuthill-McKee method to renumber the nodes in Figure 7.3 so as to decrease the bandwidth.

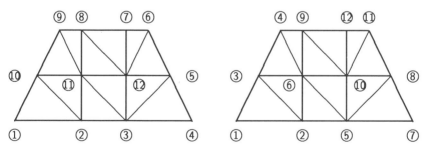

**Figure 7.3**: Renumbering of the Nodes.

The initial numbering sequence follows the perimeter and then includes the internal nodes. This may seem an unnatural or artificial numbering scheme, but is quite likely to occur if the interior nodes were generated automatically using a mesh generator. The purpose of this example is to show how, in a fairly automatic way, the algorithm can give significant improvement in the bandwidth.

Recall that the renumbering does not take the degrees of freedom at each node into account. Therefore, the connectivity matrix (introduced in Chapter 4) is very similar to a stiffness matrix if there is only one degree of freedom at each node. For the initial numbering, this matrix is

|    | 1 | 2 | 3 | 4 | 5 | 6 | 7 | 8 | 9 | 10 | 11 | 12 |
|----|---|---|---|---|---|---|---|---|---|----|----|----|
| 1  | # | + |   |   |   |   |   |   |   | +  |    |    |
| 2  | + | # | + |   |   |   |   |   |   | +  | +  |    |
| 3  |   | + | # | + | + |   |   |   |   |    | +  | +  |
| 4  |   |   | + | # | + |   |   |   |   |    |    |    |
| 5  |   |   | + | + | # | + |   |   |   |    |    | +  |
| 6  |   |   |   |   | + | # | + |   |   |    |    | +  |
| 7  |   |   |   |   |   | + | # | + |   |    |    | +  |
| 8  |   |   |   |   |   |   | + | # | + |    | +  | +  |
| 9  |   |   |   |   |   |   |   | + | # | +  | +  |    |
| 10 | + | + |   |   |   |   |   |   | + | #  | +  |    |
| 11 |   | + | + |   |   |   |   | + | + | +  | #  | +  |
| 12 |   | + |   |   | + | + | + | + |   |    | +  | #  |

This gives a semi-band $B = 10$. It must be emphasized at this stage that the above numbering is not necessarily bad; the 'goodness' of a numbering depends on the type of algorithms used to operate on the arrays. The above array is considered *sparse*, and there are routines designed explicitly to handle this type of array. The solver we present next works best with tightly banded arrays.

The first step in the renumbering is to determine the degree of each node. That is, we determine for each node the number of other nodes connected to it; this is its degree.

|          |            |
|----------|------------|
| degree 1: | None      |
| degree 2: | 1, 4      |
| degree 3: | 6, 7, 9   |
| degree 4: | 2, 5, 8, 10 |
| degree 5: | 3         |
| degree 6: | 11, 12    |

Pick Node 1 as the starting node (since it is of lowest degree) and run through each level renumbering the nodes in sequence.

|          | old number  | new number  |
|----------|-------------|-------------|
| level 1: | 1           | 1           |
| level 2: | 2, 10       | 2, 3        |
| level 3: | 9, 3, 11    | 4, 5, 6     |
| level 4: | 4, 5, 8, 12 | 7, 8, 9, 10 |
| level 5: | 6, 7        | 11, 12      |

This new numbering gives the following connectivity matrix.

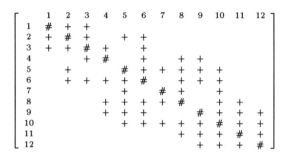

The semi-band in this new scheme is $B = 6$. This gives a reduction in storage requirements of nearly a factor of two. Note that this result nearly corresponds to what would be obtained using the wavefront scheme with Node 1 as the focal point.

It is possible that another renumbering could reduce the bandwidth even more, but the reduction will be minimal compared to what has already been achieved. Thus the main benefit of this renumbering is in the context of automatic mesh generators. That is, it makes it possible to piece the mesh (frame) together using any convenient numbering scheme; then a pass through this algorithm will give consecutive numbers with a small bandwidth. This is especially useful if the mesh is made of separate pieces with separate numbering schemes. For example, each segment could have a numbering beginning with a different multiple of 1000, say.

## 7.4   Solving Simultaneous Equations

The primary mathematical task associated with matrix analysis of frame structures consists of solving a set of $N$ simultaneous linear algebraic equations for $N$ unknowns. For small scale problems, a wide variety of schemes can be used; but for the large systems usually associated with structural analyses, great care must be exercised in choosing the correct method.

The following introduces an effective scheme which is called the *factorization method* (also referred to as the *method of decomposition*). This approach is particularly well suited for matrix analysis of structures because it provides the efficiency of the well-known Gaussian elimination process with a matrix format. (See Reference [34] for more discussion of the Gaussian elimination scheme.) Since the stiffness matrices of linearly elastic structures are always symmetric, a specialized type of factorization known as the *Cholesky method* will be developed for symmetric matrices.

## Modified Cholesky Method

To begin the discussion, let $[K]$ represent a symmetric matrix of size $[N \times N]$. We wish to factor it into the following triple product:

$$[K] = [U]^T [D][U]$$

This, in expanded form, is

$$
\begin{bmatrix}
K_{11} & K_{12} & K_{13} & \cdots & K_{1N} \\
K_{21} & K_{22} & K_{23} & \cdots & K_{2N} \\
K_{31} & K_{32} & K_{33} & \cdots & K_{3N} \\
\cdots & \cdots & \cdots & \cdots & \cdots \\
K_{N1} & K_{N2} & K_{N3} & \cdots & K_{NN}
\end{bmatrix}
=
\begin{bmatrix}
1 & 0 & 0 & \cdots & 0 \\
U_{12} & 1 & 0 & \cdots & 0 \\
U_{13} & U_{23} & 1 & \cdots & 0 \\
\cdots & \cdots & \cdots & \cdots & \cdots \\
U_{1N} & U_{2N} & U_{3N} & \cdots & 1
\end{bmatrix}
\times
$$

$$
\begin{bmatrix}
D_{11} & 0 & 0 & \cdots & 0 \\
0 & D_{22} & 0 & \cdots & 0 \\
0 & 0 & D_{23} & \cdots & 0 \\
\cdots & \cdots & \cdots & \cdots & \cdots \\
0 & 0 & 0 & \cdots & D_{NN}
\end{bmatrix}
\begin{bmatrix}
1 & U_{12} & U_{13} & \cdots & U_{1N} \\
0 & 1 & U_{23} & \cdots & U_{2N} \\
0 & 0 & 1 & \cdots & U_{3N} \\
\cdots & \cdots & \cdots & \cdots & \cdots \\
0 & 0 & 0 & \cdots & 1
\end{bmatrix}
$$

This technique is known as the *modified Cholesky method*; the 'usual' Cholesky method uses the decomposition $[K] = [U]^T[U]$ and requires all diagonal terms to be positive. We will see that the diagonal matrix $\lceil D \rfloor$ has some very interesting and useful properties.

The elements of the matrix $[K]$ essentially are obtained as inner products of the rows and the columns of $[U]$. Thus, the inner products of the first column with itself and subsequent columns produce

$$K_{11} = D_{11}, \quad K_{12} = D_{11}U_{12}, \quad K_{13} = D_{11}U_{13}, \quad \ldots, \quad K_{1N} = D_{11}U_{1N}$$

Similarly, the inner products of the second column with itself and subsequent columns are

$$K_{22} = D_{11}U_{12}^2 + D_{22}, \quad K_{23} = D_{11}U_{12}U_{13} + D_{22}U_{23}, \quad \ldots, \quad K_{2N} = D_{11}U_{12}U_{1N} + D_{22}U_{2N}$$

and so on. From this, the diagonal terms in $[K]$ can be written as

$$K_{ii} = D_{11}U_{1i}^2 + D_{22}U_{2i}^2 + D_{33}U_{3i}^2 + \ldots + D_{ii}$$

or, compactly,

$$K_{ii} = D_{ii} + \sum_{k=1}^{i-1} D_{kk}U_{ki}^2 \qquad (1 \le i \le N)$$

Similarly, off-diagonal terms in the first row and other rows are

$$K_{ij} = D_{11}U_{1i}U_{1j} + D_{22}U_{2i}U_{2j} + D_{33}U_{3i}U_{3j} + \ldots + D_{ii}U_{ij}$$

and again, compactly,

$$K_{ij} = D_{ii}U_{ij} + \sum_{k=1}^{i-1} D_{kk}U_{ki}U_{kj} \qquad (1 \le i \le N) \quad (i+1 \le j \le N)$$

Since we actually want the elements of $\lceil D \rfloor$ and $[U]$, then these can be found by rearranging the above as follows:

$$D_{ii} = K_{ii} - \sum_{k=1}^{i-1} D_{kk}U_{ki}U_{ki} \qquad (1 \le i \le N)$$

$$U_{ij} = \left(K_{ij} - \sum_{k=1}^{i-1} D_{kk}U_{ki}U_{kj}\right)/D_{ii} \qquad (1 \le i \le N) \quad (i+1 \le j \le N)$$

$$U_{ij} = 0 \qquad (i < j) \qquad\qquad (7.1)$$

These three equations constitute the recurrence formulas in an algorithm for factoring the matrix $[K]$ into $[U^T DU]$ form. In these equations, the rows are traversed as $i = 1, N$; and for each row the column is traversed as $j = i, N$.

A significant attribute of this decomposition algorithm is that it can be done *in place*. That is, the decomposed matrix is stored in the same memory locations as the original matrix. This gives the minimum storage requirements. An additional consequence is that the decomposed array inherits the same bandedness as the original array.

The diagonal coefficients $D_{ii}$ have a very useful structural interpretation which can be seen by re-arranging the system

$$[K]\{u\} = [U]^T\lceil D \rfloor[U]\{u\} = \{P\}$$

as

$$\lceil D \rfloor\{v\} = \{Q\}, \qquad \{v\} \equiv [U]\{u\}, \qquad \{Q\} \equiv [U^T]^{-1}\{P\}$$

In words, the diagonal matrix $\lceil D \rfloor$ represents the stiffness of an uncoupled system obtained by transformation from the original system by using $[U]$ as the transformation matrix. The stiffness matrix is therefore singular if at least one $D_{ii}$ is zero, since

$$\det[K] = \det([U]^T\lceil D \rfloor[U]) = \det[U]\det\lceil D \rfloor\det[U]^T = \det\lceil D \rfloor$$

The structural interpretation is that one of the $\{v\}$ degrees of freedom is unconstrained. Parenthetically, since the determinant of the system is simply the product of the diagonal terms, this helps explain why the criterion for stability involves $\det[K] = 0$. Therefore, $\lceil D \rfloor$ can be monitored to detect the presence of unusually stiff or soft elements — both could signal potential numerical difficulties.

**Example 7.2:**   The following array is typical (except for size) of those usually encountered in structural mechanics. It is desired to decompose it into its triangular form.

$$[K] = \begin{bmatrix} 1 & -1 & 0 & 0 \\ -1 & 2 & -1 & 0 \\ 0 & -1 & 3 & -2 \\ 0 & 0 & -2 & 3 \end{bmatrix} = [U^T][\, D\,][\, U\,]$$

Start with $i = 1$, and in succession vary $j = 2, 3, 4$ to obtain

$$\begin{aligned} D_{11} &= K_{11} = 1 \\ U_{12} &= (K_{12})/D_{11} = -1 \\ U_{13} &= (K_{13})/D_{11} = 0 \\ U_{14} &= (K_{14})/D_{11} = 0 \end{aligned}$$

Now increment $i$ to 2, and in succession vary $j = 3, 4$ to obtain

$$\begin{aligned} D_{22} &= (K_{22} - D_{11}U_{12}U_{12}) = 1 \\ U_{23} &= (K_{23} - D_{11}U_{12}U_{13})/D_{22} = -1 \\ U_{24} &= (K_{24} - D_{11}U_{12}U_{14})/D_{22} = 0 \end{aligned}$$

Now increment $i$ to 3, and for $j = 4$ obtain

$$\begin{aligned} D_{33} &= K_{33} - (D_{11}U_{13}U_{13} + D_{22}U_{23}U_{23}) = 3 - 1 = 2 \\ U_{34} &= [K_{34} - (D_{11}U_{12}U_{14} + D_{22}U_{23}U_{24})]/D_{33} = [-2]/2 = -1 \end{aligned}$$

Finally, increment $i$ to 4 and obtain

$$D_{44} = K_{44} - (D_{11}U_{14}U_{14} + D_{22}U_{24}U_{24} + D_{33}U_{34}U_{34}) = 3 - 2 = 1$$

This gives the decomposition as

$$[K] = [U^T][\, D\,][\, U\,] = \begin{bmatrix} 1 & 0 & 0 & 0 \\ -1 & 1 & 0 & 0 \\ 0 & -1 & 1 & 0 \\ 0 & 0 & -1 & 1 \end{bmatrix} \begin{bmatrix} 1 & 0 & 0 & 0 \\ 0 & 1 & 0 & 0 \\ 0 & 0 & 2 & 0 \\ 0 & 0 & 0 & 1 \end{bmatrix} \begin{bmatrix} 1 & -1 & 0 & 0 \\ 0 & 1 & -1 & 0 \\ 0 & 0 & 1 & -1 \\ 0 & 0 & 0 & 1 \end{bmatrix}$$

Note that all this information can be stored in-place as

$$[U^T D U] = \begin{bmatrix} 1 & -1 & 0 & 0 \\ -1 & 1 & -1 & 0 \\ 0 & -1 & 2 & -1 \\ 0 & 0 & -1 & 1 \end{bmatrix}$$

It is significant that this array has the same bandedness as the original.

## Column-wise Sequence

The above recurrence equations imply that the diagonal term $D_{ii}$ is computed first, followed by the calculation of the other terms in the $i$-th row of $[U]$. This row-wise generation of terms requires twice as many multiplications as actually needed. This increase in the number of operations can be avoided by changing to the following column-wise sequence:

$$U_{ij} = \frac{1}{D_{ii}} \left( K_{ij} - \sum_{k=1}^{i-1} D_{kk} U_{ki} U_{kj} \right) \qquad (1 \leq j \leq N) \ \ (1 \leq i \leq j)$$

$$D_{jj} = K_{jj} - \sum_{k=1}^{j-1} D_{kk} U_{kj} U_{kj} \qquad (1 \leq j \leq N)$$

Note that the product $D_{kk} U_{kj}$ appears in both equations, so let this product be

$$U_{kj}^* = D_{kk} U_{kj}$$

and perform the calculations for $U_{ij}$ and $D_{jj}$ in the following manner:

$$U_{ij}^* = K_{ij} - \sum_{k=1}^{i-1} U_{ki} U_{kj}^* \qquad (1 \leq j \leq N) \ \ (1 \leq i \leq j)$$

$$D_{jj} = K_{jj} - \sum_{k=1}^{j-1} U_{kj} U_{kj}^* \qquad (1 \leq j \leq N) \tag{7.2}$$

where $U_{kj} = U_{kj}^*/D_{kk}$ Thus, for column $j$ the intermediate product $U_{ij}^*$ is obtained for each off-diagonal term after the first. Then, the diagonal term $D_{jj}$ is computed, during which calculation the final value of each off-diagonal term is also found. By this sequence of operations, the number of multiplications is reduced to that for the usual Cholesky method; in addition, the calculation of square roots is avoided. In fact, it is slightly more efficient for that reason.

**Example 7.3:** Decompose the array of the last example, but this time use the column-wise sequence.
Start with $j = 1$, obtain

$$D_{11} = K_{11} = 1$$

Now increment $j$ to 2, and take $i = 1$ to obtain

$$U_{12}^* = K_{12} = -1$$
$$U_{12} = (U_{12}^*)/D_{11} = -1$$
$$D_{22} = (K_{22} - U_{12} U_{12}^*) = 1$$

Now increment $j$ to 3, and vary $i = 1, 2$ to obtain

$$
\begin{aligned}
U_{13}^* &= K_{13} = 0 \\
U_{23}^* &= K_{23} - U_{12}U_{13}^* = -1 \\
U_{13} &= U_{13}^*/D_{11} = -1 \\
U_{23} &= U_{23}^*/D_{22} = -1 \\
D_{33} &= K_{33} - (U_{13}U_{13}^* + U_{23}U_{23}^*) = 3 - 1 = 2
\end{aligned}
$$

Finally, increment $j$ to 4 and vary $i = 1, 2, 3$ to obtain

$$
\begin{aligned}
U_{14}^* &= K_{14} = 0 \\
U_{24}^* &= K_{24} - U_{12}U_{14}^* = 0 \\
U_{34}^* &= K_{34} - (U_{12}U_{14}^* + U_{23}U_{24}^*) = -2 \\
U_{14} &= U_{14}^*/D_{11} = 0 \\
U_{24} &= U_{24}^*/D_{22} = 0 \\
U_{34} &= U_{34}^*/D_{33} = -2/2 = -1 \\
D_{44} &= K_{44} - (U_{14}U_{14}^* + U_{24}U_{24}^* + U_{34}U_{34}^*) = 3 - 2 = 1
\end{aligned}
$$

This gives the same decomposition as before. The operations count, however, is quite different; 16 for the present example, 26 for the last example.

## Solution by Forward and Back Substitution

Assume that the following system of linear algebraic equations are to be solved:

$$[K]\{u\} = \{P\}$$

in which $\{u\}$ is a column vector of $N$ unknowns and $\{P\}$ is a column vector of $N$ known constant terms. As a preliminary step, replace $[K]$ with its decomposition to obtain

$$[U]^T\lceil D \rfloor[U]\{u\} = \{P\}$$

Now define the vector $\{y\}$ to be

$$[U]\{u\} \equiv \{y\}$$

In expanded form this expression is

$$
\begin{bmatrix}
1 & U_{12} & U_{13} & \cdots & U_{1N} \\
0 & 1 & U_{23} & \cdots & U_{2N} \\
0 & 0 & 1 & \cdots & U_{3N} \\
\cdots & \cdots & \cdots & \cdots & \cdots \\
0 & 0 & 0 & \cdots & 1
\end{bmatrix}
\begin{Bmatrix}
u_1 \\ u_2 \\ u_3 \\ \cdots \\ u_N
\end{Bmatrix}
=
\begin{Bmatrix}
y_1 \\ y_2 \\ y_3 \\ \cdots \\ y_N
\end{Bmatrix}
$$

In addition, define the vector $\{z\}$ to be

$$\lceil D \rfloor\{y\} \equiv \{z\}$$

for which the expanded form is

$$
\begin{bmatrix}
D_{11} & 0 & 0 & \cdots & 0 \\
0 & D_{22} & 0 & \cdots & 0 \\
0 & 0 & D_{33} & \cdots & 0 \\
\cdots & \cdots & \cdots & \cdots & \cdots \\
0 & 0 & 0 & \cdots & D_{NN}
\end{bmatrix}
\begin{Bmatrix}
y_1 \\ y_2 \\ y_3 \\ \cdots \\ y_N
\end{Bmatrix}
=
\begin{Bmatrix}
z_1 \\ z_2 \\ z_3 \\ \cdots \\ z_N
\end{Bmatrix}
$$

From these it is apparent that

$$ [U]^T \{z\} \equiv \{P\} $$

Or, in expanded form:

$$
\begin{bmatrix}
1 & 0 & 0 & \cdots & 0 \\
U_{12} & 1 & 0 & \cdots & 0 \\
U_{13} & U_{23} & 1 & \cdots & 0 \\
\cdots & \cdots & \cdots & \cdots & \cdots \\
U_{1N} & U_{2N} & U_{3N} & \cdots & 1
\end{bmatrix}
\begin{Bmatrix}
z_1 \\ z_2 \\ z_3 \\ \cdots \\ z_N
\end{Bmatrix}
=
\begin{Bmatrix}
P_1 \\ P_2 \\ P_3 \\ \cdots \\ P_N
\end{Bmatrix}
$$

The original vector of unknowns $\{u\}$ may now be obtained in three steps using the above three intermediate products. First solve for the vector $\{z\}$. Since $[U]^T$ is a lower triangular matrix, each element can be calculated in a series of forward substitutions. For example, the first few terms are

$$ z_1 = P_1, \qquad z_2 = P_2 - U_{12}z_1, \qquad z_3 = P_3 - U_{13}z_1 - U_{23}z_2 $$

In general, the recurrence formula becomes

$$ z_i = P_i - \sum_{k=1}^{i-1} U_{ki}z_k \qquad (1 \le i \le N) $$

The second step consists of solving for the vector $\{y\}$. Since $\lceil D \rfloor$ is a diagonal matrix, each element can be found simply by dividing terms in $\{z\}$ by the corresponding diagonal term of $\lceil D \rfloor$, as follows:

$$ y_i = \frac{z_i}{D_{ii}} \qquad (1 \le i \le N) $$

The recurrence formula can be applied in either a forward or a backward sequence. Finally, the vector $\{u\}$ can now be found. Since $[U]$ is an upper triangular matrix, each element is determined in a backward substitution procedure. The last few terms are

$$ u_N = y_N, \quad u_{N-1} = y_{N-1} - U_{N-1,N}u_N, \quad u_{N-2} = y_{N-2} - U_{N-2,N-1}u_{N-1} - U_{N-2,N}u_N $$

and so on. In general, the elements of $\{u\}$ may be calculated from the recurrence formula:

$$u_N = y_N, \qquad u_i = y_i - \sum_{k=i+1}^{N} U_{ik} u_k \qquad (N \geq i \geq 1)$$

This step completes the solution of the original equations for the unknown quantities. The solution procedure involves both a forward and a backward substitution.

It is apparent that once a decomposition has been obtained, the solution to many different load problems requires only the forward and back substitution. We will make use of this feature when solving dynamic problems; for example, in Chapter 13 we will pose transient problems as a sequence of pseudo-static problems with the load vector changing but the stiffness remaining constant.

**Example 7.4:**  Solve the following system of equations

$$\begin{bmatrix} 1 & -1 & 0 & 0 \\ -1 & 2 & -1 & 0 \\ 0 & -1 & 3 & -2 \\ 0 & 0 & -2 & 3 \end{bmatrix} \begin{Bmatrix} u_1 \\ u_2 \\ u_3 \\ u_4 \end{Bmatrix} = \begin{Bmatrix} 1 \\ 2 \\ 3 \\ 4 \end{Bmatrix}$$

This is the same array as in the previous examples, so using its $[U^T D U]$ form of decomposition, gives

$$\begin{bmatrix} 1 & 0 & 0 & 0 \\ -1 & 1 & 0 & 0 \\ 0 & -1 & 1 & 0 \\ 0 & 0 & -1 & 1 \end{bmatrix} \begin{bmatrix} 1 & 0 & 0 & 0 \\ 0 & 1 & 0 & 0 \\ 0 & 0 & 2 & 0 \\ 0 & 0 & 0 & 1 \end{bmatrix} \begin{bmatrix} 1 & -1 & 0 & 0 \\ 0 & 1 & -1 & 0 \\ 0 & 0 & 1 & -1 \\ 0 & 0 & 0 & 1 \end{bmatrix} \begin{Bmatrix} u_1 \\ u_2 \\ u_3 \\ u_4 \end{Bmatrix} = \begin{Bmatrix} 1 \\ 2 \\ 3 \\ 4 \end{Bmatrix}$$

This is broken into a series of substitution problems corresponding to the three square arrays. The first of these requires solving

$$\begin{bmatrix} 1 & 0 & 0 & 0 \\ -1 & 1 & 0 & 0 \\ 0 & -1 & 1 & 0 \\ 0 & 0 & -1 & 1 \end{bmatrix} \begin{Bmatrix} z_1 \\ z_2 \\ z_3 \\ z_4 \end{Bmatrix} = \begin{Bmatrix} 1 \\ 2 \\ 3 \\ 4 \end{Bmatrix}$$

This gives, in sequence,

$$z_1 = 1, \qquad z_2 = 3, \qquad z_3 = 6, \qquad z_4 = 10$$

These form the right hand side of the second problem

$$\begin{bmatrix} 1 & 0 & 0 & 0 \\ 0 & 1 & 0 & 0 \\ 0 & 0 & 2 & 0 \\ 0 & 0 & 0 & 1 \end{bmatrix} \begin{Bmatrix} y_1 \\ y_2 \\ y_3 \\ y_4 \end{Bmatrix} = \begin{Bmatrix} 1 \\ 3 \\ 6 \\ 10 \end{Bmatrix}$$

This simply requires dividing through by the diagonal terms to give

$$y_1 = 1, \qquad y_2 = 3, \qquad y_3 = 3, \qquad y_4 = 10$$

The final step is the back substitution in

$$
\begin{bmatrix} 1 & -1 & 0 & 0 \\ 0 & 1 & -1 & 0 \\ 0 & 0 & 1 & -1 \\ 0 & 0 & 0 & 1 \end{bmatrix}
\begin{Bmatrix} u_1 \\ u_2 \\ u_3 \\ u_4 \end{Bmatrix}
=
\begin{Bmatrix} 1 \\ 3 \\ 3 \\ 10 \end{Bmatrix}
$$

This gives, in sequence,

$$
u_4 = 10, \qquad u_3 = 13, \qquad u_2 = 16, \qquad u_1 = 17
$$

The correctness of this solution is verified by substituting it into the original system and multiply out. That is,

$$
\begin{bmatrix} 1 & -1 & 0 & 0 \\ -1 & 2 & -1 & 0 \\ 0 & -1 & 3 & -2 \\ 0 & 0 & -2 & 3 \end{bmatrix}
\begin{Bmatrix} 17 \\ 16 \\ 13 \\ 10 \end{Bmatrix}
=
\begin{Bmatrix} 17 - 16 \\ -17 + 32 - 13 \\ -16 + 39 - 20 \\ -26 + 30 \end{Bmatrix}
=
\begin{Bmatrix} 1 \\ 2 \\ 3 \\ 4 \end{Bmatrix}
$$

This process of verifying the solution is often used to assess if significant round-off errors have accumulated. Note, however, that the original matrix (and not its decomposed form) must be used, thus requiring additional storage.

## Solving Banded Systems of Equations

The final component to our system solver is to be able to take advantage of the banded nature of the arrays that usually occur in structural analyses. There are only two modifications we need do. The first involves setting limits on $(i, j)$ so that they address elements only inside the banded region. The second involves re-mapping the matrix addressing since the matrix $[\,K\,]$ is actually stored in upper banded form.
    The recursion formulas for the decomposition are

$$
U_{ij}^* \;=\; K_{ij} - \sum_{k=j1}^{i-1} U_{ki} U_{kj}^* \qquad (1 \le j \le N) \;\; (j1 \le i \le j)
$$

$$
D_{jj} \;=\; K_{jj} - \sum_{k=j1}^{j-1} U_{kj} U_{kj}^* \qquad (1 \le j \le N)
$$

where $U_{kj} = U_{kj}^* / D_{kk}$ and $j1 = \max(j - B + 1, 1)$. The corresponding recursions for the substitution sequences are

$$
z_i \;=\; B_i - \sum_{k=j1}^{i-1} U_{ki} z_k, \qquad\qquad j1 = \max(i - B + 1, 1) \;\; (1 \le i \le N)
$$

$$
y_i \;=\; \frac{z_i}{D_{ii}} \qquad\qquad\qquad\qquad (1 \le i \le N)
$$

$$
x_N = y_N, \qquad x_i = y_i - \sum_{k=i+1}^{j2} U_{ik} x_k, \qquad j2 = \min(i + B - 1, N) \;\; (N \ge i \ge 1)
$$

The mapping of the addresses is achieved simply as

$$K(i,j) = K^{ub}(i, j - B + 1)$$

where the superscript $ub$ means upper banded.

It happens that multiplication and division are far more computationally costly than addition or subtraction, hence the computational cost is usually based only on the former. The cost for the modified Cholesky method is

$$cost = \tfrac{1}{2}NB^2 - \tfrac{1}{3}B^3 + \tfrac{1}{2}B^2 - \tfrac{2}{3}B$$

For large systems this shows that the computational cost of the decomposition is about $\tfrac{1}{2}NB^2$. It is seen that if advantage is not taken of the bandedness, then the method is essentially on the order of $\tfrac{1}{6}N^3$; since $B$ is typically less than 10% of $N$ this translates in a computational difference of at least 100. The cost for the solution stage is about

$$cost = 2NB + N$$

from which it is apparent that the decomposition is the significant part of the total solution cost.

To put this computational cost into perspective, consider solving a medium sized problem (1000 degrees of freedom) on our benchmark machine of 1 MFlops. Assuming a 10% bandwidth, we have

$$cost \approx \tfrac{1}{2} \times 1000 \times 100^2 \div (1 \times 10^6) = 5 \; seconds$$

On the other hand, a large problem with 10,000 degrees of freedom would take about one hour.

## 7.5   Solving Eigenvalue Problems

Solving large eigenvalue problems is one of the most challenging problems of computational mechanics; it is time-consuming, has large memory requirements, and is fraught with many pit-falls. From the numerical point of view, it is essentially a non-linear problem and therefore requires many iterations for a solution.

The eigenvalue problems that arise in vibration and stability studies are represented, respectively, by

$$[K]\{u\} - \lambda[M]\{u\} = 0, \qquad [K]\{u\} + P[K_G]\{u\} = 0$$

For these cases, the eigenpairs $(\lambda, \{u\})$ and $(P, \{u\})$ are unknown and need to be determined. This form is called the *generalized eigenvalue problem*. The simpler *standard* form is represented by

$$[K]\{u\} - \lambda\{u\} = 0, \qquad [K]\{u\} + P\{u\} = 0$$

More details on the properties of eigenvalues and eigenvectors can be found in Chapter 11; here we use only those properties necessary to help us estimate the first few lowest eigenvalues of the system.

We will concentrate on the following form of the eigenvalue problem

$$[K]\{u\} - \lambda[M]\{u\} = 0$$

In vibration problems $\lambda = \omega^2$ (where $\omega$ is the vibration frequency) and therefore $\lambda$ is always positive. On the other hand, in stability problems $\lambda = -P$ and therefore it can be positive or negative (as we have already seen in Example 5.5). To avoid loss of generality in the following, we will therefore allow $\lambda$ to be positive or negative.

## Reduction to Standard Form

A simple way of reducing the general system to standard form is to multiply across by the inverse of the mass matrix, that is,

$$[M]^{-1}[K]\{u\} = \lambda\{u\}$$

Unfortunately, the product $[M]^{-1}[K]$ is no longer symmetric and this adds overhead to any solution algorithm. There is, however, another way that maintains the symmetry of the matrices.

Note that the mass matrix can be decomposed as

$$[M] = [U]^T[U]$$

Pre-multiply both sides of the eigenvalue problem by $[U]^{T^{-1}}$ and realizing that the product of a matrix with its inverse is the unity matrix, get

$$[U]^{T^{-1}}[K][U]^{-1}[U]\{u\} = \lambda[U]^{T^{-1}}[U]\{u\}$$

Now introduce the new arrays

$$[A] \equiv [U]^{T^{-1}}[K][U]^{-1}, \qquad \{z\} \equiv [U]\{u\}$$

This allows the problem to be written in the standard form

$$[A]\{z\} = \lambda\{z\}$$

The eigenvalues obtained in this way are identical to those of the original system. To obtain the actual eigenvectors, it is necessary to do a transformation of the form

$$\{u\} = [U]^{-1}\{z\}$$

This scheme requires pre-processing of the data which adds overhead to the computational cost. This cost is on the order of $NB^2$ for the decomposition and inversions; it is quite significant when only a few eigenvalues are to be computed. The scheme also has difficulties if there are zero masses since then the decomposition will fail.

The method to be presented next works directly on the general eigenvalue problem and is efficient when only a few of the eigenvalues and eigenvectors are required. Chapter 13 will present some other schemes also.

## Vector Iteration Methods

As has been pointed out already, in the solution of an eigenvalue problem, we need to use iteration. In the vector iteration methods, the basic idea is to guess a solution, and by repeatedly substituting it in the equations for the eigensystem, to converge to a solution.

The relation considered is

$$[K]\{\phi\} = \lambda[M]\{\phi\}$$

The aim is to satisfy this equation by directly operating upon it. Keep in mind that if a solution $\{\phi\}$ is found, then any scalar multiplication of it is also a solution. The *normalization conditions* are used to remove this ambiguity; we will find the solution $\{\phi\}$ which satisfies

$$\{\phi\}^T[K]\{\phi\} = \lambda, \qquad \{\phi\}^T[M]\{\phi\} = 1$$

Let us assume a trial vector for $\{\phi\}$, say $\{u\}_1$, and introduce on the right-hand side an equivalent load vector of

$$\{R\}_1 = [M]\{u\}_1$$

The original eigenvalue problem will now resemble an equilibrium equation as encountered in a static analysis, which we may write as

$$[K]\{u\}_2 = \lambda\{R\}_1 \quad \text{or} \quad [K]\{\tilde{u}\}_2 = \{R\}_1$$

where $\{u\}_2 = \lambda\{\tilde{u}\}_2$ is the (scaled) displacement solution corresponding to the applied forces $\{R\}_1$. (In all of the following, $\lambda$ appears just as a scaling factor.) It is intuitive that now $\{u\}_2$ should be a better approximation to an eigenvector than was the original guess $\{u\}_1$. Thus, by repeating the cycle, we obtain an increasingly better approximation to an eigenvector. This procedure is the basis of *inverse iteration*. Note that if we use the decomposed form of $[K]$ then the solution part of the iteration involves only the forward and back substitutions.

Before proceeding with the development of an algorithm, we need to establish the conditions under which convergence is achieved. Only a brief outline of the convergence discussion is given here, more details can be found in References [5, 31]. Since the eigenvectors are linearly independent of each other, any arbitrary vector can be written as a linear combination of them. (This is called the *expansion theorem*.) Doing this with our guess, we have

$$\{u\} = c_1\{\phi\}_1 + c_2\{\phi\}_2 + \ldots = \{c\}^T\{\phi\}$$

where $\{c\}^T = \{c_1, c_2, \ldots\}$ is the vector of coefficients. The load vector may now be written as

$$\begin{aligned}
\{R\} = [M]\{u\} &= c_1[M]\{\phi\}_1 + c_2[M]\{\phi\}_2 + c_3[M]\{\phi\}_3 + \ldots \\
&= \frac{1}{\lambda_1}\left[c_1[K]\{\phi\}_1 + c_2\frac{\lambda_1}{\lambda_2}[K]\{\phi\}_2 + c_3\frac{\lambda_1}{\lambda_3}[K]\{\phi\}_3 + \ldots\right]
\end{aligned}$$

Since the roots are ordered so that $\lambda_1 < \lambda_2 < \lambda_3 \ldots$ the ratios $\lambda_1/\lambda_n$ are less than or equal to unity. Hence, with each iteration each guess will become more and more dominated by the vector $\{\phi\}_1$. That is, there will be convergence to the fundamental mode. It must be realized that for each iteration, effectively new coefficients $\{c\}$ are calculated.

Convergence depends on the ratio $\lambda_1/\lambda_2$, the smaller this is, the faster the process converges. It also depends on the relative strengths of the modes in the initial guess, thus if $\{u\}_1$ has a large component of the first eigenvector $\{\phi\}_1$, i.e., $c_1 \gg c_2, c_3, \ldots$, relatively few iteration steps are required for convergence. On the other hand, if the first guess is totally deficient of the first mode, i.e., $c_1 = 0$, then theoretically the process will not find $\{\phi\}_1$. In practice, however, rounding errors arising from finite precision in the computer will eventually produce a component of the first mode in one of the subsequent guesses, and this will eventually become dominant. It is expected, therefore, that convergence will always be to the fundamental mode, although the rate depends on the initial guess. For this reason, it is usual to take the initial guess to be fully populated with ones. Repeated roots ($\lambda_2 = \lambda_1$, say) is obviously a problem but we postpone its solution until Chapter 13.

## Computer Algorithm

The iteration is begun by assuming a starting iteration vector $\{u\}_1$. The basic step in the iteration scheme is the solution of the pseudo-static problem in which we evaluate a vector $\{\tilde{u}\}_{k+1}$ from

$$[K]\{\tilde{u}\}_{k+1} = \{R\}_k = [M]\{u\}_k$$

At each step, the results are normalized by

$$\{u\}_{k+1} = \frac{\{\tilde{u}\}_{k+1}}{\left[\{\tilde{u}\}_{k+1}^T[M]\{\tilde{u}\}_{k+1}\right]^{1/2}}$$

This normalization calculation merely assures that the length of the new iteration vector is unity with respect to the mass. That is,

$$\{u\}_{k+1}^T[M]\{u\}_{k+1} = 1$$

If this scaling is not included in the iteration, the elements of the iteration vectors grow (or decrease) in each step and do not converge to $\{\phi\}_1$ but to a multiple of it. Provided that $\{u\}_1$ is not $M$-orthogonal to $\{\phi\}_1$, meaning that $\{u\}_1^T[M]\{\phi\}_1 \neq 0$, we then have the following limits

$$\{u\}_{k+1} \to \{\phi\}_1 \qquad \text{as} \quad k \to \infty$$

The steps of the algorithm may now be stated as:

**Step 1:** In preparation for solving a system of equations (the pseudo-static problem), decompose the stiffness matrix as:

$$[K] = [U]^T[D][U]$$

**Step 2:** Pick an arbitrary load vector, generally populated with ones.

$$\{R\}_1^T = \{1, 1, \ldots, 1\}$$

**Step 3:** Obtain improved vector $\{\tilde{u}\}_{k+1}$ by solving the system

$$[U]^T[D][U]\{\tilde{u}\}_{k+1} = \{R\}_k$$

**Step 4:** Form the new load vector

$$\{Q\}_{k+1} = [M]\{\tilde{u}\}_{k+1}$$

**Step 5:** Form the norm

$$\|Z\| \equiv \{\tilde{u}\}_{k+1}^T[M]\{\tilde{u}\}_{k+1} = \{\tilde{u}\}_{k+1}^T\{Q\}_{k+1}$$

The latter form is used since $\{Q\}$ is already known.

**Step 6:** Use the Rayleigh quotient to get an estimate of the eigenvalue from

$$\rho_{k+1} \equiv \frac{\{\tilde{u}\}_{k+1}^T[K]\{\tilde{u}\}_{k+1}}{\{\tilde{u}\}_{k+1}^T[M]\{\tilde{u}\}_{k+1}} = \frac{\{\tilde{u}\}_{k+1}^T\{R\}_k}{\|Z\|}$$

When $k$ is sufficiently large, $\rho_{k+1}$ converges to the lowest eigenvalue $\lambda_1$.

**Step 7:** Normalize the load by

$$\{R\}_{k+1} = \frac{\{Q\}_{k+1}}{\|Z\|^{1/2}}$$

**Step 8:** Check for convergence. Convergence can be said to have been achieved when successive values of $\rho$ satisfy

$$\frac{|\rho_{k+1} - \rho_k|}{\rho_{k+1}} \leq TOL$$

where $TOL$ is the specified tolerance. Typically, it is taken as $TOL = 10^{-2s}$, giving the eigenvalues accurate to $2s$ digits. Usually, $2s$ is taken as 10. If convergence is not achieved go back to Step 3.

**Step 9:** After convergence is achieved, the eigenvalue and eigenvectors are given by

$$\lambda = \rho_{k+1}, \qquad \{\phi\} = \frac{\{\tilde{u}\}_{k+1}}{\|Z\|^{1/2}}$$

If $q$ iterations are performed, the computational cost is about

$$cost = \tfrac{1}{2}NB^2 + q\left[2NB + 2NB + N + N + N\right] \approx \tfrac{1}{2}NB^2 + 4NBq$$

Usually about 10 iterations are required for convergence (although an upper limit of 20 is often set in case convergence is not achieved). This shows that the computational cost of solving for a *single* eigenpair is on the order of the time to solve a complete statics problem since the cost is dominated by the decomposition stage.

Now a point as regards the accuracy of the eigenvalues and vectors that are found. At any stage in the iteration, the guess can be written as

$$\{u\} = \{\phi\} + \epsilon\{v\}$$

indicating that the guess is only a small perturbation away from the true mode. Using this in the Rayleigh quotient gives an estimate for the eigenvalue as

$$\rho = \frac{\{u\}^T[K]\{u\}}{\{u\}^T[M]\{u\}} = \frac{\lambda + \epsilon\left[\{v\}^T[K]\{\phi\} + \{\phi\}^T[K]\{v\}\right] + \epsilon^2\{v\}^T[K]\{v\}}{1 + \epsilon\left[\{v\}^T[M]\{\phi\} + \{\phi\}^T[M]\{v\}\right] + \epsilon^2\{v\}^T[M]\{v\}}$$

where the normalization conditions on $\{\phi\}$ have been used. The terms inside the square brackets are also zero since $\{v\}$ is orthogonal to $\{\phi\}$. From this, we therefore get the result

$$\rho \approx \lambda + \epsilon^2(\ldots)$$

Thus, if the eigenvector is determined to an accuracy of $\epsilon$, the eigenvalues will be determined to an accuracy of $\epsilon^2$. For example, suppose the eigenvectors are known to $\pm 10^{-5}$, then the eigenvalues will be known to $\pm 10^{-10}$.

**Example 7.5:**   Determine the lowest eigenvalue and the corresponding eigenvector for the following system.

$$\left[\begin{bmatrix} 1 & -1 & 0 & 0 \\ -1 & 2 & -1 & 0 \\ 0 & -1 & 3 & -2 \\ 0 & 0 & -2 & 3 \end{bmatrix} - \lambda \begin{bmatrix} 1 & 0 & 0 & 0 \\ 0 & 2 & 0 & 0 \\ 0 & 0 & 3 & 0 \\ 0 & 0 & 0 & 4 \end{bmatrix}\right] \begin{Bmatrix} u_1 \\ u_2 \\ u_3 \\ u_4 \end{Bmatrix} = 0$$

Step 1 of the algorithm requires the decomposition of the stiffness matrix. This is the same array as in the previous example, so using its $[U^TDU]$ form of decomposition gives

$$[K] = \begin{bmatrix} 1 & 0 & 0 & 0 \\ -1 & 1 & 0 & 0 \\ 0 & -1 & 1 & 0 \\ 0 & 0 & -1 & 1 \end{bmatrix} \begin{bmatrix} 1 & 0 & 0 & 0 \\ 0 & 1 & 0 & 0 \\ 0 & 0 & 2 & 0 \\ 0 & 0 & 0 & 1 \end{bmatrix} \begin{bmatrix} 1 & -1 & 0 & 0 \\ 0 & 1 & -1 & 0 \\ 0 & 0 & 1 & -1 \\ 0 & 0 & 0 & 1 \end{bmatrix}$$

Step 2 chooses an initial guess for the effective load vector. Take

$$\{R\}_1^T = \{1, 1, 1, 1\}$$

Step 3 solves the static problem

$$[K]\{\tilde{u}\} = \{R\}$$

using the initial guess for the load vector. The solution is

$$\{\tilde{u}\} = \{8.5000, 7.5000, 5.5000, 4.0000\}$$

Step 4 computes the temporary vector

$$\{Q\} = [M]\{\tilde{u}\} = \{8.5000, 15.0000, 16.5000, 16.0000\}$$

The norm from Step 5 is now computed as

$$\|Z\| \equiv \{\tilde{u}\}^T\{Q\} = 18.4255$$

The first estimate of the eigenvalue is given by

$$\rho = \frac{\{\tilde{u}\}^T\{R\}}{\|Z\|} = 0.0751104$$

In anticipation of the need for another iteration, we compute in Step 7 another load vector

$$\{R\} = \frac{\{Q\}}{\|Z\|^{1/2}} = \{.4613, .8140, .8954, .8683\}$$

We now check for convergence according to

$$CONV = \frac{|\rho_{k+1} - \rho_k|}{\rho_{k+1}} \le TOL$$

This criterion is based on successive estimates of $\rho$, and we only have one, we must repeat the process.

Just the major numbers will now be stated.

*Iteration 2 :*
$$\{\tilde{u}\} = \{5.8614, 5.4001, 4.1247, 3.0392\}$$
$$\{Q\} = \{5.8614, 10.8002, 12.3741, 12.1570\}$$
$$\|Z\| = 13.441248$$
$$\rho = 0.0743521$$
$$\{R\} = \{.4360, .8035, .9206, .9044\}$$
$$CONV = 1.019899 \times 10^{-2}$$

*Iteration 3 :*
$$\{\tilde{u}\} = \{5.8204, 5.3843, 4.1447, 3.0646\}$$
$$\{Q\} = \{5.8204, 10.7687, 12.4342, 12.2586\}$$

$$\|Z\| = 13.452346$$
$$\rho = 0.0743352$$
$$\{R\} = \{.4326\,, .8005\,, .9243\,, .9112\}$$
$$CONV = 2.268693 \times 10^{-4}$$

*Iteration 4 :*

$$\{\tilde{u}\} = \{5.8133\,, 5.3806\,, 4.1475\,, 3.0687\}$$
$$\{Q\} = \{5.8133\,, 10.7613\,, 12.4425\,, 12.2750\}$$
$$\|Z\| = 13.452642$$
$$\rho = 0.0743348$$
$$\{R\} = \{.4321\,, .7999\,, .9249\,, .9124\}$$
$$CONV = 6.365913 \times 10^{-6}$$

*Iteration 5 :*

$$\{\tilde{u}\} = \{5.8121\,, 5.3800\,, 4.1479\,, 3.0694\}$$
$$\{Q\} = \{5.8121\,, 10.7600\,, 12.4438\,, 12.2778\}$$
$$\|Z\| = 13.452651$$
$$\rho = 0.0743347$$
$$\{R\} = \{.4320\,, .7998\,, .9250\,, .9126\}$$
$$CONV = 1.827175 \times 10^{-7}$$

*Iteration 6 :*

$$\{\tilde{u}\} = \{5.8119\,, 5.3799\,, 4.1480\,, 3.0695\}$$
$$\{Q\} = \{5.8119\,, 10.7598\,, 12.4440\,, 12.2783\}$$
$$\|Z\| = 13.452651$$
$$\rho = 0.0743347$$
$$\{R\} = \{.4320\,, .7998\,, .9250\,, .9127\}$$
$$CONV = 5.255246 \times 10^{-10}$$

*Iteration 7 :*

$$\{\tilde{u}\} = \{5.8119\,, 5.3799\,, 4.1480\,, 3.0695\}$$
$$\{Q\} = \{5.8119\,, 10.7598\,, 12.4441\,, 12.2783\}$$
$$\|Z\| = 13.452651$$
$$\rho = 0.0743347$$
$$\{R\} = \{.4320\,, .7998\,, .9250\,, .9127\}$$
$$CONV = 1.511786 \times 10^{-10}$$

*Iteration 8 :*

$$\{\tilde{u}\} = \{5.8119\,, 5.3799\,, 4.1480\,, 3.0696\}$$
$$\{Q\} = \{5.8119\,, 10.7598\,, 12.4441\,, 12.2784\}$$
$$\|Z\| = 13.452651$$
$$\rho = 0.0743347$$
$$\{R\} = \{.4320\,, .7998\,, .9250\,, .9127\}$$

$$CONV = 4.349014 \times 10^{-12}$$

Convergence has finally been achieved with the eigenvalue $\rho = 0.0743347$ and the eigenvector obtained from

$$\{\phi\} = \frac{\{\tilde{u}\}}{\|Z\|^{1/2}} = \{.4320\,,.3999\,,.3083\,,.2281\}$$

## Vector Iteration with Shifts

The inverse vector iteration scheme will always converge to the lowest eigenvalue and the corresponding eigenvector. Convergence to the second lowest eigenvalue can, however, be achieved provided that the trial vector is such that in its representation as a superposition of modes, there is no contribution of the first mode. In general, an arbitrary trial vector cannot be expected to be orthogonal to the first mode. If, however, the first mode shape has already been determined, the new trial vector can be modified to sweep away the first mode shape from it. This is called *Gram-Schmidt orthogonalization*. In practice, since each eigenvector is obtained only approximately, the reduction process gives guesses contaminated with the swept eigenvectors. This is especially true after a number of the modes have been determined. Strictly speaking, it is necessary to orthogonalize the guesses at each new iteration. This adds considerable overhead to the computational cost.

A more effective scheme for finding least dominant eigenvalues is to use vector iteration combined with shifts of the eigenvalue axis. This is most effective when a sequence of eigenvalues is required because the previous values can then be used to estimate the shifts.

Introduction of a shift $\mu$ allows the eigenvalue to be re-arranged as

$$\lambda = (\lambda - \mu) + \mu = \tilde{\lambda} + \mu$$

Substituting this into the original eigensystem now leads to

$$[K - \mu M]\{\phi\} = (\lambda - \mu)[M]\{\phi\}$$

or, equivalently,

$$[\tilde{K}]\{\phi\} = \tilde{\lambda}[M]\{\phi\}$$

Inverse iteration on this system will converge to the eigenvalue corresponding to the lowest value of $\tilde{\lambda} = (\lambda - \mu)$. This, of course, is the value of $\lambda$ closest to the shift $\mu$. The eigenvectors of the shifted system are identical to those of the original system.

It is thus evident that by choosing a shift close to the desired eigenvalue, a more accurate estimate of the eigenvalue, as well as of the corresponding eigenvector, can be obtained after a relatively small number of inverse iterations. Further, if the shift point is located between eigenvalues $\lambda_n$ and $\lambda_{n+1}$, and $(\mu - \lambda_n)$ is smaller than $(\lambda_{n+1} - \mu)$, iteration will converge to $\lambda_n$. On the other hand, if $(\lambda_{n+1} - \mu)$ is small than $(\mu - \lambda_n)$, iteration will converge to $(\lambda_{n+1} - \mu)$. Obviously, rapid convergence can be achieved if $\mu$ is located close to the desired eigenvalue.

The success of iteration with shift depends on the selection of an appropriate shift point. If the requirement is to obtain the eigenvalues nearest a specified value, the origin is shifted to that value and the selection poses no difficulties. (This situation may occur, for example, in vibration problems if the applied load has a dominant frequency; then the system may be interrogated to find the resonances closest to it.) This is also a powerful method for obtaining eigenvectors when the eigenvalues have been obtained by some other method. In these cases, convergence can occur in about 3 or 4 iterations.

For the general case, however, the shift must be determined by marching through all the lower eigenvalues. To implement this as an algorithm to determine the sequence of the lowest few eigenpairs, inverse iteration must be combined with shifting. The first few eigenvalues and mode shapes are calculated by the standard inverse iteration technique along with Gram-Schmidt orthogonalization. The same process is then employed for the next group, but first the stiffness matrix is shifted and decomposed again. The next set of eigenvalues and mode shapes are then calculated using inverse iteration with the shifted origin.

This scheme, when implemented in the manner above, has difficulties with eigen-clusters and repeated eigenvalues, and unless special precautions are taken many cigcnvalues can be missed. These issues are reconsidered in Chapter 13 when we discuss the *subspace iteration* method for the partial solution of eigenvalue problems.

**Example 7.6:** Use vector iteration with a shift to determine a second eigenvalue and corresponding eigenvector for the system of the last example.

Generally, a shift of any value can be taken; care should be taken, however, not to make the system inadvertently singular. For example, a shift of $\mu = 1$ would put zeros on the first three diagonal locations. Parenthetically, if a system is initially singular a shift can be used to remove the rigid body modes. We will use a shift of $\mu = 1.1$ giving the new system

$$
\left[ \begin{bmatrix} -0.1 & -1 & 0 & 0 \\ -1 & -0.2 & -1 & 0 \\ 0 & -1 & -0.3 & -2 \\ 0 & 0 & -2 & -1.4 \end{bmatrix} - \bar{\lambda} \begin{bmatrix} 1 & 0 & 0 & 0 \\ 0 & 2 & 0 & 0 \\ 0 & 0 & 3 & 0 \\ 0 & 0 & 0 & 4 \end{bmatrix} \right] \begin{Bmatrix} u_1 \\ u_2 \\ u_3 \\ u_4 \end{Bmatrix} = 0
$$

Step 1 of the algorithm requires the decomposition of the stiffness matrix; its $[U^T D U]$ form of decomposition is

$$
[K] = \begin{bmatrix} 1 & 0 & 0 & 0 \\ 10 & 1 & 0 & 0 \\ 0 & -0.102 & 1 & 0 \\ 0 & 0 & 4.97 & 1 \end{bmatrix} \begin{bmatrix} -0.1 & 0 & 0 & 0 \\ 0 & 9.80 & 0 & 0 \\ 0 & 0 & -0.402 & 0 \\ 0 & 0 & 0 & 8.55 \end{bmatrix} \begin{bmatrix} 1 & 10 & 0 & 0 \\ 0 & 1 & -0.102 & 0 \\ 0 & 0 & 1 & 4.97 \\ 0 & 0 & 0 & 1 \end{bmatrix}
$$

A significant point to note is the number of negative diagonal terms. The diagonal matrix, as a bi-product of the decomposition, allows the performance of a *Sturm sequence* count. Simply, a count of the number of negative diagonal terms gives the number of negative eigenvalues. In this case there are two, indicating that there are two roots less than the shift value 1.1. Hence the iteration will converge to either the second or the third root.

The iterations follow as in the last example. We will report just the major results.

*Iteration* 1 :

$$\{\tilde{u}\} = \{-.256502, -.974350, -.548628, .069469\}$$

$$\{Q\} = \{-.256502, -1.948700, -1.645885, .277877\}$$

$$\|Z\| = 1.699056097335189$$

$$\rho = -5.923568139102905E - 001$$

$$\{R\} = \{-.150967, -1.146931, -.968706, .163548\}$$

$$CONV = 1.000000000000000$$

*Iteration* 2 :

$$\{\tilde{u}\} = \{1.623872, -.011420, -.474657, .561261\}$$

$$\{Q\} = \{1.623872, -.022840, -1.423970, 2.245046\}$$

$$\|Z\| = 2.138498161978666$$

$$\rho = 6.987324945197783E - 002$$

$$\{R\} = \{.759351, -.010680, -.665874, 1.049823\}$$

$$CONV = 9.477590759785729$$

*Iteration* 3 :

$$\{\tilde{u}\} = \{1.388462, -.898198, -1.198142, .961757\}$$

$$\{Q\} = \{1.388462, -1.796395, -3.594425, 3.847028\}$$

$$\|Z\| = 3.398217240514243$$

$$\rho = 2.486525580328753E - 001$$

$$\{R\} = \{.408585, -.528629, -1.057738, 1.132072\}$$

$$CONV = 7.189924366563740 \times 10^{-1}$$

*Iteration* 4 :

$$\{\tilde{u}\} = \{1.930304, -.601616, -1.281352, 1.021880\}$$

$$\{Q\} = \{1.930304, -1.203231, -3.844057, 4.087520\}$$

$$\|Z\| = 3.681372134639745$$

$$\rho = 2.670284109674704E - 001$$

$$\{R\} = \{.524344, -.326843, -1.044191, 1.110325\}$$

$$CONV = 6.881609663937110 \times 10^{-2}$$

*Iteration* 6 :

$$\{\tilde{u}\} = \{1.822625, -.659191, -1.312835, 1.053119\}$$

$$\{Q\} = \{1.822625, -1.318381, -3.938504, 4.212475\}$$

$$\|Z\| = 3.714548057965802$$

$$\rho = 2.691383575276401E - 001$$

$$\{R\} = \{.490672, -.354924, -1.060291, 1.134048\}$$
$$CONV = 8.515886482413376 \times 10^{-4}$$

*Iteration* 14 :

$$\{\tilde{u}\} = \{1.804138, -.666049, -1.312361, 1.059767\}$$
$$\{Q\} = \{1.804138, -1.332099, -3.937084, 4.239066\}$$
$$\|Z\| = 3.715030836268836$$
$$\rho = 2.691767697366436E - 001$$
$$\{R\} = \{.485632, -.358570, -1.059772, 1.141058\}$$
$$CONV = 6.130160219766258 \times 10^{-11}$$

Convergence has finally been achieved with the eigenvalue $\lambda = \mu + \rho = 1.1 + 0.2691767697 = 1.3691767697$ and the eigenvectors obtained from

$$\{\phi\} = \frac{\{\tilde{u}\}}{\|Z\|^{1/2}} = \{.485632, -.179285, -.353257, .285265\}$$

This is the third root. The initial guess of unity for the load vector was a rather bad guess, this explains why the first few iterations give very poor convergence. But these results do show that nonetheless convergence will eventually occur.

## Problems

**7.1** Assemble the stiffness matrix for a rod, fixed at one end, modeled with two elements each of length $L$. Assume there is no load applied at the middle node, *condense* the middle degree of freedom (remove it from the system of equations) and show that the resulting stiffness relation is the same as if a single element of length $2L$ had been used.

[Reference [14], pp. 143]

**7.2** Show that the result of the previous exercise holds irrespective of the position of the middle node.

**7.3** A matrix and its inverse are given exactly by

$$[K] = \begin{bmatrix} 5 & 7 & 6 & 5 \\ 7 & 10 & 8 & 7 \\ 6 & 8 & 10 & 9 \\ 5 & 7 & 9 & 10 \end{bmatrix}, \quad [K]^{-1} = \begin{bmatrix} 68 & -41 & -17 & 10 \\ -41 & 25 & 10 & -6 \\ -17 & 10 & 5 & -3 \\ 10 & -6 & -3 & 2 \end{bmatrix}$$

Use floating point arithmetic (carrying four digits) to obtain the inverse. This is an example of an *ill-conditioned* array.

[Reference [32], pp. 126]

**7.4** An eigenvalue problem is described by

$$\left[ \begin{bmatrix} 1 & -1 & 0 \\ -1 & 2 & -1 \\ 0 & -1 & 1 \end{bmatrix} - \lambda \begin{bmatrix} 1 & 0 & 0 \\ 0 & 2 & 0 \\ 0 & 0 & 1 \end{bmatrix} \right] \left\{ \begin{array}{c} u_1 \\ u_2 \\ u_3 \end{array} \right\} = 0$$

Show that the usual inverse iteration algorithm does not work, but that after imposing a shift it does work.

**7.5** Choose two matrices [ $K$ ] and [ $M$ ] for which the eigensolution is known. Write a computer program that computes the matrix products

$$\{u\}^T[\,K\,]\{u\}\,, \qquad \{u\}^T[\,M\,]\{u\}$$

for randomly chosen vectors $\{u\}$. Show that the ratio of the products is never less than the lowest eigenvalue of the corresponding eigenvalue problem. This is known as *Rayleigh's principle*.

## Exercises

**7.1** Write a small program to subtract 1.01 from $N$, $N$ times. Vary $N$ so as to investigate round-off error.

**7.2** Show that

$$\begin{bmatrix} 4 & -6 & -3 \\ -3 & 0 & -3 \\ -1 & -2 & -1 \end{bmatrix} = \begin{bmatrix} 1 & 0 & 0 \\ -.75 & 1 & 0 \\ -.25 & .7778 & 1 \end{bmatrix} \begin{bmatrix} 4 & -6 & -3 \\ 0 & -4.5 & -5.25 \\ 0 & 0 & 2.3333 \end{bmatrix}$$

**7.3** Show that

$$\begin{bmatrix} 4 & -6 & -3 \\ -6 & 0 & -6 \\ -3 & -6 & -3 \end{bmatrix} = \begin{bmatrix} 1 & 0 & 0 \\ -1.5 & 1 & 0 \\ -.75 & 1.1667 & 1 \end{bmatrix} \begin{bmatrix} 4 & 0 & 0 \\ 0 & -9 & 0 \\ 0 & 0 & 7.0003 \end{bmatrix} \begin{bmatrix} 1 & -1.5 & -.75 \\ 0 & 1 & 1.1667 \\ 0 & 0 & 1 \end{bmatrix}$$

**7.4** An eigenvalue problem is described by

$$\left[ \begin{bmatrix} 5 & -4 & 1 & 0 \\ -4 & 6 & -4 & 1 \\ 1 & -4 & 6 & -4 \\ 0 & 1 & -4 & 5 \end{bmatrix} - \lambda \begin{bmatrix} 2 & 0 & 0 & 0 \\ 0 & 2 & 0 & 0 \\ 0 & 0 & 1 & 0 \\ 0 & 0 & 0 & 1 \end{bmatrix} \right] \left\{ \begin{array}{c} u_1 \\ u_2 \\ u_3 \\ u_4 \end{array} \right\} = 0$$

Use inverse iteration to find the lowest mode.
$$[\lambda = 0.09654\,, \quad \{u\} = \{0.3126, 0.4955, 0.4791, 0.2898\}]$$

**7.5** Use a shift of $\mu = 10.0$ for the arrays of the last example to show find the fourth mode results.     $[\lambda = 10.6385\,, \quad \{u\} = \{-0.1076, 0.2556, -0.7283, 0.5620\}]$

**7.6** A stiffness matrix of size [16×16] has a semi-bandwidth of 2. The main diagonal terms are $k_{11} = 110$, $K_{ii} = 120 + i$ for $i = 2, 15$, and $K_{1616} = 126$ with all the off-diagonal terms being $-10$. Treat this as a standard eigenvalue problem and use vector iteration to find the eigenvalues and eigenvectors.
$$[\lambda_1 = 102.2, \lambda_4 = 113.4, \lambda_8 = 124.7, \lambda_{12} = 139.2, \lambda_{16} = 151.5]$$

# Chapter 8

# Dynamics of Elastic Systems

This chapter gives an introduction to the dynamics of elastic systems. We restrict the emphasis to those concepts that will be used directly in later chapters. The response of a simple spring-mass system will be used to motivate these concepts. References [25, 40, 44] are good sources for additional details on the material covered.

The concept of vibration is fundamental to understanding the dynamics of structures. The study of vibration is concerned with the oscillatory motion of bodies; all bodies with elasticity and mass are capable of exhibiting it. Resonant (or natural) frequencies are those frequencies at which the structure exhibits relatively large vibration responses for relatively small inputs. Even if the excitation forces are not sinusoidal, these frequencies will tend to dominate the response.

In practice, large resonant responses are mitigated by the presence of damping, and so both resonance and damping must be considered simultaneously. It is difficult to estimate the damping forces in a real structure with the same accuracy as the elastic and inertia forces; consequently a rigorous computer simulation is seldom possible. So for convenience, the form of damping is usually chosen so that it is conducive to easy mathematical manipulation. Two forms of damping models are discussed.

Allied to the idea of natural frequencies associated with a structure, is the idea that every arbitrary loading history can be considered as composed of a spectrum of frequencies; we therefore introduce spectral analysis because it provides a powerful tool for simplifying the analysis of continuous systems.

## 8.1   Harmonic Motion and Vibration

A vibration executed without the presence of external forces is called a *free vibration*. A simple pendulum is a typical example. Vibration that takes place under the excitation of periodic external forces is called a *forced* vibration. An example of forced vibration is that due to unbalance in rotating machinery.

A vibration motion such as

$$u(t) = A \sin \omega t$$

is called *simple harmonic* motion with an *amplitude A* and an *angular frequency ω*. A plot of this function is shown in Figure 8.1. The time for the response to repeat itself is $T = 2\pi/\omega$ and is called the *period*. The rate of repetition is called the frequency $f = 1/T$. The relation between displacement, velocity and acceleration for the point undergoing harmonic motion is obtained simply by differentiation, that is,

$$
\begin{aligned}
\text{displacement:} \quad u &= A\sin(\omega t) \\
\text{velocity:} \quad \dot{u} &= \omega A\cos(\omega t) \quad = \quad \omega A\sin(\omega t - \pi/2) \\
\text{acceleration:} \quad \ddot{u} &= -\omega^2 A\sin(\omega t) \quad = \quad \omega^2 A\sin(\omega t - \pi)
\end{aligned}
$$

We use the notation of a super dot to mean derivative with respect to time. The behavior of all three responses is harmonic and is shown (scaled) in Figure 8.1. It is obvious that they all have the same shape. What is different is their phase — how much they need to be moved relative to each other so as to overlap exactly. In the above case, for example, the velocity is 90 degrees ($\pi/2$ radians) out of phase with the displacement. Phase plays are very important role in the analysis of vibrating systems. It is apparent from this that a general expression for harmonic motion is $u(t) = A\sin(\omega t + \phi)$ where $\phi$ is a phase shift.

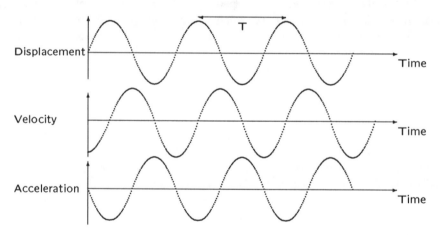

**Figure 8.1**: Simple harmonic motion.

The description of the dynamic response of elastic systems will be motivated by considering the simple case of a single spring-mass system. Consider the free body diagram of the mass attached to the spring of stiffness $k$ as shown in Figure 8.2. We identify four forces acting on the displaced mass. The applied force $P$ is the agent causing the displacement, the elastic force $ku$ attempts to return the mass to its original position, the inertia force $-m\ddot{u}$ acts so as to keep the mass where it is, and finally the damping force $F^d$ also attempts to retard the motion. The equation of motion for the mass is therefore

$$ku + m\ddot{u} + F^d = P(t) \tag{8.1}$$

All real structures experience some sort of dissipation of energy (or damping) when set in motion. This is due to such factors as friction with the surrounding air, and internal friction of the material itself. The scientific nature of friction is still not well known, therefore its treatment in vibration is approached from the point of view of convenience. We will consider the question in more detail later, but as a first attempt at modeling damping, we will look at *viscous* damping as represented mechanically by the dash-pot. The dashpot exerts a retarding force which is proportional to the instantaneous velocity. Thus we write $F^d = c\dot{u}$ where $c$ is the damping constant. The equation of motion that we will mostly discuss is therefore

$$ku + c\dot{u} + m\ddot{u} = P(t) \tag{8.2}$$

where we seek to find $u(t)$ when $P(t)$ is specified.

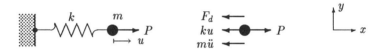

**Figure 8.2**: Simple spring-mass system.

**Example 8.1:**   Determine the motion of the mass in Figure 8.2 after it is displaced from its initial position and released. Assume no damping.

The differential equation of motion (after release) reduces to

$$ku + m\ddot{u} = 0$$

This is a second order differential equation with constant coefficients. We expect solutions of the form

$$u(t) = A\cos \alpha t + B\sin \alpha t$$

where $A$ and $B$ are the constants of integration, and $\alpha$ is, so far, an undetermined constant. Substitute the assumed solution into the differential equation to get

$$[k - m\alpha^2]A\cos \alpha t + [k - m\alpha^2]B\sin \alpha t = 0$$

Since the differential equation must be satisfied for any value of time, then we must have that

$$k - m\alpha^2 = 0$$

This specifies $\alpha$ to be

$$\alpha = \pm\sqrt{\frac{k}{m}} = \pm\omega_o$$

and gives the general solution as

$$u(t) = A \cos \omega_o t + B \sin \omega_o t$$

The arbitrary constants $A$ and $B$ are determined from the *initial conditions*. The problem as stated says that initially the mass is displaced and then released from rest. The initial conditions are therefore that at $t = 0$ we have

$$u(0) = u_o, \qquad \dot{u}(0) = 0$$

This gives $A = u_o$ and $B = 0$, and the solution

$$u(t) = u_o \cos \omega_o t$$

This is shown plotted in Figure 8.3. The system is exhibiting an harmonic motion of frequency $\omega = \omega_o = \sqrt{k/m}$. This value is called the *natural* frequency. A single degree of freedom system, when set in free vibration motion, vibrates at only one frequency, and that frequency depends only on the material properties of the system.

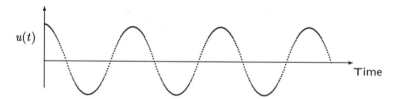

**Figure 8.3**: Free vibration response.

**Example 8.2:**   Suppose the mass of the last example is already in motion, then at a particular instant in time $t = 0$, let the initial conditions be $u(0) = u_o$ and $\dot{u}(0) = v_o$. Determine the time history of the mass.

Irrespective of how the motion is initiated, a free vibration is described as the sum of a sine and cosine term in the form

$$u(t) = A \cos \omega_o t + B \sin \omega_o t$$

Using the initial conditions gives that $A = u_o$ and $B = v_o/\omega_o$ allowing the time history to be written as

$$u(t) = u_o \cos \omega_o t + \frac{v_o}{\omega_o} \sin \omega_o t = C \cos(\omega_o t - \phi)$$

where the amplitude and phase are given by, respectively,

$$C \equiv \sqrt{u_o^2 + v_o^2/\omega_o^2}, \qquad \phi \equiv \tan^{-1}(u_o \omega_o / v_o)$$

Such motion is also periodic and of frequency $\omega_o$.

**Example 8.3:**   Determine the response of the mass of Figure 8.2 to a sinusoidally varying applied load $P(t) = \hat{P} \sin \omega t$. Neglect damping.

Under this circumstance the equation of motion becomes

$$ku + m\ddot{u} = P(t) = \hat{P} \sin \omega t$$

where $P(t)$ is called the *forcing function*. This differential equation is *inhomogeneous* because of the non-zero on the right hand side. Thus the solution will comprise two parts; the general solution obtained after setting $P = 0$, and the particular (or complementary) solution obtained so as to give $P$.

We already know that the homogeneous solution is given by

$$u_h(t) = A \cos \omega_o t + B \sin \omega_o t$$

Look for particular solutions of similar form, that is,

$$u_p(t) = C \cos \alpha t + D \sin \alpha t$$

where $\alpha$ is an as yet unspecified frequency and $C, D$ are arbitrary constants. On substituting into the differential equation, get

$$[k - m\alpha^2]C \cos \alpha t + [k - m\alpha^2]D \sin \alpha t = \hat{P} \sin \omega t$$

This must be true at any value of time; hence separately equating the terms associated with the sines and cosines gives

$$C = 0, \qquad D = \frac{\hat{P}}{k - \omega^2 m}, \qquad \alpha = \omega$$

The total displacement response can therefore be written as

$$u(t) = A \cos \omega_o t + B \sin \omega_o t + \frac{\hat{P}}{k - \omega^2 m} \sin \omega t$$

Again, the coefficients are obtained from the initial conditions. Using the initial conditions of the last example gives the complete solution as

$$u(t) = u_o \cos \omega_o t + \frac{v_o}{\omega_o} \sin \omega_o t - \frac{\hat{P}\omega/\omega_o}{k - \omega^2 m} \sin \omega_o t + \frac{\hat{P}}{k - \omega^2 m} \sin \omega t$$

The first three terms carry the natural frequency $\omega_o$ while the last term carries the forcing frequency $\omega$. In any real system, where some slight damping always exists, the only motion that will persist is the motion described by the last term. Hence we call the last term the *steady state* response, while the rest are the *transients*.

An interesting feature of this solution is observed when the forcing frequency is varied; it is seen that the amplitude of the response changes. Indeed, when

$$\omega^2 = \frac{k}{m} = \omega_o^2$$

the response is infinite, even for very small values of excitation force. This situation is called *resonance*. Figure 8.4 shows how the steady state amplitude ratio

$$\frac{\hat{u}}{\hat{P}/k} = \frac{1}{1 - \omega^2 m/k} = \frac{1}{1 - \omega^2/\omega_o^2}$$

varies as a function of frequency. We will show later that in practical situations there is always some damping and therefore an infinite response is never achieved as shown in the figure.

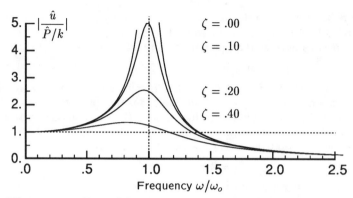

**Figure 8.4**: Forced frequency response of spring-mass system.

## 8.2 Complex Notation

The use of complex algebra facilitates the mathematical analysis of vibration especially when we deal with phase shifts. We therefore find it propitious to introduce it at this stage.

A complex quantity is written as

$$z = a + ib, \qquad i \equiv \sqrt{-1}$$

This can be thought of as a vector with components $a$ and $b$ as shown in Figure 8.5; $a$ is the *real part*, $b$ is the *imaginary part*. The magnitude and orientation is given by

$$|z| = \sqrt{a^2 + b^2} = A, \qquad \phi = \tan^{-1}(a/b)$$

Consequently, an alternative form for the complex number is

$$z = A(\cos\phi + i\sin\phi)$$

We can put this in a convenient form by noting the following relation for exponential functions. The Taylor series expansion of the exponential function $e^{ix}$ is

$$\begin{aligned} e^{ix} &\approx 1 + (ix) + \tfrac{1}{2}(ix)^2 + \tfrac{1}{6}(ix)^3 + \tfrac{1}{24}(ix)^4 + \tfrac{1}{120}(ix)^5 + \tfrac{1}{720}(ix)^6 + \ldots \\ &= [1 - \tfrac{1}{2}x^2 + \tfrac{1}{24}x^4 - \tfrac{1}{720}x^6 + \ldots] + i[x - \tfrac{1}{6}x^3 + \tfrac{1}{120}x^5 + \ldots] \\ &= \cos x + i\sin x \end{aligned}$$

We can now write the complex number as

$$z = Ae^{i\phi}$$

Some other relations of interest are

$$\cos\phi = Re\{e^{i\phi}\} = [e^{i\phi} + e^{-i\phi}]/(2), \qquad \sin\phi = Im\{e^{i\phi}\} = [e^{i\phi} - e^{-i\phi}]/(2i)$$

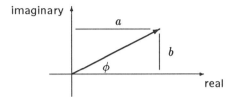

imaginary

$a$

$b$

$\phi$

real

**Figure 8.5**: Complex number as a vector.

where $Re$ and $Im$ stand for real part and imaginary part, respectively.

The addition, multiplication, and so on, of complex numbers follows the usual rules of vector algebra. For example, suppose we have two complex numbers

$$z_1 = a + ib = A_1 e^{i\phi_1}, \qquad z_2 = c + id = A_2 e^{i\phi_2}$$

Then addition is achieved by adding the components

$$z_1 + z_2 = (a + c) + i(b + d) = A_1 e^{i\phi_1} + A_2 e^{i\phi_2}$$

Multiplication is given by

$$z_1 z_2 = (ac - bd) + i(ad + bc) = A_1 A_2 e^{i(\phi_1 + \phi_2)}$$

The exponential form makes multiplication very simple.

To show how these ideas can help to simplify the description of harmonic motions, consider the equation of motion

$$kv + c\dot{v} + m\ddot{v} = P_o \cos(\omega t + \delta)$$

where all terms are real. Now introduce the complementary equation of motion

$$kw + c\dot{w} + m\ddot{w} = P_o \sin(\omega t + \delta)$$

Multiply the second equation by $i$ and add it to the first. The result shows that the complex variable $u \equiv v + iw$ must satisfy the following differential equation

$$ku + c\dot{u} + m\ddot{u} = P_o e^{i(\omega t + \delta)} = \hat{P} e^{i\omega t}$$

In the last form for the load, we have incorporated the phase with the applied load so that $\hat{P}$, in general, is a complex quantity. If we solve this equation for $u$, then we can recover both $v$ or $w$ from

$$v = Re\{u\}, \qquad w = Im\{u\}$$

respectively. We emphasize that working with the complex variable $u$ is equivalent to working with the real variable $v$; no information is gained or lost, it is just a matter of convenience.

The solution for harmonic motion is written simply as

$$u(t) = \hat{u}e^{i\omega t}$$

In the following, the super hat notation will designate the complex amplitude of each frequency component; these components are called the *spectral amplitude*. It is understood that when the actual displacement is required then the above is combined with its complex conjugate to give a real response.

**Example 8.4:** Find the forced frequency response for the spring-mass system of Figure 8.2.

The equation of motion for forced single frequency sinusoidal excitation may be written as

$$ku + c\dot{u} + m\ddot{u} = \hat{P}e^{i\omega t}$$

where $\hat{P}$ is the excitation force and $\omega$ is the excitation frequency. Using a trial solution of the form

$$u(t) = \hat{u}\,e^{i\omega t}$$

gives the velocity and acceleration as

$$
\begin{aligned}
\dot{u}(t) &= i\omega\,\hat{u}e^{i\omega t} = i\omega\,u \\
\ddot{u}(t) &= (i\omega)^2\,\hat{u}e^{i\omega t} = -\omega^2\,\hat{u}e^{i\omega t} = -\omega^2\,u
\end{aligned}
$$

This shows that differentiation is accomplished by multiplying by $i\omega$. Therefore by substituting for $u(t)$ and canceling the common time factors, we get

$$[k + i\omega\,c - \omega^2\,m]\hat{u} = \hat{P}$$

This is solved to give

$$\hat{u} = \frac{\hat{P}}{[k - \omega^2 m + i\omega c]} = \frac{\omega_o^2\,\hat{P}/k}{[\omega_o^2 - \omega^2 + i2\zeta\omega\omega_o]} = \frac{\hat{P}/k}{[1 - (\omega/\omega_o)^2 + i2\zeta\omega/\omega_o^2]}$$

where $\omega_o \equiv \sqrt{k/m}$ is the undamped natural frequency, $\zeta = c/2m\omega_0$ is the dimensionless damping ratio, and $\hat{P}/k$ is the static extension of the spring caused by the force.

We also write this solution as

$$u(t) = \hat{u}e^{i\omega t} = \left[\frac{1}{1 - (\omega/\omega_0)^2 + i2\zeta\omega/\omega_0}\right]\frac{\hat{P}}{k}e^{i\omega t} \equiv H(\omega)\hat{P}e^{i\omega t}$$

It can be seen that the displacement is proportional to the applied force, and the proportionality factor $H(\omega)$ is called the *frequency response function* (FRF); it is complex and depends on frequency. Figure 8.4 can also be used as a plot of the magnitude of $kH(\omega)$. We will have more to say about this function later.

## 8.3  Damping

All real structures experience some sort of energy dissipation or damping when set in motion. This is due to such factors as friction with the surrounding air, and internal friction of the material itself. This section considers some of the consequences of this on the motion.

There are two simple mathematical models for damping in a vibrating structure; the damping may be viscous or hysteretic. In the first, energy dissipation per cycle is proportional to the forcing frequency, while in hysteretic damping, it is independent of frequency. Mathematically, the two types are very similar; we shall therefore give a brief comparison of their effects but concentrate on the viscous damping.

### Critical Damping

Before we proceed with discussing the effects of damping, we would first like to get a measure of what is meant by small amounts of damping. To this end, consider the free vibration of the system with viscous damping. The equation of motion is

$$ku + c\dot{u} + m\ddot{u} = 0$$

Look for particular solutions of this in the form $u(t) = Ae^{i\alpha t}$. Substitute into the differential equation and get the characteristic equation

$$A[k + ic\alpha - m\alpha^2] = 0$$

The value of $\alpha$ that satisfies this is obtained by solving the quadratic equation and is

$$\alpha = \frac{ic}{2m} \pm \frac{1}{2m}\sqrt{4mk - c^2}$$

The time response of the solution is affected by the sign of the radical term as

$$
\begin{aligned}
c^2 &> 4mk; &&\text{overdamped} \\
c^2 &= 4mk; &&\text{critical damping} \\
c^2 &< 4mk; &&\text{underdamped}
\end{aligned}
$$

Let the critical damping be given by

$$c_c \equiv \sqrt{4mk} = 2m\omega_o$$

then the characteristic values of $\alpha$ are given by

$$\alpha = \omega_o[i\zeta \pm \sqrt{1 - \zeta^2}]$$

where $\zeta \equiv c/c_c$ is the ratio of the damping to critical damping. The free vibration solutions are

$$u(t) = e^{-\zeta\omega_o t}[Ae^{-i\omega_d t} + Be^{+i\omega_d t}]$$

where $\omega_d \equiv \omega_o\sqrt{1 - \zeta^2}$ is called the *damped natural frequency*. The critical point occurs when $\zeta = 1$, thus we say that the structure is lightly damped when $\zeta \ll 1$. This is the situation of most interest to us in structural analysis.

**Example 8.5:**   Determine the motion of the mass of Figure 8.2 after it is displaced from its initial position and released. Assume the system is lightly damped.
    We use as initial conditions that at $t = 0$

$$u(0) = u_o, \qquad \dot{u}(0) = 0$$

This gives the solution

$$u(t) = \tfrac{1}{2}u_o e^{-\omega_o\zeta t}[(1 + \frac{i\omega_o\zeta}{\omega_d})e^{-i\omega_d t} + (1 - \frac{i\omega_o\zeta}{\omega_d})e^{+i\omega_d t}]$$

which is shown plotted in Figure 8.6. Note that it eventually decreases to zero, but oscillates as it does so. The frequency of oscillation is $\omega_d = \omega_o\sqrt{1 - \zeta^2} \approx \omega_o(1 - \tfrac{1}{2}\zeta^2)$. Hence, for small amounts of damping this is essentially the undamped natural frequency. The rate of decay is dictated by the term $e^{-\omega_o\zeta t} = e^{-ct/2m}$.

**Figure 8.6**: Damped response.

## Viscous and Hysteretic Damping

We shall compare the forced frequency response of the system for both viscous and hysteretic (structural) damping; in both cases we assume that the system is only lightly damped. We repeat some of the results for the viscous model so as to have a direct comparison with the hysteretic case.

    The equation of motion for forced single frequency sinusoidal excitation of the system with viscous damping may be written as

$$ku + c\dot{u} + m\ddot{u} = \hat{P}e^{i\omega t}$$

where $\hat{P}$ is the excitation force and $\omega$ is the excitation frequency. Using a trial solution of the form

$$u(t) = \hat{u}e^{i\omega t}$$

we can show by differentiation and substitution that

$$\hat{u} = \frac{\hat{P}}{k - \omega^2 m + i\omega c} = \frac{\omega_o^2 \hat{P}/k}{\omega_0^2 - \omega^2 + i2\zeta\omega_o\omega}$$

where $\omega_o = \sqrt{k/m}$ is the undamped natural frequency, $\zeta = c/2m\omega_o$ is the dimensionless damping ratio, and $\hat{P}/k$ is the extension in the spring caused by the force alone. Thus, the displacement history is

$$u(t) = \hat{u}e^{i\omega t} = \left[\frac{1}{1 - (\omega/\omega_o)^2 + i2\zeta\omega/\omega_o}\right]\frac{\hat{P}}{k}e^{i\omega t} = H(\omega)\hat{P}e^{i\omega t}$$

It can be seen that the displacement is proportional to the applied force, and the proportionality factor $H(\omega)$ is called the *frequency response function* (FRF). It is complex and depends on frequency. The amplitude of the displacement is given by

$$|\hat{u}| = |H(\omega)\hat{P}| = \left[\frac{1}{\sqrt{\{1 - (\omega/\omega_o)^2\}^2 + (2\zeta\omega/\omega_o)^2}}\right]\frac{\hat{P}}{k}$$

and the phase of the response lags behind the applied force by an angle $\Delta$

$$\Delta = \tan^{-1}\frac{2\zeta\omega/\omega_o}{[1 - (\omega/\omega_o)^2]}$$

The solution can therefore, alternatively, be written in the form

$$u(t) = \left[\frac{1}{\sqrt{\{1 - (\omega/\omega_o)^2\}^2 + (2\zeta\omega/\omega_o)^2}}\right]\frac{\hat{P}}{k}e^{i(\omega t - \Delta)} = |H(\omega)|\,\hat{P}e^{i(\omega t - \Delta)}$$

which emphasizes the separate effects of amplitude and phase. The amplitude response is shown in Figure 8.4 for different values of damping.

Many materials, when subjected to cyclic strain, generate internal friction that dissipates energy per cycle that is relatively independent of the strain rate. In the present context, this means the damping force is taken as

$$F^d = h\frac{\dot{u}}{\omega}, \qquad \hat{F}^d = ih\,\hat{u}$$

where $h$ is the damping constant. It is important to realize that the hysteretic damping idealization is restricted to the forced frequency case because otherwise the frequency in its definition is undefined. If we take the frequency as $\omega_o$, the natural frequency, then this damping reduces to the viscous case.

The equation of motion for a single degree of freedom system with structural damping is written in the time domain as

$$ku + \frac{h}{\omega}\dot{u} + m\ddot{u} = P(t)$$

and in the spectral form as

$$\left[k(1 + i\gamma) - \omega^2 m\right]\hat{u} = \hat{P} \qquad \text{or} \qquad \hat{u} = \frac{\hat{P}}{k}\left[\frac{1}{1 - (\omega/\omega_o)^2 + i\gamma}\right]$$

where $\gamma \equiv h/k$ is called the structural damping factor. The frequency response function is obtained from

$$u(t) = \hat{u}e^{iwt} = H(\omega)\hat{P}e^{iwt} = \left[\frac{1}{\sqrt{\{1 - (\omega/\omega_o)^2\}^2 + \gamma^2}}\right]\frac{\hat{P}}{k}e^{i(\omega t - \Delta)}$$

where the displacement lags behind the force by

$$\Delta = \tan^{-1}\left[\frac{\gamma}{1 - (\omega/\omega_o)^2}\right]$$

For hysteretic damping, the maximum response occurs exactly at $\omega/\omega_o = 1$, independent of the damping. At very low frequencies, the response depends on the amount of damping, unlike the viscous case. When the system is vibrating at the natural frequency with $\omega/\omega_o = 1$, both the viscous and hysteretic models give the same results if we have $\gamma = 2\zeta$.

## Effects of Damping

The frequency response function, $H(\omega)$, can be interpreted as a magnification factor between the input force and the output response. Figure 8.4 shows the absolute value of this as a function of the frequency ratio $\omega/\omega_o$ for various values of the damping ratio $\zeta$. We can see that increasing the damping diminishes the peak amplitudes. Further, there is a shift of these peaks to the left of $\omega/\omega_o = 1$. In fact, the peaks occur at frequencies given by

$$\omega = \omega_o\sqrt{(1 - 2\zeta^2)}$$

and the peak value of $|H(\omega)|$ is given by

$$|H(\omega)| = \frac{1/k}{2\zeta\sqrt{1 - \zeta^2}} \approx \frac{1/k}{2\zeta}$$

This last relation is for light damping ($\zeta \leq 0.1$) and shows the sensitivity of the peak to damping. The points where the amplitude of $|H(\omega)|$ reduces to $1/\sqrt{2}$ of its peak value are called the *half power* points. The difference in the frequencies at the half power points for light damping can be shown to be

$$\Delta\omega = \omega_2 - \omega_1 = 2\zeta\omega_o$$

For this reason, the term $2\zeta$ is sometimes called the *Loss Factor*.

Since the frequency response function is a complex quantity, it can therefore be broken up into its real and imaginary parts by multiplying the numerator and the denominator by its complex conjugate. Thus

$$\begin{aligned} H(\omega) &= \left[\frac{1 - (\omega/\omega_o)^2}{\{1 - (\omega/\omega_o)^2\}^2 + (2\zeta\omega/\omega_o)^2} - \frac{i2\zeta\omega/\omega_o}{\{1 - (\omega/\omega_o)^2\}^2 + (2\zeta\omega/\omega_o)^2}\right]\frac{1}{k} \\ &= H_R + iH_I \end{aligned}$$

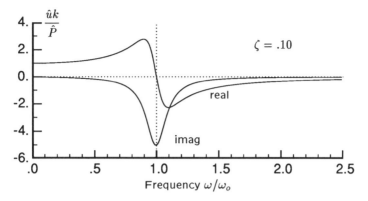

**Figure 8.7**: Real and imaginary components of the FRF.

As shown in Figure 8.7, the real component of the FRF has a zero at $\omega/\omega_o = 1$, independent of damping and exhibits maxima at frequencies given by

$$\omega_1 = \omega_o\sqrt{1 - 2\zeta}, \qquad \omega_2 = \omega_o\sqrt{1 + 2\zeta}$$

These frequencies are often used to estimate the damping of the system from

$$2\zeta = \frac{1 - (\omega_1/\omega_2)^2}{1 + (\omega_1/\omega_2)^2}$$

The plot of the imaginary part of the FRF has a peak close to $\omega/\omega_o = 1$ which is sharper than that of the magnitude of $[H(\omega)]$.

It must be kept in mind that real structures exhibit neither viscous nor structural damping in its pure form. More likely, they exhibit a combination of both, with the proportion of each probably depending on the frequency range. Additionally, much of the damping in structures comes from the joints and the interaction with attachments. As a consequence, the damping is not a material "constant" like the stiffness or density that can be determined by component testing. Because we deal with lightly damped structures, it is sufficient that we consider just the viscous model.

## 8.4   Forced Response

From the discussions of the previous sections, we will limit the following analysis only to the viscous damping case. Under this circumstance the equation of motion becomes

$$ku + c\dot{u} + m\ddot{u} = P(t)$$

where $P(t)$ is the known arbitrary forcing function. This differential equation is called *inhomogeneous* because of the non-zero on the right hand side. Thus the solution will comprise two parts; the general solution obtained after setting $P = 0$, and the particular (or complementary) solution obtained so as to give $P$.

We have already shown that for the lightly damped case ($\zeta \ll 1$) there is the general homogeneous solution

$$u(t) = e^{-\omega_o \zeta t}[Ae^{-i\omega_d t} + Be^{+i\omega_d t}]$$

with $\omega_d = \omega_o\sqrt{1 - \zeta^2}$. We now wish to consider the particular solutions.

## Duhamel's Integral

An *impulsive* force is a large force that acts over a short period of time. The time integral of the force is referred to as the *impulse* of the force. We can obtain the transient response to an arbitrary force history $P(t)$ by considering the force to be the sum of a sequence of impulses.

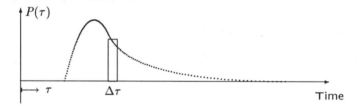

**Figure 8.8**: Impulsive loads.

Specifically, consider an arbitrary force history $P(t)$ as shown in Figure 8.8 with one of the impulses indicated. Each impulse is

$$P\Delta\tau$$

The action of this impulse on the mass is to cause a change of momentum given by

$$m\Delta\dot{u} = P\Delta\tau \qquad \text{or} \qquad \Delta\dot{u} = \frac{P\Delta\tau}{m}$$

If the mass is initially at rest then the change in velocity is the initial velocity for the motion. That is, we have

$$u(0) = u_o = 0\,, \qquad \dot{u}(0) = v_o = \frac{P\Delta\tau}{m}$$

The response to this impulse is

$$u(t) = \frac{P\Delta\tau}{m\omega_d}e^{-\zeta\omega_o(t-\tau)}\sin\omega_d(t - \tau)$$

The term $(t - \tau)$ takes into account the fact that the pulse occurs at time $\tau$ and not time zero. The actual force history is a series of these impulses at different times $\tau$, hence the cumulative effect is obtained by letting $\Delta\tau$ become very small and replacing the summation by an integral over the full time to give

$$u(t) = \frac{1}{\omega_d m} \int_0^t P(\tau) e^{-\zeta\omega_o(t-\tau)} \sin[\omega_d(t - \tau)] \, d\tau$$

This is called *Duhamel's integral* and represents a particular solution of the differential equation of motion subjected to an arbitrary forcing function.

   If the initial conditions are not zero, then the homogeneous solution must be added to complete the solution. Since it can be shown that the particular solution and its derivative is zero at $t = 0$, then we can determine the constants of integration as before. This gives

$$u(t) = e^{-\zeta\omega_o t}\left(u_o \cos \omega_d t + \frac{\zeta\omega_o u_o + v_o}{\omega_d} \sin \omega_d t\right) + \frac{1}{\omega_d m} \int_0^t P(\tau) e^{-\zeta\omega_o(t-\tau)} \sin[\omega_d(t-\tau)] \, d\tau$$

For simple forcing functions (for example stepped loading) the integration may be performed exactly, but generally it must be done numerically. To this end, it is useful to rearrange the response relation as

$$u(t) = \frac{1}{\omega_d} e^{-\zeta\omega_o t} \left[A(t) \sin \omega_d t - B(t) \cos \omega_d t\right]$$

where the two time functions are given by

$$A(t) = \zeta\omega_o u_o + v_o + \frac{1}{\omega_d m} \int_0^t P(\tau) e^{\zeta\omega_o \tau} \cos \omega_d \tau \, d\tau$$

$$B(t) = u_o + \frac{1}{\omega_d m} \int_0^t P(\tau) e^{\zeta\omega_o \tau} \sin \omega_d \tau \, d\tau$$

In this way, the integral can be accomplished in an incremental step by step fashion. If, however, the damping is such that the exponential inside the integral becomes large (and produces numerical problems) then the corresponding $e^{-\zeta\omega_o t}$ term must be brought inside the integral. This can be a significant disadvantage since the computational effort 'pyramids', that is, at each new time increment the integration is performed anew from time zero. For this reason other integration schemes are developed in Chapter 13.

   **Example 8.6:**   The spring-mass system of Figure 8.2 is initially at rest. Find the response to the following step loading

$$\begin{aligned} P(t) &= 0 & t < 0 \\ P(t) &= P_o & t > 0 \end{aligned}$$

The initial conditions are such that $u_o = 0$ and $v_o = 0$. The solution is obtained by substituting for the force into Duhamel's integral to get

$$u(t) = \frac{1}{\omega_d m} \int_0^t P(\tau) e^{-\zeta \omega_o dt} \sin[\omega_d(t - \tau)] \, d\tau = \frac{P_o}{k} \left[ 1 - e^{-\zeta \omega_o dt} (\cos \omega_d t + \frac{\omega_o \zeta}{\omega_d} \sin \omega_d t) \right]$$

This response is shown in Figure 8.9 for two values of damping. Note that the response oscillates about the static deflection position.

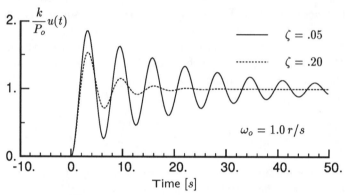

**Figure 8.9**: Response to step loading.

## Spectral Analysis Method

An alternative approach for finding the forced response is to work mostly in the frequency domain. Appendix B shows that complicated responses can be viewed as a superposition of many sinusoids each of different frequency. That is,

$$u(t) = \sum_{n=o}^{N-1} \hat{u}_n e^{+i\omega_n t}$$

where $\hat{u}_n$ is the spectrum of amplitudes (one for each frequency). Representing arbitrary time signals in this fashion is called *spectral analysis*. Summing such a series would ordinarily be a very time consuming task except for the availability of the fast Fourier transform algorithm (FFT) to do the job. This algorithm is so efficient that it has revolutionized the whole area of spectral analysis. Therefore, in the subsequent analysis, we will assume that any arbitrary time input or response can be represented in the spectral form

$$F(t) = \sum_{n=o}^{N-1} C_n e^{+i\omega_n t}$$

and the tasks of performing the summations are accomplished with a computer program.

In the spectral analysis approach to dynamic problems, we replace the actual force with its spectral representation

$$P(t) = \sum \hat{P} e^{i\omega t}$$

and look for particular solutions of the differential equation in a similar form

$$u(t) = \sum \hat{u} e^{i\omega t}$$

where $\hat{u}$ is the unknown response spectrum. On substituting into the differential equation, and rearranging, get

$$\sum \left[ k + i\omega c - \omega^2 m \right] \hat{u} e^{i\omega t} = \sum \hat{P} e^{i\omega t}$$

Since this must be true at any time, then we can remove the summations and the time factors on both sides of the equation to give

$$[k + i\omega c - \omega^2 m]\hat{u} = \hat{P}$$

This has the straight-forward solution that

$$\hat{u} = \frac{\hat{P}}{[k + i\omega c - \omega^2 m]} = H(\omega)\hat{P}$$

The displacement response can therefore be written as

$$u(t) = \sum H(\omega)\hat{P} e^{i\omega t} = \sum \frac{\hat{P} e^{i\omega t}}{[k + i\omega c - \omega^2 m]}$$

If we treat the product $H(\omega)\hat{P}$ as the complex components of the transform, then an FFT inverse transform can be performed to give the time response. Numerous examples of doing this are given in Reference [14].

**Example 8.7:**   Determine the particular solution when $P(t)$ is a rectangular pulse.
   For a rectangular pulse we have that

$$\hat{P}(\omega) = P_o a \left\{ \frac{\sin(\omega a/2)}{\omega a/2} \right\} e^{-i\omega(t_o + a/2)}$$

Assume the mass is completely at rest before the application of the force. The resulting displacement is given by

$$u(t) = \sum H(\omega) P_o a \left\{ \frac{\sin(\omega a/2)}{\omega a/2} \right\} e^{-i\omega(t_o + a/2)} e^{i\omega t}$$

This is a rather complicated expression, so it is expected that numerical methods are used for its evaluation.

# Problems

**8.1** Show that by choosing a reference frame located at the static deflection of a spring-mass system, that gravity has no effect on the free vibration.

[Reference [40], pp. 14]

**8.2** Show that the free response of an overdamped system does nor exhibit any oscillation.

[Reference [40], pp. 29]

**8.3** Consider the forced frequency response of an undamped system near resonance, that is, $\omega - \omega_o = 2\Delta\omega$. Show that the response is

$$u(t) = \frac{\omega_o^2 P_o/k}{\Delta\omega(\omega_o + \omega)} \sin \Delta\omega \sin(\omega_o + \omega)t/2$$

Plot this function and observe the 'beating' effect.

**8.4** Show that the general solution for critical damping is

$$u(x) = [A + Bt]e^{\alpha t}$$

[Reference [23], pp. 204]

**8.5** Show that the difference in frequency at the half power points is given by

$$\Delta\omega = \omega_2 - \omega_1 = 2\zeta\omega_o$$

**8.6** Show that for a ramp loading $P(t) = at$, the Duhamel's integral evaluates to

$$\frac{a}{\omega_o^2}[\omega_o t - \sin \omega_o t]$$

The judicious combination of ramps can be used to synthesize a variety of loading conditions including triangular.

[Reference [35], pp. 352]

**8.7** Plot the real part of the frequency response function (FRF) against its imaginary part. Show that it forms a circle.

[Reference [50], pp. 59]

# Exercises

**8.1** Show that the division of two complex numbers is given by

$$\frac{z_1}{z_2} = \frac{a + ib}{c + id} = \frac{ac + bd}{c^2 + d^2} + i\frac{bc - ad}{c^2 + d^2}$$

**8.2** A point moves in such a way that its position at time $t$ is given by

$$z = r_1 e^{i\omega t} + r_2 e^{-i\omega t}$$

Show that the locus of the point is an ellipse with major and minor axes of $r_1 + r_2$, $r_1 - r_2$, respectively.

**8.3** The period and maximum displacement of the sinusoidal motion of a point on a structure are $0.125\,s$ and $0.25\,in$, respectively. For that point determine the frequency in Hertz, the maximum velocity, and the maximum acceleration.

$$[\ddot{u}_{max} = 1.64\,g]$$

**8.4** Measurements on a building indicate a maximum sinusoidal displacement of $0.005\,in$ and a maximum velocity of $0.5\,in/s$. Determine the frequency of vibration and the maximum acceleration.                                   $[\ddot{u}_{max} = 0.13\,g]$

**8.5** A vehicle for landing on the moon has a mass of $4500\,kg$. The damped spring-undercarriage system of the vehicle has a stiffness of $450\,kN/m$ and a damping ratio of 0.20. The rocket lift engines are cut off when the vehicle is hovering at an altitude of $10\,m$ Calculate the maximum deflection of the undercarriage system when the vehicle hits the surface. Neglect deflection due to self weight and assume $g = 1.6\,m/s^2$.                                                      $[0.428\,m]$

**8.6** A wooden block of mass $3\,kg$ is restrained by a spring of stiffness $2\,N/mm$. A bullet of mass $0.2\,kg$ is fired at a speed of $20\,m/s$ into the block. Obtain the maximum displacement of the block neglecting damping, and if damping is 10% of critical.                                                          $[50\,mm, 43\,mm]$

**8.7** A water tower has a mass of $24000\,kg$ which can be assumed to be lumped at the tank center. The lateral stiffness of the supporting frame is $5000\,kN/m$. The tank is subjected to a lateral load that varies as a half sine wave of amplitude $200\,kN$ and period $1.0\,s$. Obtain the response during the application of the force.                   $[u(t) = 0.0494(\sin 6.283t - 0.4353\sin 14.43t)]$

**8.8** A hydrometer float is used to measure the specific gravity of fluids. The mass is $0.0372\,kg$ and the diameter protruding above the liquid surface is $6.4\,mm$. Determine the period of oscillation in a fluid of specific gravity 1.20.    $[1.97\,s]$

**8.9** A mass of $0.907\,kg$ is attached to the end of a spring of stiffness $7\,N/cm$. Determine the critical damping.

**8.10** To calibrate a dashpot, the velocity of the plunger is measured when a given force is applied. If a $0.5\,lb$ weight produced a constant velocity of $1.2\,in/s$. Determine the damping when used with the previous system.          $[\xi = 1.45]$

**8.11** A vibrating system consists of a $4.534\,kg$ mass, a spring of stiffness $35\,N/cm$, and a dashpot with damping $0.1243\,N/(cm \cdot s)$. Find the damping factor and the ratio of any two consecutive amplitudes.

**8.12** A weight attached to a spring of stiffness $525\,N/m$ has a viscous device. When the weight is displaced and released, the period of vibration is found to be $1.80\,s$, and the ration of consecutive amplitudes is $4.2 : 1.0$. Determine the amplitude when a force $P = 2\sin 3t$ acts on the system.          $[7.97\,mm]$

# Chapter 9

# Vibration of Rod Structures

This chapter deals with our first application of the matrix methods to the motion of continuous systems. These systems can exhibit quite complicated behavior because the responses are functions of both space and time. We show, however, that spectral analysis can be used to reduce these problems to a series of pseudo-static problems and thus make them amenable to the solution procedures already established. As a special case, we establish the free vibration problem as an eigenvalue problem.

Systems with a continuous mass distribution have an infinite number of resonances and mode shapes or, equivalently, an infinite number of degrees of freedom. However, to achieve a matrix formulation for the dynamics of continuous systems, it is necessary to discretize the mass distribution. We develop two schemes for doing this. These approximations result in a finite number of degrees of freedom, and consequently a finite number of resonances also.

## 9.1   Rod Theory

We first establish some exact results for the dynamics of rods. This will aid in the later comparison with the approximate matrix methods. As done in Chapter 2, we consider a uniform rod; that is, both the mass and stiffness of the rod are assumed uniformly distributed along its length. Additionally, assumptions such as small deformations, linear behavior, and so on, are assumed to be still applicable.

**Figure 9.1**: Rod with infinitesimal rod element.

## Equation of Motion

Consider the balance of forces on the rod segment shown in Figure 9.1; by the process already demonstrated in Chapter 2, we get the following dynamic equilibrium equation

$$\frac{\partial F}{\partial x} = \rho A \ddot{u} - q + \eta \dot{u}$$

where $\rho A$ is the mass per unit length and $\eta$ is the damping per unit length. Use of Hooke's law combined with the strain-displacement relation gives

$$\frac{F}{A} = \sigma = E\epsilon = E\frac{\partial u}{\partial x}$$

We now substitute for $F$ into the equilibrium equation to get

$$\frac{\partial}{\partial x}[EA\frac{\partial u}{\partial x}] = \rho A\frac{\partial^2 u}{\partial t^2} - q + \eta\frac{\partial u}{\partial t}$$

This is the general form of the equation of motion when the section properties can change slightly. In the case when the properties are uniform, we have

$$EA\frac{\partial^2 u}{\partial x^2} = \rho A\frac{\partial^2 u}{\partial t^2} - q + \eta\frac{\partial u}{\partial t}$$

This is the governing equation of motion for the dynamic response of a rod. Note that it has space derivatives in addition to the time derivatives. As a result, the solution $u(x,t)$ is a function of both space and time. These problems are very difficult to solve, in general, and so the spectral analysis approach introduced in Chapter 8 will be further developed as a tool for their solution.

## A Note on Spectral Analysis

The idea of representing the time variation of a function by a summation of harmonic functions is extended here to representing arbitrary functions of time and position resulting from the solution of wave equations. The approach is to remove the time variation by using the spectral representation of the solution. This leaves a new differential equation for the coefficients which, in many cases, can be integrated directly.

Consider the time variation of the solution at a particular point in space; it has the spectral representation

$$u(x_1, y_1, t) = f_1(t) = \sum C_{1n}e^{i\omega_n t}$$

At another point, the solution behaves as a second time function $f_2(t)$ and is represented by the Fourier coefficients $C_{2n}$. That is, the coefficients are different at each spatial point. Thus, the solution at an arbitrary position has the following spectral representation

$$u(x, y, t) = \sum \hat{u}_n(x, y, \omega_n)e^{i\omega_n t}$$

where $\hat{u}_n(x,y)$ are the spatially dependent Fourier coefficients. Notice that these coefficients are functions of frequency $\omega$, and thus there is no reduction in the total number of independent variables.

For shorthand, the summation and subscripts will often be understood and the function will be given the representation

$$u(x,y,t) \qquad \Rightarrow \qquad \hat{u}_n(x,y,\omega_n) \quad \text{or} \quad \hat{u}(x,y,\omega)$$

Sometimes, we will write the representation simply as $\hat{u}$.

The differential equations are in terms of both space and time derivatives. Since these equations are linear, then it is possible to apply the spectral representation to each term appearing. Thus, the spectral representation for the time derivative is

$$\frac{\partial u}{\partial t} = \frac{\partial}{\partial t} \sum \hat{u}_n e^{i\omega_n t} = \sum i\omega_n \hat{u}_n e^{i\omega_n t}$$

In shorthand, this becomes

$$\frac{\partial u}{\partial t} \qquad \Rightarrow \qquad i\omega_n \hat{u}_n \quad \text{or} \quad i\omega\,\hat{u}$$

In fact, time derivatives of general order have the representation

$$\frac{\partial^m u}{\partial t^m} \qquad \Rightarrow \qquad (i\omega_n)^m\,\hat{u}_n \quad \text{or} \quad (i\omega)^m\,\hat{u}$$

Herein lies the advantage of the spectral approach to solving differential equations — time derivatives are replaced by algebraic expressions in the Fourier coefficients. That is, there is a reduction in the number of derivatives occurring.

Similarly, the spatial derivatives are represented by

$$\frac{\partial u}{\partial x} = \frac{\partial}{\partial x} \sum \hat{u}_n e^{i\omega_n t} = \sum \frac{\partial \hat{u}_n}{\partial x} e^{i\omega_n t}$$

and in shorthand notation

$$\frac{\partial u}{\partial x} \qquad \Rightarrow \qquad \frac{\partial \hat{u}_n}{\partial x} \quad \text{or} \quad \frac{\partial \hat{u}}{\partial x}$$

In this case there does not appear to be any reduction; as will be seen later, with the removal of time as an independent variable, these derivatives often become ordinary derivatives, and thus more amenable to integration.

## Spectral Solution

Based on the previous developments, it is sufficient for us to study motions of the form

$$u(x,t) = \sum \hat{u}(x) e^{i\omega t}$$

where we understand that the solution to a general problem would require the super-position of many terms. Note that in contrast to the previous chapter, the spectral amplitude $\hat{u}(x)$ is considered a function of $x$. Substituting this representation for $u(x,t)$ into the equation of motion and canceling the common time factor, we obtain

$$EA\frac{d^2\hat{u}}{dx^2} = -\rho A\omega^2\hat{u} - \hat{q} + i\omega\eta\hat{u}$$

This is an ordinary differential equation with constant coefficients where the frequency appears as a parameter. We can rewrite the differential equation as

$$EA\frac{d^2\hat{u}}{dx^2} + k^2\hat{u} = -\hat{q}\,, \qquad\qquad k^2 \equiv (\rho A\omega^2 - i\omega\eta)$$

In this form, the governing equation for the rod resembles the governing equations for beam buckling already encountered in Section 5.4. The advantage of the spectral approach to dynamic problems is that it essentially reduces the problem to a pseudo-static one. Consequently, all the principles and approaches established for static problems are applicable here also. Indeed, a summary of the relationships for the structural quantities

$$\text{Displacement}: \qquad \hat{u} = \hat{u}(x)$$

$$\text{Force}: \qquad \hat{F} = +EA\frac{d\hat{u}}{dx} \qquad\qquad (9.1)$$

$$\text{Loading}: \qquad \hat{q} = -EA\frac{d^2\hat{u}}{dx^2} + (\omega^2\rho A - i\omega\eta)\hat{u} \qquad (9.2)$$

shows that the only difference in comparison with the rod equations of Chapter 2 is the addition of the mass and damping related terms in the expression for the loading. Since $\omega$ is treated as a parameter in the spectral analysis, we see once again that the displacement $\hat{u}(x)$ (via its derivatives) can be viewed as the fundamental unknown of interest.

## Viscoelastic Rod

The previous derivation included the effect of damping through the external term $-\eta\dot{u}$. Damping can also manifest itself through a time dependent constitutive relation; this is called *viscoelasticity*. We give only a brief review of it here; our main intention is to show that it does not affect the spectral formulation of the problem.

The derivation of the equation of motion follows the same procedure as before except that the stress-strain behavior is time dependent. While there are many ways of expressing this relationship (Reference [8], for example, demonstrates the use of hereditary integrals), the following form is adequate for present purposes:

$$\sum_p a_p\frac{d^p\sigma}{dt^p} = \sum_q b_q\frac{d^q e}{dt^q}\,, \qquad\qquad p, q = 0, 1, 2, \ldots \qquad (9.3)$$

Here the stress and strain are related through multiple derivatives in time. We can express this relation in the spectral form as

$$\{\sum_p a_p(i\omega)^p\}\hat{\sigma} = \{\sum_q b_q(i\omega)^q\}\hat{\epsilon}$$

or simply

$$\hat{\sigma} = E(\omega)\hat{\epsilon}, \qquad E(\omega) \equiv \frac{\{\sum_q b_q(i\omega)^q\}}{\{\sum_p a_p(i\omega)^p\}} \tag{9.4}$$

This resembles the linear elastic relation. Indeed, if we view the Young's modulus of the previous section as being frequency dependent, then the equation of motion derived there is applicable to the viscoelastic case. One further note: All of the material coefficients in Equation(9.4) are real even though the relationship itself is complex; if, however, we allow the coefficients to be complex then our relationship also includes the effect of hysteretic damping.

To amplify on these results, consider, for concreteness, the special case of the *standard linear solid*. This is sometimes visualized as a mechanical model consisting of a parallel spring and dashpot in series with another spring. The constitutive relation is described by

$$\dot{\sigma} + \frac{\sigma}{\eta}[E_1 + E_2] = E_1\dot{\epsilon} + \frac{E_1 E_2}{\eta}\epsilon$$

Unlike the linear elastic material, this requires three material properties $E_1, E_2, \eta$ to describe it. The viscoelastic modulus is

$$E(\omega) = E_1\frac{E_2 + i\omega\eta}{E_1 + E_2 + i\omega\eta} \tag{9.5}$$

and has the very slow and very fast behavior limits of

$$E(0) \approx \frac{E_1 E_2}{E_1 + E_2}, \qquad E(\infty) \approx E_1$$

Note that both of these limits are elastic. Consequently, the viscoelastic energy dissipation occurs only in the middle range. Reference [43] discusses this dissipation.

**Example 9.1:**   Obtain the solution to the differential equation for the rod. Assume there is no applied loading.

We begin by assuming a trial solution of the form

$$\hat{u}(x) = \mathbf{A}e^{-i\alpha x}$$

where $\mathbf{A}$ is a complex, frequency dependent, amplitude. Substitute this to get

$$EA(-i\alpha)^2\mathbf{A}e^{-i\alpha x} - k^2\mathbf{A}e^{-i\alpha x} = 0$$

On canceling the common terms, the solution of this equation requires that

$$-\alpha^2 EA + k^2 = 0 \qquad \Rightarrow \qquad \alpha = \pm k = \pm\sqrt{(\omega^2\rho A - i\omega\eta)/EA}$$

Two values are obtained for $\alpha$, and therefore we can write the total solution as

$$\hat{u}(x) = \mathbf{A}e^{-ikx} + \mathbf{B}e^{+ikx}, \qquad k \equiv \sqrt{(\omega^2\rho A - i\omega\eta)/EA} \qquad (9.6)$$

where $\mathbf{A}, \mathbf{B}$ are complex, frequency dependent, constants to be determined by the boundary conditions. Sometimes, it is more convenient to consider the solution in the equivalent form

$$\hat{u}(x) = c_1 \cos kx + c_2 \sin kx \qquad (9.7)$$

As to which is preferable depends on the particular boundary value problem; generally, when damping is present the wavenumber $k$ is complex and the exponential form is then preferable.

It is worth keeping in mind that while the above solution satisfies the differential equation, it does not necessarily satisfy the boundary conditions of any particular problem. It is the choice of particular constants of integration that achieve this.

**Example 9.2:**   Consider the forced response of the fixed-free rod shown in Figure 9.2.

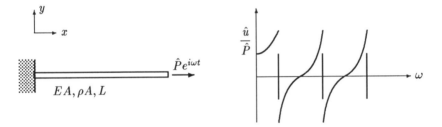

**Figure 9.2**: Rod with applied load.

The approach we use is essentially the same as what we did for the static problems of Chapter 2 except that now the boundary conditions are time dependent. Specifically, we impose at $x = 0$ that $u(t) = 0$ and at $x = L$ that $P(t) = \hat{P}e^{i\omega t}$. The spectral form of these become

$$\hat{u}(0) = 0 \;=\; c_1 + 0$$

$$\hat{P} = EA\frac{d\hat{u}(L)}{dx} = 0 \;=\; EA[-c_1 k \sin kL + c_2 k \cos kL] = 0$$

The solution for the coefficients is

$$c_1 = 0, \qquad c_2 = \frac{\hat{P}}{EA\,k\cos kL}$$

The response at a general point is now seen to be

$$\hat{u}(x) = c_2 \sin kx = \frac{\sin kx}{EA\,k\cos kL}\,\hat{P}$$

Note that this solution is valid at any frequency; however, it exhibits special behavior at certain frequencies.

Let us consider the displacement of the tip of the rod. The unit response as a function of frequency is shown in Figure 9.2. At certain frequencies, the response becomes infinite. This means that even very small applied loads would generate very large responses. It is instructive to compare these results with the corresponding results for the stability of beams in Section 5.4. If we write the above solution in the form of a stiffness relation for the tip, that is,

$$\frac{EA}{L}\left[\frac{kL\,\cos kL}{\sin kL}\right]\hat{u}(L) = \hat{P}$$

then the infinite responses occur when the stiffness goes to zero. These problems appear to be governed by a frequency dependent stiffness relation.

The critical values of frequency occur at

$$\cos kL = 0 \qquad \text{or} \qquad kL = \tfrac{1}{2}\pi, \tfrac{3}{2}\pi, \cdots, \tfrac{1}{2}(2n+1)\pi$$

leading to the resonant frequencies

$$\omega_n = \frac{(n+\tfrac{1}{2})\pi}{L}\sqrt{\frac{EA}{\rho A}}$$

The special frequencies are independent of the applied load.

**Example 9.3:**  Consider the free vibrations of the fixed-fixed rod shown in Figure 9.3. By *free vibration* is meant that the rod moves without any external driving force.

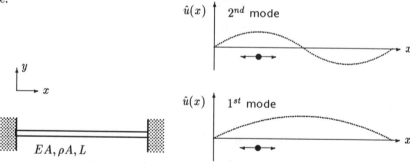

**Figure 9.3**: Fixed-fixed rod.

The approach is to use the above general solution and to impose the boundary conditions to determine the particular values for the coefficients. This is essentially the same as what is done in static problems. The fixed-fixed conditions are that $u(t) = 0$ at $x = 0$ and $x = L$, requiring, in the spectral form, that

$$\hat{u}(0) = 0 \ = \ c_1 + 0$$
$$\hat{u}(L) = 0 \ = \ c_1 \cos kL + c_2 \sin kL$$

The first equation gives that $c_1 = 0$ and therefore the second equation leads to

$$c_2 \sin kL = 0$$

In general, this has the only solution $c_2 = 0$, which is a trivial solution corresponding to no motion. There is a special circumstance, however, under which a non-trivial solution is obtained; when

$$\sin kL = 0 \quad \text{or} \quad kL = n\pi, \quad n = 0, 1, 2, 3, \cdots$$

The frequencies corresponding to these are obtained as

$$\omega_n = k\sqrt{\frac{EA}{\rho A}} = \frac{n\pi}{L}\sqrt{\frac{EA}{\rho A}}$$

It is only at these frequencies that the above forms for $\hat{u}(x)$ is non-trivial. Note that there is an infinite number of frequencies, but they occur at discrete values.

The *mode shapes*, i.e., the solution shape at these special frequencies are

$$\hat{u}(x) = c_2 \sin k_n x = c_2 \sin\left(n\pi\frac{x}{L}\right)$$

There is a unique shape at each frequency as shown in Figure 9.3.

**Example 9.4:** Consider the free vibrations of the free-free rod shown in Figure 9.4. This problem is unconstrained, and from the statics point of view it should give difficulties. However, the mass acts as an inertia force and therefore makes the system stable.

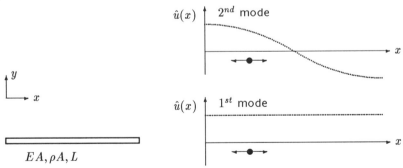

**Figure 9.4:** Free-free rod.

The approach is essentially the same as what was done in the previous problem except that the free-free conditions of $P(t) = 0$ at $x = 0$ and $x = L$ are imposed. The spectral form of these become

$$EA\frac{d\hat{u}(0)}{dx} = 0 = EA[c_2k]$$

$$EA\frac{d\hat{u}(L)}{dx} = 0 = EA[-c_1k\sin kL + c_2k\cos kL] = 0$$

The first of these gives that $c_2 = 0$ and therefore the second equation leads to

$$c_1\, k\sin kL = 0$$

The non-trivial solution is obtained from

$$k\sin kL = 0 \qquad \text{or} \qquad kL = n\pi, \qquad n = 0,1,2,3,\cdots$$

The frequencies corresponding to these are obtained as

$$\omega_n = k\sqrt{\frac{EA}{\rho A}} = \frac{n\pi}{L}\sqrt{\frac{EA}{\rho A}}$$

The mode shapes for the solution are

$$\hat{u}(x) = c_1\cos k_n x = c_1\cos(n\pi\frac{x}{L})$$

These are shown plotted in Figure 9.4. For $n = 0$, we have the mode shape $\hat{u}(x) = c_1 = constant$. This corresponds to a *rigid body mode* since there is no deformation.

We give a summary of the above results in Table 9.1. The frequencies are related to the factor $(kL)$ by

$$\omega_n = \frac{(kL)_n}{L}\sqrt{\frac{EA}{\rho A}}$$

These will be used later in the chapter for comparison with approximate solutions.

| BC | $(kL)_1$ | $(kL)_2$ | $(kL)_3$ | Frequency Equation | |
|---|---|---|---|---|---|
| fixed-fixed | 3.142 | 6.283 | 9.425 | $\sin kL = 0$, | $kL = n\pi$ |
| fixed-free | 1.571 | 4.712 | 7.854 | $\cos kL = 0$, | $kL = (n+\frac{1}{2})\pi$ |
| free-free | 3.142 | 6.283 | 9.425 | $\sin kL = 0$, | $kL = n\pi$ |

Table 9.1: Frequency factor $kL$ for the first three non-rigid body modes.

These last few examples show that the solution for free vibrations of a rod involves solving and eigenvalue problem. More specifically, it involves solving a *transcendental* eigenvalue problem because the unknown eigenvalues (which are related to the frequencies) appear as arguments of circular and hyperbolic functions. Solving this type of problem is very difficult especially when multiple rod segments are connected together. The remainder of this chapter will therefore be devoted to transitioning from the exact dynamic analysis to an approximate matrix formulation. Our main achievement will be the ability to state the free vibration problem as an *algebraic* eigenvalue problem and therefore have access to the numerical solution procedures of Chapter 7.

## 9.2   Structural Connections

We now look at what happens when multiple rod sections are connected together to form a structure. The approach, conceptually, is the same as for the static structures; we write the solution for each segment in terms of the constants of integration, and apply both compatibility and equilibrium conditions at the joints in order to determine the constants. In fact, the analogy is almost exact when we use the spectral approach. It will be seen that a system of equations ensue, and these can be put in matrix form. The vanishing of the determinant then gives the conditions for resonance.

Before we look at the structure, we investigate a situation present in dynamic problems but not at all in the static case. We refer to the effect of concentrated masses.

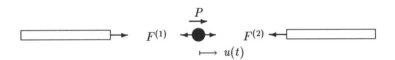

**Figure 9.5**: Rod connection.

Consider two rod segments connected by a rigid joint as shown in Figure 9.5. Let the joint have a concentrated mass $m_c$. Compatibility between the two segments requires that

$$u^{(1)}(t) = u^{(2)}(t) = u(t)$$

The spectral form for this looks almost identical

$$\hat{u}^{(1)} = \hat{u}^{(2)} = \hat{u}$$

Dynamic equilibrium at the joint gives

$$-F^{(1)}(t) + F^{(2)}(t) + P(t) = m_c \ddot{u}(t)$$

The spectral form of this becomes

$$-\hat{F}^{(1)} + \hat{F}^{(2)} + \hat{P} = -\omega^2 m_c \hat{u}$$

From this it is apparent that a concentrated mass affects the structure as an effective force of amount $\omega^2 m_c \hat{u}$. This force, however, is frequency dependent.

**Example 9.5:**   Consider the free vibration of a rod, fixed at one end and with a concentrated mass at the other end, as shown in Figure 9.6.

**Figure 9.6**: Rod with concentrated mass.

The equation of motion of the rod section is the same as before and therefore we can use the general solution of Equation(9.7). To impose the boundary conditions, we separate the concentrated mass from the rod as shown in Figure 9.6. The boundary conditions require that

$$\text{at}\quad x = 0: \qquad u = 0$$
$$\text{at}\quad x = L: \qquad F = EA\frac{\partial u}{\partial x} = -m_c\frac{\partial^2 u}{\partial t^2}$$

as shown by the free body diagram of Figure 9.6. The spectral form of these equations are

$$\text{at}\quad x = 0: \qquad \hat{u} = 0$$
$$\text{at}\quad x = L: \qquad \hat{F} = EA\frac{d\hat{u}}{dx} = m_c\omega^2\hat{u}$$

Using the general solution for the rod section

$$\hat{u}(x) = c_1 \cos kx + c_2 \sin kx$$

gives the system of equations representing the boundary conditions as

$$c_1 = 0$$
$$EA[-c_1 k \sin kL + c_2 k \cos kL] = m_c\omega^2[c_1 \cos kL + c_2 \sin kL]$$

The first equation gives that $c_1 = 0$; to have a solution that is non-trivial, we require that

$$\cos kL - \gamma kL \sin kL = 0$$

where $\gamma = m_c/\rho AL$, is the ratio of the concentrated mass to the total mass of the rod. This frequency equation is transcendental and is therefore difficult to solve in simple form. Note, however, the following limiting cases

$$\text{if}\quad \gamma = 0 \quad \Rightarrow \quad kL = \tfrac{1}{2}\pi, \tfrac{3}{2}\pi, \cdots, (n+\tfrac{1}{2})\pi$$
$$\text{if}\quad \gamma = \infty \quad \Rightarrow \quad kL = 0, \pi, 2\pi, \cdots, n\pi$$

These correspond to the free and fixed end conditions, respectively. That is, a very large mass has the effect of fixing the response at that point.

For arbitrary values of mass ratio the situation is more complicated; there are multiple frequencies at which the above transcendental equation is satisfied. Some values for the first mode are shown in the following chart.

| $\gamma$ | 0 | 3.0 | 1.0 | .70 | .30 | .10 | .01 | $\infty$ |
|---|---|---|---|---|---|---|---|---|
| $kL$ | $\pi/2$ | 1.20 | .86 | .75 | .52 | .32 | .10 | 0 |
| | fixed | | | | | | | free |

These were obtained by solving

$$\tan \xi = \frac{1}{\gamma \xi}$$

for different values of *gamma*. Figure 9.7 shows the graphical scheme for solving this. The figure also shows that increasing the concentrated mass decreases the resonant frequency for a given mode. In general, the addition of mass decreases the resonant frequencies.

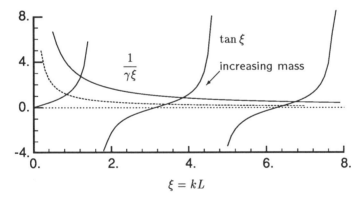

**Figure 9.7**: Graphical scheme for finding roots.

**Example 9.6:**  It would appear that the limit $\gamma = \infty$ can be achieved either by letting $m_c \longrightarrow \infty$ or having $\rho AL \longrightarrow 0$. The latter would correspond to a concentrated mass attached to a massless spring which we know gives only a single resonant frequency and not the multiple ones as indicated above. Show the proper way to take the limit to a massless rod.

The density of the rod appears in two places; in the definition of $\gamma = m_c/\rho AL$, and in the definition $k = \omega\sqrt{\rho A/EA}$. Therefore, we cannot take the limit on $\gamma$

without simultaneously considering its effect on the characteristic equation. Consider $\xi \equiv kL = \omega L\sqrt{\rho A/EA}$ as being small, the expansion of the characteristic equation then gives

$$[1 - \tfrac{1}{2}\xi^2 + \ldots] - \gamma\xi[\xi - \tfrac{1}{6}\xi^3 + \ldots] = 0$$

Retaining only powers of $\xi^2$ this becomes

$$1 - (\gamma + \tfrac{1}{2})\xi^2 = 0$$

Hence for small $\rho A$, we get

$$\xi = \frac{1}{\sqrt{\gamma + \tfrac{1}{2}}} \qquad \text{or} \qquad \omega = \frac{1}{L}\sqrt{\frac{EA}{m_c + \tfrac{1}{2}\rho AL}}$$

It is apparent that we have only one resonant frequency, and that when $\rho A = 0$ this describes a spring mass system with a spring stiffness of $EA/L$.

**Example 9.7:**   Determine the natural frequency for the rod structure made of two segments as shown in Figure 9.8. Let the material of both segments be the same, but take the areas and lengths as being different.

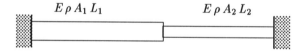

$E\rho A_1 L_1$               $E\rho A_2 L_2$

**Figure 9.8**: Simple connected rod structure.

We use the general solution for both rod segments

$$\hat{u}(x) = c_1 \cos kx + c_2 \sin kx$$

For the first section, we have at $x = 0$

$$\hat{u} = 0 \quad \Rightarrow \quad 0 = c_1$$

From this we get that the deflected shape is

$$\hat{u}(x) = c_2 \sin kx$$

where for this section $k_1^2 = \rho A_1/EA_1 = k^2$.
    For the second section, we have at $x = L_2$ (we use $x$ to mean the local distance for the section)

$$\hat{u} = 0 \quad \Rightarrow \quad 0 = c_1^* \cos kL_2 + c_2^* \sin kL_2$$

where $k_2^2 = \rho A_2/EA_2 = k^2$ is the same as for the first section. From this boundary condition, we get for the deflected shape

$$\hat{u}(x) = c_1^* \cos kx + c_2^* \sin kx\,, \qquad c_1^* \cos kL_2 + c_2^* \sin kL_2 = 0$$

We do not substitute for $c_2^*$ in terms of $c_1^*$ just in case $\sin kL$ could be zero.

The compatibility and equilibrium equations must be satisfied at the junction of the two rod segments. Continuity of displacement across the two sections gives

$$\hat{u}^{(1)} = \hat{u}^{(2)} \quad \Rightarrow \quad c_2 \sin kL_1 = c_1^*$$

The second equation to be imposed is equilibrium at the junction. With reference to the free body diagram in Figure 9.5, we have

$$\hat{F}^{(1)} = \hat{F}^{(2)} \quad \Rightarrow \quad EA_1[kc_2 \cos kL_1] = EA_2[kc_2^*]$$

These equations can be put into a matrix form as

$$\begin{bmatrix} \sin kL_1 & -1 & 0 \\ EA_1 k \cos kL_1 & 0 & -EA_2 k \\ 0 & \cos kL_2 & \sin kL_2 \end{bmatrix} \left\{ \begin{array}{c} c_2 \\ c_1^* \\ c_2^* \end{array} \right\} = 0$$

The determinant of this is

$$\det = EA_1 k \cos kL_1 \sin kL_2 + EA_2 k \sin kL_1 \cos kL_2 = 0$$

This equation cannot be solved explicitly for $k$. There are multiple modes but computing them requires solving a transcendental equation.

A special case to test the result is when $L_1 = L_2 = L/2$, then the characteristic equation reduces to

$$\det = (EA_1 + EA_2)k \sin kL = 0$$

This is the characteristic equation for the fixed-fixed rod and the resonant frequencies do not depend on the areas.

## 9.3 Exact Dynamic Stiffness Matrix

We will use the governing differential equation to derive a matrix formulation for the dynamics of rod structures. The basic idea in establishing the stiffness relation is the same as used in Chapter 2: having integrated the governing differential equations, we rewrite the constants of integration in terms of the nodal degrees of freedom. Using the displacement function, we are then in a position to determine the member force distribution. This, in turn, can be related to the nodal loads.

We assume that the loading is zero over the element length, then the general deflected shape is

$$\hat{u}(x) = c_1 \cos kx + c_2 \sin kx\,, \qquad k = k(\omega)$$

We write the coefficients $c_1$, $c_2$ in terms of the nodal degrees of freedom. That is, at $x = 0$ and $x = L$,

$$\hat{u}(0) \equiv \hat{u}_1 = c_1, \qquad \hat{u}(L) \equiv \hat{u}_2 = c_1 \cos kL + c_2 \sin kL$$

The solution for $c_1$ and $c_2$ is now given as

$$c_1 = \frac{1}{\Delta}[\sin kL\, \hat{u}_1], \qquad c_2 = \frac{1}{\Delta}[-\cos kL\, \hat{u}_1 + \hat{u}_2], \qquad \Delta \equiv \sin kL$$

At this stage, we rewrite the deflection function in terms of just the nodal degrees of freedom. Specifically,

$$\hat{u}(x) = \frac{1}{\Delta}[\sin k(L - x)\, \hat{u}_1 + \sin kx\, \hat{u}_2] \equiv \hat{f}_1(x)\, \hat{u}_1 + \hat{f}_2(x)\, \hat{u}_2$$

It is interesting to compare the spectral rod shape functions $\hat{f}_1$ and $\hat{f}_2$ with their static counterparts. We have

$$\frac{\sin k(L - x)}{\sin kL} \Leftrightarrow \frac{(L - x)}{L}, \qquad \frac{\sin kx}{\sin kL} \Leftrightarrow \frac{x}{L}$$

We can easily show that equality is attained in the limit as $kL \approx 0$; in other words, if the element is very small or the frequency is very low.

The force distribution is obtained from the displacement as

$$\hat{F}(x) = EA\frac{d\hat{u}}{dx} = EA[\hat{f}_1'(x)\, \hat{u}_1 + \hat{f}_2'(x)\, \hat{u}_2] = \frac{EAk}{\Delta}[-\cos k(L - x)\, \hat{u}_1 + \cos kx\, \hat{u}_2]$$

The values at $x = 0$ and $x = L$ are, respectively,

$$\hat{F}_1 = -\hat{F}(0) = \frac{EAk}{\Delta}[\cos kL\, \hat{u}_1 - \hat{u}_2], \qquad \hat{F}_2 = \hat{F}(L) = \frac{EAk}{\Delta}[-\hat{u}_1 + \cos kL\, \hat{u}_2]$$

We can now put these in the form of a stiffness relation

$$\begin{Bmatrix} \hat{F}_1 \\ \hat{F}_2 \end{Bmatrix} = \frac{EA}{L}\frac{kL}{\sin kL}\begin{bmatrix} \cos kL & -1 \\ -1 & \cos kL \end{bmatrix}\begin{Bmatrix} \hat{u}_1 \\ \hat{u}_2 \end{Bmatrix} \qquad \text{or} \qquad \{\hat{F}\} = [\,\hat{k}\,]\{\hat{u}\} \qquad (9.8)$$

We will call the matrix $[\,\hat{k}\,]$ a *dynamic stiffness* matrix. It is symmetric and, depending on the wavenumber $k$, could also be complex. Aside from this latter characteristic, we treat the dynamic stiffness in identical fashion as the static stiffness. For example, the assemblage procedure is identical to that done for the static rod structures and results in a global dynamic stiffness relation of

$$[\hat{K}]\{\hat{u}\} = \{\hat{P}\}$$

The structural dynamic stiffness matrix $[\hat{K}]$ enjoys the same symmetry and bandedness properties of the static elastic stiffness.

**Example 9.8:**   Determine the frequency equation for the fixed-free rod.

The procedure we follow is the same as for other matrix methods. Number the nodes with Node 1 at the fixed end and Node 2 at the other end. The unknown degree of freedom is

$$\{\hat{u}_u\} = \{\hat{u}_2\}$$

There is only one element, hence the reduced dynamic structural stiffness matrix is just the fourth quadrant of the element stiffness. The stiffness relation is therefore

$$\frac{EA}{L}\frac{kL}{\sin kL}\left[\cos kL\right]\hat{u}_2 = 0$$

Setting the determinant to zero gives

$$\frac{kL}{\sin kL}\cos kL = 0$$

That is, we have

$$kL = \tfrac{1}{2}\pi, \tfrac{3}{2}\pi, \ldots, (n+\tfrac{1}{2})\pi$$

This is the same as previously obtained. Note that $kL = 0$ is not a root.

# 9.4   Approximate Matrix Formulation

The dynamic stiffness matrix as developed is exact; however, the unknown resonant frequencies are deeply imbedded in its transcendental form. What we are now interested in is a simplification of this matrix to allow easier determination of the resonant frequencies. We first approximate the dynamic stiffness by considering the limit when the element size is small or, equivalently, when the frequencies are low. We also look at the lumped approach where we simply replace the distributed mass as a set of concentrated masses at the ends of the element.

## Approximate Dynamic Stiffness

The frequency appears in the trigonometric terms $\cos kL$ and $\sin kL$. We will perform a Taylor series expansion of these with respect to $\xi \equiv kL$. This will give a polynomial, but we will retain only enough terms so that the stiffness, ultimately, contains the frequency up to $\omega^2$ terms. Note that this expansion on small $\xi$ is equivalent to requiring a small element length $L$ or restricting the analysis to low frequencies..

We will start with the determinant. Using the series expansion for the sine term, we get

$$\Delta = \sin\xi \approx \xi - \tfrac{1}{6}\xi^3 + \cdots = \xi[1 - \tfrac{1}{6}\xi^2 + \cdots]$$

It is important to realize that the series has been truncated precisely as indicated above because that is the first non-trivial point at which the determinant can exhibit a zero. We will need the reciprocal of the determinant and this is approximated as

$$\frac{1}{\Delta} \approx \frac{1}{\xi}[1 + \tfrac{1}{6}\xi^2 + \cdots]$$

We now do the expansion on the stiffness terms out to the same order. For example,

$$\hat{k}_{11} \approx \frac{EA}{L}[1 - \tfrac{1}{2}\xi^2 + \cdots]\frac{\xi}{\xi}[1 + \tfrac{1}{6}\xi^2 + \cdots] = \frac{EA}{L}[1 - \tfrac{1}{3}\xi^2 + \cdots]$$

Similar expansions on the other terms gives

$$[\hat{k}] \approx \frac{EA}{L}\left[\begin{bmatrix} 1 & -1 \\ -1 & 1 \end{bmatrix} - \frac{\xi^2}{6}\begin{bmatrix} 2 & 1 \\ 1 & 2 \end{bmatrix} + \cdots\right]$$

This approximation neglects terms of $\xi^4$ and higher. The first term we recognize as the rod element stiffness; the second term contains $\omega$ and we suspect that it is associated with the dynamic behavior of the rod.

Consider the particular case when there is external damping, then

$$\frac{EA}{L}\xi^2 = \frac{EA}{L}k^2 L^2 = \omega^2 \rho A L - i\omega\eta L$$

The expansion for the dynamic stiffness matrix may now be written as

$$[\hat{k}] \approx \frac{EA}{L}\begin{bmatrix} 1 & -1 \\ -1 & 1 \end{bmatrix} + i\omega\frac{\eta L}{6}\begin{bmatrix} 2 & 1 \\ 1 & 2 \end{bmatrix} - \omega^2\frac{\rho AL}{6}\begin{bmatrix} 2 & 1 \\ 1 & 2 \end{bmatrix} + \cdots \qquad (9.9)$$

We saw earlier that the inertia of a concentrated mass is given by

$$m_c \ddot{u} \qquad \text{or} \qquad -\omega^2 m_c \hat{u}$$

and similarly, that the spectral representation of the velocity is $\dot{u} \Leftrightarrow i\omega$. This suggests that we can write the approximate dynamic stiffness relation as

$$[\hat{k}] = [k] + i\omega[c] - \omega^2[m]$$

and make the interpretations that $[k]$ is the element elastic stiffness matrix, $[c]$ is the element damping matrix, and $[m]$ is the element mass matrix. It also suggests that we can write the approximate dynamic stiffness relation *in the time domain* as

$$\{F\} = [k]\{u\} + [c]\{\dot{u}\} + [m]\{\ddot{u}\}$$

In this form all matrices are real and constant.

There are a few interesting points worth noting about the developments just completed. First, if we compare the diagonal terms of the approximate dynamic stiffness matrix (for no damping) with the exact values, we see they go through a zero only once as $\omega$ is varied. Therefore, at most, only two resonant frequencies are obtained for this element. Contrast this with the infinite number obtainable from the exact dynamic stiffness matrix. Thus we conclude that to improve the approximate solution we must use many elements for a given member length. Second, the matrix form for the damping is same as for the mass. Actually, we can write

$$[\,c\,] = \frac{\eta}{\rho A}[\,m\,]$$

This shows that the damping matrix is *proportional* to the mass matrix. Without going into the details, we can show that for the viscoelastic rod where $E = E(\omega)$ the damping can be proportional to the stiffness matrix. Therefore, the representation of the damping matrix as

$$[\,c\,] = \alpha[\,m\,] + \beta[\,k\,]$$

where $\alpha$ and $\beta$ are constants is justified by our developments. This damping representation is called *Rayleigh damping*.

## Lumped Mass Approach

It is instructive at this stage to take a different approach to establishing the approximate matrix formulation. We will use the so-called *lumped mass* method; that is, the distributed mass is replaced by concentrated masses located at the nodes. Specifically, we first replace the distributed mass of the rod by a concentrated one, and then simply put half of it at Node 1 and the other half at Node 2. Figure 9.9 indicates the free body diagrams for the element replaced with concentrated masses.

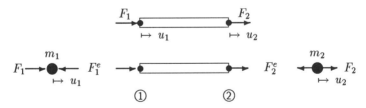

**Figure 9.9**: Rod element replaced with lumped mass.

Since the distributed mass is assumed to be concentrated at the nodes, then the equation of motion for the massless middle section becomes

$$EA\frac{\partial^2 u}{\partial x^2} = 0 \qquad \frac{\partial^2 u}{\partial t^2} = 0$$

This resembles the equation for the static case, except that the displacement shape $u(x,t)$ is a function of space *and* time. This displacement shape is determined (as before) to be

$$u(x,t) = (1 - \frac{x}{L})u_1(t) + (\frac{x}{L})u_2(t) = f_1(x)u_1(t) + f_2(x)u_2(t)$$

This is time-dependent insofar as the nodal values themselves depend on time but the rod shape functions are independent of time. The elastic forces $F_1^e$ and $F_2^e$ are related to the nodal displacements $u_1$ and $u_2$ as

$$\left\{ \begin{array}{c} F_1^e \\ F_2^e \end{array} \right\} = \frac{EA}{L} \left[ \begin{array}{cc} 1 & -1 \\ -1 & 1 \end{array} \right] \left\{ \begin{array}{c} u_1 \\ u_2 \end{array} \right\}$$

This, of course, is the same stiffness relation for the rod as obtained in Chapter 2, but again, a reminder that it is time dependent through the nodal degrees of freedom.

Dynamic equilibrium of the concentrated masses now gives

$$\begin{array}{llll} \text{at Node 1:} & F_1 - F_1^e - F_1^d & = & m_1 \ddot{u}_1 \equiv \frac{1}{2}\rho A L \ddot{u}_1 \\ \text{at Node 2:} & F_2 - F_2^e - F_2^d & = & m_2 \ddot{u}_2 \equiv \frac{1}{2}\rho A L \ddot{u}_2 \end{array}$$

In matrix form

$$\left\{ \begin{array}{c} F_1 \\ F_2 \end{array} \right\} = \left\{ \begin{array}{c} F_1^e \\ F_2^e \end{array} \right\} + \left\{ \begin{array}{c} F_1^d \\ F_2^d \end{array} \right\} + \frac{\rho A L}{2} \left[ \begin{array}{cc} 1 & 0 \\ 0 & 1 \end{array} \right] \left\{ \begin{array}{c} \ddot{u}_1 \\ \ddot{u}_2 \end{array} \right\} = [\, k \,]\{u\} + [\, c \,]\{\dot{u}\} + [\, m \,]\{\ddot{u}\}$$

where the mass and damping matrices are defined as

$$[\, m \,] \equiv \frac{\rho A L}{2} \left[ \begin{array}{cc} 1 & 0 \\ 0 & 1 \end{array} \right], \qquad [\, c \,] \equiv \frac{\eta L}{2} \left[ \begin{array}{cc} 1 & 0 \\ 0 & 1 \end{array} \right]$$

These matrices are called the *lumped mass* and *lumped damping* matrices, respectively. The most significant characteristic of a lumped matrix is that it is diagonal. The advantage in using this matrix is that the assembled structural mass matrix is always diagonal which leads to significantly fewer computations and less computer storage requirements. In contrast, the mass and damping matrices obtained in the previous section are called *consistent* matrices for reason that will become apparent in Chapter 12. Some of the examples to follow will compare the performance of the lumped mass matrix against the consistent mass model.

## 9.5   Matrix Form of Dynamic Problems

The matrix methods are most powerful in those cases where it is difficult to integrate the differential equations directly. The presence of concentrated masses is an example of such a case. The following are some examples to demonstrate the ability to analyze the free vibration of structures.

The assemblage process under dynamic circumstances is nearly identical to that of Chapter 2. Thus in reference to Figure 2.6 we get the same equilibrium equations; the only slight modification is that we replace the nodal load vector $\{F\}$ for each element with its expansion in terms of the stiffness, damping, and mass matrices. The result is that we get the equations of motion of the structures as a whole to be

$$[K]\{u\} + [C]\{\dot{u}\} + [M]\{\ddot{u}\} = \{P\}$$

where $[K]$ is the elastic structural stiffness matrix, $[C]$ is the assembled damping matrix, and $[M]$ is the structural mass matrix. These latter two matrices are assembled in exactly the same manner as for the elastic stiffness. As a result, the mass and damping matrices will exhibit all the symmetry and bandedness properties of the stiffness matrix.

When the structural joints have mass, we need only amend the structural mass matrix as follows

$$[M] = \sum_i [m^{(i)}] + \lceil M_c \rfloor$$

where $\lceil M_c \rfloor$ is the collection of joint concentrated masses. This is a diagonal matrix. For the proportional damping case,

$$[C] = \alpha[M] + \beta[K]$$

where $\alpha$ and $\beta$ are constants.

The computer solution of the structural equations of motion is discussed in more detail in a later chapter, but it is of value now to consider some of the major problem types originating from this set. The above equations of motion are to be interpreted as a system of differential equations in time for the unknown nodal displacements $\{u\}$, subject to the known forcing histories $\{P\}$, and a set of boundary and initial conditions. Generally, these require some numerical scheme for integration over time. Therefore, for *transient* dynamic problems, the matrix method approach becomes computationally intensive in two respects. First, a substantial increase in the number of elements must be used in order to model the mass distribution accurately. The other is that the complete system of equations must be solved at each time increment. These issues are dealt with in Chapter 13.

For the special case when the excitation force is harmonic, that is,

$$\{P\} = \{\hat{P}\} e^{i\omega t} \quad \text{or} \quad \left\{ \begin{array}{c} P_1 \\ P_2 \\ \vdots \\ P_n \end{array} \right\} = \left\{ \begin{array}{c} \hat{P}_1 \\ \hat{P}_2 \\ \vdots \\ \hat{P}_n \end{array} \right\} e^{i\omega t}$$

(note that many of the $P_n$ could be zero) then the response is also harmonic and given by

$$\{u\} = \{\hat{u}\}\, e^{i\omega t} \qquad \text{or} \qquad \left\{ \begin{array}{c} u_1 \\ u_2 \\ \vdots \\ u_n \end{array} \right\} = \left\{ \begin{array}{c} \hat{u}_1 \\ \hat{u}_2 \\ \vdots \\ \hat{u}_n \end{array} \right\} e^{i\omega t}$$

This type of analysis is referred to as *forced frequency* analysis. Substituting these forms into the differential equations gives

$$[\,K\,]\{\hat{u}\}e^{i\omega t} + i\omega[\,C\,]\{\hat{u}\}e^{i\omega t} - \omega^2[\,M\,]\{\hat{u}\}e^{i\omega t} = \{\hat{P}\}e^{i\omega t}$$

or, after canceling through the common time factor,

$$\big[[\,K\,] + i\omega[\,C\,] - \omega^2[\,M\,]\big]\{\hat{u}\} = \{\hat{P}\} \qquad \text{or} \qquad [\,\hat{K}\,]\{\hat{u}\} = \{\hat{P}\}$$

Thus, the solution can be obtained analogous to the static problem; the difference is that the stiffness matrix is modified by the inertia term $\omega^2[\,M\,]$ and the complex damping term $i\omega[\,C\,]$. This is the discrete approximation of the dynamic structural stiffness. It is therefore frequency dependent as well as being complex. This system of equations is now recognized as the spectral form of the equations of motion of the structure. One approach, then, to transient problems is to evaluate the above at each frequency and use the FFT [9] for time domain reconstructions. This is feasible but a more full fledged spectral approach based on the exact dynamic stiffness is developed in References [14, 15].

A case of very special interest is that of free vibrations. When the damping is zero this case gives the mode shapes that are very important in a modal analysis (as we shall see in Chapter 11). For free vibrations of the system, the applied loads $\{P\}$ are zero giving the equations of motion as

$$\big[[\,K\,] - \omega^2[\,M\,]\big]\{\hat{u}\} = 0$$

This is a system of homogeneous equations for the nodal displacements $\{\hat{u}\}$. For a nontrivial solution, the determinant of the matrix of coefficients must be zero. We thus conclude that this is an eigenvalue problem, $\omega^2$ are the eigenvalues and the corresponding $\{\hat{u}\}$ the eigenvectors of the problem. Note that the larger the number of elements (for a given structure), the larger the system of equations; consequently, the more eigenvalues we can obtain.

**Example 9.9:**  Consider the free vibration of the fixed-fixed rod shown in Figure 9.10. Use two elements to find an approximate solution. This problem was already treated exactly earlier in the chapter and so will act as a good point for comparison.

The element stiffness matrices are

$$[k^{(12)}] = \frac{EA}{L/2}\begin{bmatrix} 1 & -1 \\ -1 & 1 \end{bmatrix}, \qquad [k^{(23)}] = \frac{EA}{L/2}\begin{bmatrix} 1 & -1 \\ -1 & 1 \end{bmatrix}$$

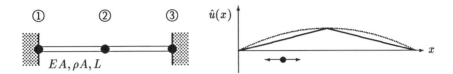

**Figure 9.10**: Fixed-fixed rod with two elements.

giving the full assembled structural stiffness matrix as

$$[K] = \frac{2EA}{L} \begin{bmatrix} 1 & -1 & 0 \\ -1 & 2 & -1 \\ 0 & -1 & 1 \end{bmatrix}$$

Note that this is the same as if it were a static problem. The element mass matrices (using the consistent mass matrix) are

$$[m^{(12)}] = \frac{\rho AL/2}{6} \begin{bmatrix} 2 & 1 \\ 1 & 2 \end{bmatrix}, \qquad [m^{(23)}] = \frac{\rho AL/2}{6} \begin{bmatrix} 2 & 1 \\ 1 & 2 \end{bmatrix}$$

giving the full assembled structural mass matrix as

$$[M] = \frac{\rho AL}{12} \begin{bmatrix} 2 & 1 & 0 \\ 1 & 4 & 1 \\ 0 & 1 & 2 \end{bmatrix}$$

If, on the other hand, the lumped mass model is used then the structural mass matrix becomes

$$[M] = \frac{\rho AL}{4} \begin{bmatrix} 1 & 0 & 0 \\ 0 & 2 & 0 \\ 0 & 0 & 1 \end{bmatrix}$$

We will compare the performance of this against the consistent mass model.
The equations of motion in full form for the free vibration of the structure are

$$\left[ \frac{2EA}{L} \begin{bmatrix} 1 & -1 & 0 \\ -1 & 2 & -1 \\ 0 & -1 & 1 \end{bmatrix} - \omega^2 \frac{\rho AL}{12} \begin{bmatrix} 2 & 1 & 0 \\ 1 & 4 & 1 \\ 0 & 1 & 2 \end{bmatrix} \right] \left\{ \begin{array}{c} \hat{u}_1 \\ \hat{u}_2 \\ \hat{u}_3 \end{array} \right\} = 0$$

This is reduced in the usual manner by removing the fixed degrees of freedom. That is, the boundary conditions are used to determine the unknown degrees of freedom as

$$u_1 = u_3 = 0, \quad \ddot{u}_1 = \ddot{u}_3 = 0 \quad \Rightarrow \quad \{u_u\} = \{u_2\}, \quad \{\ddot{u}_u\} = \{\ddot{u}_2\}$$

Consequently, the reduced structural matrices are

$$[K^*] = \frac{2EA}{L}[\,2\,], \qquad [M^*] = \frac{\rho AL}{12}[\,4\,]$$

The eigenvalue problem now simply becomes

$$\left[\frac{2EA}{L}2 - \omega^2\frac{\rho AL}{12}4\right]\hat{u}_2 = 0$$

allowing the resonant frequency to be obtained as

$$\omega = \frac{\sqrt{12}}{L}\sqrt{\frac{EA}{\rho A}} \simeq \frac{3.46}{L}\sqrt{\frac{EA}{\rho A}}, \qquad \omega_{exact} = \frac{\pi}{L}\sqrt{\frac{EA}{\rho A}}$$

In comparison with the exact solution previously obtained, there is a difference of about 10%. However, there is only one value computed — the two element formulation is incapable of giving any higher resonances.

The corresponding lumped mass result is

$$\omega = \frac{\sqrt{8}}{L}\sqrt{\frac{EA}{\rho A}} \simeq \frac{2.83}{L}\sqrt{\frac{EA}{\rho A}}$$

This value is an underestimate by about the same amount that the consistent mass is an overestimate. Thus it appears, from an accuracy point of view, there is no significant difference between the two approaches.

The mode shape for this solution is simply $\{0, 1, 0\}$. This corresponds to the first symmetric mode of the exact solution as shown in Figure 9.10.

**Example 9.10:**   Consider the same problem as above, but this time use three elements.

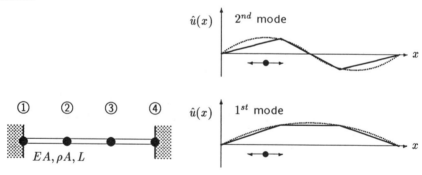

**Figure 9.11**: Fixed-fixed rod with three elements.

Number the nodes as shown in Figure 9.11. Only consider the reduced matrices; the unknown displacements and known forces are, respectively,

$$\{u_u\} = \begin{Bmatrix} u_2 \\ u_3 \end{Bmatrix}, \qquad \{P_k\} = \begin{Bmatrix} 0 \\ 0 \end{Bmatrix}, \qquad \{IDbc\} = \{0; 1; 2; 0\}$$

The reduced element stiffnesses are

$$[k^{*(12)}] = \frac{3EA}{L}[\ 1\ ], \qquad [k^{*(23)}] = \frac{3EA}{L}\begin{bmatrix} 1 & -1 \\ -1 & 1 \end{bmatrix}, \qquad [k^{*(34)}] = \frac{3EA}{L}[\ 1\ ]$$

giving the reduced structural stiffness matrix as

$$[K^*] = \frac{3EA}{L}\begin{bmatrix} 2 & -1 \\ -1 & 2 \end{bmatrix}$$

The reduced element mass matrices are

$$[m^{*(12)}] = \frac{\rho AL}{18}[\ 2\ ], \qquad [m^{*(23)}] = \frac{\rho AL}{18}\begin{bmatrix} 2 & 1 \\ 1 & 2 \end{bmatrix}, \qquad [m^{*(34)}] = \frac{\rho AL}{18}[\ 2\ ]$$

giving the reduced structural mass matrix as

$$[M^*] = \frac{\rho AL}{18}\begin{bmatrix} 4 & 1 \\ 1 & 4 \end{bmatrix}$$

The corresponding lumped mass matrix is

$$[M^*] = \frac{\rho AL}{3}\begin{bmatrix} 1 & 0 \\ 0 & 1 \end{bmatrix}$$

The eigenvalue system of equations becomes

$$\left[\frac{3EA}{L}\begin{bmatrix} 2 & -1 \\ -1 & 2 \end{bmatrix} - \omega^2\frac{\rho AL}{18}\begin{bmatrix} 4 & 1 \\ 1 & 4 \end{bmatrix}\right]\left\{\begin{matrix} \hat{u}_2 \\ \hat{u}_3 \end{matrix}\right\} = 0$$

These equations can now be solved to obtain the eigenvalues. That is, the frequency equation is obtained by multiplying the determinant out, and rearranging to get

$$\left(\frac{54EA}{5L^2\rho A} - \omega^2\right)\left(\frac{54EA}{L^2\rho A} - \omega^2\right) = 0$$

The solutions of this are

First mode :  $\quad \omega = \dfrac{\sqrt{54/5}}{L}\sqrt{\dfrac{EA}{\rho A}} \simeq \dfrac{3.29}{L}\sqrt{\dfrac{EA}{\rho A}}$

Second mode:  $\quad \omega = \dfrac{\sqrt{54}}{L}\sqrt{\dfrac{EA}{\rho A}} \simeq \dfrac{7.35}{L}\sqrt{\dfrac{EA}{\rho A}}$

The accuracy of the first mode is improved, but also, an estimate of the second frequency is obtained. (Recall that for this problem, the numerical factor for the theoretical solution varies as $n\pi$.) The lumped mass results are also improved giving

First mode :  $\quad \omega = \dfrac{\sqrt{9}}{L}\sqrt{\dfrac{EA}{\rho A}} \simeq \dfrac{3.0}{L}\sqrt{\dfrac{EA}{\rho A}}$

Second mode:  $\quad \omega = \dfrac{\sqrt{27}}{L}\sqrt{\dfrac{EA}{\rho A}} \simeq \dfrac{5.20}{L}\sqrt{\dfrac{EA}{\rho A}}$

Again, these lumped frequencies are on the lower side of theory by about the same amount that the consistent results are higher.

The corresponding mode shapes are (irrespective of the mass matrix)

$$\{\hat{u}\}_1 = \{0,1,1,0\}, \qquad \{\hat{u}\}_2 = \{0,1,-1,0\}$$

It is apparent that these are estimates for the first symmetric and first anti-symmetric mode shapes, respectively, as shown in Figure 9.11.

**Example 9.11:**   For the rod shown in Figure 9.12, determine the effect of the concentrated mass at the end of a rod on the resonant frequency.

**Figure 9.12**: Two element solution for rod with concentrated mass.

We will use a two element solution for demonstration. The reduced structural matrices are

$$[K^*] = \frac{EA2}{L} \begin{bmatrix} 2 & -1 \\ -1 & 1 \end{bmatrix}, \qquad [M^*] = \frac{\rho AL}{12} \begin{bmatrix} 4 & 1 \\ 1 & 2 \end{bmatrix} + \begin{bmatrix} 0 & 0 \\ 0 & m_c \end{bmatrix} = \frac{\rho AL}{12} \begin{bmatrix} 4 & 1 \\ 1 & 2+12\gamma \end{bmatrix}$$

where $\gamma = m_c/\rho AL$. The frequency equation is obtained by setting the determinant of the system of equations to zero. We get

$$(7 + 48\gamma)\lambda^2 - 24(10 + 24\gamma)\lambda + 576 = 0$$

where $\lambda^2 \equiv \omega^2 \rho AL/EA$. The solution of this quadratic equation is

$$\omega L \sqrt{\frac{\rho A}{EA}} = kL = \sqrt{12 \frac{(10 + 24\gamma) \pm \sqrt{72 + 266\gamma + 576\gamma^2}}{(7 + 56\gamma)}}$$

If $\gamma = 100$ , then $kL \simeq 0.1$, which agrees very well with the exact solution. Some other limits are

$$\begin{array}{llll}
\text{if} & \gamma = 0 & kL = 1.61 & \text{compare} \quad kL = \pi/2 = 1.57 \\
\text{if} & \gamma = \infty & kL = 1/\sqrt{\gamma} = 0 & \text{compare} \quad kL = 0
\end{array}$$

The full behavior is shown in the Figure 9.12 for the first mode. The corresponding result for the lumped case is

$$\omega L \sqrt{\frac{\rho A}{EA}} = kL = \sqrt{4 \frac{(2 + 4\gamma) \pm \sqrt{2 + 8\gamma + 16\gamma^2}}{(1 + 4\gamma)}}$$

This is also shown in the figure. Again the lumped mass result consistently underestimates by the amount that the consistent mass overestimates.

**Example 9.12:**   Consider the free vibration of the fixed-fixed rod structure shown in Figure 9.8. Use two elements to find an approximate solution.

Number the nodes from left to right as 1, 2, 3, respectively. The boundary conditions are used to determine the unknown degrees of freedom as

$$u_1 = u_3 = 0, \quad \ddot{u}_1 = \ddot{u}_3 = 0 \quad \Rightarrow \quad \{u_u\} = \{u_2\}, \quad \{\ddot{u}_u\} = \{\ddot{u}_2\}$$

The element stiffness matrices are

$$[k^{(12)}] = \frac{EA_1}{L_1} \begin{bmatrix} 1 & -1 \\ -1 & 1 \end{bmatrix}, \quad [k^{(23)}] = \frac{EA_2}{L_2} \begin{bmatrix} 1 & -1 \\ -1 & 1 \end{bmatrix}$$

giving the reduced assembled structural stiffness matrix as

$$[K^*] = \left[ \frac{EA_1}{L_1} + \frac{EA_2}{L_2} \right]$$

The element mass matrices (using the consistent mass matrix) are

$$[m^{(12)}] = \frac{\rho A_1 L_1}{6} \begin{bmatrix} 2 & 1 \\ 1 & 2 \end{bmatrix}, \quad [m^{(23)}] = \frac{\rho A_2 L_2}{6} \begin{bmatrix} 2 & 1 \\ 1 & 2 \end{bmatrix}$$

giving the reduced assembled structural mass matrix as

$$[M^*] = \tfrac{1}{3} [\rho A_1 L_1 + \rho A_2 L_2]$$

If, on the other hand, the lumped mass model is used then the reduced structural mass matrix becomes

$$[M^*] = \tfrac{1}{2} [\rho A_1 L_1 + \rho A_2 L_2]$$

The eigenvalue problem is

$$\left[ (\frac{EA_1}{L_1} + \frac{EA_2}{L_2}) - \omega^2 (\rho A_1 L_1 + \rho A_2 L_2)/2 \right] \hat{u}_2 = 0$$

allowing the resonant frequency to be obtained as

$$\omega = \sqrt{3} \sqrt{\frac{EA_1/L_1 + EA_2/L_2}{\rho A_1 L_1 + \rho A_2 L_2}}$$

In the special case where $L_1 = L_2 = L/2$, we get

$$\omega = \frac{\sqrt{12}}{L} \sqrt{\frac{EA_1 + EA_2}{\rho A_1 + \rho A_2}} = \frac{\sqrt{12}}{L} \sqrt{\frac{E}{\rho}}$$

which is independent of the areas as shown earlier.

## Problems

**9.1** Plot the dynamic stiffness $\hat{k}_{11}$ as a function of frequency.

[Reference [14], pp. 142]

**9.2** Compare the $\hat{k}_{11}$ term from the approximate dynamic stiffness with the exact value. At what frequency do they differ by 5%?

**9.3** Show that if the area of the rod varies as $A(x) = A_o(1 + x/a)^m$ the differential equation of motion is

$$\frac{d^2\hat{u}}{dx^2} + \left(\frac{m}{a+x}\right)\frac{d\hat{u}}{dx} + \beta^2\hat{u} = 0, \qquad \beta = \omega\sqrt{\frac{\rho A_o}{EA_o}}$$

This is a form of Bessel's equation; show that the solutions are

$$\hat{u}(x) = c_1 z^\gamma J_\gamma(z) + c_2 z^\gamma Y_\gamma(z), \qquad \gamma = \tfrac{1}{2}(1-m), \qquad z = \beta(a+x)$$

where $J(z)$ and $Y(z)$ are Bessel functions of the first and second kind, respectively.

[Reference [14], pp. 80]

**9.4** Show that for the special case of $m = 1$, then

$$\hat{u}(x) = \mathbf{A}J_o(z) + \mathbf{B}Y_o(z)$$

and that the asymptotic form of this result when $x \gg a$ is

$$\hat{u}(x) \;=\; c_1\sqrt{\frac{2}{\pi\beta x}}\cos\left(\beta x - \frac{\pi}{4}\right) + c_1\sqrt{\frac{2}{\pi\beta x}}\sin(\beta x - \frac{\pi}{4}) = \frac{1}{\sqrt{\beta x}}\{\bar{c}_1 e^{-i\beta x} + \bar{c}_2 e^{i\beta x}\}$$

which is similar in form to that for the uniform rod except for the decreasing amplitude.

[Reference [14], pp. 81]

**9.5** Show that if the dynamic stiffness is expanded further, that the next term is

$$-\omega^4(\rho AL)^2\frac{L}{360EA}\begin{bmatrix} 8 & 7 \\ 7 & 8 \end{bmatrix}$$

[Reference [35], pp. 283]

**9.6** Use the previous expansion to obtain the frequency relation for the fixed free and free-free rod. Use one element.

**9.7** Show that the equation of motion for the twisting of a uniform shaft is

$$GJ\frac{\partial^2\phi}{\partial x^2} = \rho I\frac{\partial^2\phi}{\partial t^2}$$

Show that the all the results for the shaft can be recovered from those of the rod by making the associations

$$\phi \leftrightarrow u, \quad GJ \leftrightarrow EA, \quad \rho I \leftrightarrow \rho A, \quad T \leftrightarrow F,$$

[Reference [6], pp. 245]

## Exercises

**9.1** Consider the free vibrations of a fixed-free rod. Show that the frequency equation is

$$\cos kL = 0$$

leading to the resonant frequencies $\omega_n L = (n + \frac{1}{2})\pi\sqrt{EA/\rho A}$. Also show that the corresponding mode shapes are

$$\hat{u}(x) = c_2 \sin[(n + \frac{1}{2})\pi\frac{x}{L}]$$

**9.2** Since the consistent mass matrix overestimates the frequency while the lumped mass underestimates it, show that an inconsistent mass matrix defined by

$$[\,m\,] = \frac{1}{2}\frac{\rho AL}{6}\begin{bmatrix} 2 & 1 \\ 1 & 2 \end{bmatrix} + \frac{1}{2}\frac{\rho AL}{2}\begin{bmatrix} 1 & 0 \\ 0 & 1 \end{bmatrix} = \frac{\rho AL}{12}\begin{bmatrix} 5 & 1 \\ 1 & 5 \end{bmatrix}$$

gives very good results.

**9.3** A steel composite rod is made of lengths $L_1$, $L_2$, $L_3$ and areas $A_1$, $A_2$, $A_3$. Show that the resonant frequencies for longitudinal vibration are obtained from

$$A_3 T_3 (A_1 T_1 + A_2 T_2) - A_2(A_1 - A_2 T_1 T_2) = 0, \qquad T_i \equiv \tan \omega L_i \sqrt{\rho A / EA}$$

**9.4** A steel rod of length $2\,m$ is clamped at both ends. The cross-section is square of side $50\,mm$, find the first three natural frequencies. Use four elements.

[8256, 17748, 28842]

**9.5** An aluminum rod of length $10\,in$ and diameter $0.5\,in$ is freely supported. Use the minimum number of elements to compute the first three resonant frequencies.

[0, 66667, 133330]

**9.6** Determine the expression for the natural frequency of torsional oscillations of a uniform rod (shaft) clamped at the middle and free at the ends.

$$[\omega_n = (2n - 1)\pi\sqrt{G/rho}/L]$$

**9.7** Determine the natural frequencies of a torsional system consisting of a uniform shaft of mass moment of inertia $J_s$ with a disk of inertia $J_o$ attached to each end. Check the fundamental frequency by reducing the uniform shaft to a torsional spring with end masses.  $[[(kJ_o/J_s)^2 - 1]\tan k = 2(kJ_o/J_s), \ k = \omega L\sqrt{\rho/G}]$

Chapter 10

# Vibration of Beam Structures

We treat the dynamics of beams in a manner similar to that for rods. The sequence of analysis followed is: derive the governing differential equations, use spectral analysis to obtain general solutions, obtain an exact matrix formulation, obtain an approximate matrix formulation.

We emphasize the approximate matrix formulation because it is in a convenient form for eigenanalysis. Consequently, an issue of importance is the accuracy of the various approximations.

## 10.1    Spectral Analysis of Beams

We carry over most of the assumptions used in Chapter 3 as regards the beam behavior; in particular, we assume that the dynamic response of the beam is characterized by the transverse displacement of the centerline $v(x,t)$. We add two load terms that are relevant to dynamic problems; the first is inertia associated with the mass per unit length $\rho A$, the other is dissipation associated with the viscous damping per unit length $\eta$.

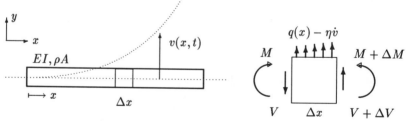

**Figure 10.1**: Flexural motion of infinitesimal beam element.

The equations of motion for the small beam segment of Figure 10.1 are

$$\frac{\partial M}{\partial x} + V = I\ddot{\phi}, \qquad \frac{\partial V}{\partial x} = \rho A \ddot{v} - q + \eta \dot{v}$$

We make the further assumptions that the rotational inertia $I\ddot{\phi}$ is negligible, and that the elastic property of the beam, as described by the moment deflection relation,

$$M = EI\frac{\partial^2 v}{\partial x^2}$$

is still valid in the dynamic case. These allow us to simplify the equations of motion for the beam to

$$EI\frac{\partial^4 v}{\partial x^4} + \eta\frac{\partial v}{\partial t} + \rho A\frac{\partial^2 v}{\partial t^2} = q(x,t) \qquad (10.1)$$

(All properties are taken as constant along the length.) Note that this equation is not the same wave equation as for the rod — while the inertia and damping terms are very similar, there are four space derivatives here. The consequence of this is that the mode shapes are spatially dependent in a different manner.

Assume that the applied load and the beam response have the spectral representations

$$q(x,t) = \sum \hat{q}(x)e^{i\omega t}, \qquad v(x,t) = \sum \hat{v}(x)e^{i\omega t}$$

Substitute these into the equation of motion and cancel the common time factor. The resulting differential equation for the spectral amplitude is

$$EI\frac{d^4\hat{v}}{dx^4} + i\omega\eta\hat{v} - \omega^2\rho A\hat{v} = \hat{q}$$

This can be written in the compact form

$$\frac{d^4\hat{v}}{dx^4} - k^4\hat{v} = \frac{\hat{q}}{EI}, \qquad k^4 \equiv (\omega^2\rho A - i\eta\omega)/EI$$

This is an ordinary differential equation with constant coefficients. If we consider the Young's modulus to be frequency dependent, then this is also the governing differential equation for a viscoelastic beam.

We are now in a position to summarize the relationships for the structural quantities as

$$\text{Displacement:} \qquad \hat{v} = \hat{v}(x)$$

$$\text{Slope:} \qquad \hat{\phi} = \frac{d\hat{v}}{dx} \qquad (10.2)$$

$$\text{Moment:} \qquad \hat{M} = +EI\frac{d^2\hat{v}}{dx^2} \qquad (10.3)$$

$$\text{Shear:} \qquad \hat{V} = -EI\frac{d^3\hat{v}}{dx^3} \qquad (10.4)$$

$$\text{Loading:} \qquad \hat{q} = +EI\frac{d^4\hat{v}}{dx^4} - (\omega^2\rho A - i\omega\eta)\hat{v} \qquad (10.5)$$

The only difference in comparison with the beam equations of Chapter 3 is the addition of the mass and damping related terms in the expression for the loading. Since

$\omega$ is treated as a parameter in the spectral analysis, it is seen from these that the transverse displacement $\hat{v}(x)$ (via its derivatives) can be viewed as the fundamental unknown of interest. It is emphasized that the spectral approach to dynamics essentially reduces the problem to a pseudo-static one. Thus, all the principles and approaches established for static problems are applicable here also.

When solving these beam equations, information may be given at any of the five levels above, thus requiring integrations (or differentiations) to obtain the other quantities. Integration gives rise to constants of integration which must be found from the boundary and compatibility conditions. In the general case, there are four constants of integration (since the highest derivative is four). As an example, we will integrate the loading relation. To simplify matters, assume that in addition to the material properties $EI$, $\rho A$, $\eta$, the loading $\hat{q}$ is also constant. The loading relation is then an inhomogeneous fourth order differential equation in the deflection $\hat{v}(x)$. The complete solution is the sum of the homogeneous solution and a particular solution. It is easy to show that the particular solution is

$$\hat{v}_p(x) = \frac{-\hat{q}}{EIk^4}$$

To obtain the homogeneous solution, begin by assuming a trial solution of the form

$$\hat{v}_h(x) = Ae^{-i\alpha x}$$

where $\alpha$ is an unknown constant. Substitute this into the differential equation to get

$$(-i\alpha)^4 Ae^{-i\alpha x} - k^4 Ae^{-i\alpha x} = 0$$

Since the exponential term is never zero, it can be canceled giving

$$+\alpha^4 - k^4 = 0 \qquad \Rightarrow \qquad \alpha^2 = \pm k^2 \qquad \Rightarrow \qquad \alpha = \pm k, \pm ik$$

Thus, four possible values are obtained for $\alpha$, giving the general homogeneous solution in the form

$$\hat{v}_h(x) = \mathbf{A}e^{-ikx} + \mathbf{B}e^{-kx} + \mathbf{C}e^{+ikx} + \mathbf{D}e^{+kx}$$

where $\mathbf{A}, \mathbf{B}, \mathbf{C}, \mathbf{D}$ are complex constants to be determined by the boundary conditions. We can now write the total solution as

$$\hat{v}(x) = \mathbf{A}e^{-ikx} + \mathbf{B}e^{-kx} + \mathbf{C}e^{+ikx} + \mathbf{D}e^{+kx} - \frac{\hat{q}}{EIk^4} \qquad (10.6)$$

Note that the material properties and frequency appear in the spectrum relation as represented by

$$k \equiv [\omega^2 - i\eta\omega/\rho A]^{\frac{1}{4}}[\frac{\rho A}{EI}]^{\frac{1}{4}} = k(\omega)$$

Therefore, although we keep the same functional form for the solution at each frequency, the actual shapes are quite different.

Sometimes it is more convenient to consider the solution in the equivalent form

$$\hat{v}(x) = c_1 \cos kx + c_2 \sin kx + c_3 \cosh kx + c_4 \sinh kx - \frac{\hat{q}}{EIk^4} \qquad (10.7)$$

As to which is preferable depends on the particular boundary value problem. As was pointed out with rods; generally, when damping is present in the system the exponential form is preferred, when damping is zero then the above form is easier to work with because none of the terms are complex.

**Example 10.1:**   Analyze the free vibrations of the simply-supported beam shown in Figure 10.2. Neglect damping.

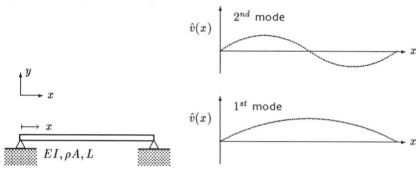

**Figure 10.2**: Simply supported beam.

The solution approach is to use the boundary conditions to determine the coefficients in the general solution for the deflected shape. Since there is no damping, as well as no applied load, we take this general solution as

$$\hat{v}(x) = c_1 \cos kx + c_2 \sin kx + c_3 \cosh kx + c_4 \sinh kx$$

The boundary conditions are that the deflection and moment are zero at each end. The spectral form of these boundary conditions are

$$\text{at} \quad x = 0\,, L: \qquad \hat{v} = 0$$

$$\text{at} \quad x = 0\,, L: \qquad EI\frac{d^2\hat{v}}{dx^2} = 0$$

These four conditions give, after substituting for the displacement function,

$$\begin{aligned} c_1 + c_3 &= 0 \\ c_1 \cos kL + c_2 \sin kL + c_3 \cosh kL + c_4 \sinh kL &= 0 \\ c_1 - c_3 &= 0 \\ -c_1 \cos kL - c_2 \sin kL + c_3 \cosh kL + c_4 \sinh kL &= 0 \end{aligned}$$

It is obvious that $c_1 = c_3 = 0$ and that $c_2$ and $c_4$ must satisfy

$$\begin{bmatrix} \sin kL & \sinh kL \\ -\sin kL & \sinh kL \end{bmatrix} \begin{Bmatrix} c_2 \\ c_4 \end{Bmatrix} = 0$$

For a non-trivial solution, this requires that the determinant of the matrix associated with this system be zero. That is

$$\det \begin{vmatrix} \sin kL & \sinh kL \\ -\sin kL & \sinh kL \end{vmatrix} = 0 \qquad \text{or} \qquad 2\sin kL \sinh kL = 0$$

Since the hyperbolic term $\sinh kL$ is zero only when $kL = 0$, then it is required that

$$\sin kL = 0 \qquad \text{or} \qquad kL = n\pi, \qquad n = 1,2,3,\ldots$$

Expand $k^4$ in terms of frequency to give

$$k^4 = \frac{\rho A \omega^2}{EI} = \frac{n^4 \pi^4}{L^4} \qquad \text{or} \qquad \omega = \frac{n^2 \pi^2}{L^2} \sqrt{\frac{EI}{\rho A}}$$

These are the resonant frequencies; just as for the rod, there is an infinity of discrete resonant frequencies.

The mode shapes corresponding to the resonant frequencies are determined by evaluating $c_2$ and $c_4$. We find that when $kL = n\pi$ then $c_4 = 0$ and $c_2$ is arbitrary. Hence the mode shapes are

$$\hat{v}(x) = c_2 \sin\!\left(n\pi \frac{x}{L}\right)$$

The first two of these are shown plotted in Figure 10.2. Since the coefficient $c_2$ is arbitrary, it is usual to set it to unity.

**Example 10.2:**  Analyze the free vibrations of the fixed-fixed beam shown Figure 10.3. Neglect damping.

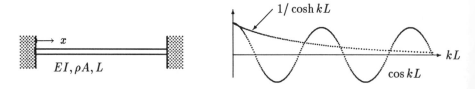

**Figure 10.3**: Fixed-fixed beam.

The fixed-fixed conditions are that the deflection and slope at both ends of the beam are zero. The spectral form of these conditions leads to the following homogeneous set of equations

$$\begin{bmatrix} 1 & 0 & 1 & 0 \\ \cos kL & \sin kL & \cosh kL & \sinh kL \\ 0 & 1 & 0 & 1 \\ -\sin kL & \cos kL & \sinh kL & \cosh kL \end{bmatrix} \begin{Bmatrix} c_1 \\ c_2 \\ c_3 \\ c_4 \end{Bmatrix} = 0$$

For a non-trivial solution for the coefficients, the determinant of this system must be zero. This determinant, after expanding and simplifying, gives the following frequency equation

$$\cos kL \cosh kL - 1 = 0 \qquad \text{or} \qquad \cos kL = \frac{1}{\cosh kL}$$

This is a transcendental equation and is therefore difficult to solve. Figure 10.3 gives a graphical idea of how the roots are determined: whenever the plot of $f_1(\omega) = \cos kL$ intersects the plot of $f_2(\omega) = 1/\cosh kL$, we find a root. The sequence of roots so determined for this problem is

$$kL = 4.73\,, 7.85\,, 10.99\,, \cdots$$

Note that $kL = 0$ is an intercept, but it is of no interest since then $\hat{v}(x) = 0$. The frequencies are obtained as

$$\omega_n = \frac{(kL)_n^2}{L^2}\sqrt{\frac{EI}{\rho A}}\,, \qquad (kL)^2 = 22.37\,, 61.67\,, 120.90\,, \cdots$$

There are multiple resonant frequencies.

The mode shapes for this problem are

$$\hat{v}(x) = (\sin kL - \sinh kL)(\cos kx - \cosh kx) - (\cos kL - \cosh kL)(\sin kx - \sinh kx)$$

Notice that they are rather complicated shapes. In fact, for most beam problems the mode shapes are complicated like this. We will therefore not make direct use of the mode shapes.

For future reference, Table 10.1 gives results for some other boundary conditions. The frequency is given by

$$\omega_n = \frac{(kL)_n^2}{L^2}\sqrt{\frac{EI}{\rho A}}$$

Note that the free-free and pinned-free boundary conditions also have rigid body modes. A curious feature of this table is the fact that apparently different physical situations (such as the fixed-fixed and free-free beams) can have the same frequency equation. This comes about because the specification of the boundary conditions are in terms of derivatives of $\hat{v}(x)$; since this function comprises the trigonometric and hyperbolic functions, then we see that it is possible, on differentiation, for similar boundary conditions to appear.

# 10.2  Structural Connections

We now look at the necessary conditions to impose when multiple beams are connected together to form a beam structure. Our approach, conceptually, is the same as for the static structures; we write the solution for each beam segment in terms of the constants of integration, and then apply both compatibility and equilibrium conditions at the joints in order to determine these constants. The only difference is that for the dynamic problem we work in terms of the spectral form of the equations.

Consider two beam segments connected by a rigid joint as shown in Figure 10.4. To add some generality, let the joint have a concentrated mass $m_c$. Compatibility between the two beam segments requires that

$$v^{(1)}(t) = v^{(2)}(t) = v(t)\,, \qquad \phi^{(1)}(t) = \phi^{(2)}(t) = \phi(t)$$

| BC | $(kL)_1^2$ | $(kL)_2^2$ | $(kL)_3^2$ | Frequency Equation |
|---|---|---|---|---|
| cantilever | 3.516 | 22.03 | 61.70 | $\cos kL \cosh kL + 1 = 0$ |
| pinned-pinned | 9.870 | 39.48 | 88.83 | $\sin kL = 0$ |
| fixed-pinned | 15.42 | 49.97 | 104.2 | $\sin kL \cosh kL - \cos kL \sinh kL = 0$ |
| pinned-free | 15.42 | 49.97 | 104.2 | $\sin kL \cosh kL - \cos kL \sinh kL = 0$ |
| fixed-fixed | 22.37 | 61.67 | 120.9 | $\cos kL \cosh kL - 1 = 0$ |
| free-free | 22.37 | 61.67 | 120.9 | $\cos kL \cosh kL - 1 = 0$ |

Table 10.1: Frequency factor $(kL)^2$ for the first three non-rigid body modes.

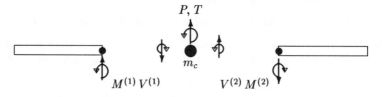

**Figure 10.4**: Beam connection.

The spectral forms for these are

$$\hat{v}^{(1)} = \hat{v}^{(2)} = \hat{v}, \qquad \hat{\phi}^{(1)} = \hat{\phi}^{(2)} = \hat{\phi} \tag{10.8}$$

Dynamic equilibrium at the joint gives

$$-V^{(1)}(t) + V^{(2)}(t) + P(t) = m_c \ddot{v}(t), \qquad -M^{(1)}(t) + M^{(2)}(t) + T(t) = 0$$

The spectral forms of these equations are

$$- \hat{V}^{(1)} + \hat{V}^{(2)} + \hat{P} + \omega^2 m_c \hat{v} = 0, \qquad -\hat{M}^{(1)} + \hat{M}^{(2)} + \hat{T} = 0 \tag{10.9}$$

There are a total of four equations that must be imposed at each junction. It is apparent that a concentrated mass affects the structure as an effective force of amount $\omega^2 m_c \hat{v}$. This force, however, is frequency dependent. It is interesting to observe that at a particular frequency, the force exerted by the mass is similar to that of a spring of stiffness $-\omega^2 m_c$.

In like manner, if the beam segments are connected by a pinned joint, then we impose continuity of deflection only. The remaining three equations come from equilibrium; since the moment transmitted is zero, we have

$$-\hat{V}^{(1)} + \hat{V}^{(2)} + \hat{P} + \omega^2 m_c \hat{v} = 0, \qquad -\hat{M}^{(1)} = 0, \qquad \hat{M}^{(2)} = 0$$

Other connectivity conditions are handled in a similar way.

**Example 10.3:** Investigate the effect of a concentrated mass on the frequency equation of the pinned-pinned beam shown in Figure 10.5. Neglect damping.

**Figure 10.5**: Beam with concentrated mass.

The generic spectral solution for an arbitrary beam segment is

$$\hat{v}(x) = c_1 \cos kx + c_2 \sin kx + c_3 \cosh kx + c_4 \sinh kx$$

We will use this in both beam segments. The pinned boundary condition at $x = 0$ for the first segment requires that

$$\hat{v} = 0 \implies 0 = c_1 + c_3$$
$$EI\frac{d^2\hat{v}}{dx^2} = 0 \implies 0 = -k^2 c_1 + k^2 c_3$$

This gives the deflection function as

$$\hat{v}(x) = c_2 \sin kx + c_4 \sinh kx$$

The pinned boundary condition at $x = L$ for the second member leads to (note that we are using $x$ are a local distance)

$$\hat{v} = 0 \implies 0 = c_1^* C + c_2^* S + c_3^* C_h + c_4^* S_h$$
$$EI\frac{d^2\hat{v}}{dx^2} = 0 \implies 0 = k^2[-c_1^* C - c_2^* S + c_3^* C_h + c_4^* S_h]$$

where, for convenience, we have used the notations $C \equiv \cos kL$, $S \equiv \sin kL$, $C_h \equiv \cosh kL$, $S_h \equiv \sinh kL$. These two conditions can be simplified to

$$c_1^* C + c_2^* S = 0, \qquad c_3^* C_h + c_4^* S_h = 0$$

The compatibility conditions of Equations(10.8) lead to

$$c_2 S + c_4 S_h = c_1^* + c_3^*, \qquad k[c_2 C + c_4 C_h] = k[c_2^* + c_4^*]$$

The equilibrium conditions of Equations(10.9) lead to

$$EIk^3[-c_2 C + c_4 C_h] - EIk^3[-c_2^* + c_4^*] + \omega^2 m_c[c_1^* + c_3^*] = 0$$
$$-EIk^2[-c_2 S + c_4 S_h] + EIk^2[-c_1^* + c_3^*] = 0$$

There are six equations to determine the six coefficients. We will first simplify the equations. From the third and sixth equation we get

$$c_1^* = c_2 S, \qquad c_3^* = c_4 S_h$$

These combined with the second equation gives

$$c_4^* = -c_4 C_h$$

The remaining equations may now be put in the matrix form

$$\begin{bmatrix} CS & 0 & S \\ C & 2C_h & -1 \\ -C + \alpha S & 2C_h + \alpha S_h & 1 \end{bmatrix} \begin{Bmatrix} c_2 \\ c_4 \\ c_2^* \end{Bmatrix} = 0$$

where $\alpha = \gamma kL$, and $\gamma = m_c/\rho AL$ is the ratio of the concentrated mass to the mass of half of the beam. The determinant of the matrix becomes, after some simplification,

$$4\cos kL \sin kL \cosh kL + \alpha \sin kL(\cos kL \sinh kL - \sin kL \cosh kL) = 0$$

Not surprising, this frequency equation is transcendental and therefore cannot be solved in closed form. We recognize, with the help of Table 10.1, the following limiting cases:

if $\gamma = 0 = \alpha$ $\Rightarrow$ pinned-pinned, length $= 2L$
if $\gamma = \infty = \alpha$ $\Rightarrow$ pinned-fixed, and pinned-pinned, length $= L$

For an arbitrary value of $\gamma$, there are multiple frequencies at which the above equation is satisfied. These values can be obtained numerically using the scheme outlined in the last example.

## 10.3   Exact Matrix Formulation

We will use the general solution of the governing differential equations to derive an exact matrix formulation for the dynamics of beams. The resulting stiffness matrix is frequency dependent. It is shown that the resonances are obtained by solving the transcendental eigenvalue problem. This makes the exact formulation unsuitable for eigenanalysis, however, it forms the basis of an approximate scheme to be presented in the next section.

The basic idea in establishing the stiffness relation is the same as used in Chapter 3: having integrated the governing differential equations, we rewrite the constants of integration in terms of the nodal degrees of freedom. Using the displacement function, we are then in a position to determine the member moment and shear distributions. These, in turn, can be related to the nodal loads. The only difference is that we use the spectral form of the displacement solution.

We assume that the loading is zero over the element length. Hence the general deflected shape is

$$\hat{v}(x) = c_1 \cos kx + c_2 \sin kx + c_3 \cosh kx + c_4 \sinh kx, \qquad k = k(\omega)$$

Write the coefficients $c_n$ in terms of the nodal degrees of freedom. Thus, at $x = 0$,

$$\hat{v}(0) \equiv v_1 = c_1 + c_3, \qquad \frac{d\hat{v}(0)}{dx} \equiv \hat{\phi}_1 = kc_2 + kc_4$$

Consequently, the deflected shape is

$$k\hat{v}(x) = c_1 k[\cos kx - \cosh kx] + c_2 k[\sin kx - \sinh kx] + \hat{v}_1 k \cosh kx + \hat{\phi}_1 \sinh kx$$

Similarly, at the other node we write

$$\begin{aligned}
k\hat{v}_2 &= c_1 k[\cos kL - \cosh kL] + c_2 k[\sin kL - \sinh kL] + \hat{v}_1 k \cosh kL + \hat{\phi}_1 \sinh kL \\
\hat{\phi}_2 &= -kc_1[\sin kL + \sinh kL] + kc_2[\cos kL - \cosh kL] + \hat{v}_1 k \sinh kL + \hat{\phi}_1 \cosh kL
\end{aligned}$$

The system of equations to determine the remaining coefficients can now be arranged as

$$\begin{bmatrix} (C - C_h) & (S - S_h) \\ -(S + S_h) & (C - C_h) \end{bmatrix} \begin{Bmatrix} c_1 \\ c_2 \end{Bmatrix} = \begin{Bmatrix} -\hat{v}_1 \xi C_h - \hat{\phi}_1 L S_h + \hat{v}_2 \xi \\ -\hat{v}_1 \xi S_h - \hat{\phi}_1 L C_h + \hat{\phi}_2 L \end{Bmatrix} \frac{1}{\xi}$$

where we have used the notations $\xi \equiv kL$, and $C \equiv \cos\xi$, $S \equiv \sin\xi$, $C_h \equiv \cosh\xi$, $S_h \equiv \sinh\xi$. We will use Cramer's rule to solve this. First we get the determinant and rearrange it as

$$\Delta = (C - C_h)^2 + (S + S_h)(S - S_h) = 2[1 - CC_h]$$

The solution for $c_1$ is now given as

$$c_1 = \det \begin{vmatrix} -\hat{v}_1 \xi C_h - \hat{\phi}_1 L S_h + \hat{v}_2 \xi & \xi(S - S_h) \\ -\hat{v}_1 \xi S_h - \hat{\phi}_1 L C_h + \hat{\phi}_2 L & \xi(C - C_h) \end{vmatrix} \frac{1}{\xi\Delta}$$

Multiplying this out gives

$$c_1 = [v_1 \xi(1 + SS_h - CC_h) + \phi_1 L(SC_h - CS_h) + v_2 \xi(C - C_h) + \phi_2 L(S_h - S)]\frac{1}{\xi\Delta}$$

We can obtain the other coefficient in a similar manner

$$c_2 = [-v_1 \xi(CS_h + SC_h) + \phi_1 L(1 - CC_h - SS_h) + v_2 \xi(S + S_h) + \phi_2 L(C - C_h)]\frac{1}{\xi\Delta}$$

At this stage we can rewrite the deflection function in terms of the nodal degrees of freedom. That is, we have similar to Equation(3.6),

$$\hat{v}(x) = \hat{g}_1(x)\hat{v}_1 + \hat{g}_2(x)L\hat{\phi}_1 + \hat{g}_3(x)\hat{v}_2 + \hat{g}_4(x)L\hat{\phi}_2$$

where

$$\begin{aligned}
\hat{g}_1(x) &\equiv [1 - CC_h]\hat{h}_0(x) + [SS_h]\hat{h}_2(x) - [CS_h + SC_h]\hat{h}_3(x) \\
\hat{g}_2(x) &\equiv [1 - CC_h]\hat{h}_1(x) + [SC_h - SS_h]\hat{h}_2(x) - [SS_h]\hat{h}_3(x) \\
\hat{g}_3(x) &\equiv [C - C_h]\hat{h}_2(x) + [S + S_h]\hat{h}_3(x) \\
\hat{g}_4(x) &\equiv [S_h - S]\hat{h}_2(x) + [C - C_h]\hat{h}_3(x)
\end{aligned} \qquad (10.10)$$

The functions $\hat{h}_n(x)$, defined as

$$\hat{h}_0(x) \equiv \cos kx + \cosh kx \qquad \hat{h}_1(x) \equiv \sin kx + \sinh kx$$
$$\hat{h}_2(x) \equiv \cos kx - \cosh kx \qquad \hat{h}_3(x) \equiv \sin kx - \sinh kx$$

are arranged in increasing powers of $x$. That is, in the limit of small $k$ they are $1$, $x$, $x^2$, $x^3$, respectively. The functions $\hat{g}_n(x)$ are the dynamic shape functions for the beam; they are frequency dependent.

Now that we have the deflection function in terms of the nodal degrees of freedom, we can write the nodal forces and moments also in terms of the nodal degrees of freedom. For example, the moment distribution is

$$\hat{M}(x) = EI\frac{d^2\hat{v}}{dx^2} = EI[\hat{g}_1''(x)\hat{v}_1 + \hat{g}_2''(x)L\hat{\phi}_1 + \hat{g}_3''(x)\hat{v}_2 + \hat{g}_4''(x)L\hat{\phi}_2]$$

The moment at $x = 0$ is

$$\hat{M}_1 = -\hat{M}(0) = EI\frac{2\xi}{L^2}[v_1\xi SS_h + \phi_1 L(SC_h - CS_h) + v_2\xi(C - C_h) + \phi_2 L(S_h - S)]\frac{1}{\Delta}$$

The shear force distribution is

$$\hat{V}(x) = -EI\frac{d^3\hat{v}}{dx^3} = -EI[\hat{g}_1'''(x)\hat{v}_1 + \hat{g}_2'''(x)L\hat{\phi}_1 + \hat{g}_3'''(x)\hat{v}_2 + \hat{g}_4'''(x)L\hat{\phi}_2]$$

giving a value at $x = 0$ of

$$\hat{V}_1 = -\hat{V}(0) = EI\frac{\xi^2}{L^3}[v_1\xi(CS_h + SC_h) + \phi_1 L(SS_h) - v_2\xi(S + S_h) - \phi_2 L(C - C_h)]\frac{1}{\Delta}$$

Similar expressions can be obtained for the nodal loads $\hat{M}_2$ and $\hat{V}_2$, from which the stiffness relation can be established. The complete stiffness matrix is

$$\left\{\begin{array}{c} \hat{V}_1 \\ \hat{M}_1 \\ \hat{V}_2 \\ \hat{M}_2 \end{array}\right\} = \left[\quad \hat{k}(\omega) \quad \right] \left\{\begin{array}{c} \hat{v}_1 \\ \hat{\phi}_1 \\ \hat{v}_2 \\ \hat{\phi}_2 \end{array}\right\} \qquad \text{or} \qquad \{\hat{F}\} = [\ \hat{k}\ ]\{\hat{u}\}$$

where the dynamic stiffness $[\ \hat{k}\ ]$ is given as

$$\frac{EI}{L^3}\frac{2}{\Delta}\left[\begin{array}{cccc} (SC_h + CS_h)\xi^3 & SS_h\xi^2 L & -(S + S_h)\xi^3 & (C_h - C)\xi^2 L \\ & (SC_h - CS_h)\xi L^2 & -(C_h - C)\xi^2 L & -(S - S_h)\xi L^2 \\ & & (SC_h + CS_h)\xi^3 & -SS_h\xi^2 L \\ & \text{sym} & & (SC_h - CS_h)\xi L^2 \end{array}\right]$$

$$(10.11)$$

This is a symmetric matrix and exhibits many of the same repetitions as the static beam element stiffness.

It is apparent that since we can view the spectral analysis as a series of pseudo-static problems, that the assemblage proceeds identically to that of the static case. This leads to the structural system

$$[\hat{K}]\{\hat{u}\} = \{\hat{P}\}$$

This equations looks like the static equation, but the 'hat' notation is a reminder that all the terms are frequency dependent.

**Example 10.4:**   Use the dynamic stiffness matrix to determine the frequency equation for a vibrating cantilever.

The procedure we follow is the same as for other matrix methods. Number the nodes with 1 at the fixed end and 2 at the other, then the unknown degrees of freedom are

$$\{\hat{u}_u\} = \{\hat{v}_2, \hat{\phi}_2\}$$

There is only one element, hence the reduced structural stiffness matrix is just the fourth quadrant of the element stiffness. The reduced stiffness relation is therefore

$$\frac{EI}{L^3}\frac{\xi^3}{\Delta} \begin{bmatrix} (SC_h + CS_h)\xi^3 & -SS_h\xi^2 L \\ -SS_h\xi^2 L & (SC_h - CS_h)\xi L^2 \end{bmatrix} \begin{Bmatrix} \hat{v}_2 \\ \hat{\phi}_2 \end{Bmatrix} = 0$$

The determinant simplifies to

$$(SC_h + CS_h)(SC_h - CS_h) - S^2 S_h^2 = 0 \qquad \text{or} \qquad 1 + CC_h = 0$$

This is the same as is listed in Table 10.1.

**Example 10.5:**   A dynamic load is applied to the center of a fixed-fixed beam of length $2L$ as shown in Figure 10.6. Determine the frequency response function.

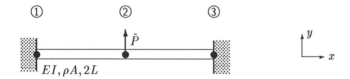

**Figure 10.6:** Central impact of a clamped beam.

Number the nodes as shown in the figure, then the unknown degrees of freedom and known forces are

$$\{\hat{u}_u\} = \{\hat{v}_2, \hat{\phi}_2\}, \qquad \{\hat{P}_k\} = \{\hat{P}, 0\}$$

The reduced element stiffness matrices are

$$[\hat{k}^{*(12)}] = \frac{EI2}{L^3\Delta} \begin{bmatrix} (SC_h + CS_h)\xi^3 & -SS_h\xi^2 L \\ -SS_h\xi^2 L & (SC_h - CS_h)\xi L^2 \end{bmatrix}$$

and

$$[\hat{k}^{*(23)}] = \frac{EI2}{L^3\Delta} \left[ \begin{array}{cc} (SC_h + CS_h)\xi^3 & +SS_h\xi^2 L \\ +SS_h\xi^2 L & (SC_h - CS_h)\xi L^2 \end{array} \right]$$

The assembled structural stiffness relation is therefore

$$\frac{4EI}{L^3\Delta} \left[ \begin{array}{cc} (SC_h + CS_h)\xi^3 & 0 \\ 0 & (SC_h - CS_h)\xi L^2 \end{array} \right] \left\{ \begin{array}{c} \hat{v}_2 \\ \hat{\phi}_2 \end{array} \right\} = \left\{ \begin{array}{c} \hat{P} \\ 0 \end{array} \right\}$$

It is obvious that $\hat{\phi}_2 = 0$ implying symmetry of the deformation. Solving the remaining equation gives

$$\hat{v}_2 = \frac{L^3}{4EI}\frac{\Delta}{\xi^3}\frac{\hat{P}}{[CS_h + SC_h]}$$

A time response can be obtained by performing an inverse FFT on the spectrum $\hat{v}_2$.

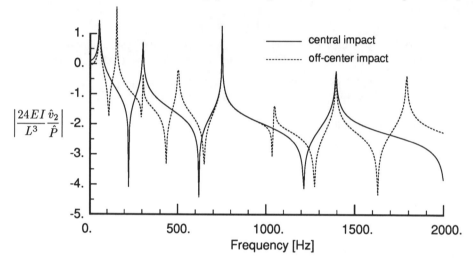

**Figure 10.7**: Log of frequency response function for impact at two locations.

The log of the frequency response function is shown plotted in Figure 10.7. There are two characteristics to note. First, the response has very sharp spectral peaks occurring at the resonances, and therefore the time domain response will be dominated by these frequencies. Second, the impact at the one-third position creates many more spectral peaks; in other words it excites more of the resonant frequencies.

## 10.4   Approximate Matrix Formulation

The dynamic stiffness matrix as developed is exact; however, the unknown frequency in an eigenanalysis is deeply imbedded in its transcendental form. That is, when we solve for the resonant frequencies, we must solve a transcendental eigenvalue problem.

There are no standard methods for solving the transcendental eigenvalue problem (although References [36, 47] discuss some successful schemes), and this makes direct use of the exact dynamic stiffness matrix difficult for vibration problems. In contrast, there are many standard procedures for solving the algebraic eigenvalue problem as already shown in Chapter 5. What we are now interested in is a simplification of the dynamic stiffness matrix to allow easier determination of the resonances. The price is that the stiffness relation is no longer exact.

We look at two schemes. The first uses the exact matrix and by use of a Taylor series expansion obtains an algebraic form. The other method lumps the distributed properties. This is not as accurate but its derivation is instructive.

## Approximate Dynamic Stiffness

The frequency appears in the term $\xi \equiv kL$ and in the trigonometric terms $C \equiv \cos \xi$, $S \equiv \sin \xi$, $C_h \equiv \cosh \xi$, $S_h \equiv \sinh \xi$. We will make a Taylor series expansion of these with respect to $\xi$. Note that this expansion on small $\xi$ is equivalent to requiring a small element length $L$ or restricting the analysis to low frequencies.

We will start with the determinant. Using the series expansion for the cosine and cosine hyperbolic terms, we get

$$
\begin{aligned}
\Delta &= 2[1 - CC_h] \\
&\approx 2[1 - (1 - \tfrac{1}{2}\xi^2 + \tfrac{1}{24}\xi^4 - \tfrac{1}{720}\xi^6 - \tfrac{1}{8!}\xi^8 + \cdots)(1 + \tfrac{1}{2}\xi^2 + \tfrac{1}{24}\xi^4 + \tfrac{1}{720}\xi^6 + \tfrac{1}{8!}\xi^8 + \cdots)] \\
&\approx \frac{2\xi^4}{6}[1 - \tfrac{1}{420}\xi^4 + O(\xi^8)]
\end{aligned}
$$

We have truncated the series precisely as indicated above because that is the first non-trivial point at which the determinant can exhibit a zero. We will need the reciprocal of the determinant and this is approximated as

$$
\frac{2}{\Delta} \approx \frac{6}{\xi^4}[1 + \tfrac{1}{420}\xi^4 + O(\xi^8)]
$$

We now do the expansion on the stiffness terms out to the same order. For example,

$$
SS_h \approx [\xi - \tfrac{1}{6}\xi^3 + \tfrac{1}{120}\xi^5 - \tfrac{1}{7!}\xi^7 + \cdots][\xi + \tfrac{1}{6}\xi^3 + \tfrac{1}{120}\xi^5 + \tfrac{1}{7!}\xi^7 + \cdots] = \xi^2[1 - \tfrac{1}{90}\xi^4 + O(\xi^8)]
$$

Hence the second stiffness term is

$$
\hat{k}_{12} \approx \frac{EI}{L^3}\frac{6}{\xi^4}[1 + \tfrac{1}{420}\xi^4 + O(\xi^8)]\xi^2[1 - \tfrac{1}{90}\xi^4 + O(\xi^8)]\xi^2 L = \frac{EI}{L^3}[6L - \xi^4\frac{22L}{420} + O(\xi^8)]
$$

In like manner, we can expand for all the stiffness terms to finally get the complete approximate element stiffness matrix as

$$
[\hat{k}] = \frac{EI}{L^3}\left[
\begin{bmatrix}
12 & 6L & -12 & 6L \\
6L & 4L^2 & -6L & 2L^2 \\
-12 & -6L & 12 & -6L \\
6L & 2L^2 & -6L & 4L^2
\end{bmatrix}
- \frac{\xi^4}{420}
\begin{bmatrix}
156 & 22L & 54 & -13L \\
22L & 4L^2 & 13L & -3L^2 \\
54 & 13L & 156 & -22L \\
-13L & -3L^2 & -22L & 4L^2
\end{bmatrix}
\right]
$$

This approximation neglects terms of order $\zeta^8$ and higher. We recognize the first matrix as the elastic stiffness for beams. Replacing $\zeta$ in terms of the density and damping, that is,

$$\zeta^4 = (kL)^4 = (\omega^2 \rho A - i\omega\eta)\frac{L^4}{EI}$$

we see that the leading coefficient for the second matrix is proportional to $\omega^2$ and $i\omega$. Recall from Chapter 8 that the inertia force and damping force have the following associations

$$-\omega^2 m\hat{u} \longleftrightarrow m\ddot{u}, \qquad i\omega c\hat{u} \longleftrightarrow c\dot{u}$$

Therefore, we can introduce a mass matrix defined as

$$[\,m\,] \equiv \frac{\rho AL}{420}\begin{bmatrix} 156 & 22L & 54 & -13L \\ 22L & 4L^2 & 13L & -3L^2 \\ 54 & 13L & 156 & -22L \\ -13L & -3L^2 & -22L & 4L^2 \end{bmatrix}$$

This symmetric matrix is called the *consistent mass matrix* for beams. Note that these masses do not necessarily have any simple interpretation of masses at nodes. It is apparent that we can also introduce a *consistent damping matrix* for beams as

$$[\,c\,] = \frac{\eta}{\rho A}[\,m\,]$$

This is an example of the damping matrix being proportional to the mass matrix.

The beam stiffness relation is now written as

$$\{\hat{F}\} = [\,\hat{k}\,]\{\hat{u}\} \approx [[\,k\,] + i\omega[\,c\,] - \omega^2[\,m\,]]\{\hat{u}\}$$

If we compare the undamped diagonal terms of the approximate dynamic stiffness matrix with the exact values, we see they go through a zero only once as $\omega$ is varied. Therefore, at most, only four resonant frequencies are obtained for this element. Contrast this with the infinite number obtainable from the exact matrix. We conclude that to improve the approximate solution we must use many elements for a given member length.

## Lumped Matrix Formulation for Beams

We will now take a different approach to establishing the approximate stiffness relation: the mass and damping distributions will be treated by the lumped method. That is, we replace the distributed quantities by concentrated values at Nodes 1 and 2. The damping follows the mass, so we will show the development only for the mass.

First consider an element of length $L$ as shown in Figure 10.8. Let the mass distribution be replaced by the concentrated masses $m_1$ and $m_2$ at the nodes, then the equation of motion for the massless middle section is

$$EI\frac{\partial^4 v}{\partial x^4} = 0$$

This is just the equation for the static deflection shape. Recall that the deflection of the element can be expressed in terms of the four nodal degrees of freedom $v_1, \phi_1, v_2$, and $\phi_2$ as (from Chapter 3)

$$v(x,t) = g_1(x)v_1(t) + g_2(x)L\phi_1(t) + g_3(x)v_2(t) + g_4(x)L\phi_2(t)$$

This is time-dependent only insofar as the nodal values depend on time. The elastic restoring forces and moments are related to the nodal displacements as

$$\left\{ \begin{array}{c} V_1^e \\ M_1^e \\ V_2^e \\ M_2^e \end{array} \right\} = \frac{EI}{L^3} \left[ \begin{array}{cccc} 12 & 6L & -12 & 6L \\ 6L & 4L^2 & -6L & 2L^2 \\ -12 & -6L & 12 & -6L \\ 6L & 2L^2 & -6L & 4L^2 \end{array} \right] \left\{ \begin{array}{c} v_1 \\ \phi_1 \\ v_2 \\ \phi_2 \end{array} \right\}$$

This is obviously the same stiffness relation for the beam as obtained in Chapter 3. Keep in mind that this is time dependent because the nodal values are time dependent.

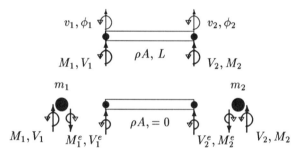

**Figure 10.8**: Beam element replaced with lumped mass model.

To estimate the lumped parameters, we consider only rigid body motions. When $v_1(t) = v_2(t)$ then the total mass $\rho AL$ only displaces and we get the same effect if we place half of the total mass at each end. When the beam rotates such that $\phi_1(t) = \phi_2(t)$ and $v_1(t) = -v_2(t)$, that is, the entire beam is rotating about its center, then the moment of inertia is $\rho AL^3/12$. But the contribution from the lumped masses (at the ends) is

$$2(\frac{\rho AL}{2})(\frac{L}{2})^2 = \frac{3}{12}\rho AL^3$$

This is already larger than the total as estimated from the distributed inertia, hence we will take the additional contribution to the rotational inertia as being zero. Dynamic equilibrium of the concentrated masses now gives

at Node 1: $\quad V_1 - V_1^e - V_1^d = m_1\ddot{v}_1 = \frac{1}{2}\rho AL\ddot{v}_1, \quad M_1 - M_1^e = I_1\ddot{\phi}_1 = 0$

at Node 2: $\quad V_2 - V_2^e - V_1^d = m_2\ddot{v}_2 = \frac{1}{2}\rho AL\ddot{v}_2, \quad M_2 - M_2^e = I_2\ddot{\phi}_2 = 0$

We use a similar argument for the damping, realizing that the total damping on the element is $\eta L$ and it is lumped at both nodes. The above equations in matrix form are

$$
\left\{\begin{array}{c} V_1 \\ M_1 \\ V_2 \\ M_2 \end{array}\right\} = \left\{\begin{array}{c} V_1^e \\ M_1^e \\ V_2^e \\ M_2^e \end{array}\right\} + \frac{\rho AL}{2}\left[\begin{array}{cccc} 1 & 0 & 0 & 0 \\ 0 & 0 & 0 & 0 \\ 0 & 0 & 1 & 0 \\ 0 & 0 & 0 & 0 \end{array}\right]\left\{\begin{array}{c} \ddot{v}_1 \\ \ddot{\phi}_1 \\ \ddot{v}_2 \\ \ddot{\phi}_2 \end{array}\right\} + \frac{\eta L}{2}\left[\begin{array}{cccc} 1 & 0 & 0 & 0 \\ 0 & 0 & 0 & 0 \\ 0 & 0 & 1 & 0 \\ 0 & 0 & 0 & 0 \end{array}\right]\left\{\begin{array}{c} \dot{v}_1 \\ \dot{\phi}_1 \\ \dot{v}_2 \\ \dot{\phi}_2 \end{array}\right\}
$$

$$
= [\,k\,]\{u\} + [\,m\,]\{\ddot{u}\} + [\,c\,]\{\dot{u}\}
$$

The matrix $[\,m\,]$ is called the *lumped mass matrix* for beams. It is diagonal. Note that to avoid numerical difficulties, it is usual to replace the zero $m_{22}$ and $m_{44}$ terms with a value given by $\alpha \rho AL$ where $\alpha$ is a very small number.

## Comparison of Consistent and Lumped Masses

The lumped mass has some very appealing properties. For one, it is easy to associate a physical model with it. From a practical point of view, the more important property is that it is diagonal. This means that after assemblage the structural mass matrix is also diagonal. This leads to significantly fewer computations and less computer storage requirements.

In contrast, the components of the consistent mass matrix do not suggest any physical association (at least, not according to the derivation we used.) Further, after assemblage the structural mass matrix will be as populated as the stiffness matrix.

A pertinent question to ask is: Since both mass models are approximate, does one nonetheless perform better than the other? Table 10.2 shows the calculations of the first two resonant frequencies for beams with various boundary conditions. In each case four elements were used. It is apparent that the consistent mass is superior in every case by a significant margin. Therefore, all things being equal, the consistent mass is recommended for flexural motions.

Note that in every case the consistent model overestimates the frequency whereas the lumped model underestimates it.

## 10.5   Beam Structures Problems

Irrespective of the derivation of the element matrices, the approximate stiffness relation for the element can be written as

$$
\{F\} = [\,k\,]\{u\} + [\,c\,]\{\dot{u}\} + [\,m\,]\{\ddot{u}\}
$$

Each of the quantities on the right hand side are to be viewed as 'interior' to the element, that is, during the assemblage process only the nodal force vector $\{F\}$ interacts with the other elements.

| BC | Mode | Exact | Consistent | Lumped |
|---|---|---|---|---|
| cantilever | 1 | 3.516 | 3.516 (+0.00%) | 3.418 (-2.79%) |
| | 2 | 22.03 | 22.06 (+0.14%) | 20.09 (-8.80%) |
| pinned-pinned | 1 | 9.872 | 9.872 (+0.02%) | 9.867 (-0.04%) |
| | 2 | 39.48 | 39.63 (+0.39%) | 39.19 (-0.73%) |
| fixed-fixed | 1 | 22.37 | 22.40 (+0.15%) | 22.30 (-0.30%) |
| | 2 | 61.67 | 62.25 (+0.93%) | 59.25 (-3.92%) |
| free-free | 1 | 22.37 | 22.40 (+0.13%) | 18.91 (-15.5%) |
| | 2 | 61.67 | 62.06 (+0.62%) | 48.00 (-22.2%) |

Table 10.2: Performance of two mass models. Errors are in parentheses.

The assemblage is performed as illustrated previously. Imposing dynamic equilibrium at each nodal point in terms of $\{F\}$ and substituting for the respective force vectors in terms of the stiffness, mass and damping matrices, gives

$$[K]\{u\} + [C]\{\dot{u}\} + [M]\{\ddot{u}\} = \{P\}$$

where $[K]$ is the structural stiffness matrix, $[C]$ is the assembled damping matrix, and $[M]$ is the structural mass matrix. These latter matrices are assembled in identical fashion as done for the elastic stiffness; consequently they enjoy all of the same symmetry and bandedness properties.

When the beam joints have inertias, then these are added to the total mass simply by

$$[M] = \sum_i [m^{(m)}] + \lceil M_c \rfloor$$

where $\lceil M_c \rfloor$ is the diagonal matrix of concentrated inertias (note that this could also include concentrated rotational inertias). Damping can be either specified or taken as

$$[C] = \alpha[M] + \beta[K]$$

A special case of this has $\beta = 0$ and $\alpha = \eta/\rho A$ where $\eta$ is the damping per unit length of beam. This will give correspondence with solutions obtained for the viscously damped beam. Note that this is exact for structures only when the damping of each element is the same.

**Example 10.6:**   Use one element to determine the resonant frequencies of the simply-supported beam shown in Figure 10.9. Neglect damping.

Number the nodes as shown in the figure. The boundary conditions are

$$v_1 = v_2 = 0$$

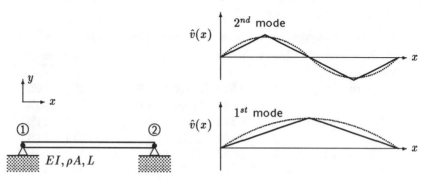

**Figure 10.9**: Simply supported beam.

Consequently, the unknown displacements and known forces are

$$\{u_u\} = \left\{ \begin{matrix} \phi_1 \\ \phi_2 \end{matrix} \right\}, \qquad \{P_k\} = \left\{ \begin{matrix} 0 \\ 0 \end{matrix} \right\}$$

Thus, the pertinent reduced matrices associated with the degrees of freedom $\phi_1$ and $\phi_2$ are (using consistent mass)

$$[K^*] = \frac{EI}{L^3} \begin{bmatrix} 4L^2 & 2L^2 \\ 2L^2 & 4L^2 \end{bmatrix} \qquad [M^*] = \frac{\rho AL}{420} \begin{bmatrix} 4L^2 & -3L^2 \\ -3L^2 & 4L^2 \end{bmatrix}$$

The governing equation for the free vibration of the beam is

$$\frac{EI}{L^3} \begin{bmatrix} 4 & 2 \\ 2 & 4 \end{bmatrix} \left\{ \begin{matrix} \phi_1 \\ \phi_2 \end{matrix} \right\} + \frac{\rho AL}{420} \begin{bmatrix} 4 & -3 \\ -3 & 4 \end{bmatrix} \left\{ \begin{matrix} \ddot{\phi}_1 \\ \ddot{\phi}_2 \end{matrix} \right\} = \left\{ \begin{matrix} 0 \\ 0 \end{matrix} \right\}$$

The spectral form of this equation is

$$\left[ \frac{EI}{L^3} \begin{bmatrix} 4 & 2 \\ 2 & 4 \end{bmatrix} - \omega^2 \frac{\rho AL}{420} \begin{bmatrix} 4 & -3 \\ -3 & 4 \end{bmatrix} \right] \left\{ \begin{matrix} \hat{\phi}_1 \\ \hat{\phi}_2 \end{matrix} \right\} = \left\{ \begin{matrix} 0 \\ 0 \end{matrix} \right\}$$

The resonant frequency $\omega$ can be obtained by setting the determinant to zero. Introduce the parameter $\lambda^2 \equiv EI/(L^4 \rho A)$, then the determinant becomes

$$(4\lambda^2 - \omega^2 4/420)^2 - (2\lambda^2 + \omega^2 3/420)^2 = 0$$

Take the square root of this

$$(4\lambda^2 - \omega^2 4/420) = \pm(2\lambda^2 + \omega^2 3/420)$$

Hence

$$\omega^2 = 120\,\lambda^2 \qquad \text{and} \qquad 2520\,\lambda^2$$

Substituting for $\lambda$, gives the fundamental frequency as

$$\omega = \frac{\sqrt{120}}{L^2} \sqrt{\frac{EI}{\rho A}} \simeq \frac{11}{L^2} \sqrt{\frac{EI}{\rho A}}, \qquad \omega_{exact} = \frac{\pi^2}{L^2} \sqrt{\frac{EI}{\rho A}} \simeq \frac{10}{L^2} \sqrt{\frac{EI}{\rho A}}$$

This compares well with the exact solution giving a difference of about 10%. The frequency for the second mode is

$$\omega = \sqrt{\frac{2520}{L^4}\frac{EI}{\rho A}} = \frac{50.2}{L^2}\sqrt{\frac{EI}{\rho A}}, \qquad \omega_{exact} = \frac{4\pi^2}{L^2}\sqrt{\frac{EI}{\rho A}} = \frac{39.5}{L^2}\sqrt{\frac{EI}{\rho A}}$$

The comparison with the exact solution shows a much larger difference for this mode.

   If we take the values of the frequencies and substitute them into the spectral form of the equations of motion, we obtain respectively

$$\hat{\phi}_2 = -\hat{\phi}_1, \qquad \hat{\phi}_2 = \hat{\phi}_1$$

Thus the first mode is symmetric and the second mode is anti-symmetric as shown in Figure 10.9. Note that to plot the mode shape, it is not sufficient to just plot the nodal displacements — the slope information must also be used. However, when many elements are used, it is sufficient to just plot the displacement.

   These results show that if only one element is used then only two resonances can be obtained and no information at all is obtained about the higher modes. In fact, we get one resonance for each degree of freedom — if we want many modes we must have many degrees of freedom in our model.

**Example 10.7:**  Use two elements to find an approximate solution for the free flexural vibration of the fixed-fixed beam shown in Figure 10.10. Note that two elements are required so as to give some non-zero degrees of freedom.

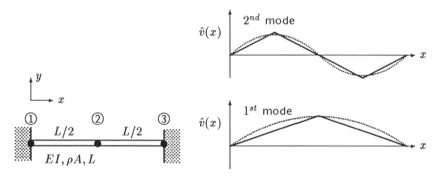

**Figure 10.10**: Fixed-fixed beam.

   Number the nodes as shown in the figure. The boundary conditions are used to determine the unknown degrees of freedom as

$$v_1 = v_3 = \phi_1 = \phi_3 = 0 \qquad \Rightarrow \qquad \{u_u\} = \{v_2, \phi_2\}, \qquad \{\ddot{u}_u\} = \{\ddot{v}_2, \ddot{\phi}_2\}$$

The reduced element matrices are (noting that each element is of length $L/2$)

$$[k^{*(12)}] = \frac{EI}{L^3/8}\begin{bmatrix} 12 & -6L/2 \\ -6L/2 & 4L^2/4 \end{bmatrix}, \qquad [k^{*(23)}] = \frac{EI}{L^3/8}\begin{bmatrix} 12 & 6L/2 \\ 6L/2 & 4L^2/4 \end{bmatrix}$$

giving the assembled structural stiffness matrix as

$$[K^*] = \frac{16EI}{L^3} \begin{bmatrix} 12 & 0 \\ 0 & L^2 \end{bmatrix}$$

The element mass matrices (using consistent masses) are

$$[m^{*(12)}] = \frac{\rho AL/2}{420} \begin{bmatrix} 156 & -22L/2 \\ -22L/2 & 4L^2/4 \end{bmatrix}, \qquad [m^{*(23)}] = \frac{\rho AL/2}{420} \begin{bmatrix} 156 & 22L/2 \\ 22L/2 & 4L^2/4 \end{bmatrix}$$

giving the reduced assembled structural mass matrix as

$$[M^*] = \frac{\rho AL}{420} \begin{bmatrix} 156 & 0 \\ 0 & L^2 \end{bmatrix}$$

Note that the mass matrices also assemble into diagonal form.

The equations of motion for the free vibration of the beam are

$$\left[ \frac{16EI}{L^3} \begin{bmatrix} 12 & 0 \\ 0 & L^2 \end{bmatrix} - \omega^2 \frac{\rho AL}{420} \begin{bmatrix} 156 & 0 \\ 0 & L^2 \end{bmatrix} \right] \left\{ \begin{matrix} \hat{v}_2 \\ \hat{\phi}_2 \end{matrix} \right\} = 0$$

The eigenvalue problem now leads to the following frequency equation

$$(64\frac{105}{L^4}\frac{EI}{\rho A} - \omega^2 13)(\frac{105}{L^4}\frac{EI}{\rho A} - \omega^2) = 0$$

allowing the frequencies to be obtained as

First mode :    $\omega_1 = \dfrac{8}{L^2}\sqrt{\dfrac{105}{13}}\sqrt{\dfrac{EI}{\rho A}} \simeq \dfrac{24.3}{L^2}\sqrt{\dfrac{EI}{\rho A}}$,    $\omega_{exact} = \dfrac{22.37}{L^2}\sqrt{\dfrac{EI}{\rho A}}$

Second mode:   $\omega_2 = \dfrac{8}{L^2}\sqrt{105}\sqrt{\dfrac{EI}{\rho A}} \simeq \dfrac{82.0}{L^2}\sqrt{\dfrac{EI}{\rho A}}$,    $\omega_{exact} = \dfrac{61.67}{L^2}\sqrt{\dfrac{EI}{\rho A}}$

In comparison with the exact solution previously obtained, there is a difference of about 10% for the first mode. The mode shapes for these solutions are simply $\{v_1, \phi_1\} = \{1, 0\}, \{0, 1\}$, respectively. This corresponds to the first symmetric mode and first anti-symmetric mode of the exact solution.

The corresponding equations of motion using the lumped mass model are

$$\left[ \frac{16EI}{L^3} \begin{bmatrix} 12 & 0 \\ 0 & L^2 \end{bmatrix} - \omega^2 \frac{\rho AL}{48} \begin{bmatrix} 24 & 0 \\ 0 & 0 \end{bmatrix} \right] \left\{ \begin{matrix} \hat{v}_2 \\ \hat{\phi}_2 \end{matrix} \right\} = 0$$

The determinant of the eigenvalue problem now simply becomes

$$(64\frac{6}{L^4}\frac{EI}{\rho A} - \omega^2)(64\frac{24}{L^4}\frac{EI}{\rho A}) = 0$$

allowing the frequency for the first mode to be obtained as

First mode :    $\omega_1 = \dfrac{8}{L^2}\sqrt{6}\sqrt{\dfrac{EI}{\rho A}} \simeq \dfrac{19.6}{L^2}\sqrt{\dfrac{EI}{\rho A}}$

This is on the low side by more than the consistent is on the high side but nonetheless a reasonable estimate. This apparent accuracy comes from the fact that for the first mode the rotational degrees of freedom are zero and therefore the lumped model's inadequate modeling of the rotational inertia is not noticeable. However, for the second mode (which has only a rotation at the center) the frequency estimate is grossly in error. Actually, the second frequency is estimated to be infinite.

**Example 10.8:**   A concentrated mass is located at the end of a cantilever beam as shown in Figure 10.11. Investigate the effect the mass has on the free vibrations of the beam.

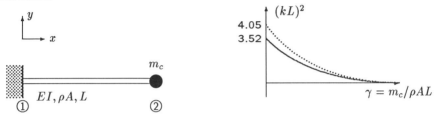

**Figure 10.11**: Cantilever beam with concentrated mass.

We will use one element to model the beam. Numbering the nodes as shown in the figure, we can obtain the reduced structural matrices as

$$[K^*] = \frac{EI}{L^3}\begin{bmatrix} 12 & -6L \\ -6L & 4L^2 \end{bmatrix}$$

$$[M^*] = \frac{\rho AL}{420}\begin{bmatrix} 156 & -22L \\ -22L & 4L^2 \end{bmatrix} + m_c\begin{bmatrix} 1 & 0 \\ 0 & 0 \end{bmatrix} = \frac{\rho AL}{420}\begin{bmatrix} 156 + 420\gamma & -22L \\ -22L & 4L^2 \end{bmatrix}$$

where $\gamma \equiv m_c/\rho AL$ is the ratio of the concentrated mass to the total mass of the beam. The frequency equation is obtained from the spectral form of the equations of motion

$$\left[\lambda^2\begin{bmatrix} 12 & -6L \\ -6L & 4L^2 \end{bmatrix} - \frac{\omega^2}{420}\begin{bmatrix} (156+420\gamma) & -22L \\ -22L & 4L^2 \end{bmatrix}\right]\begin{Bmatrix} \hat{v}_2 \\ \hat{\phi}_2 \end{Bmatrix} = 0, \qquad \lambda^2 \equiv \frac{1}{L^4}\frac{EI}{\rho A}$$

by setting the determinant to zero. That is

$$\omega^4(1 + 12\gamma) - \omega^2\lambda^2 12(78 + 420\gamma) + 15120\lambda^4 = 0$$

If $\gamma = 100$, say, then $\omega \simeq 0.187\lambda$, which agrees very well with the exact solution. Some other limits are

$$\begin{array}{llll} \text{if} & \gamma = 0 & \omega = 4.05\lambda & \text{compare} \quad 3.52\lambda \\ \text{if} & \gamma = \infty & \omega = 0 & \text{compare} \quad 0 \end{array}$$

The full behavior for $(kL)^2 = \omega L^2\sqrt{\rho A/EI}$ for the first mode is shown in Figure 10.11.

## Problems

**10.1** Show that the general equation of motion for a beam with mildly varying section or material properties is given by

$$\frac{\partial^2}{\partial x^2}[EI\frac{\partial^2 v}{\partial x^2}] + \eta\frac{\partial v}{\partial t} + \rho A\frac{\partial^2 v}{\partial t^2} = q(x,t)$$

**10.2** Show that for a beam on an elastic foundation, that the wavenumber is given by

$$k^4 = [\omega^2\rho A - \alpha]/EI$$

where $\alpha$ is the elasticity of the foundation.

[Reference [14], pp. 116]

**10.3** Use the Taylor series expansion of the exact stiffness matrix to show that the matrix representation of the elastic foundation is proportional to the mass matrix.

**10.4** Consider the vibrations of a beam with an axial compressive load $P$. Show that

$$v(x,t) = Ae^{ikx - \omega t)}, \qquad 4k = \pm\sqrt{\lambda \pm \sqrt{\lambda^2 + 4\alpha^2}}$$

where $\lambda^2 = P/EI$ and $\alpha^2 = \omega^2\rho A/EI$ is a solution. Show for a simply supported beam of length $L$ that the resonant frequencies are

$$\omega_n = \frac{n\pi}{L\sqrt{\rho A}}\sqrt{\frac{n^2\pi^2 EI}{L^2} - P}$$

Note that the frequency goes to zero when the axial load approaches the buckling load.

[Reference [16], pp. 325]

**10.5** Prove that in the limit as $kL$ becomes very small, that the spectral shape functions $\hat{g}_n(x)$ approach those of Chapter 3.

**10.6** Choose an arbitrary frequency and plot the spectral shape functions $\hat{g}_n(x)$ as a function of $x$.

**10.7** Choose an arbitrary frequency and plot the shear force and bending moment diagrams for the cantilever beam.

**10.8** Compare the $\hat{k}_{11}$ term of the exact stiffness with the corresponding term of the approximate stiffness using the consistent mass. (Neglect damping). At what value of $kL$ do they differ by 5%?

[Reference [14], pp. 148]

**10.9** Show that the frequency equation for a cantilever beam with a lumped mass at its tip is

$$(1 + \cos kL \sin kL \cosh kL) + \gamma(\cos kL \sinh kL - \sin kL \cosh kL) = 0$$

**10.10** Show that for the cantilever beam with a concentrated mass at the tip, that in the limit as the beam density goes to zero we have only one resonant frequency given by

$$\omega = \sqrt{\frac{3EI}{m_c L^3}} = \frac{1}{L^2}\sqrt{\frac{EI}{\rho A}}\sqrt{\frac{3}{\gamma}}$$

## Exercises

**10.1** The end of the tailpipe of an automobile is observed to vibrate significantly when the engine is idling. Consider the tailpipe as a beam, discuss the feasibility of reducing the vibration by adding an extra bracket, by repositioning the existing bracket, or by adding a mass at the end of the pipe.

**10.2** A steel shaft $4\,ft$ long is carried in bearings at each end. The middle $1\,ft$ is $2\,in$ in diameter with the remainder being $1\,in$ in diameter. Determine the fundamental frequency. $\qquad [\omega = 195\,r/s]$

**10.3** An aluminum cantilever beam of uniform thickness $0.482\,in$ tapers from a width of $0.871\,in$ to $0.194\,in$ over its length of $16\,in$. Determine an approximation for the first natural frequency. $\qquad [f = 86.89\,Hz]$

**10.4** A steel beam of length $2\,m$ is clamped at both ends. The cross-section is square of side $50\,mm$, find the first few natural frequencies. Use four elements. $\qquad [414, 1150, 2281, 4316]$

**10.5** An aluminum beam of length $10\,in$ and diameter $0.5\,in$ is freely supported. Compute the first few resonant frequencies. $\qquad [0, 0, 12432, 38909]$

**10.6** Owing to a slight eccentricity of its motor, a rocket has developed lateral vibrations in its free flight after burn-out. Analyze the rocket by considering it as a uniform beam modeled with two elements. Let $L = 200\,in$, $EI = 10 \times 10^9\,lb\cdot in^2$ and $\rho A = 0.042\,lb\cdot s^2/in^2$. $\qquad [\omega = 273.5\,r/s]$

**10.7** The center and both ends of a beam of length $2L$ are simply supported. Obtain the first few resonant frequencies if $L = 200\,in$, $EI = 10 \times 10^9\,lb\cdot in^2$ and $\rho A = 0.042\,lb\cdot s^2/in^2$. $\qquad [\omega = 534.5, 1000, 2450\,r/s]$

**10.8** A concrete beam $2 \times 2 \times 12\,in$ is cross-section, supported at two points $0.224L$ from the ends, was found to resonate at $1690\,Hz$. If the density of concrete is $153\,lb/ft^3$, determine the modulus of elasticity. $\qquad [3.48\,msi]$

Chapter 11

# Modal Analysis of Frames

In the previous two chapters, we obtained an exact formulation for the dynamics of continuous systems. For vibration problems, the natural frequencies and mode shapes are determined by solving a transcendental eigenvalue problem. Consequently, we found the approximate formulations easier to use because they lead to an algebraic eigenvalue problem. In this chapter, we take the approximate formulation one step further by utilizing properties of algebraic eigensystems to introduce the concept of the modal model. This model involves a transformation of the description of dynamic systems in terms of stiffness and mass to a new set of equivalent variables in terms of the structural modes of vibration. This provides a scheme for analyzing the dynamics of complicated structures in terms of more useful quantities. References [17, 50] give very readable introductions to modal analysis.

Our interest is in 3-D frame structures, so we first complete the work of the previous chapters by showing the exact formulation for frames but almost immediately move on to the approximate formulation. This approach shows the correct transformation of the approximate damping and mass matrices.

## 11.1  Dynamic Stiffness for Space Frames

The approach to establishing the dynamic stiffness matrix for the space frame is essentially the same as used in Chapter 4 for establishing the elastic stiffness. That is, we consider each loading system to be independent of each other. Therefore, it is simply a matter of augmenting the separate stiffnesses for rods, beams and shafts to the full degrees of freedom of the frame, and then superposing them.

Once this is achieved, we will obtain the results for the approximate dynamic stiffness. This in turn will be used to obtain the results for plane frames.

### Assemblage of the Exact Dynamic Stiffness

The displacement of each node of a space frame is described by three translational and three rotational components of displacement, giving six degrees of freedom at each unrestrained node. In local coordinates, these forces and displacements are related to each other through a $[12 \times 12]$ matrix

$$\{\bar{F}\} = [\,\bar{k}\,]\{\bar{u}\}$$

where the bar indicates local coordinates. The stiffness matrix $[\,\bar{k}\,]$ is constructed similar to Equation(4.3) by augmenting the individual behaviors to $[12 \times 12]$.

In global coordinates, the element nodal displacement vector is composed of four separate vectors, namely,

$$\{\hat{u}\} = \left\{ \{\hat{u}_1,\, \hat{v}_1,\, \hat{w}_1\};\, \{\hat{\phi}_{x1},\, \hat{\phi}_{y1},\, \hat{\phi}_{z1}\};\, \{\hat{u}_2,\, \hat{v}_2,\, \hat{w}_2\};\, \{\hat{\phi}_{x2},\, \hat{\phi}_{y2},\, \hat{\phi}_{z2}\} \right\}$$

and the corresponding nodal force vector by

$$\{\hat{F}\} = \left\{ \{\hat{F}_{x1},\, \hat{F}_{y1},\, \hat{F}_{z1}\};\, \{\hat{M}_{x1},\, \hat{M}_{y1},\, \hat{M}_{z1}\};\, \{\hat{F}_{x2},\, \hat{F}_{y2},\, \hat{F}_{z2}\};\, \{\hat{M}_{x2},\, \hat{M}_{y2},\, \hat{M}_{z2}\} \right\}$$

Each of these vectors are separately transformed by the $[3 \times 3]$ rotation matrix $[\,R\,]$. Hence the complete transformation from global to local coordinates is given by

$$\{\bar{F}\} = [\,T\,]\{\hat{F}\}, \qquad \{\bar{u}\} = [\,T\,]\{\hat{u}\}$$

where

$$[\,T\,] \equiv \begin{bmatrix} R & 0 & 0 & 0 \\ 0 & R & 0 & 0 \\ 0 & 0 & R & 0 \\ 0 & 0 & 0 & R \end{bmatrix}$$

is a $[12 \times 12]$ matrix. This matrix is identical to that used in Chapter 4. Substituting for the barred vectors into the element stiffness relation allows us to obtain the global stiffness of the member as

$$[\,\hat{k}\,] = [\,T\,]^T [\,\bar{k}\,][\,T\,]$$

This is formally the same relation as obtained for the static analysis.

The assemblage process follows that of the other structures already encountered; each member stiffness is rotated to the global coordinate system and then augmented to the system size. The structural stiffness matrix is then

$$[\,\hat{K}\,] = \sum_m [\,T\,]_m^T [\,\bar{k}\,]_m [\,T\,]_m$$

where the summation is over each member. It is important to realize that the transformation occurs before the assemblage and therefore each member is transformed individually. We are now in a position to write the dynamic stiffness relation for the structure as

$$[\,\hat{K}\,]\{\hat{u}\} = \{\hat{P}\}$$

If there are concentrated masses at the nodes, then we make the following replacement

$$[\hat{K}] \longrightarrow [\hat{K}] - \omega^2 \lceil M_c \rfloor$$

where $\lceil M_c \rfloor$ is the diagonal matrix of concentrated masses. Note that this matrix can also include rotational inertias.

## Assemblage of the Approximate Dynamic Stiffness

We have already shown in Chapters 9 and 10 that the dynamic stiffness of each member has the expansion

$$[\hat{\bar{k}}] \approx [\bar{k}] + i\omega [\bar{c}] - \omega^2 [\bar{m}]$$

Therefore, we conclude that the global assembled form of the approximate stiffness is

$$[\hat{K}] \approx [K] + i\omega [C] - \omega^2 [M]$$

where the separate matrices are obtained as

$$[K] = \sum_m [T]_m^T [\bar{k}]_m [T]_m, \qquad [M] = \sum_m [T]_m^T [\bar{m}]_m [T]_m$$

with a similar transformation to obtain the matrix $[C]$. Finally, we have already made the following associations

$$i\omega \hat{u} \longrightarrow \dot{u}(t), \qquad -\omega^2 \hat{u} \longrightarrow \ddot{u}(t)$$

Therefore, the dynamics of the structure are described by the following equations of motion

$$[K]\{u\} + [C]\{\dot{u}\} + [M]\{\ddot{u}\} = \{P\}$$

If there are concentrated masses we replace the mass matrix with

$$[M] \longrightarrow [M] + \lceil M_c \rfloor$$

The equations of motion for the general frame is seen to be described by a system of second order ordinary differential equations.

The global stiffness for an arbitrary frame element in two-dimensions was already given in Equation(4.4); here we state the corresponding mass matrix for a member oriented at an angle $\theta$ to the global $x$-axis

$$[m] = \frac{\rho A L}{420} \begin{bmatrix} 156S^2 & & & & & \\ -156CS & 156C^2 & & & sym & \\ -22LS & 22LC & 4L^2 & & & \\ 54S^2 & -54CS & -13LS & 156S^2 & & \\ -54CS & 54C^2 & 13LC & -156CS & 156C^2 & \\ 13LS & -13LC & -3L^2 & 22LS & -22LC & 4L^2 \end{bmatrix} \qquad (11.1)$$

where the abbreviations $C \equiv \cos\theta$, $S \equiv \sin\theta$ are used. The damping matrix transforms similar to the mass matrix. Note that if the lumped mass model is used, then the mass matrix remains diagonal under a rotation. In that case it is easy to write it directly in the global coordinates.

**Example 11.1:**   Use the exact dynamic stiffness matrix to determine the frequency equation for the frame shown in Figure 11.1. Each member has the same material and section properties. Neglect damping.

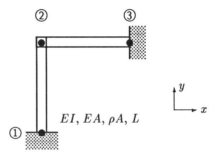

**Figure 11.1**: Frame with fixed-fixed supports.

Numbering the nodes as shown, the total degrees of freedom are

$$\{\hat{u}\} = \{\hat{u}_1, \hat{v}_1, \hat{\phi}_1; \hat{u}_2, \hat{v}_2, \hat{\phi}_2; \hat{u}_3, \hat{v}_3, \hat{\phi}_3\}$$

The fixed boundary conditions require that

$$\hat{u}_1 = \hat{v}_1 = \hat{\phi}_1 = 0 \qquad \text{and} \qquad \hat{u}_3 = \hat{v}_3 = \hat{\phi}_3 = 0$$

giving the reduced degrees of freedom as

$$\{\hat{u}_u\} = \{\hat{u}_2, \hat{v}_2, \hat{\phi}_2\}$$

The reduced assembled dynamic stiffness can be shown to be

$$\left[ \frac{EA}{L} \frac{\xi}{S} \begin{bmatrix} C & 0 & 0 \\ 0 & C & 0 \\ 0 & 0 & 0 \end{bmatrix}_{rod} \right.$$
$$\left. + \frac{EI}{L^3} \frac{2}{\Delta} \begin{bmatrix} (SC_h + CS_h)\xi^2 & 0 & SS_h\xi^2 L \\ 0 & (SC_h + CS_h)\xi^2 & SS_h\xi^2 L \\ SS_h\xi^2 L & SS_h\xi^2 L & 2(SC_h - CS_h)\xi L^2 \end{bmatrix}_{beam} \right] \begin{Bmatrix} \hat{u}_2 \\ \hat{v}_2 \\ \hat{\phi}_2 \end{Bmatrix} = 0$$

The eigenvalues are obtained by setting the determinant to zero. The resulting frequency equation is obviously very complicated and must be solved numerically.

We can simplify the equations by assuming the members to be inextensible. That is, we impose the further constraints that

$$\hat{u}_2 = 0 \quad \text{and} \quad \hat{v}_2 = 0$$

These conditions give the reduced structural dynamic stiffness as

$$\frac{EI}{L^3} \frac{4\xi L^2}{\Delta} \left[ \sin kL \cosh kL - \cos kL \sinh kL \right] \hat{\phi}_2 = 0$$

This is a transcendental eigenvalue problem and the frequency equation is

$$\sin kL \cosh kL - \cos kL \sinh kL = 0$$

We recognize this (from Table 10.1) as the frequency equation for the vibration of a fixed-pinned beam. We see that even in this very simple case the frequency equation turns out to be relatively complicated and difficult to solve.

**Example 11.2:** Use the approximate matrix formulation to find the first few resonant frequencies for the plane frame shown in Figure 11.1. Each member has the same material and section properties. Neglect damping.

We will use two elements to model the problem. Numbering the nodes as shown, the total degrees of freedom are

$$\{u\} = \{u_1, v_1, \phi_1; u_2, v_2, \phi_2; u_3, v_3, \phi_3\}$$

The fixed boundary conditions require that

$$u_1 = v_1 = \phi_1 = 0 \quad \text{and} \quad u_3 = v_3 = \phi_3 = 0$$

for all time. This gives the reduced system as

$$\{u_u\} = \{u_2, v_2, \phi_2\}, \quad \{\ddot{u}_u\} = \{\ddot{u}_2, \ddot{v}_2, \ddot{\phi}_2\}$$

The reduced element stiffness matrices for both members are for the non-zero degree of freedoms $\{u_2, v_2, \phi_2\}$. For Member 1-2 with connectivity 1 to 2, the orientation is $\theta = 90^\circ$, giving

$$[k^{*(12)}] = \frac{EA}{L} \begin{bmatrix} 0 & 0 & 0 \\ 0 & 1 & 0 \\ 0 & 0 & 0 \end{bmatrix} + \frac{EI}{L^3} \begin{bmatrix} 12 & 0 & 6L \\ 0 & 0 & 0 \\ 6L & 0 & 4L^2 \end{bmatrix}$$

For Member 2-3 with connectivity 2 to 3, the orientation is $\theta = 0^\circ$, giving

$$[k^{*(23)}] = \frac{EA}{L} \begin{bmatrix} 1 & 0 & 0 \\ 0 & 0 & 0 \\ 0 & 0 & 0 \end{bmatrix} + \frac{EI}{L^3} \begin{bmatrix} 0 & 0 & 0 \\ 0 & 12 & 6L \\ 0 & 6L & 4L^2 \end{bmatrix}$$

The corresponding mass matrices are

$$[m^{*(12)}] = \frac{\rho AL}{420} \begin{bmatrix} 156 & 0 & 22L \\ 0 & 0 & 0 \\ 22L & 0 & 4L^2 \end{bmatrix}, \qquad [m^{*(23)}] = \frac{\rho AL}{420} \begin{bmatrix} 0 & 0 & 0 \\ 0 & 156 & 22L \\ 0 & 22L & 4L^2 \end{bmatrix}$$

The reduced structural stiffness and mass matrices can therefore be assembled. This results in the equations of motion of the reduced system being

$$
\left[ \frac{EA}{L} \begin{bmatrix} 1 & 0 & 0 \\ 0 & 1 & 0 \\ 0 & 0 & 0 \end{bmatrix} + \frac{EI}{L^3} \begin{bmatrix} 12 & 0 & 6L \\ 0 & 12 & 6L \\ 6L & 6L & 8L^2 \end{bmatrix} \right] \left\{ \begin{matrix} u_2 \\ v_2 \\ \phi_2 \end{matrix} \right\}
$$

$$
- \frac{\rho AL}{420} \begin{bmatrix} 156 & 0 & 22L \\ 0 & 156 & 22L \\ 22L & 22L & 8L^2 \end{bmatrix} \left\{ \begin{matrix} \ddot{u}_2 \\ \ddot{v}_2 \\ \ddot{\phi}_2 \end{matrix} \right\} = \left\{ \begin{matrix} 0 \\ 0 \\ 0 \end{matrix} \right\}
$$

Introducing the notations

$$
\alpha \equiv \frac{EA}{L}, \qquad \beta \equiv \frac{EI}{L^3}, \qquad \lambda \equiv \omega^2 \frac{\rho AL}{420}
$$

we get the eigenvalue problem as

$$
\left[ \begin{bmatrix} \alpha + 12\beta & 0 & \beta 6L \\ 0 & \alpha + 12\beta & \beta 6L \\ \beta 6L & \beta 6L & \beta 8L^2 \end{bmatrix} - \lambda \begin{bmatrix} 156 & 0 & 22L \\ 0 & 156 & 22L \\ 22L & 22L & 8L^2 \end{bmatrix} \right] \left\{ \begin{matrix} \hat{u}_2 \\ \hat{v}_2 \\ \hat{\phi}_2 \end{matrix} \right\} = 0
$$

The characteristic equation is

$$
(\alpha + 12\beta - 156\lambda)[35\lambda^2 - \lambda(\alpha + 90\beta) + \beta(\alpha + 3\beta)] = 0
$$

One of the roots gives

$$
\lambda = \frac{\alpha + 12\beta}{156} \qquad \text{or} \qquad \omega = \sqrt{\frac{420}{156} \frac{EA/L + EI/L^3}{\rho AL}}
$$

Substituting this into the equations of the eigenvalue problem, we get that

$$
\hat{\phi}_2 = 0 \qquad \text{and} \qquad \hat{u}_2 = -\hat{v}_2
$$

This is the symmetric mode. In the inextensible case ($\alpha = EA/L$ very large) we get

$$
\omega = \frac{1.64}{L} \sqrt{\frac{EA}{\rho A}}
$$

which shows that the resonance goes to infinity as $EA$ increases.
    The other two resonances occur at

$$
\lambda = \frac{1}{70} \left[ (\alpha + 90\beta) \pm \sqrt{\alpha^2 + 40\alpha\beta + 7680\beta^2} \right]
$$

Again, suppose that the axial stiffness is very large, then

$$
\omega = \frac{20.5}{L^2} \sqrt{\frac{EI}{\rho A}} \qquad \text{and} \qquad \frac{3.46}{L} \sqrt{\frac{EA}{\rho A}}
$$

The second of these resonances approaches infinity as $EA$ increases, but the first has a finite value. The first mode corresponds to

$$\hat{u}_2 = \hat{v}_2 = 0, \qquad \hat{\phi}_2 \neq 0$$

This deformation is the same as if the structure is pinned at Node 2.

These examples clearly show that even though the geometry is symmetric, that the anti-symmetric vibration could be the dominant one. Also, the use of the inextensibility assumption can profoundly influence the prediction of the resonances. Thus in the above problem, of the three possible modes, only one is predicted to occur.

## 11.2   Modal Matrix

It is apparent that the analysis of complicated structures will involve systems that have very many degrees of freedom and therefore are described by a large number of equations This is all the more true since the use of the approximate stiffness requires subdividing a given member into many small elements. This section develops some of the concepts which form the basis for the treatment and understanding of the dynamical behavior of large systems. Central to this development is the concept of the *modal matrix* because through it the system can be transformed into a set of uncoupled equations.

### Orthogonality of Mode Shapes

When an undamped system is excited, it will continue to vibrate long after the initial disturbance is gone. Further, it vibrates with a characteristic shape (called the *mode shape*) governed by the following system of equations

$$[K]\{u\} + [M]\{\ddot{u}\} = 0$$

Since the motion is harmonic, then $\{u(t)\} = \{\hat{u}\}e^{i\omega t}$, and the characteristic shape $\{\hat{u}\}$ satisfies the algebraic system of equations

$$\left[[K] - \lambda[M]\right]\{\hat{u}\} = 0 \tag{11.2}$$

where $\lambda = \omega^2$. These equations are homogeneous, hence the solutions, in general, are zero. The only time a non-trivial solution is obtained is when the determinant of the coefficients is zero. Thus Equation(11.2) is recognized as the familiar eigenvalue problem; $\lambda$ are the eigenvalues and $\{\hat{u}\}$ are the eigenvectors. There are as many eigenvalues as the order of the system of equations. That is, the solution yields $N$ eigenvalues $\lambda_m$ and $N$ corresponding eigenvectors $\{\hat{u}\}_m$.

It is apparent that if $\{\hat{u}\}$ is a solution, then $\alpha\{\hat{u}\}$ is also a solution when $\alpha$ is a scalar constant. As a reminder that the mode shapes are some sort of normalized version of the displacements $\{\hat{u}\}$, the notation

$$\{\phi\} \equiv normalized\,\{\hat{u}\}$$

will be used. We will discuss some particular forms of normalization later.

Consider two arbitrary, non-null vectors $\{v\}_1$ and $\{v\}_2$. For the square matrix $[\,A\,]$ to be *positive definite*, we must have that the triple product

$$\{v\}_1^T[\,A\,]\{v\}_1 = constant$$

be greater than zero. If the matrix $[\,A\,]$ is symmetric, we also have that

$$\{v\}_1^T[\,A\,]\{v\}_2 = \{v\}_2^T[\,A\,]\{v\}_1$$

We will use these two important results to establish some properties of the mode shapes.

Each mode shape will satisfy the equation of motion, that is, when substituted into Equation(11.2) they give

$$[\,K\,]\{\phi\}_i = \lambda_i[\,M\,]\{\phi\}_i$$

Pre-multiply this by the transpose of another mode shape $\{\phi\}_j$

$$\{\phi\}_j^T[\,K\,]\{\phi\}_i = \lambda_i\{\phi\}_j^T[\,M\,]\{\phi\}_i$$

Now write the equation for the $j^{th}$ mode and pre-multiply this by the transpose of the $i^{th}$ mode; that is,

$$\{\phi\}_i^T[\,K\,]\{\phi\}_j = \lambda_j\{\phi\}_i^T[\,M\,]\{\phi\}_j$$

Subtract these, and since the mass and stiffness matrices are symmetric, then obtain

$$0 = (\lambda_i - \lambda_j)\{\phi\}_i^T[\,M\,]\{\phi\}_j$$

We chose the mode shapes to be at two different natural frequencies, therefore $\lambda_i \neq \lambda_j$ resulting in

$$\{\phi\}_i^T[\,M\,]\{\phi\}_j = 0$$

This is a statement of the *orthogonality* property of the mode shapes with respect to the mass matrix. By analogy to vector algebra, it means that the eigenvectors are perpendicular (orthogonal) to each other, and their vector dot product is therefore zero. It is emphasized, however, that in the present case we have a weighting factor $[\,M\,]$. In a similar manner, it can be seen that

$$\{\phi\}_i^T[\,K\,]\{\phi\}_j = 0$$

There are cases of *repeated roots*; that is, the system has different modes at the same frequency. The above development only shows that these modes are orthogonal to all other modes but not necessarily to each other. Actually, the eigenvectors are not unique and a linear combination of them may also satisfy the equations of motion. In these circumstances we will prescribe that the mode shapes associated with repeated roots be orthogonal to each other.

## Modal Mass and Stiffness Matrices

If we set $i = j = m$ in the analysis of the previous section, then the two mode shapes we are dealing with are the same, and therefore the triple product is equal to some non-zero constant. That is,

$$\{\phi\}_m^T[M]\{\phi\}_m = \tilde{M}_{mm}$$
$$\{\phi\}_m^T[K]\{\phi\}_m = \tilde{K}_{mm} = \lambda_m \tilde{M}_{mm} = \omega_m^2 \tilde{M}_{mm}$$

$\tilde{M}_{mm}$ and $\tilde{K}_{mm}$ are called the *modal mass* and *modal stiffness* of the $m^{th}$ mode, respectively.

These relations show that the mass and stiffness matrix can be converted to a single constant, one for each mode, by multiplying by the mode shapes. Thus, construct a square matrix $[\Phi]$ whose columns are the normalized mode shape vectors as

$$[\Phi] \equiv \left[ \left\{ \begin{matrix} \phi_1 \\ \phi_2 \\ \vdots \\ \phi_N \end{matrix} \right\}_1 \left\{ \begin{matrix} \phi_1 \\ \phi_2 \\ \vdots \\ \phi_N \end{matrix} \right\}_2 \cdots \left\{ \begin{matrix} \phi_1 \\ \phi_2 \\ \vdots \\ \phi_N \end{matrix} \right\}_N \right]$$

$[\Phi]$ is referred to as the *modal matrix*. It is a fully populated matrix of order $[N \times N]$ and typically is not symmetric.

The orthogonal properties of the mode shapes and the definition of the modal mass and stiffness, can now be expressed in matrix form as

$$[\Phi]^T[M][\Phi] = \lceil \tilde{M} \rfloor, \qquad [\Phi]^T[K][\Phi] = \lceil \tilde{K} \rfloor$$

where $\lceil \tilde{M} \rfloor$ and $\lceil \tilde{K} \rfloor$ are diagonal matrices of order $[N \times N]$.

## Normalization of Mode Shapes

The modal vector represents a shape rather than the absolute deflection of the structure; that is, the ratio of the elements of the modal vector are fixed not their absolute value. If, however, one of the values is fixed then the eigenvector becomes unique in an absolute sense also. The process of scaling the elements of the mode shape is called *normalization*; the resulting scaled modes are called *orthonormal* modes. There are several methods available for doing this, the following is a partial list:

1. The largest element is set to unity.
2. The length of the mode vector is set to unity.
3. A particular, physically significant, element is set to unity.
4. The modal mass is set to unity such that $\{\phi\}_m^T[M]\{\phi\}_m = \tilde{M}_{mm} = 1$

The first three of these are useful when the mode shapes are to be plotted. The last of the scaling schemes simplifies some relationships; for example, the modal stiffness is now numerically equal to the eigenvalues since

$$\tilde{K}_{mm} = \lambda_m \tilde{M}_{mm} = \lambda_m = \omega_m^2$$

A caution is in order here; one of the disadvantages of normalizing the mode shapes is that it is easy to loose track of the physical units of quantities. For example, in the previous equation it is tempting to say that the stiffness has units of $[rad/sec]^2$. Essentially, the method of scaling determines the numeric value of the modal mass, therefore whenever needed we will state this value.

**Example 11.3:**   The simple frame shown in Figure 11.2 is modeled with two degrees of freedom. Determine the eigenvalues and eigenvectors associated with this system if the governing equations of motion are

$$\begin{bmatrix} 4 & -2 \\ -2 & 6 \end{bmatrix} \begin{Bmatrix} u_1 \\ u_2 \end{Bmatrix} + \begin{bmatrix} 1 & 0 \\ 0 & 2 \end{bmatrix} \begin{Bmatrix} \ddot{u}_1 \\ \ddot{u}_2 \end{Bmatrix} = \begin{Bmatrix} P_1 \\ P_2 \end{Bmatrix}$$

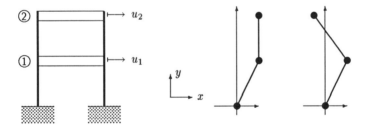

**Figure 11.2**: Two degree of freedom system.

The spectral form of these equations for free vibration leads to the system

$$\begin{bmatrix} 4-\lambda & -2 \\ -2 & 6-2\lambda \end{bmatrix} \begin{Bmatrix} \hat{u}_1 \\ \hat{u}_2 \end{Bmatrix} = \begin{Bmatrix} 0 \\ 0 \end{Bmatrix}, \qquad \lambda = \omega^2$$

The determinant must be zero for a non-trivial solution; thus on multiplying out and rearranging, we get

$$(4-\lambda)(6-2\lambda) - 4 = 0 \quad \text{or} \quad \lambda^2 - 7\lambda + 10 = 0$$

Since this is quadratic, then the roots are

$$\lambda_{1,2} = \frac{7 \pm \sqrt{49-40}}{2} = \frac{7}{2} \pm \frac{3}{2} = 2, 5$$

Thus the ordered eigenvalues are $\lambda_1 = 2$ and $\lambda_2 = 5$. The two natural frequencies are

$$\omega_1 = \sqrt{\lambda_1} = \sqrt{2} \quad \text{and} \quad \omega_2 = \sqrt{\lambda_2} = \sqrt{5}$$

The mode shape for the first mode is obtained by substituting $\lambda_1$ into the original system to give

$$\begin{bmatrix} 4-2 & -2 \\ -2 & 6-4 \end{bmatrix} \begin{Bmatrix} \hat{u}_1 \\ \hat{u}_2 \end{Bmatrix}_1 = \begin{Bmatrix} 0 \\ 0 \end{Bmatrix}$$

and these become, when written out separately,

$$\begin{aligned} 2\hat{u}_1 - 2\hat{u}_2 &= 0 \\ -2\hat{u}_1 + 2\hat{u}_2 &= 0 \end{aligned}$$

From both equations we have that $\hat{u}_1 = \hat{u}_2$, thus the first mode shape is

$$\{\hat{u}\}_1 = \hat{u}_1\{\phi\}_1 = \hat{u}_1 \begin{Bmatrix} 1 \\ 1 \end{Bmatrix}$$

where the magnitude of $\hat{u}_1$ is arbitrary. Similarly, for the second mode we get after substituting for $\lambda_2$

$$\begin{bmatrix} 4-5 & -2 \\ -2 & 6-10 \end{bmatrix} \begin{Bmatrix} \hat{u}_1 \\ \hat{u}_2 \end{Bmatrix}_2 = \begin{Bmatrix} 0 \\ 0 \end{Bmatrix}$$

giving as separate equations

$$\begin{aligned} -\hat{u}_1 - 2\hat{u}_2 &= 0 \\ -2\hat{u}_1 - 4\hat{u}_2 &= 0 \end{aligned}$$

Both of these equations give $\hat{u}_2 = -\frac{1}{2}\hat{u}_1$. Thus the second mode shape is

$$\{\hat{u}\}_2 = \hat{u}_1\{\phi\}_2 = \hat{u}_1 \begin{Bmatrix} 1 \\ -\frac{1}{2} \end{Bmatrix}$$

Again, this has been normalized to the first displacement.

These mode shapes are shown plotted in Figure 11.2. Strictly speaking, we should only plot the values of $\hat{u}_1$ and $\hat{u}_2$, but since the orthogonality properties are not affected by augmenting the eigenvectors with zeros, then a clearer picture is obtained by incorporating the zero displacement at the attachment.

**Example 11.4:**   Show that the eigenvectors of the last example are orthogonal.
We will consider only the normalized forms

$$\{\phi\}_1 = \begin{Bmatrix} 1 \\ 1 \end{Bmatrix} \quad \text{and} \quad \{\phi\}_2 = \begin{Bmatrix} 1 \\ -\frac{1}{2} \end{Bmatrix}$$

First note that these are not orthogonal in the simple vector dot product sense, since

$$\begin{Bmatrix} 1 \\ 1 \end{Bmatrix}^T \begin{Bmatrix} 1 \\ -\frac{1}{2} \end{Bmatrix} = \{1 - \tfrac{1}{2}\} \neq 0$$

The eigenvectors are orthogonal *with respect to* the mass and stiffness matrices.
For the mass matrix, we have

$$\begin{Bmatrix} 1 \\ 1 \end{Bmatrix}^T \begin{bmatrix} 1 & 0 \\ 0 & 2 \end{bmatrix} \begin{Bmatrix} 1 \\ -\frac{1}{2} \end{Bmatrix} = \begin{Bmatrix} 1 \\ 1 \end{Bmatrix}^T \begin{Bmatrix} 1 \\ -1 \end{Bmatrix} = 0$$

And for the stiffness matrix

$$\begin{Bmatrix} 1 \\ 1 \end{Bmatrix}^T \begin{bmatrix} 4 & -2 \\ -2 & 6 \end{bmatrix} \begin{Bmatrix} 1 \\ -\frac{1}{2} \end{Bmatrix} = \begin{Bmatrix} 1 \\ 1 \end{Bmatrix}^T \begin{Bmatrix} 5 \\ -5 \end{Bmatrix} = 0$$

**Example 11.5:**   Determine the modal mass and modal stiffness for the system of
Figure 11.2.
Recalling that

$$\{\phi\}_1 = \begin{Bmatrix} 1 \\ 1 \end{Bmatrix} \qquad \text{and} \qquad \{\phi\}_2 = \begin{Bmatrix} 1 \\ -\frac{1}{2} \end{Bmatrix}$$

then the generalized mass for the first mode is

$$\tilde{M}_{11} = \begin{Bmatrix} 1 \\ 1 \end{Bmatrix}^T \begin{bmatrix} 1 & 0 \\ 0 & 2 \end{bmatrix} \begin{Bmatrix} 1 \\ 1 \end{Bmatrix} = 3$$

and for the second mode

$$\tilde{M}_{22} = \begin{Bmatrix} 1 \\ -\frac{1}{2} \end{Bmatrix}^T \begin{bmatrix} 1 & 0 \\ 0 & 2 \end{bmatrix} \begin{Bmatrix} 1 \\ -\frac{1}{2} \end{Bmatrix} = \frac{3}{2}$$

Note that there is only one mass for each mode. The modal stiffnesses are

$$\tilde{K}_{11} = \begin{Bmatrix} 1 \\ 1 \end{Bmatrix}^T \begin{bmatrix} 4 & -2 \\ -2 & 6 \end{bmatrix} \begin{Bmatrix} 1 \\ 1 \end{Bmatrix} = 6, \qquad \tilde{K}_{22} = \begin{Bmatrix} 1 \\ -\frac{1}{2} \end{Bmatrix}^T \begin{bmatrix} 4 & -2 \\ -2 & 6 \end{bmatrix} \begin{Bmatrix} 1 \\ -\frac{1}{2} \end{Bmatrix} = \frac{15}{2}$$

It is useful to note that these results for the stiffness could also be obtained by using
the relationship involving the resonant frequency, that is,

$$\tilde{K}_{11} = \omega_1^2 \tilde{M}_{11} = 2 \times 3 = 6 \qquad \text{and} \qquad \tilde{K}_{22} = \omega_2^2 \tilde{M}_{22} = 5 \times 3/2 = 15/2$$

The exact same results are also obtained when the full modal matrix is used.
First establish the modal matrix as

$$[\Phi] = \begin{bmatrix} 1 & 1 \\ 1 & -\frac{1}{2} \end{bmatrix}$$

Now pre- and post-multiply the mass and stiffness matrices by this modal matrix to
get for the mass

$$\lceil \tilde{M} \rfloor = \begin{bmatrix} 1 & 1 \\ 1 & -\frac{1}{2} \end{bmatrix} \begin{bmatrix} 1 & 0 \\ 0 & 2 \end{bmatrix} \begin{bmatrix} 1 & 1 \\ 1 & -\frac{1}{2} \end{bmatrix} = \begin{bmatrix} 3 & 0 \\ 0 & \frac{3}{2} \end{bmatrix}$$

and for the stiffness

$$\lceil \tilde{K} \rfloor = \begin{bmatrix} 1 & 1 \\ 1 & -\frac{1}{2} \end{bmatrix} \begin{bmatrix} 4 & -2 \\ -2 & 6 \end{bmatrix} \begin{bmatrix} 1 & 1 \\ 1 & -\frac{1}{2} \end{bmatrix} = \begin{bmatrix} 6 & 0 \\ 0 & \frac{15}{2} \end{bmatrix}$$

These results emphasize the diagonal nature of the modal mass and stiffness matrices. They also show that if the mode shapes are normalized in a different manner then different numerical values will be obtained for the modal stiffness and mass.

**Example 11.6:**   Determine the set of othonormalized eigenvectors for the above problem.

We had determined a set of eigenvectors as

$$\{\phi\}_1 = \begin{Bmatrix} 1 \\ 1 \end{Bmatrix} \qquad \text{and} \qquad \{\phi\}_2 = \begin{Bmatrix} 1 \\ -\frac{1}{2} \end{Bmatrix}$$

From these we computed a set of modal masses as

$$\tilde{M}_{11} = 3, \qquad \tilde{M}_{22} = \frac{3}{2}$$

We introduce a new set of modal vectors given by

$$\{\phi\}_1 \equiv \frac{1}{\sqrt{3}} \begin{Bmatrix} 1 \\ 1 \end{Bmatrix} = \begin{Bmatrix} 0.577 \\ 0.577 \end{Bmatrix}, \qquad \{\phi\}_2 \equiv \frac{1}{\sqrt{\frac{3}{2}}} \begin{Bmatrix} 1 \\ -\frac{1}{2} \end{Bmatrix} = \begin{Bmatrix} 0.816 \\ -0.408 \end{Bmatrix}$$

These modal vectors result in a unit modal mass matrix.

# 11.3   Transformation to Principal Coordinates

Consider a system described by the following coupled equations of motion

$$\begin{bmatrix} k_{11} & k_{12} \\ k_{12} & k_{22} \end{bmatrix} \begin{Bmatrix} u_1 \\ u_2 \end{Bmatrix} + \begin{bmatrix} m_1 & 0 \\ 0 & m_2 \end{bmatrix} \begin{Bmatrix} \ddot{u}_1 \\ \ddot{u}_2 \end{Bmatrix} = \begin{Bmatrix} P_1 \\ P_2 \end{Bmatrix}$$

In these equations, the coupling is due to the fact that the stiffness matrix is not diagonal. This is called *elastic* or *static* coupling. When the mass matrix is not diagonal; the coupling is termed *inertial* or *dynamic* coupling. If we obtained diagonal mass and stiffness matrices simultaneously, then the system would be uncoupled and each equation would be similar to that of a single degree of freedom system. These could then be solved independently of each other. Such a transformation will be shown here.

Consider the transformation of the displacements to new values by the equation

$$\{u\} = [\ \Phi\ ]\{\eta\}$$

where $[\ \Phi\ ]$ is the modal matrix. Here $\{\eta\}$ are called *principal coordinates* or *normal coordinates*. The equations of motion in terms of these new coordinates are

$$[\ K\ ][\ \Phi\ ]\{\eta\} + [\ M\ ][\ \Phi\ ]\{\ddot{\eta}\} = \{P\}$$

Pre-multiply this by the transpose of the modal matrix to get

$$[\ \Phi\ ]^T[\ K\ ][\ \Phi\ ]\{\eta\} + [\ \Phi\ ]^T[\ M\ ][\ \Phi\ ]\{\ddot{\eta}\} = [\ \Phi\ ]^T\{P\}$$

Because of the orthogonal properties of the mode shapes and the definition of the modal mass and stiffness, the equations of motion become

$$\lceil \tilde{K} \rfloor \{\eta\} + \lceil \tilde{M} \rfloor \{\ddot{\eta}\} = [\ \Phi\ ]^T \{P\}$$

This represents $N$ uncoupled equations of the form

$$\tilde{K}_{mm}\eta_m + \tilde{M}_{mm}\ddot{\eta}_m = \{\phi\}^T_m\{P\} = \tilde{P}_m$$

where $\{\phi\}_m$ is the $m^{th}$ mode shape, $\tilde{M}_{mm}$ and $\tilde{K}_{mm}$ are the $m^{th}$ modal mass and stiffness, respectively. Each equation above is the equation of motion for a single degree of freedom system, and since $\tilde{K}_{mm} = \omega^2_m \tilde{M}_{mm}$, can be written as

$$\omega^2_m\eta_m + \ddot{\eta}_m = \frac{\tilde{P}_m}{\tilde{M}_{mm}} = \frac{\{\phi\}^T_m\{P\}}{\{\phi\}^T_m[M]\{\phi\}_m} \tag{11.3}$$

This can be integrated directly to give the generalized response as

$$\eta_i(t) = \eta_m(0)\cos\omega_m t + \frac{1}{\omega_m}\dot{\eta}_m(0)\sin\omega_m t + \frac{1}{\omega_m \tilde{M}_{mm}}\int_0^t \tilde{P}_m(\tau)\sin[\omega_m(t-\tau)]\,d\tau$$

This is *Duhamel's Integral* we have already encountered in Chapter 8. For simple forcing functions (for example stepped loading) the integration may be performed analytically, but generally it must be done numerically.

Once time responses for all the $\eta$'s are obtained, the solution in terms of the original coordinates can be obtained by simply transforming back to the physical coordinates according to

$$\{u\} = [\ \Phi\ ]\{\eta\}$$

In general, for an $N$ degree of freedom system, the responses are

$$\begin{Bmatrix} u_1 \\ u_2 \\ \vdots \\ u_N \end{Bmatrix} = \begin{Bmatrix} \phi_1 \\ \phi_2 \\ \vdots \\ \phi_N \end{Bmatrix}_1 \eta_1(t) + \begin{Bmatrix} \phi_1 \\ \phi_2 \\ \vdots \\ \phi_N \end{Bmatrix}_2 \eta_2(t) + \cdots \begin{Bmatrix} \phi_1 \\ \phi_2 \\ \vdots \\ \phi_N \end{Bmatrix}_M \eta_M(t)$$

This can also be written as

$$\{u(t)\} = \sum_m^M \{\phi\}_m\eta_m(t)$$

This is a fundamental relation in the dynamics of structures. It shows that the response of any complicated system can be conceived as the superposition of the responses of the natural vibration modes. Further, the summation need not extend to $N$ (the system size) but can be truncated at $M < N$. Indeed, for many practical problems (such as earthquake analysis) $M$ is significantly less than $N$ and generally on the order of 1 to 5.

**Example 11.7:**  Consider again the simple system described by the equations of motion

$$\begin{bmatrix} 4 & -2 \\ -2 & 6 \end{bmatrix} \begin{Bmatrix} u_1 \\ u_2 \end{Bmatrix} + \begin{bmatrix} 1 & 0 \\ 0 & 2 \end{bmatrix} \begin{Bmatrix} \ddot{u}_1 \\ \ddot{u}_2 \end{Bmatrix} = \begin{Bmatrix} P_1 \\ P_2 \end{Bmatrix}$$

Obtain the equations of motion in principal coordinates. Also, determine the free vibration characteristics of the system.

We have already shown that the modal matrix $[\ \Phi\ ]$ is assembled as

$$[\ \Phi\ ] = \begin{bmatrix} 1 & 1 \\ 1 & -\frac{1}{2} \end{bmatrix}$$

The coordinate transformation is therefore given by

$$\begin{Bmatrix} u_1 \\ u_2 \end{Bmatrix} = \begin{bmatrix} 1 & 1 \\ 1 & -\frac{1}{2} \end{bmatrix} \begin{Bmatrix} \eta_1 \\ \eta_2 \end{Bmatrix}$$

Applying this to the equation of motion and pre-multiplying by $[\ \Phi\ ]^T$ gives

$$[\ \Phi\ ]^T \begin{bmatrix} 4 & -2 \\ -2 & 6 \end{bmatrix} [\ \Phi\ ] \begin{Bmatrix} \eta_1 \\ \eta_2 \end{Bmatrix} + [\ \Phi\ ]^T \begin{bmatrix} 1 & 0 \\ 0 & 2 \end{bmatrix} [\ \Phi\ ] \begin{Bmatrix} \ddot{\eta}_1 \\ \ddot{\eta}_2 \end{Bmatrix} = [\ \Phi\ ]^T \begin{Bmatrix} P_1 \\ P_2 \end{Bmatrix}$$

Multiplying out, get

$$\begin{bmatrix} 6 & 0 \\ 0 & 15/2 \end{bmatrix} \begin{Bmatrix} \eta_1 \\ \eta_2 \end{Bmatrix} + \begin{bmatrix} 3 & 0 \\ 0 & 3/2 \end{bmatrix} \begin{Bmatrix} \ddot{\eta}_1 \\ \ddot{\eta}_2 \end{Bmatrix} = \begin{bmatrix} 1 & 1 \\ 1 & -1/2 \end{bmatrix} \begin{Bmatrix} P_1 \\ P_2 \end{Bmatrix}$$

Note how the mass and stiffness matrices have been diagonalized. The equations of motion can be separated as

$$6\eta_1 + 3\ddot{\eta}_1 = P_1 + P_2$$
$$\tfrac{15}{2}\eta_2 + \tfrac{3}{2}\ddot{\eta}_2 = P_1 - \tfrac{1}{2}P_2$$

It is worth observing that while the coordinates are uncoupled, all the applied forces now act at each generalized node as shown in Figure 11.3.

**Figure 11.3:** The modal representation of the undamped 2-DoF system.

**Example 11.8:**  Consider a special case of the last example when $P_2 = 0$ and $P_1$ is a stepped loading of magnitude $P_o$. Obtain the solution if the system is initially at rest.

The differential equations after $t = 0$ are simplified to

$$\ddot{\eta}_1 + 2\eta_1 = \tfrac{1}{3}P_o$$
$$\ddot{\eta}_2 + 5\eta_2 = \tfrac{2}{3}P_o$$

Integrate this using the Duhamel's integral. The initial conditions are that $\eta(0)$ and $\dot{\eta}(0)$ are zero for both modes, leading to the modal responses

$$\eta_1(t) = \tfrac{1}{6}P_o[1 - \cos\omega_1 t], \qquad \omega_1 = \sqrt{2}$$
$$\eta_2(t) = \tfrac{4}{45}P_o[1 - \cos\omega_2 t], \qquad \omega_1 = \sqrt{5}$$

Performing the transformation back to physical coordinates, gives the total response as

$$u_1(t) = \eta_1(t) + \eta_2(t) = \tfrac{1}{6}P_o[1 - \cos(\omega_1 t)] + \tfrac{4}{45}P_o[1 - \cos(\omega_2 t)]$$
$$u_2(t) = \eta_1(t) - \tfrac{1}{2}\eta_2(t) = \tfrac{1}{6}P_o[1 - \cos(\omega_1 t)] - \tfrac{2}{45}P_o[1 - \tfrac{1}{2}\cos(\omega_2 t)]$$

This is shown plotted in Figure 11.4. Note that the response for $u_1(t)$ appears somewhat random even though it is made up of only two sinusoids.

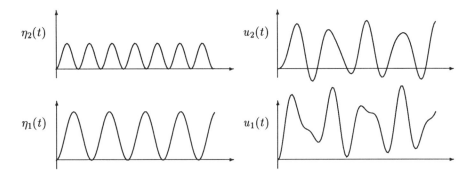

**Figure 11.4**: Transient response of the 2-DoF system.

**Example 11.9:**   With reference to the last two examples, what set of applied forces will excite only the second mode of vibration?

If the applied force for the first mode is zero then there will be no response in that mode. For the present system we have that

$$\tilde{P}_1 = P_1 + P_2$$

Hence by choosing $P_2(t) = -P_1(t)$ then this mode is not excited. That is, if equal but opposite forces are applied at the two masses, then only the second mode is excited.

## 11.4   Forced Damped Motion

The equations of motion of a multiple degree of freedom system with viscous damping present can be written in matrix form as

$$[K]\{u\} + [C]\{\dot{u}\} + [M]\{\ddot{u}\} = \{P\}$$

A simplifying assumption usually introduced is that the distribution of damping throughout the structure is proportional. That is

$$[C] = \alpha[M] + \beta[K]$$

where $\alpha$ and $\beta$ are constants; this is usually referred to as *Rayleigh damping.* We have already seen a special case of this in Chapter 9 as the consistent damping matrix where $\beta = 0$ and $\alpha = \eta\rho A$.

With proportional damping, the coordinate transformation using the modal matrix will diagonalize the damping matrix, in addition to diagonalizing the mass and stiffness matrices. Thus the coupled equations can be uncoupled to single degree of freedom systems as

$$\lceil \tilde{K} \rfloor\{\eta\} + \lceil \tilde{C} \rfloor\{\dot{\eta}\} + \lceil \tilde{M} \rfloor\{\ddot{\eta}\} = [\Phi]^T\{P\}$$

This represents an uncoupled set of equations for damped single degree of freedom systems. The $m^{th}$ equation is

$$\tilde{K}_{mm}\eta_m + \tilde{C}_{mm}\dot{\eta}_m + \tilde{M}_{mm}\ddot{\eta}_m = \{\phi\}_m^T\{P\} = \tilde{P}_m$$

Since $\tilde{K}_{mm} = \omega_m^2 \tilde{M}_{mm}$, this can be rewritten as

$$\omega_m^2\eta_m + 2\zeta_m\omega_m\dot{\eta}_m + \ddot{\eta}_m = \frac{\{\phi\}_m^T\{P\}}{\{\phi\}_m^T[M]\{\phi\}_m} = \frac{\tilde{P}_m}{\tilde{M}_{mm}} \qquad (11.4)$$

where the damping ratio is defined as

$$\zeta_m \equiv \frac{\tilde{C}_{mm}}{2\sqrt{\tilde{K}_{mm}\tilde{M}_{mm}}} = \frac{\tilde{C}_{mm}}{2\omega_m\tilde{M}_{mm}} = \frac{\alpha + \beta\omega_{mm}^2}{2\omega_m}$$

This is a very interesting relation because it shows how the modal damping is affected differently by the mass and stiffness matrices. When the damping is proportional to the mass it decreases at the higher frequencies; on the other hand, stiffness proportional damping increases at the higher frequencies. We see that although the damping matrix is "constant" the effect it has on the response is different at each frequency.

The above system of equations can be integrated according to the methods of Chapter 8 and the physical response again given by

$$\{u(t)\} = \sum_{m}^{M}\{\phi\}_m\eta_m(t)$$

Even in the damped case, we can view the response of a complicated structure as the superposition of modes.

## Forced Frequency Response

As shown in Chapter 8, it is often more convenient to analyze damped systems in the frequency domain. That is, we are interested in the forced frequency response of the structure when subjected to frequencies that are not necessarily the undamped resonant values.

The uncoupled equations of motion are transformed into the frequency domain to give

$$\tilde{K}_{mm}\hat{\eta}_m + i\omega\tilde{C}_{mm}\hat{\eta}_m - \omega^2\tilde{M}_{mm}\hat{\eta}_m = \hat{\tilde{P}}_m$$

This is solved directly to give the response

$$\hat{\eta}_m(\omega) = \frac{\hat{\tilde{P}}_m}{[\tilde{K}_{mm} + i\omega\tilde{C}_{mm} - \omega^2\tilde{M}_{mm}]} = \frac{\hat{\tilde{P}}_m}{[\omega_m^2 + 2i\zeta_m\omega\omega_m - \omega^2]\tilde{M}_{mm}}$$

Let

$$\hat{\eta}_m(\omega) = \hat{h}_m(\omega)\hat{\tilde{P}}_m , \qquad \hat{h}_m(\omega) \equiv \frac{1}{[\omega_m^2 + 2i\zeta_m\omega_m\omega - \omega^2]\tilde{M}_{mm}}$$

and since the modal matrix is time independent, then we also have that

$$\{\hat{u}\} = [\ \Phi\ ]\{\hat{\eta}\}$$

Consequently, we can write the spectral response in physical coordinates as

$$\{\hat{u}\} = \sum_m^M \{\phi\}_m\{\hat{\eta}\}_m = \sum_m^M \{\phi\}_m\hat{h}_m(\omega)\hat{\tilde{P}}_m = \sum_m^M \{\phi\}_m\{\phi\}_m^T\hat{h}_m(\omega)\{\hat{P}\}$$

Notice that the load vector $\{\hat{P}\}$ is independent of the mode, hence we can write

$$\{\hat{u}\} = \left[\hat{H}(\omega)\right]\{\hat{P}\}$$

where we have defined a square matrix $\left[\hat{H}(\omega)\right]$ as

$$[\hat{H}(\omega)] \equiv \sum_m^M \{\phi\}_m\{\phi\}_m^T\hat{h}_m = \sum_m^M \frac{\{\phi\}_m\{\phi\}_m^T}{[\omega_m^2 + 2i\zeta_m\omega_m\omega - \omega^2]\tilde{M}_{mm}} \qquad (11.5)$$

Keep in mind that $\{\phi\}_m\{\phi\}_m^T$ is a square array of size $[M \times M]$. It is apparent that $[\hat{H}(\omega)]$ is the reciprocal of the dynamic stiffness $[\hat{K}]$. It has various names including the *frequency response matrix*, the *receptance matrix*, and the *admittance matrix*.

Note that the generalized force $\tilde{P}_i$ is actually a combination of all the applied forces. Obviously, the combination is different for each mode so it is reasonable to inquire as to how the individual force components participate for each mode. Assume a spectral representation so that $\{P(t)\} = \{\hat{P}\}e^{i\omega t}$, then the term

$$\frac{\hat{\tilde{P}}_m}{\tilde{M}_{mm}} = \frac{\{\phi\}_m^T\{\hat{P}\}}{\{\phi\}_m^T[M]\{\phi\}_m}$$

is called the *mode participation factor*.

**Example 11.10:** Determine the frequency response matrix for the two degree of freedom system of Figure 11.2. Assume modal damping of $\zeta = 0.001$.

First note that

$$\{\phi\}_1\{\phi\}_1^T = \left\{\begin{matrix}1\\1\end{matrix}\right\}\left\{\begin{matrix}1\\1\end{matrix}\right\}^T = \begin{bmatrix}1 & 1\\1 & 1\end{bmatrix}, \qquad \{\phi\}_2\{\phi\}_2^T = \left\{\begin{matrix}1\\-\frac{1}{2}\end{matrix}\right\}\left\{\begin{matrix}1\\-\frac{1}{2}\end{matrix}\right\}^T = \frac{1}{4}\begin{bmatrix}4 & -2\\-2 & 1\end{bmatrix}$$

The frequency response matrix is now determined as

$$[\hat{H}(\omega)] = \begin{bmatrix}1 & 1\\1 & 1\end{bmatrix}\frac{1}{[2 - \omega^2 + .002\sqrt{2}i\omega]3} + \frac{1}{4}\begin{bmatrix}4 & -2\\-2 & 1\end{bmatrix}\frac{1}{[5 - \omega^2 + .002\sqrt{5}i\omega]\frac{15}{2}}$$

Plots of these are shown in Figure 11.5. Note that all the peaks are at the same frequencies. The responses are obtained from

$$\left\{\begin{matrix}\hat{u}_1\\\hat{u}_2\end{matrix}\right\} = [\hat{H}(\omega)]\left\{\begin{matrix}\hat{P}_1\\\hat{P}_2\end{matrix}\right\} = \begin{bmatrix}\hat{H}_{11}(\omega) & \hat{H}_{12}(\omega)\\\hat{H}_{21}(\omega) & \hat{H}_{22}(\omega)\end{bmatrix}\left\{\begin{matrix}\hat{P}_1\\\hat{P}_2\end{matrix}\right\}$$

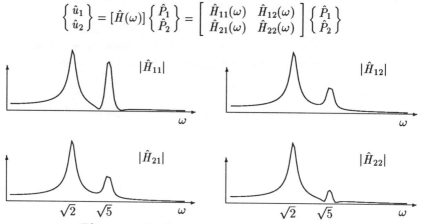

**Figure 11.5:** Frequency response matrix.

**Example 11.11:** Show that the frequency response matrix of the last example can also be obtained from the dynamic stiffness matrix. Neglect damping.

The spectral form the two degree of freedom system is

$$\begin{bmatrix}(4 - \omega^2) & -2\\-2 & (6 - 2\omega^2)\end{bmatrix}\left\{\begin{matrix}\hat{u}_1\\\hat{u}_2\end{matrix}\right\} = \left\{\begin{matrix}\hat{P}_1\\\hat{P}_2\end{matrix}\right\}$$

By use of Cramer's rule this gives

$$\left\{\begin{matrix}\hat{u}_1\\\hat{u}_2\end{matrix}\right\} = \frac{1}{\Delta}\begin{bmatrix}(6 - 2\omega^2) & 2\\2 & (4 - \omega^2)\end{bmatrix}\left\{\begin{matrix}\hat{P}_1\\\hat{P}_2\end{matrix}\right\}$$

where

$$\Delta = (6 - 2\omega^2)(4 - \omega^2) - 4 = 2(\omega^2 - 2)(\omega^2 - 5)$$

The scheme of partial fractions will show an exact correspondence between this relation and the frequency response matrix. What can be observed here is that the determinant $\Delta$ goes through zeros at precisely the values of the natural frequencies.

## 11.5   The Modal Model

We have seen that based on a structure's material properties, geometry and boundary conditions, an analytical model can be formulated which describes dynamic structural properties in terms of mass, damping and stiffness distributions. Equations of motion developed from this model are a set of simultaneous, second order, ordinary differential equations written as

$$[K]\{u\} + [C]\{\dot{u}\} + [M]\{\ddot{u}\} = \{P\}$$

Because these equations describe the response of the structure as a function of time, this description of the structure is called a *time domain* model.

We also developed an alternative model of the structure which describes the dynamic properties as a function of frequency. This involved defining each of the structure's modes of vibration. Each mode is defined by four parameters: an undamped natural frequency, a measure of energy dissipation or damping, a characteristic mode shape, and a measure of mode shape inertial scaling (modal mass). These quantities comprise the structure's *modal model* or *frequency domain model*. The structural response in the time domain and in the frequency domain are represented by

$$\{u(t)\} = \sum_{m}^{M}\{\phi\}_m \eta_m(t)\,, \qquad \{\hat{u}(\omega)\} = \sum_{m}^{M}\{\phi\}_m \hat{\eta}_m = \sum_{m}^{M}\{\phi\}_m\{\phi\}_m^T \hat{h}_m(\omega)\{\hat{P}\}$$

The modal model and the time domain model are equivalent descriptions of a structure's dynamic properties since they are obtained as a transformation of one to the other.

Modal analysis is the process of defining a structure's dynamic properties by its modes of vibration. This can be done either analytically or from test data; although both approaches can yield a common result, for practical purposes each method has advantages and disadvantages and these are discussed next.

### Analytical Modal Analysis

Analytical modal analysis is a procedure used to uncouple the structural equations of motion by use of the modal matrix transformation. It takes as its beginning point the existence of the stiffness and mass matrices from which the modal matrix is computed. The modal response of the structure is then found by summing the respective modal responses, in accordance with their degree of participation in the structural motion.

An analytical model is the only way to investigate a structure's dynamic response before any prototype has been built; they can be used to evaluate different design concepts. This capability — to do pre-prototype investigations of a structure's dynamic behavior — is perhaps the single most important advantage of mathematical modeling. Analytical methods are easier to use for design optimization because it is usually much easier to make a change to a computer model and rerun the analysis than to change a physical model. Additionally, the structure or component may operate in a hostile environment such as zero gravity, high centrifugal force, cryogenic, high temperature, high or low pressure, chemically corrosive or radioactive, all of which may cause testing problems, but analysis can simulate these conditions rather easily. Finally, analytical modal analysis produces results in greater detail than is normally done in testing; an unexpected mode is less likely to be missed than with test methods.

One weakness of analytical methods is that the damping distribution in a structure is very difficult to prescribe. A commonly used approach is to assume that the structure is proportionally damped. The mode shapes and natural frequencies are then identical to those of the undamped case and can be calculated from equations which do not include the damping matrix. This arbitrary assignment of the damping means that eventually the model will have to be verified experimentally.

## Modal Testing

The process of extracting the modal model of a structure from experimental data is called *modal testing*. Although the modal model of a structure can be obtained either by mathematical modeling or by modal testing, generally the term "modal analysis" is used to refer to the process of extracting modal parameters form test data rather than analytically.

Modal testing is a formalized method for identification of natural frequencies and mode shapes of structures. It utilizes dedicated modal test equipment, and requires a formalized procedure for disturbing the structure into motion, and then recording the distribution of the resulting motions throughout the structure. The end results of a modal test are the various natural frequencies, mode shapes, and damping data of the structure. The accuracy of the dynamic model obtained through modal testing is constrained by the quality of the test data; the modal model of the test structure is obtained directly by curve fitting the measured frequency response data. Because of this direct curve fitting, modal analysis can be used to construct a dynamic model of structures too difficult or costly to model analytically.

Modal testing provides the structural matrices which govern the modes and natural frequencies. Thus the basic structural dynamic data, when obtained accurately from a valid test, also provides a true identification of the structural properties for the modes of interest. These derived matrices are based on the measured participation of the mass, stiffness and damping properties in the modes of interest, for the actual boundary conditions which the structure is experiencing. For a design which has already reached the prototype stage, an experimental modal survey may actually be cheaper and quicker than an analytical model for some purposes.

A difficult class of problems where modal analysis is invaluable involves the case where a vibration is observed but the nature of the forcing function is unknown. Analytical techniques can be employed to alter the dynamic characteristics but, without knowledge of the frequency content of the forcing function, the vibration may become worse.

## Integrating Analysis and Modal Testing

In many problems, both analysis and testing can be used to complement each other. If both techniques are used on the same project, the accuracy of the analytical model and the test procedure can be validated. Often the modal test results indicate where refinement of the analytical model may be necessary. For example, a cursory modal test may be run to routinely confirm or suggest improvements in a finite element model. Modes may be measured selectively or all modes in a given frequency band can be extracted. The modal data (including damping) can then be used directly in a finite element model for the structure or component, for subsequent problem solving, or re–designing the equipment for more optimum dynamic response.

Testing and analysis can be merged in other ways. There are many problems in which different phases of the project are best analyzed with one method or the other. For example, assume a vibration is noticed in a structure but the cause is unknown. Testing methods can be used to determine the amplitude and frequency content of the vibration which are then used as applied displacements in an analytical model of the structure. This model can then be used to predict the stresses to determine whether a fatigue failure is likely. Also, the model can be used to determine structural modifications to shift the natural frequencies. After the changes are made, testing can verify that the amplitudes have indeed been reduced.

In large scale projects, it may prove convenient for the structure to be handled by separation into components with some described by finite element models, and the remainder described by their experimentally measured modal properties.

# 11.6 Dynamic Structural Testing

The experimental measurement of the modal properties for a complex structure has been sufficiently difficult that, until recently, only very expensive structures such as aerospace vehicles have been thoroughly tested to verify predicted normal modes. However, recent advancements in minicomputer based equipment and software have substantially reduced the labor, capital, and time required for a typical modal test. In fact, the change in equipment and technique has been so radical that the term "modal analysis" which would logically be synonymous with linear structural dynamic analysis is now used to mean modal property measurement. This inseparation of modal analysis and modal testing is what prompts this section; however, we can only give a brief review of its essentials. Most of the missing details can be found in References [17, 50].

The basic difficulty in modal measurement is that when a structure is dynamically loaded (by any conveniently available means), the response will usually be composed of contributions from many different modes. Resolving the shape or deformation pattern characteristic of a particular mode requires that its contribution to the response at each monitored point be somehow separated out of the total. This is essentially the reverse of the process demonstrated in Figure 11.4.

## Procedure

Modal testing has been used as an effective engineering tool for almost fifty years. Historically, the experimental determination of modal parameters was referred to as *resonance testing*. When a structure is lightly damped and its natural frequencies are well separated, the structure behaves as a single degree of freedom system; its natural frequencies are very distinct, observable and can be easily isolated. Such structures are referred to as simple structures. The terms simple and complex structures, when used in connection with modal testing, are not related to the shape, material, geometry, cost or mission of the structure but rather to the density of natural frequencies and the levels of damping. The simplicity or complexity here is a function of the modal characteristics and the level of testing sophistication needed to extract these parameters.

With the advances in both the technology and applications of structures, especially in the aerospace industry, the inadequacy of resonance testing became more noticeable especially when dealing with complex structures.

The dynamic properties of a structure can be determined by subjecting the structure to a known excitation and measuring the resulting responses. Conceptually, this can be shown by considering

$$\begin{array}{ccccc} \text{Excitation} & & \text{Structure} & & \text{Response} \\ P(t) & \longrightarrow & H(t) & \longrightarrow & u(t) \end{array}$$

Any arbitrary set of time–varying excitations or inputs $P(t)$ interact with the structure's dynamic properties $H(t)$ to yield responses $u(t)$. Obviously, if we know the dynamic characteristics $H(t)$, we can predict the response caused by any excitation.

The relationship in the time domain is difficult to use, but in the frequency domain it is quite simple. The input-output relationship can be written as a simple algebraic equation:

$$\hat{u}(\omega) = \hat{H}(\omega) \times \hat{P}(\omega), \qquad \hat{H}(\omega) = \hat{u}(\omega) \div \hat{P}(\omega)$$

$\hat{H}(\omega)$ is called the *transfer function* and expresses the dynamic characteristics of the structure as a function of the frequency $\omega$. By measuring the excitation $P(t)$ and response $u(t)$, and computing their Fourier transforms, the transfer function is obtained from the quotient $\hat{u}(\omega)/\hat{P}(\omega)$. This is known as the *frequency response function* (FRF) and can be measured directly with modern multichannel spectrum analyzers. To help fix ideas, it is worth keeping Figure 10.7 in mind as a sample frequency response function. This figure shows how an impact at two locations can be used to excite the different modes and give independent information.

The usual method of testing involves exciting the structure with an input having energy over a frequency range wide enough to include the natural frequencies of all modes of interest. Signals proportional to input (usually force) and response (usually acceleration) are digitized and stored. A fast Fourier transform algorithm is used to obtain frequency domain representations of the raw data and these in turn are used with appropriate averaging to obtain frequency response measurements with adequate statistical reliability from

$$\hat{H}_{jk}(\omega) = \frac{\hat{u}_j(\omega)}{\hat{P}_k(\omega)} = \frac{\hat{u}_j(\omega)\hat{P}_k^*(\omega)}{|\hat{P}_k(\omega)|^2}$$

where $\hat{u}_j(\omega)$ is the discrete Fourier transform of the response at degree of freedom $i$, and $\hat{P}_k(\omega)$ is the discrete Fourier transform of input to degree of freedom $k$. Curve fitting routines are applied to the frequency response data to identify the natural frequencies within the given frequency range. The corresponding mode shapes are extracted from the digitized amplitude data at the natural frequencies.

## Curve Fitting Frequency Response Data

Modal analysis by curve fitting in the frequency domain has become very popular. The procedure involves systematic selection of modal parameters in analytic functions such that they are best matched to the measured frequency response functions.

The FRF between two degrees of freedom in a structure is written as:

$$\hat{H}_{jk}(\omega) = \sum_{m=1}^{M} \left[ \frac{A_{jk}^{(m)}}{(i\omega - \omega_m p_m)(i\omega - \omega_m p_m^*)} \right]$$

where

$$A_{jk}^{(m)} = \frac{\phi_j^{(m)} \phi_k^{(m)}}{\tilde{M}_{mm}}, \qquad p_m = \zeta_m + i\sqrt{1 - \zeta_m^2}$$

The term $\hat{H}_{jk}$ is a single element of the array $[\hat{H}(\omega)]$ of Equation(11.5), and can be thought of as one of the sub-figures of Figure 11.5. During the curve fitting, the test provides $\hat{H}_{jk}$ and $\omega$ and the unknowns are $A_{jk}^{(m)}$, $\zeta_m$ and $\omega_m$. When experimental data in the vicinity of only a single mode is used then this is called single degree of freedom (SDOF) curve fitting. For closely spaced or heavily damped modes multiple degree of freedom (MDOF) curve fitting must be used.

In practice, the FRFs are measured to define a single row or a column of the frequency response matrix. Curve fitting each of these will yield the corresponding row or column of the residue matrices $[A^{(m)}]$ from which the modal vectors $\{\phi\}_m$ can be obtained. More advanced methods exploit the redundancy of the matrix to gain statistical improvement in the vector estimates.

As the process is repeated on functions $H_{jk}(\omega)$ for different values of $j$ and $k$, an entire column of the frequency response matrix is gradually built up. If $N$ degrees of freedom are identified as significant then an $[N \times N]$ matrix of frequency response functions can be measured. However, it can be shown that the information contained in the matrix is highly redundant and that (in theory) only a single column or row of functions need be measured. The residues $A_{jk}^{(m)}$ obtained from the driving point response can be used to normalize the mode vectors such that they are orthonormal with respect to the mass matrix and convenient for response simulation.

## 11.7  Structural Modification

Computing the eigenvalues of a large structural system can be very expensive. This becomes all the more noticeable in iterative situations such as design, when essentially the same problem is solved many times but with small design changes. We conclude this chapter by using some ideas from modal analysis to develop a scheme for calculating how minor changes in a structure affect the frequencies of the whole structure. This is done without solving the complete eigenvalue problem.

From the foregoing sections, it is apparent that our typical structural system (with viscous damping) can be described by a set of uncoupled equations of the form

$$\tilde{K}_{mm}\eta_m + \tilde{C}_{mm}\dot{\eta}_m + \tilde{M}_{mm}\ddot{\eta}_m = \{\phi\}_m^T\{P\} = \tilde{P}_m$$

where $\tilde{K}_{mm}$, $\tilde{M}_{mm}$ and $\tilde{C}_{mm}$ are the modal stiffness, mass and damping, respectively. These are obtained by modal transformations of the respective matrices. The undamped resonant frequencies of such a system are obtained from

$$\omega_m^2 = \frac{\tilde{K}_{mm}}{\tilde{M}_{mm}}$$

Imagine making a small change in the structure so as to cause a small change in the stiffness and mass. Then to first order terms, the change in frequency can be estimated as (by differentiation)

$$2\omega_m\Delta\omega_m = \frac{\Delta\tilde{K}_{mm}}{\tilde{M}_{mm}} - \frac{\tilde{K}_{mm}\Delta\tilde{M}_{mm}}{\tilde{M}_{mm}^2}$$

Divide both sides of this relation by the previous equation to give

$$\frac{\Delta\omega_m}{\omega_m} = \frac{\Delta\tilde{K}_{mm}}{2\tilde{K}_{mm}} - \frac{\Delta\tilde{M}_{mm}}{2\tilde{M}_{mm}}$$

The changes in stiffness and mass are estimated by

$$\Delta\tilde{K}_{mm} \equiv \{\phi\}_m^T[\Delta K][\phi]_m, \qquad \Delta\tilde{M}_{mm} \equiv [\Phi]_m^T[\Delta M][\Phi]_m$$

where it is the original modal matrix that is used. Once the modal matrix for a structure has been established, then this gives an inexpensive method for estimating the effect of small structural changes. Implicit in this scheme is the assumption that the mode shapes are relatively insensitive to small structural changes.

A similar development can be done for the damping matrix.

**Example 11.12:** A small accelerometer is mounted to the system of Figure 11.2 in order to measure the frequency response. Estimate the effect the mass of the accelerometer has on the response.

Recall from before that the modal description is summarized as

$$\lceil \tilde{K} \rfloor = \begin{bmatrix} 6 & 0 \\ 0 & \frac{15}{2} \end{bmatrix}, \qquad \lceil \tilde{M} \rfloor = \begin{bmatrix} 3 & 0 \\ 0 & \frac{3}{2} \end{bmatrix}, \qquad \{\phi\}_1 = \begin{Bmatrix} 1 \\ 1 \end{Bmatrix}, \qquad \{\phi\}_2 = \begin{Bmatrix} 1 \\ -\frac{1}{2} \end{Bmatrix}$$

First consider putting the accelerometer on mass 1, then

$$[\Delta M] = \begin{bmatrix} m_a & 0 \\ 0 & 0 \end{bmatrix}$$

and the changes in modal masses are

$$\Delta \tilde{M}_{11} = \begin{Bmatrix} 1 \\ 1 \end{Bmatrix}^T \begin{bmatrix} m_a & 0 \\ 0 & 0 \end{bmatrix} \begin{Bmatrix} 1 \\ 1 \end{Bmatrix} = m_a, \qquad \Delta \tilde{M}_{22} = \begin{Bmatrix} 1 \\ -\frac{1}{2} \end{Bmatrix}^T \begin{bmatrix} m_a & 0 \\ 0 & 0 \end{bmatrix} \begin{Bmatrix} 1 \\ -\frac{1}{2} \end{Bmatrix} = m_a$$

Note that both modal masses are affected, hence both modal frequencies will be affected. That is

$$\frac{\Delta \omega_1}{\omega_1} = 0 - \frac{m_a}{2 \times 3} = -\frac{1}{6} m_a, \qquad \frac{\Delta \omega_2}{\omega_2} = 0 - \frac{m_a}{2 \times \frac{3}{2}} = -\frac{1}{3} m_a$$

In both cases, the resonance is decreased, but the second mode by a greater amount. That is, the effective mass is greater at the higher frequency.

Now consider putting the accelerometer on mass 2, then

$$[\Delta M] = \begin{bmatrix} 0 & 0 \\ 0 & m_a \end{bmatrix}$$

and the changes in modal masses are

$$\Delta \tilde{M}_{11} = \begin{Bmatrix} 1 \\ 1 \end{Bmatrix}^T \begin{bmatrix} 0 & 0 \\ 0 & m_a \end{bmatrix} \begin{Bmatrix} 1 \\ 1 \end{Bmatrix} = m_a, \qquad \Delta \tilde{M}_{22} = \begin{Bmatrix} 1 \\ -\frac{1}{2} \end{Bmatrix}^T \begin{bmatrix} 0 & 0 \\ 0 & m_a \end{bmatrix} \begin{Bmatrix} 1 \\ -\frac{1}{2} \end{Bmatrix} = \frac{1}{4} m_a$$

Note that this time both modal masses are affected differently. The changes in frequencies are

$$\frac{\Delta \omega_1}{\omega_1} = 0 - \frac{m_a}{2 \times 3} = -\frac{1}{6} m_a, \qquad \frac{\Delta \omega_2}{\omega_2} = 0 - \frac{\frac{1}{4} m_a}{2 \times \frac{3}{2}} = -\frac{1}{12} m_a$$

Again, the resonance is decreased, but this time the second mode shows less effect.

## Problems

**11.1** Show that if an eigenvector is augmented with zeros, and that the mass and stiffness matrices are augmented with arbitrary values, that the modal mass and stiffness are unchanged.

**11.2** Show that interchanging the degree of freedoms $i$ and $j$ and the corresponding rows and columns of the system matrices, does not affect the orthogonality properties.

**11.3** Suppose a structure has two rigid body modes characterized by the eigenpairs $(0, \{\phi\}_1)$ and $(0, \{\phi\}_2)$. Show that the mode shapes are orthogonal to all other modes and that

$$[K]\{\phi\}_1 = 0, \qquad [K]\{\phi\}_2 = 0$$

[Reference [28], pp. 68]

**11.4** It is desired to do modal testing on a two degree of freedom system. What are the minimum number of tests required?

**11.5** Let the modal damping be specified as $2\zeta_m\omega_m$ on the diagonal and zero elsewhere. Show that the damping matrix is fully populated and given by

$$[C] = [\Phi^T]^{-1}[\tilde{C}][\Phi] = [M][\sum_m 2\zeta_m\omega_m\{\phi\}_m\{\phi\}_m^T][M]$$

[Reference [23], pp. 452]

**11.6** Generalize the orthogonality relations to

$$\{\phi\}_i^T[KM^{-1}]^n[K]\{\phi\}_j = \lambda^{n+1}, \qquad \{\phi\}_i^T[M][M^{-1}K]^n\{\phi\}_j = \lambda^{n+1}$$

where $n$ is any positive or negative integer.

[Reference [23], pp. 447]

**11.7** Show that if the damping matrix is taken in the following proportional form

$$[C] = \sum_{n=0} \alpha_n[M][M^{-1}K]^n$$

that it leads to a diagonal modal damping matrix.

[Reference [23], pp. 451]

**11.8** Given the following pairs $(\omega_m, \zeta_m)$ of modal damping: $(2, .002)$, $(3, .030)$, $(7, .040)$, $(15, .100)$, $(19, .140)$; show that a best curve-fit for the Rayleigh damping coefficients gives

$$[C] = 0.01498[M] + 0.01405[K]$$

[Reference [5], pp. 529]

**11.9** Integrate the equations of motion as $\int_t^{t+\epsilon}$ where $\epsilon$ is a short time over which an impulsive force acts. Show that the equations of motion shortly after the impulse become

$$[M]\{\dot{u}\} = \int^\epsilon \{P\}dt = \{I\}$$

where $\{I\}$ is the vector of impulses.

[Reference [6], pp. 549]

## Exercises

**11.1** A $48\,in$ long steel bar of circular cross-section of $1.5\,in$ diameter is bent into an L shape and fixed to a rigid support at one end. Using the lumped mass approach determine the natural frequencies and mode shapes.
$[f_1 = 20.56\,Hz,\ f_2 = 41.51\,Hz]$

**11.2** Consider a steel frame similar to that in Figure 11.1 except that the vertical member is $2\,m$ while the horizontal one is $1\,m$. If it has a cross-section of $24\,mm \times 24\,mm$, use twelve elements to determine the first few resonances.
$[27.17\,r/s,\ 74.59\,r/s]$

**11.3** Find the first few resonances of the aluminum truss of Figure 4.4. Let $h = 10\,m$, $A = 100\,mm^2$ and $\theta = 45^\circ$. $[359.46,\ 566.46\ r/s]$

**11.4** A steel truss similar to Figure 4.5 has a length of $5\,m$ and cross-sectional area of $250\,mm^2$. Use one element per member to obtain the vibration mode shapes. What would happen if more elements were used?   $[362.52,\ 642.79,\ 869.99\ r/s]$

**11.5** An aluminum frame similar to Figure 5.16 but with cross-section $1\,in \times 1\,in$ and length $100\,in$, is modeled with four elements. What are the resonances and mode shapes? $[86.421,\ 126.24,\ 324.20\ r/s]$

**11.6** A steel portal frame (two vertical members of length $L$ with a cross member of length $2L$). Use four elements to determine the resonances. The cross-section is $1\,in \times 1\,in$ and length $10\,ft$. $[8.8474,\ 16.049,\ 51.491\ r/s]$

**11.7** Add a diagonal member (same properties) to the portal frame of the previous problem and determine the resonances. What has happened?
$[16029,\ 33.814,\ 63.664\ r/s]$

**11.8** A three-story building frame is to be modeled as a shear building with the following data: $m_1 = m_2 = 3$, $m_3 = 2\ kip \cdot s^2/ft$; $k_1 = 1000$, $k_2 = 800$, $k_3 = 600$, $kip/ft$. Determine the natural frequencies and mode shapes.
$[8.7,\ 21.0,\ 29.2,\ r/s]$

**11.9** Suppose the building of the previous problem is subjected to a suddenly applied constant horizontal force of $50\,kip$ at the second floor. Plot the deflection of the top floor up to the first peak of response. $[u_{max} = 2.83\,in]$

# Chapter 12

# General Structural Principles II

In this chapter, we will develop alternative descriptions of the equations of motion. We are motivated to do this because in complex structures involving distributed mass and elasticity, the direct vectorial method of the last few chapters may be difficult to apply.

We extend the work and energy ideas of Chapter 6 by introducing the scalar kinetic energy function to account for the inertial effects. The principle of stationary potential energy is extended to Hamilton's principle which states that the difference of the kinetic and potential energies considered over any time interval is a stationary value. This formulation has the advantage of dealing only with purely scalar energy quantities, and is independent of any particular coordinate system. The variational and Ritz analysis ideas of this chapter are covered extensively in References [5, 29].

Consistent with our approach in Chapter 6, we use the general structural principles to develop approximate solutions. We illustrate this in two ways. We first use Hamilton's principle to obtain improved differential descriptions of the structural models. This is important in structural dynamics because it is not always clear how to establish the governing equations in a direct manner — this is especially true in determining the correct statement of the boundary conditions. As a practical matter, computer solutions require the problem to be posed in the form of a finite number of discrete unknowns. To this end, we combine the Ritz method with Hamilton's principle to obtain Lagrange's equation of motion. This recovers the approximate matrix formulations of the last few chapters but, in addition, it also shows the way for treating more complicated systems (such as variable cross-section or distributed loads).

## 12.1 Elements of Analytical Dynamics

This section briefly reviews some of the elements of analytical dynamics. This is necessary preparation for developing the general principles for analyzing the dynamic response of continuous structures. We are particularly interested in making clear the types of systems to which our equations apply. In the literature (Lanczos' book [29] is an excellent reference), it is usual to find dynamic problems classified as:

- Holonomic or Non-holonomic
- Conservative or Non-conservative
- Scleronomic or Rheonomic

These terms will now be explained.

To describe the dynamic behavior of a system, we select the smallest possible number of variables. These are called the *generalized coordinates* of the system and are denoted by $u_1, u_2, \ldots, u_N$. We can consider these generalized coordinates as the coordinates of a 'point' in an N-dimensional space. Then we can view

$$u_1(t), \quad u_2(t), \quad \ldots$$

as a 'curve' in this space. This geometrical picture can be a great aid to our thinking; no matter how numerous the particles constituting a given mechanical system may be, or how complicated the relations existing between them, the entire mechanical system (at a particular instant in time) is pictured as a single point of a many-dimensional space. This space is called the *configuration space* of the system.

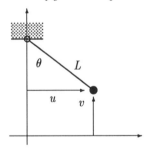

Figure 12.1: Simple pendulum.

Consider a simple pendulum for which the string length $L$ is constant, as shown in Figure 12.1. Using rectangular coordinates $u$ and $v$ we have the constraint relation

$$u^2 + (L - v)^2 = L^2$$

Let the support point move in the vertical direction in a prescribed manner; the constraint is now

$$u^2 + (L - A \sin \omega t - v)^2 = L^2$$

In both cases the constraint is expressed in 'finite' form; our treatment of dynamic problems will be limited to systems in which the constraints are expressible by means of equations containing the coordinates and the time, but not containing any time derivatives of the coordinates. Such a system is called *holonomic*. In a holonomic system, we can give arbitrary independent variations to the generalized coordinates without violating the constraints; in the case of a non-holonomic system this cannot be done. Rolling motion might provide the only real mechanical illustration of non-holonomic constraints, consequently little is lost on ignoring non-holonomic systems in this book. Therefore, for the systems we will consider, the generalized coordinates satisfy the following:

- Their values determine the configuration of the system.

- They may be varied arbitrarily and independently without violating the con-straints of the system.

When the forces acting on a system are such that the work done by them, in the passage of the system from one configuration to another is independent of the way in which this passage is carried out, then the system is said to be *conservative*. The work done by the forces is called the *potential energy* of the system. Examples of conservative forces are: gravity forces between masses, forces due to all types of springs and elastic bodies (assuming "perfectly elastic" material). Non-conservative forces include those of friction, drag, and various types which depend on time and velocity. In a conservative system the generalized forces are derivable from a potential energy function $V$ as

$$P_i = -\frac{\partial V}{\partial u_i}$$

In a *scleronomic* system, a configuration of the system is given when the values of a set of generalized coordinates are assigned; in a *rheonomic* system it is necessary to assign also the value of the time. In other words, a scleronomic system is one which has only 'fixed' constraints, whereas a rheonomic system has 'moving' constraints. For example, the pendulum with fixed support is scleronomic; the pendulum for which the point of support is given an assigned motion is rheonomic. Systems that are subjected to time-dependent constraints are generally non-conservative in the sense that the total mechanical energy varies with time, since the constraining forces perform work on the system. (Keep in mind that energy would be conserved if the *whole* system were considered; we are referring here to the system defined as the mass and the string.) The essential difference between a scleronomic and a rheonomic system is that in the former the 'total energy' remains constant during the motion; rheonomic systems do not satisfy any conservation laws but scleronomic systems do. We see that the law of the conservation of energy has to be restricted to systems which are scleronomic in both the work function and kinematical conditions.

Systems which are holonomic, conservative and scleronomic are often referred to as 'simple' systems.

There are forces associated with the kinematical constraints. To illustrate, con-sider the simple cantilever beam shown in Figure 5.7; when we view it as a free body diagram disconnected from the built-in end, we have the loads $P$ and $Q$ and the reac-tions. To distinguish $P$ and $Q$ from the latter, we call them *applied forces* and we call the latter reactions *forces of constraint*. In the present case, the forces of constraint are an axial force, a shear force, and a moment. For any system with 'workless constraints' (the forces of constraint do no work in the virtual displacement satisfying the instantaneous constraints), all forces other than the reactions of constraint are called applied forces.

The virtual work done by a general system is

$$\delta W = P_1 \delta u_1 + P_2 \delta u_2 + \cdots$$

For a system with workless constraints we have

$$\delta W = 0$$

That is, the work done by the applied forces is zero. We can give a striking geometrical interpretation of the principle of virtual work; since work is the scalar product of force and virtual displacement, the vanishing of it means the force is perpendicular to any possible virtual displacement. Suppose the given mechanical system is free of any constraints, then the principle requires that the force shall vanish because there is no vector which can be perpendicular to all directions in space. On the other hand, if there are kinematical constraints, then the principle does not require the force to vanish but only that it be perpendicular to the subspace of constraints. Consider again the 2-dimensional example of the pendulum of Figure 12.1. The kinematic constraint is such that the mass is constrained to move in a circle of radius $L$, thus the subspace of constraints is the displacement in the radial direction. Two possible components of force could be applied to the mass; one component acts perpendicular to the motion (in the radial direction) while the other acts tangential to the motion. The principle of virtual work requires that the generalized force be in the tangential direction only.

## 12.2   Hamilton's Principle

We are now interested in applying the ideas of the last section to continuous or distributed systems. To make the developments concrete, we initially analyze the one-dimensional rod; the extension to multiple dimensions will follow easily since we work with scalar quantities. We have already established that the principle of virtual work for the rod is given by

$$\delta W = \delta W^s + \delta W^b - \delta U = \int_S f^s \delta u \, dS + \int_V f^b \delta u \, dV - \int_V \sigma \, \delta \epsilon \, dV = 0$$

where the three contributions are from the surface tractions, the body forces, and the strain energy, respectively. In dynamic problems, we need to account for two new aspects: first is the presence of inertia forces, and second is that all quantities are functions of time. We will take care of the former by use of d'Alembert's principle and the latter by time averaging.

## D'Alembert's Principle and Kinetic Energy

D'Alembert's principle converts a dynamics problem into an equivalent problem of static equilibrium by treating the inertia as a body force. For example, for a rod we treat the total body force (per unit length) as comprised of $f^b - \rho A \ddot{u}$ where $-\rho A \ddot{u}$ is the inertia of an infinitesimal volume.

The virtual work of the body force is

$$\delta W^b = \int_V f^b \delta u \, dV - \int_V \rho A \ddot{u} \delta u \, dV$$

In writing this relation, we suppose that the performance of the virtual displacement consumes no time; that is, the real motion of the system is stopped while the virtual displacement is performed. Consequently, the time variable is conceived to remain constant while the virtual displacement is executed. We must suppose, however, that the inertia forces corresponding to the real motion persist during the virtual displacement since we wish to calculate the virtual work of these forces.

Another important consideration is the admissibility of certain displacements as virtual displacements; the real displacements take place over real time and may no longer be admissible as virtual displacements. As an example of a situation in which the real displacements are not admissible as virtual displacements, consider a simple pendulum in which the string is being shortened at a constant rate. That is, the distance from the point of suspension to the mass reduces with time. The virtual displacement given to the system at that time is along the tangent to an arc. (Since this displacement is perpendicular to the force of constraint exerted by the string, the string tension does no work on it.) The real displacement that takes place in a short time has an additional component along the string and therefore the force of constraint will, in fact, do work on the real displacement.

We will concentrate on the inertia term in the above virtual work expression. Noting that

$$\frac{d}{dt}(\dot{u} \, \delta u) = \ddot{u} \, \delta u + \dot{u} \, \delta \dot{u}$$

we can write the inertia term as

$$\int \rho A \ddot{u} \, \delta u dV = \int \rho A \frac{d}{dt}(\dot{u} \, \delta u) dV - \int \rho A \dot{u} \, \delta \dot{u} dV = \int \rho A \frac{d}{dt}(\dot{u} \, \delta u) dV - \int \delta T \, dV$$

where we have introduced the concept of *kinetic energy* defined as

$$T \equiv \tfrac{1}{2} \int_V \rho A \dot{u}^2 dV \qquad \text{such that} \qquad \delta T \equiv \int_V \rho A \dot{u} \, \delta \dot{u} \, dV$$

At this stage, the principle of virtual work becomes

$$\delta W = \delta W^s + \delta W^b - \delta U + \delta T - \frac{d}{dt} \int_V (\dot{u} \delta u) dV = 0$$

It remains now to remove the last integral term.

## Hamilton's Formulation

Hamilton refined the concept that a motion can be viewed as a path in configuration space; he showed that for a system with given configurations at times $t_1$ and $t_2$, of all the possible configurations between these two times the actual that occurs satisfies a stationary principle. This essentially geometric idea is illustrated in Figure 12.2 where a varied path in both space and time are shown. To fix ideas, consider a beam in motion with its position described by $v(x,t)$. At a particular instant in time, $t$, we can imagine a varied deformation shape shown as the dotted line in (a). The end constraints at $x = 0$ and $L$ are not varied however. Now consider a particular point on the beam and plot its position over time; this gives the solid line of (b) and it represents the 'Newtonian path' of the point. The addition of the virtual displacement $\delta u(x,t)$ gives a path that may look like the dotted line in (b); again there is no variations at the extreme times $t_1$ and $t_2$.

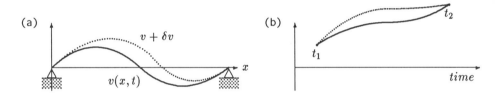

**Figure 12.2**: Hamilton's configuration space for a beam.

Hamilton disposed of the last term in the virtual work relation by integrating the equation over time between the limits of the 1 and 2 configurations. The last term is a time derivative and so may be integrated explicitly to give

$$-\int \dot{u}\delta u \, d\forall \Big|_{t_1}^{t_2}$$

By assumption, the configuration has no variations at the extreme times and hence the term is zero. Consequently, the virtual work relation becomes

$$\int_{t_1}^{t_2} [\delta W^s + \delta W^b + \delta T - \delta U]dt = 0 \tag{12.1}$$

This equation is generally known as the *extended Hamilton's principle*. An important feature of this principle is that it is formulated without reference to any particular system of coordinates. That is, it holds true for constrained as well as generalized coordinates.

In the special case when the applied forces can be derived from a scalar potential function $V$, the variations become complete variations and we can write

$$\delta \int_{t_1}^{t_2} [T - (U + V)]dt = 0 \quad \text{or} \quad \delta \int_{t_1}^{t_2} L \, dt = 0 \tag{12.2}$$

where $L = T - (U+V)$ is called the *Lagrangian*. This equation is Hamilton's principle, and it may be stated as follows:

> *Among all motions that will carry a conservative system from a given configuration at time $t_1$ to a second given configuration at time $t_2$, that which actually occurs provides a stationary value of the integral.*

Hamilton's principle is a variational principle and it is as general as Newton's second law. Note that the integral is essentially a time-averaged quantity. Also note that the work of the body forces is due to all of the body forces *except inertia*; the inertia of the system is accounted for through the kinetic energy term.

Hamilton's principle occupies an important position in analytical mechanics because it reduces the formulation of a dynamic problem to the variation of two scalar quantities: the work function and the kinetic energy, and because it is invariant under coordinate transformation.

**Example 12.1:**  Consider a particle of mass $m$ subject to a force $P$. Show that Newton's law governing the motion of the particle can be recovered from Hamilton's principle.

The kinetic energy, strain energy, and potential of the applied force are given by, respectively,

$$T = \tfrac{1}{2}m\dot{u}^2, \qquad U = 0, \qquad V = -Pu$$

Hamilton's principle for the particle is

$$\delta \int_{t_1}^{t_2} [T - (U+V)]dt = \delta \int_{t_1}^{t_2} [\tfrac{1}{2}m\dot{u}^2 - (0 - Pu)]dt = \int_{t_1}^{t_2} [m\dot{u}\,\delta\dot{u} + P\delta u]dt = 0$$

Noting that

$$\dot{u}\,\delta\dot{u}\,dt = \dot{u}\frac{d(\delta u)}{dt}dt = \dot{u}d(\delta u)$$

then we can integrate the first term in the integral by parts to give

$$m\dot{u}\delta u\Big|_{t_1}^{t_2} + \int_{t_1}^{t_2} [-m\ddot{u} + P]\delta u\,dt = 0$$

By assumption, the variation $\delta u$ at the times $t_1$ and $t_2$ are zero, then the first term is also zero. Since the time limits of integration are arbitrary, and since the variations between these limits can be arbitrary, then we conclude that the integrand must be zero. This gives

$$-m\ddot{u} + P = 0 \qquad \text{or} \qquad P = m\ddot{u}$$

This is Newton's second law, consequently, Hamilton's principle is as general as Newton's equation.

## 12.3   Approximate Structural Theories

One of the very successful applications of the energy methods is in aiding the derivation of approximate structural theories. The approach will be demonstrated in connection with higher order rod and beam theories. We pay special attention to the specification of the boundary conditions since these are not always obvious in dynamic problems.

### Modified Rod Theory

As the rod deforms longitudinally, it also contracts due to the Poisson's ratio effect. Thus, each particle of the rod also has a transverse component of velocity. We will now add this to the kinetic energy term in order to have a more accurate accounting of the energy. We will then use Hamilton's principle to derive the governing differential equations and associated boundary conditions. The modified rod theory thus developed is referred to as *Love's rod theory*.

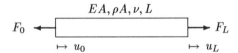

**Figure 12.3**: Rod with end loads.

The transverse strain is related to the axial strain by $\epsilon_t = -\nu\epsilon$, therefore, the transverse velocity is given by

$$\dot{u}_t = r\dot{\epsilon}_t = -\nu r\dot{\epsilon} = -\nu r\frac{\partial \dot{u}}{\partial x}$$

In this, we have assumed that the transverse displacement is proportional to the distance $r$ from the centroid of the cross-section. The total kinetic energy of the rod is readily found to be

$$T = \int_V \tfrac{1}{2}\rho[\dot{u}(x,t)^2 + \dot{u}_t(x,t)^2]d\forall = \tfrac{1}{2}\int_0^L \int_A \rho[\dot{u}^2 + \nu^2 r^2(\frac{\partial \dot{u}}{\partial x})^2]dA\,dx$$

Since $\dot{u}$ is a function only of $x$ (and time) then we can perform the integration with respect to the cross-section to give

$$T = \tfrac{1}{2}\int_0^L [\rho A\dot{u}^2 + \nu^2 \rho J(\frac{\partial \dot{u}}{\partial x})^2]dx\,, \qquad J \equiv \int_A r^2\,dA$$

$J$ is the polar moment of area. The total strain energy is given by

$$U = \tfrac{1}{2} \int_0^L EA(\frac{\partial u}{\partial x})^2 dx$$

The final term we need is the potential of the applied forces. Assume there are no distributed loads — only end loads as shown in Figure 12.3, then we have

$$V = -(-F_0 u_0 + F_L u_L) = F_0 u_0 - F_L u_L$$

We now substitute these energies into Hamilton's principle to get

$$\delta \int_{t_1}^{t_2} \left[ \tfrac{1}{2} \int_0^L [\rho A \dot{u}^2 + \nu^2 \rho J(\frac{\partial \dot{u}}{\partial x})^2] dx - \tfrac{1}{2} \int_0^L EA(\frac{\partial u}{\partial x})^2 dx - F_0 \delta u_0 + F_L \delta u_L \right] dt = 0$$

Take the variation inside

$$\int_{t_1}^{t_2} \left[ \int_0^L \left\{ \rho A \dot{u} \delta \dot{u} + \nu^2 \rho J(\frac{\partial \dot{u}}{\partial x})(\frac{\partial \delta \dot{u}}{\partial x}) - EA(\frac{\partial u}{\partial x})(\frac{\partial \delta u}{\partial x}) \right\} dx - F_0 u_0 + F_L u_L \right] dt = 0$$

We will now use integration by parts so as to have all terms multiplied by a common variation $\delta u$. For example, the time integration of the first term can be rewritten as

$$\int_{t_1}^{t_2} \rho A \dot{u} \, \delta \dot{u} \, dt = \int_{t_1}^{t_2} \rho A \dot{u} \, d(\delta u) = \rho A \dot{u} \, \delta u \Big|_{t_1}^{t_2} - \int_{t_1}^{t_2} \rho A \ddot{u} \, \delta u \, dt$$

By assumption there is no variation of the configuration at times $t_1$ and $t_2$, hence the term evaluated at these limits is zero. For the third term, we have for the space integration

$$\int_0^L EA(\frac{\partial u}{\partial x})(\frac{\partial \delta u}{\partial x}) dx = \int_0^L EA(\frac{\partial u}{\partial x}) d(\delta u) = EA\frac{\partial u}{\partial x} \delta u \Big|_0^L - \int_0^L EA\frac{\partial^2 u}{\partial x^2} dx \, \delta u$$

The middle term requires both space and time integration and this is done as a combination of the previous two terms. The result is

$$\int_{t_1}^{t_2} \int_0^L \nu^2 \rho J(\frac{\partial \dot{u}}{\partial x})(\frac{\partial \delta \dot{u}}{\partial x}) dx \, dt = - \int_{t_1}^{t_2} \nu^2 \rho J \frac{\partial \ddot{u}}{\partial x} \Big|_0^L \delta u \, dt + \int_{t_1}^{t_2} \int_0^L \nu^2 \rho J \frac{\partial^2 \ddot{u}}{\partial x^2} \delta u \, dx \, dt$$

Add all these terms together to get

$$\int_{t_1}^{t_2} \left\{ \int_0^L \left[ EA\frac{\partial^2 u}{\partial x^2} + \nu^2 \rho J \frac{\partial^2 \ddot{u}}{\partial x^2} - \rho A \ddot{u} \right] \delta u \, dx - (EA\frac{\partial u}{\partial x} + \nu^2 \rho J \frac{\partial \ddot{u}}{\partial x} - F)\delta u \Big|_0^L \right\} dt = 0$$

Since the time limits and space limits in the integrations are arbitrary, then the first integrand is zero giving the governing differential equation as

$$EA\frac{\partial^2 u}{\partial x^2} + \nu^2 \rho J \frac{\partial^2 \ddot{u}}{\partial x^2} - \rho A \frac{\partial^2 u}{\partial t^2} = 0 \qquad (12.3)$$

It is obvious that if either Poisson's ratio or the polar moment of area is negligibly small then we recover the rod equation already developed in Chapter 9.

The remaining terms must also be zero and thereby specify the boundary conditions; at either end of the rod, we specify

$$u \quad \text{or} \quad F = EA\frac{\partial u}{\partial x} + \nu^2 \rho J\frac{\partial \ddot{u}}{\partial x} \tag{12.4}$$

The natural boundary condition is a rather surprising result. We recognize the first term of it as the usual axial force relation; does the presence of the second term mean that this relation is no longer valid? Recall the discussion of the Ritz method in Chapter 6 were it was stated that we must satisfy the geometric boundary conditions explicitly but that the natural boundary conditions are implied in the potential function. Actually, both the differential equation and the boundary conditions are implied in the potential. Because we started with an approximation for the potential function we derived a governing differential equation and a set of boundary conditions most consistent with that approximation. We can imagine, therefore, proposing a different potential and having a natural boundary condition that is actually the same as the axial force relation. In fact, such a situation arises in the higher order rod theory referred to as *Mindlin-Herrmann rod theory* [14].

**Example 12.2:** Use the Love theory of rods to determine the resonant frequencies of a fixed-fixed rod.

Spectral analysis of Equation(12.3) gives

$$(1 - \omega^2\nu^2\frac{\rho J}{EA})EA\frac{d^2\hat{u}}{dx^2} + \omega^2\rho A\,\hat{u} = 0$$

This differential equation has constant coefficients and hence has the solution

$$\hat{u}(x) = c_1\cos kx + c_2\sin kx\,, \quad k = \sqrt{\frac{\omega^2\rho A}{(EA - \omega^2\nu^2\rho J)}}$$

The boundary conditions require that $\hat{u}(0) = 0$ and $\hat{u}(L) = 0$; thus a non-trivial solution can be found only if

$$\sin kL = 0 \quad \text{or} \quad k = n\pi/L$$

This is the same result as for the elementary theory of Chapter 9; the difference here is that the wavenumber $k$ is a different function of frequency. Substituting for the frequency we get the resonances as

$$\omega = \frac{n\pi}{L}\sqrt{\frac{EA}{\rho A}}\Big/\sqrt{1 + (\frac{n\pi}{L})^2\frac{\rho J}{\rho A}}$$

We recognize the leading term as the resonant frequencies predicted by the elementary theory, hence the improved theory gives lower values.

**Example 12.3:** Estimate the allowable dimensions of a rod in order that the elementary rod theory gives results within 5% of the improved theory.

We require that

$$\omega_{elem} \leq (1 + \alpha)\omega_{love}$$

where $\alpha = 0.05$. Noting that $J/A = D^2/8$ where $D$ is the diameter of the rod, then we can rearrange the criterion for the fixed-fixed rod as

$$\sqrt{1 + (\frac{n\pi}{L})^2 \frac{\rho J}{\rho A}} \leq (1 + \alpha) \qquad \text{or} \qquad \frac{nD}{L} \leq \sqrt{16\alpha + 8\alpha^2}$$

The 5% criterion can be approximated as

$$\frac{nD}{L} \leq 1$$

For the first mode we can have the length of the rod as short as the diameter. However, for the tenth mode we must have the length at least ten times the diameter.

## Timoshenko Beam Theory

We will now develop a beam theory that takes the shear deformation into account. To show the generality of the approach, we will arrive at the structural theory by way of some two-dimensional concepts. Consider a rectangular beam of length $L$, thickness $h$, and width $b$, as shown in Figure 12.4. If $b$ is small, then the beam can be regarded as in a state of plane stress.

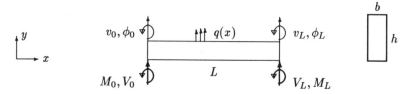

**Figure 12.4:** Timoshenko beam with end loads.

We begin by expanding the displacements in the beam by a Taylor series about the mid-plane displacements $\bar{u}(x,0)$ and $\bar{v}(x,0)$ as

$$\bar{u}(x,y) \approx \bar{u}(x,0) + y\frac{\partial \bar{u}}{\partial y}\Big|_{y=0} + \cdots = u(x) - y\phi(x) + \cdots$$

$$\bar{v}(x,y) \approx \bar{v}(x,0) + y\frac{\partial \bar{v}}{\partial y}\Big|_{y=0} + \cdots = v(x) + y\psi(x) + \cdots$$

where we have used the notations

$$u(x) = \bar{u}(x,0), \quad v(x) = \bar{v}(x,0), \quad \phi(x) = -\frac{\partial \bar{u}}{\partial y}\Big|_{y=0}, \quad \psi(x) = \frac{\partial \bar{v}}{\partial y}\Big|_{y=0}$$

Since we are interested in flexural deformations, we set $\bar{u}(x,0) = 0$. For illustration purposes, we retain only one term in each expansion and obtain the approximate displacements as

$$\bar{u}(x,y) \approx -y\phi(x), \qquad \bar{v}(x,y) \approx v(x)$$

This says that the deformation is governed by two independent functions, $v(x)$ and $\phi(x)$, that depend only on the position along the centerline. We could, of course, retain more terms in the expansion and develop an even more refined theory. We make the assumption that these kinematic representations do not change under dynamic conditions.

The axial and shear strains corresponding to the above deformation are

$$\epsilon_{xx} = \frac{\partial \bar{u}}{\partial x} = -y\frac{\partial \phi}{\partial x}, \qquad \gamma_{xy} = \frac{\partial \bar{u}}{\partial y} + \frac{\partial \bar{v}}{\partial x} = \left(-\phi + \frac{\partial v}{\partial x}\right)$$

For a slender beam undergoing flexural deformation we would expect that $\sigma_{yy} \ll \sigma_{xx}$, so that we can set $\sigma_{yy} = 0$. Substituting this relation in the strain energy density function, we have

$$U = \int_V [\tfrac{1}{2}\sigma_{xx}\epsilon_{xx} + \tfrac{1}{2}\sigma_{xy}\gamma_{xy}]dV = \int_V [\tfrac{1}{2}E\epsilon_{xx}^2 + \tfrac{1}{2}G\gamma_{xy}^2]dV$$

Substitute for the strain to get the total strain energy as

$$U = \tfrac{1}{2}\int_o^L \int_{-h/2}^{h/2} \left[Ey^2(\frac{\partial \phi}{\partial x})^2 + G(\phi - \frac{\partial v}{\partial x})^2\right] b\,dy\,dx = \tfrac{1}{2}\int_o^L \left[EI(\frac{\partial \phi}{\partial x})^2 + GA(\phi - \frac{\partial v}{\partial x})^2\right] dx$$

The total kinetic energy is

$$T = \int_V \tfrac{1}{2}\rho[\dot{u}(x,t)^2 + \dot{v}(x,t)^2]dV = \tfrac{1}{2}\int_0^L \int_A \rho[y^2\dot{\phi}^2 + \dot{v}^2]dA\,dx = \tfrac{1}{2}\int_0^L [\rho A\dot{v}^2 + \rho I\dot{\phi}^2]\,dx$$

If the applied surface tractions and loads on the beam are as shown in Figure 12.4, then the potential of these loads is

$$V = -\int_o^L q(x)v\,dx - M_L\phi_L + M_0\phi_0 - V_Lv_L + V_0v_0$$

Hamilton's principle for the beam may now be stated as

$$\delta \int_{t_1}^{t_2} \left\{ \int_0^L \left[\tfrac{1}{2}[\rho A\dot{v}^2 + \rho I\dot{\phi}^2] - \tfrac{1}{2}[EI(\frac{\partial \phi}{\partial x})^2 + GA(-\phi + \frac{\partial v}{\partial x})^2] + qv\right] dx \right.$$
$$\left. +M\phi\Big|_0^L + Vv\Big|_0^L \right\} dt = 0$$

Taking the variation inside the integrals and using integration by parts, we get

$$\int_{t_1}^{t_2} \left\{ \int_0^L \left[ GA[-\phi + \frac{\partial v}{\partial x}] - EI\frac{\partial^2 \phi}{\partial x^2} + \rho I\ddot{\phi} \right] \delta\phi \, dx \right.$$

$$- \int_0^L \left[ GA\frac{\partial}{\partial x}[-\phi + \frac{\partial v}{\partial x}] - \rho A\ddot{v} + q \right] \delta v \, dx$$

$$\left. + \ [GA\frac{\partial \phi}{\partial x} - M]\delta\phi\Big|_0^L + [GA(-\phi + \frac{\partial v}{\partial x}) - V]\delta v\Big|_0^L \right\} dt = 0 \qquad (12.5)$$

Using the fact that the variations $\delta v$ and $\delta\phi$ can be varied separately and arbitrarily, and that the limits on the integrals are also arbitrary, then we get

$$GA\frac{\partial}{\partial x}\left[\frac{\partial v}{\partial x} - \phi\right] + q \ = \ \rho A\ddot{v}$$

$$EI\frac{\partial^2 \phi}{\partial x^2} + GA\left[\frac{\partial v}{\partial x} - \phi\right] \ = \ \rho I\ddot{\phi} \qquad (12.6)$$

These are the *Timoshenko equations* of motion for a beam. In comparison to the elementary Bernoulli-Euler beam theory, this theory accounts for the shear deformation as well as the rotational inertia. The associated boundary conditions (at each end of the beam) are specified in terms of any pair of conditions selected from the following groups

$$\left\{ v \quad \text{or} \quad V = GA[\frac{\partial v}{\partial x} - \phi] \right\}, \qquad \left\{ \phi \quad \text{or} \quad M = \frac{\partial \phi}{\partial x} \right\} \qquad (12.7)$$

Thus a free boundary is specified as

$$V = 0, \qquad M = 0$$

An inadmissible set of boundary conditions are

$$v = 0, \quad V = 0 \qquad \text{or} \qquad \phi = 0, \quad M = 0$$

If either of these are imposed then there is no guarantee that the remaining term in Equation(12.5) is zero.

In order to account for the truncation error of the expansion $\bar{u} \approx -y\phi$, we can add a correction coefficient $K$ such that $\bar{u} \approx -Ky\phi$. This coefficient can be evaluated many ways; for a rectangular cross-section it usually is taken to be $2/3$ in static problems and $\pi^2/12$ for dynamic problems. It manifests itself in the differential equations and boundary conditions by replacing the shear coefficient $GA$ with $GAK$.

**Example 12.4:**   Recover the elementary beam theory from the Timoshenko equations.

We proceed in two separate steps. First we neglect the rotational inertia by setting the term $\rho I \ddot{\phi}$ to zero, from which we get

$$GA[\frac{\partial v}{\partial x} - \phi] = -EI\frac{\partial^2 \phi}{\partial x^2}$$

Substituting this into the first equation gives

$$EI\frac{\partial^3 \phi}{\partial x^2} + q = \rho A \ddot{v}$$

As a second step, we say that the shear coefficient $GA$ is very large so that

$$-\phi + \frac{\partial v}{\partial x} = 0 \quad \text{or} \quad \phi = \frac{\partial v}{\partial x}$$

When this is substituted into the previous equation we get the elementary theory. It is important to realize that the product

$$V = GA[\frac{\partial v}{\partial x} - \phi]$$

remains constant during this limiting process.

The expansion $\bar{u} = -y\phi$ assures that plane sections remain plane after deformation. The assumption $\phi = \partial v/\partial x$ further requires that the plane section remains normal to the neutral axis. Therefore, $\phi$ can be regarded as the rotation of the cross-section.

# 12.4 Lagrange's Equation

Hamilton's principle provides a complete formulation of a dynamical problem; however, to obtain solutions to problems the Hamilton integral formulation must be converted into one or more differential equations of motion in a manner as shown in the last section. For computer solution, these must be further reduced to equations using discrete unknowns. That is, we introduce some generalized coordinates (or degrees of freedom with the constrained degrees removed). At present, we will not be explicit about which coordinates we are considering, but it may be helpful to think in terms of those we introduced in association with the Ritz method in Chapter 6.

## Systems with Finite Degrees of Freedom

For holonomic constraints, the physical coordinates of the system can be expressed in terms of the generalized coordinates. Thus, if we have a function given by

$$u = u(u_1, u_2, \ldots, u_N)$$

where $u_i$ are generalized coordinates, then the time derivative of this function is

$$\dot{u} = \sum_{j=1}^{N} \frac{\partial u}{\partial u_j} \dot{u}_j$$

Consequently, we see that the kinetic energy is a function of the following form

$$T = T(u_1, u_2, \ldots, u_N; \dot{u}_1, \dot{u}_2, \ldots, \dot{u}_N)$$

The variation in the kinetic energy is given by

$$\delta T = \sum_{j=1}^{N} \frac{\partial T}{\partial u_j} \delta u_j + \sum_{j=1}^{N} \frac{\partial T}{\partial \dot{u}_j} \delta \dot{u}_j$$

We can use integration by parts on the second term to obtain

$$\int_{t_1}^{t_2} \delta T = \int_{t_1}^{t_2} \sum_{j=1}^{N} \left\{ \frac{\partial T}{\partial u_j} - \frac{d}{dt} \frac{\partial T}{\partial \dot{u}_j} \right\} \delta u_j \, dt$$

where we used the fact that the variations at the extreme times are zero.

The total potential of the conservative forces is a function of the form

$$U + V = \Pi = \Pi(u_1, u_2, \ldots, u_N)$$

and its variation is given by

$$\delta \Pi = \sum_{j=1}^{N} \frac{\partial \Pi}{\partial u_j} \delta u_j$$

Additionally, we have that the virtual work of the non-conservative forces is given by

$$\delta W^d = \sum_{j=1}^{N} Q_j \delta u_j$$

The Hamilton's integral can now be reduced to the following form

$$\int_{t_1}^{t_2} \sum_{j=1}^{N} \left\{ -\frac{d}{dt} \left( \frac{\partial T}{\partial \dot{u}_j} \right) + \frac{\partial T}{\partial u_j} - \frac{\partial (U + V)}{\partial u_j} + Q_j \right\} \delta u_j dt = 0$$

Since the virtual displacements $\delta u_j$ are independent and arbitrary, and since the time limits are arbitrary then each integrand is zero. This leads to the so-called *Lagrange's equation of motion*:

$$\frac{d}{dt} \left( \frac{\partial T}{\partial \dot{u}_j} \right) - \frac{\partial T}{\partial u_j} + \frac{\partial (U + V)}{\partial u_j} - Q_j = 0 \qquad \text{for } j = 1, 2, \ldots, N$$

It is apparent from this equation that if the system is not in motion then we recover the principle of stationary potential energy expressed in terms of generalized coordinates.

We emphasize that the transition from Hamilton's principle to Lagrange's equation was possible only by identifying $u_j$ as generalized coordinates. That is, Hamilton's principle holds true for constrained as well as generalized coordinates but Lagrange's equation is valid only for the latter.

## Small Motions

We shall form the equations of motion for small oscillations of a system having $N$ degrees of freedom. That is, we linearize Lagrange's equation; this will allow us to connect the results of the present chapter with those of the earlier chapters.

Consider small motions about an equilibrium position defined by $u_i = 0$ for all $i$. We can perform a Taylor series expansion on the potential function to give

$$U(u_1, u_2, \ldots) = U(0) + \sum_i \frac{\partial U(0)}{\partial u_i} u_i + \frac{1}{2} \sum_i \sum_j \frac{\partial^2 U(0)}{\partial u_i \partial u_j} u_i u_j + \ldots$$

The first term in this expansion is irrelevant and the second term is zero since, by assumption, the origin is an equilibrium position. We therefore have the representation of the potential as

$$U(u_1, u_2, \ldots) \approx \frac{1}{2} \sum_i \sum_j K_{ij} u_i u_j, \qquad K_{ij} \equiv \frac{\partial^2 U(0)}{\partial u_i \partial u_j}$$

We, of course, recognize $[\,K\,]$ as the structural stiffness matrix.

We can do a similar expansion for the kinetic energy. In this case, however, we also assume that the system is linear in such a way that $T$ is a function only of the velocities $\dot{u}_j$. We therefore get

$$T(\dot{u}_1, \dot{u}_2, \ldots) \approx \frac{1}{2} \sum_i \sum_j M_{ij} \dot{u}_i \dot{u}_j, \qquad M_{ij} \equiv \frac{\partial^2 T(0)}{\partial \dot{u}_i \partial \dot{u}_j}$$

We will show shortly that $[\,M\,]$ is the structural mass matrix.

The potential of the conservative forces also has an expansion similar to that for $U$, but we retain only the linear terms in $u_j$ such that

$$V = -\sum_j P_j u_j, \qquad P_j \equiv -\frac{\partial V(0)}{\partial u_j}$$

Finally, assume that the non-conservative forces are of the viscous type such that the virtual work is

$$\delta W^d = Q^d \delta u = -c\dot{u}\,\delta u$$

This suggests the introduction of a function analogous to the potential for the conservative forces

$$Q_j^d = -\frac{\partial D}{\partial \dot{u}_j} \qquad \text{where} \qquad D = D(\dot{u}_1, \dot{u}_2, \ldots, \dot{u}_N)$$

For small deformations we can write

$$D(\dot{u}_1, \dot{u}_2, \ldots, \dot{u}_N) \approx \frac{1}{2} \sum_i \sum_j C_{ij} \dot{u}_i \dot{u}_j, \qquad C_{ij} \equiv \frac{\partial^2 D(0)}{\partial \dot{u}_i \partial \dot{u}_j}$$

The function $D$ is called the *Rayleigh dissipation function.*

If we now substitute these forms for $U$, $V$, $T$ and $D$ into the Lagrange's equation we get

$$\sum_j \{K_{ij}u_j + C_{ij}\dot{u}_j + M_{ij}\ddot{u}_j\} = P_i \qquad i = 1, 2, \ldots, N$$

This is put in the familiar matrix form as

$$[K]\{u\} + [C]\{\dot{u}\} + [M]\{\ddot{u}\} = \{P\}$$

In comparison to the last few chapters, we have the meaning of $[M]$ and $[C]$ as the (generalized) structural mass and damping matrices, respectively. We now have a discrete representation of the equations of motion. As yet, we have not said how the actual coefficients can be obtained or the actual meaning of the generalized coordinates. In fact, there are many ways of establishing both; in the next few sections we will illustrate the Ritz approach as was done in Chapter 6.

**Example 12.5:**  Consider a concentrated mass $m$ attached to a massless pendulum of length $L$ as shown in Figure 12.1. Illustrate the use of Lagrange's equation in obtaining the equations of motion for the mass.

The kinetic energy, strain energy, and potential of the applied force are given by, respectively,

$$T = \tfrac{1}{2}m[\dot{u}^2 + \dot{v}^2], \qquad U = 0, \qquad V = -(-mgv) = mgv$$

where $g$ is the gravitational constant and the gravitational force is taken to act opposite the displacement $v$. In order to use Lagrange's equation, we must utilize generalized coordinates. In the present context, that means we cannot use $u$ and $v$ as independent coordinates since they are subjected to the following constraint relation

$$u^2 + v^2 = L^2 \qquad \text{or} \qquad u\dot{u} + v\dot{v} = 0$$

We could eliminate either $u$ or $v$ and then proceed. Instead, we will introduce an entirely different coordinate; namely, the rotation angle $\theta$ of the pendulum. The position of the mass is given by

$$u = L\sin\theta, \qquad v = L - L\cos\theta$$

The velocities are

$$\dot{u} = L\cos\theta\,\dot{\theta}, \qquad \dot{v} = L\sin\theta\,\dot{\theta}$$

The kinetic energy and potential of the applied force are now given by, respectively,

$$T = \tfrac{1}{2}mL^2\dot{\theta}^2, \qquad V = mg(L - L\cos\theta)$$

The partial derivatives required in the Lagrange equation are

$$\frac{\partial T}{\partial \dot{\theta}} = mL^2\dot{\theta}, \qquad \frac{\partial T}{\partial \theta} = 0, \qquad \frac{\partial V}{\partial \theta} = mgL\sin\theta$$

The equation of motion is therefore determined to be

$$mL^2\ddot{\theta} + mg\,L\sin\theta = 0$$

Note that this is a non-linear equation because of the occurrence of $\sin\theta$. If we restrict the motion such that $\theta$ is small then we get

$$mL^2\ddot{\theta} + mg\,L\theta = 0 \qquad \text{or} \qquad \ddot{\theta} + \frac{g}{L}\theta = 0$$

This is the familiar second order system governing simple harmonic motion that we have been dealing with in the last few chapters.

## 12.5   The Ritz Method

The previous section has shown that we can reduce all linear dynamic problems to a system of second order ordinary differential equations in time. The coefficients are obtained from the total energies of the system by

$$K_{ij} \equiv \frac{\partial^2 U}{\partial u_i \partial u_j}, \qquad U = \tfrac{1}{2}\int_V \{\sigma\}^T \{\epsilon\}\, d\!V \tag{12.8}$$

$$M_{ij} \equiv \frac{\partial^2 T}{\partial \dot{u}_i \partial \dot{u}_j}, \qquad T = \tfrac{1}{2}\int_V \rho\{\dot{u}\}^T \{\dot{u}\}\, d\!V \tag{12.9}$$

$$C_{ij} \equiv \frac{\partial^2 D}{\partial \dot{u}_i \partial \dot{u}_j}, \qquad D = \tfrac{1}{2}\int_V c\{\dot{u}\}^T \{\dot{u}\}\, d\!V \tag{12.10}$$

To complete the formulation, we now use the Ritz approach to establish explicit forms for the energies in terms of the generalized coordinates.

Recall from Chapter 6 that we assume an expansion for the displacements in the form, for example,

$$u(x,y,z) = \sum_i^N a_i g_i(x,y,z)$$

where the $g_i$ are linearly independent trial functions and the $a_i$ are generalized coordinates. We do the same here except that the $a_i$ are taken as functions of time. To ensure that the characteristics of the system are taken into consideration, the trial functions must satisfy the essential (geometric) boundary conditions of the problem.

**Example 12.6:**  Consider the system shown in Figure 12.5, consisting of a concentrated mass $m_c$ attached to the end of a uniform rod. Determine the equations of motion of the system. Neglect damping.

Because the rod possesses a distributed mass, the system has an infinite number of degrees of freedom. However, if we make the assumption that the displacement of the system is given by

$$u(x,t) = a(t)f(x)$$

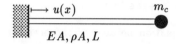

**Figure 12.5**: Rod with concentrated mass.

where $a(t)$ is an unknown generalized coordinate, and $f(x)$ is a shape function (yet to be determined), the system reduces to a single degree of freedom system. That is, once $a(t)$ is determined for a chosen $f(x)$, the displaced shape of the entire system is known over all time.

The total strain energy is given by

$$U = \tfrac{1}{2} \int_0^L EA \left(\frac{\partial u}{\partial x}\right)^2 dx = \tfrac{1}{2} a^2 \int_0^L EA \left(f'\right)^2 dx$$

The corresponding stiffness is given by

$$K = \frac{\partial^2 U}{\partial a^2} = \int_0^L EA \left(f'\right)^2 dx$$

The total kinetic energy of the rod and the attached mass is readily found to be

$$T = \int_0^L \tfrac{1}{2} \rho A \dot{u}(x,t)^2 dx + \tfrac{1}{2} m_c \dot{u}(L,t)^2 = \tfrac{1}{2} \dot{a}^2 \int_0^L \rho A f(x)^2 dx + \tfrac{1}{2} m_c \dot{a}^2 f(L)^2$$

Therefore, the effective mass of this single degree of freedom system is

$$M = \frac{\partial^2 U}{\partial \dot{a}^2} = \int_0^L \rho A f^2 dx + m_c f(L)^2$$

The equations of motion are

$$K\,a + M\,\ddot{a} = 0$$

It remains now to determine some explicit forms for $f(x)$. It is important to realize that there is nothing unique about $f(x)$. As a first choice, assume

$$a(t)f(x) = a(t)[b_0 + b_1 x]$$

where $b_0$ and $b_1$ are constants. The essential boundary condition requires $u(0,t) = 0$, hence $b_0 = 0$. Although not necessary, let us also require that $f(x)$ be non-dimensional by setting $b_1 = 1/L$. We now have

$$f(x) = x/L, \qquad f'(x) = 1/L$$

As a result, we determine the effective stiffness and mass for a uniform rod to be

$$K = \frac{EA}{L}, \qquad M = \frac{\rho AL}{3} + m_c$$

Assuming simple harmonic motion we get an estimate for the fundamental frequency as

$$\omega = \sqrt{\frac{EA/L}{m_c + \rho AL/3}}$$

It is interesting to note that this is the result obtained when the problem is modeled with a single rod element.

As a alternative trial function, choose

$$a(t)f(x) = a(t)\sin(\pi x/2L)$$

This satisfies the essential boundary condition and gives the stiffness and mass as

$$K = \frac{EA}{L}\frac{\pi^2}{8}, \qquad M = \frac{\rho AL}{2} + m_c$$

A second estimate for the fundamental frequency is therefore found to be

$$\omega = \frac{\pi}{2L}\sqrt{\frac{EA/L}{\rho A + 2m_c/L}}$$

Note that if the attached mass $m_c$ is zero, this frequency turns out to be exactly equal to the true frequency of vibration of a uniform rod. This is because the sine function $\sin(\pi x/2L)$ happens to be the true vibration shape for this case. On the other hand, if the mass of the rod is negligible then our first estimation of the frequency is the correct value.

On comparing the two masses and stiffness, we see that in the second case both the mass and stiffness have increased.

**Example 12.7:**   Consider the cantilevered beam shown in Figure 12.6 with a spring of stiffness $\alpha$ attached half-way along its length. Determine the mass and stiffness associated with this structure for a vibration in the vertical direction only. Neglect damping.

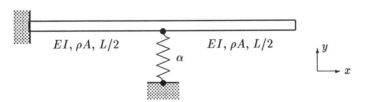

EI, ρA, L/2    α    EI, ρA, L/2    y    x

**Figure 12.6:** Beam with spring.

In this example, the essential boundary conditions at $x = 0$ are

$$v(0,t) = 0, \qquad \phi(0,t) = 0$$

Choose a Ritz expansion as

$$v(x,t) = a_0(t) + a_1(t)[\frac{x}{L}] + a_2(t)[\frac{x}{L}]^2$$

This second order displacement function permits the beam to deform into a parabola while it vibrates. The essential boundary conditions require that $a_0 = 0$, $a_1 = 0$. The total strain energy is given by

$$U = \frac{1}{2}\int_0^L EI(\frac{\partial^2 v}{\partial x^2})^2 dx + \frac{1}{2}\alpha v(L/2)^2 = \frac{1}{2}a_2^2 \int_0^L EI(\frac{2}{L^2})^2 dx + \frac{1}{2}\alpha a_2^2(\frac{1}{2})^4 = \frac{1}{2}a_2^2(\frac{4EI}{L^3} + \frac{\alpha}{16})$$

The associated stiffness is given by

$$K = \frac{\partial^2 U}{\partial a_2^2} = \frac{4EI}{L^3} + \frac{\alpha}{16}$$

The total kinetic energy of the beam is readily found to be

$$T = \int_0^L \frac{1}{2}\rho A \dot{v}^2 dx = \frac{1}{2}\dot{a}_2^2 \int_0^L \rho A(\frac{x}{L})^4 dx = \frac{1}{2}\dot{a}_2^2 \frac{\rho AL}{5}$$

Therefore the effective mass of this single degree of freedom system is

$$M = \frac{\partial^2 U}{\partial \dot{a}_2^2} = \frac{\rho AL}{5}$$

The equation of motion is

$$(\frac{4EI}{L^3} + \frac{\alpha}{16}) a + (\frac{\rho AL}{5}) \ddot{a} = 0$$

A more accurate model of the deformed vibration shape has a third order displacement term. That is, choose the Ritz expansion as

$$v(x,t) = a_0(t) + a_1(t)[\frac{x}{L}] + a_2(t)[\frac{x}{L}]^2 + a_3(t)[\frac{x}{L}]^3$$

The essential boundary conditions require that $a_0 = 0$, $a_1 = 0$, hence write the expansion as

$$v(x,t) = a_1(t)f_1(x) + a_2(t)f_2(x), \qquad f_1(x) \equiv [\frac{x}{L}]^2, \quad f_2(x) \equiv [\frac{x}{L}]^3$$

The total kinetic energy of the beam is

$$T = \int_0^L \frac{1}{2}\rho A \dot{v}^2 dx = \frac{1}{2}\int_0^L \rho A[\dot{a}_1 f_1 + \dot{a}_2 f_2]^2 dx$$

Therefore the individual masses are

$$M_{11} = \frac{\partial^2 T}{\partial \dot{a}_1^2} = \int_0^L \rho A f_1 f_1 dx = \int \rho A(\frac{x}{L})^4 = \frac{\rho AL}{5}$$

Another mass is given by

$$M_{12} = \frac{\partial^2 T}{\partial \dot{a}_1 \partial \dot{a}_2} = \int_0^L \rho A f_1 f_2 dx = \int_0^L \rho A (\frac{x}{L})^5 = \frac{\rho AL}{6}$$

Working similarly on the other terms leads to

$$[M] = \frac{\rho AL}{210} \begin{bmatrix} 42 & 35 \\ 35 & 30 \end{bmatrix}$$

The total strain energy is given by

$$U = \frac{1}{2} \int_0^L EI[a_1 f_1'' + a_2 f_2'']^2 dx + \frac{1}{2}\alpha[a_1 f_1(L/2) + a_2 f_2(L/2)]^2$$

The individual stiffnesses are

$$K_{11} = \frac{\partial^2 U}{\partial a_1^2} = \int_0^L EI f_1'' f_1'' dx + \alpha f_1(L/2) f_1(L/2) = \frac{4EI}{L^3} + \frac{\alpha}{16}$$

Similar expressions for the other terms leads to

$$[K] = \frac{EI}{L^3} \begin{bmatrix} 4 & 6 \\ 6 & 12 \end{bmatrix} + \frac{\alpha}{64} \begin{bmatrix} 4 & 2 \\ 2 & 1 \end{bmatrix}$$

Note that the $K_{11}$ stiffness and $M_{11}$ mass are the same as already obtained for the single degree of freedom case.

## Finite Element Analysis

As shown in Chapter 6, we can view the finite element method as an application of the Ritz method where instead of the trial functions spanning over the complete domain, the individual functions span only over subdomains (the finite elements) of the complete region. We can apply the same reasoning to the dynamic problem; indeed, the only new terms we need to consider are the inertia and the damping.

In order that a finite element solution be a Ritz analysis, the essential boundary conditions must be completely satisfied by the finite element nodal point displacements. Additionally, in the selection of the displacement functions, no special attention need be given to the natural boundary conditions because these conditions are imposed with the load vector and are satisfied approximately in the Ritz solution. The accuracy with which these natural boundary conditions are satisfied depends on the specific trial functions employed, and on the number of elements used to model the problem.

The approach to the application of the Ritz method to developing the element mass relation is to view the element itself as a structure with essential boundary conditions at the nodal points. An application of Hamilton's principle then gives the approximate dynamic equilibrium conditions that the nodal loads satisfy. We have already done that in general terms and were lead to Lagrange's equation. Thus, the only new aspect that we need consider here is obtaining the mass matrix from the kinetic energy term.

Let the distributed displacement fields $\{\mathcal{U}(x, y, z, t)\}$ be represented by expressions of the form

$$u(x, y, z, t) = \lfloor g(x, y, z) \rfloor \{u(t)\}$$

where $\lfloor g(x, y, z) \rfloor$ is a set of known admissible functions of the coordinates, and $\{u\}$ is a set of generalized coordinates. Let us insert this into the expression for the kinetic energy for a general elastic body

$$T = \tfrac{1}{2} \int_V \{\dot{\mathcal{U}}\}^T \{\dot{\mathcal{U}}\} dV = \tfrac{1}{2} \int_V \rho \{\dot{u}\}^T \lfloor g \rfloor^T \lfloor g \rfloor \{\dot{u}\} dV = \tfrac{1}{2} \{\dot{u}\}^T [m] \{\dot{u}\}$$

where

$$[m] \equiv \int_V \rho \lfloor g \rfloor^T \lfloor g \rfloor dV$$

Since, for a given structure, everything within this integral is a known function of the coordinates, the indicated integration can be carried out explicitly.

### General Rod and Beam Masses

We will begin by recovering the element mass matrix for the simple rod used in the earlier chapters. Recall that the displacements for the rod segment can be written in terms of the nodal values as

$$u(x) = (1 - \frac{x}{L}) u_1 + (\frac{x}{L}) u_2 \equiv f_1(x) u_1 + f_2(x) u_2$$

This displacement function satisfies the essential boundary conditions in that they give the displacements at both nodes. We therefore identify $f_1(x)$ and $f_2(x)$ as the Ritz functions. The velocity of each particle in the element is

$$\dot{u} = \frac{x}{L} (\dot{u}_2 - \dot{u}_1)$$

giving the kinetic energy of the element as

$$T = \tfrac{1}{2} \int_0^L \rho A (\dot{u})^2 \, dx = \frac{\rho A L}{6} (\dot{u}_1^2 + \dot{u}_2^2 + \dot{u}_1 \dot{u}_2)$$

We can now obtain the mass matrix from the relation

$$m_{ij} \equiv \frac{\partial^2 T}{\partial \dot{u}_i \partial \dot{u}_j}$$

We need to differentiate with respect to the parameters $\dot{u}_1$ and $\dot{u}_2$. That is,

$$\frac{\partial T}{\partial \dot{u}_1} = \frac{\rho A L}{6} (2\dot{u}_1 + \dot{u}_2), \qquad \frac{\partial T}{\partial \dot{u}_2} = \frac{\rho A L}{6} (2\dot{u}_2 + \dot{u}_1)$$

Further differentiation gives

$$[m] = \frac{\rho A L}{6} \begin{bmatrix} 2 & 1 \\ 1 & 2 \end{bmatrix}$$

We recognize this as the consistent mass matrix obtained by expansion in Chapter 9. The general expression can also be used to derive the general form of the rod mass. From the shape functions we get that

$$[g(x)] = \lfloor f_1(x), f_2(x) \rfloor$$

The element masses are therefore given by

$$m_{ij} = \int_0^L \rho A f_i(x) f_j(x) \, dx$$

In particular, we have such relations as

$$m_{11} = \int_0^L \rho A f_1(x) f_1(x) \, dx, \qquad m_{12} = \int_0^L \rho A f_1(x) f_2(x) \, dx$$

and so on. Evaluating these integrals for constant $\rho A$ gives the mass matrix already obtained above, but in this form they can be used when the density or cross-section varies along the length.

The procedure for determining the element mass matrix for beams proceeds as for the rod. Recall that the deflection can be represented in terms of the nodal values as

$$
\begin{aligned}
v(x) &= \left[ 1 - 3(\frac{x}{L})^2 + 2(\frac{x}{L})^3 \right] v_1 + (\frac{x}{L}) \left[ 1 - 2(\frac{x}{L}) + (\frac{x}{L})^2 \right] L\phi_1 \\
&\quad + (\frac{x}{L})^2 \left[ 3 - 2(\frac{x}{L}) \right] v_2 + (\frac{x}{L})^2 \left[ -1 + (\frac{x}{L}) \right] L\phi_2 \\
&= g_1(x) v_1 + g_2(x) L\phi_1 + g_3(x) v_2 + g_4(x) L\phi_2
\end{aligned}
$$

This satisfies the essential boundary conditions. The kinetic energy in the beam element is

$$T = \frac{1}{2} \int_0^L \rho A [\dot{v}]^2 \, dx = \frac{1}{2} \int_0^L \rho A [g_1(x)\dot{v}_1 + g_2(x)L\dot{\phi}_1 + g_3(x)\dot{v}_2 + g_4(x)L\dot{\phi}_2]^2 dx$$

The entities in the mass matrix can be obtained by differentiating this expression with respect to the nodal velocities to give

$$m_{ij} = \frac{\partial^2 T}{\partial \dot{u}_i \partial \dot{u}_j} = \int_0^L \rho A g_i(x) g_j(x) \, dx$$

Carrying out these integrations under the condition of constant $\rho A$ gives the consistent beam mass matrix as already obtained. In the present form they are applicable to non-uniform sections. Note that it is the symmetry of the terms $g_i(x)g_j(x)$ that insures the symmetry of the mass matrices.

**Element Damping Matrix**

Let the rod and beam have a linear damping term $c\dot{u}$ that occurs in the differential equation. We introduce the Rayleigh dissipation function given by

$$D = \tfrac{1}{2} \int_{\forall} c\dot{u}^2 d\forall$$

Hence, by analogy with the kinetic energy term we get

$$c_{ij} = \frac{\partial^2 D}{\partial \dot{u}_i \partial \dot{u}_j} = \int_0^L cAg_i(x)g_j(x)\, dx$$

For uniform damping $cA$ is constant and we see that the matrix is the same as for the mass matrix. That is, we obtain proportional damping.

## 12.6   Ritz Method Applied to Discrete Systems

We saw how we can reduce an infinite degree of freedom system to a finite system by use of the Ritz approach combined with a stationary principle. It is reasonable therefore to expect that we can reduce a finite system to an even smaller one by the same approach. This is a very important consideration because the solution cost for large dynamic problems can be quite substantial.

We use the Ritz procedure to represent the displacement shape of a discrete multiple degree of freedom system by the superposition of a few appropriately selected shape functions. That is, the displacements of the system are expressed as a superposition of several different independent shape vectors, known as Ritz vectors, each weighted by its own generalized coordinate. Thus

$$\{u\}(t) = a_1(t)\{\psi\}_1 + a_2(t)\{\psi\}_2 + \cdots + a_M(t)\{\psi\}_M = [\,\Psi\,]\{a\}$$

where $[\,\Psi\,]$ is an $[N \times M]$ matrix of Ritz shape function vectors and $\{a\}$ is the vector of $M$ generalized coordinates. We note that instead of being continuous function of spatial coordinates, the $\{\psi\}_m$ are vectors of size $N$. The energy functions are given by

$$
\begin{aligned}
U &= \tfrac{1}{2}\{u\}^T[\,K\,]\{u\} = \tfrac{1}{2}\{a\}^T[\,\Psi\,]^T[\,K\,][\,\Psi\,]\{a\} \\
T &= \tfrac{1}{2}\{\dot{u}\}^T[\,M\,]\{\dot{u}\} = \tfrac{1}{2}\{\dot{a}\}^T[\,\Psi\,]^T[\,M\,][\,\Psi\,]\{\dot{a}\} \\
D &= \tfrac{1}{2}\{\dot{u}\}^T[\,C\,]\{\dot{u}\} = \tfrac{1}{2}\{\dot{a}\}^T[\,\Psi\,]^T[\,C\,][\,\Psi\,]\{\dot{a}\}
\end{aligned}
$$

which leads to reduced matrices defined as

$$[\,\tilde{K}\,] \equiv [\,\Psi\,]^T[\,K\,][\,\Psi\,], \qquad [\,\tilde{M}\,] \equiv [\,\Psi\,]^T[\,M\,][\,\Psi\,], \qquad [\,\tilde{C}\,] \equiv [\,\Psi\,]^T[\,C\,][\,\Psi\,]$$

The load term becomes

$$V = -\{u\}^T\{P\} = -\{a\}^T[\,\Psi\,]^T\{P\}$$

Hence, the reduced load is

$$\{\tilde{P}\} \equiv [\, \Psi \,]^T \{P\}$$

The close similarity between these relations and those of modal analysis (if $[\, \Psi \,]$ is replaced by $[\, \Phi \,]$) is noted. However, these reduced matrices are not diagonal, in general.

The usefulness of the Ritz procedure is most evident when only a few shape function vectors are adequate to represent the response. In such a case, $M$ is much smaller than $N$, and the transformed property matrices are of size $[M \times M]$. The transformed equations of motion are

$$[\, \tilde{K} \,]\{a\} + [\, \tilde{C} \,]\{\dot{a}\} + [\, \tilde{M} \,]\{\ddot{a}\} = \{\tilde{P}\}$$

These equations are of order $M$ and because $M \ll N$, the reduced problem is of a significantly smaller size than the original. However, keep in mind that the reduced matrices are not necessarily diagonal (unlike the modal equations), or even banded as in the case of the original equations.

One of the great difficulties in a Ritz analysis is choosing the Ritz vectors. This is taken up in greater detail in the next chapter when we consider the subspace iteration scheme, but now we wish to consider a few basic aspects.

**Example 12.8:**   The three-story building frame shown in Figure 12.7 is modeled as a shear type building; that is, all the mass is lumped at the floor levels and the walls are assumed axially inextensible. Estimate the fundamental frequency of the building frame.

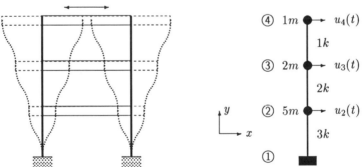

**Figure 12.7:** Three story structure.

The shear model of the building is, in fact, a three degree of freedom model as shown in the Figure. The stiffness matrix and mass matrix of the frame are

$$[\, K \,] = k \begin{bmatrix} 5 & -2 & 0 \\ -2 & 3 & -1 \\ 0 & -1 & 1 \end{bmatrix}, \qquad \lceil M \rfloor = m \begin{bmatrix} 5 & & \\ & 2 & \\ & & 1 \end{bmatrix}$$

The exact eigenvalues are easily determined to be

$$\omega^2 = \{0.269, 1.000, 2.231\} \times \frac{k}{m} \quad \text{or} \quad \omega = \{0.519, 1.000, 1.494\} \times \sqrt{\frac{k}{m}}$$

The corresponding mode shapes are

$$[\, \Phi \,] = \begin{bmatrix} .236 & .333 & .182 \\ .432 & .000 & -.560 \\ .590 & .667 & .455 \end{bmatrix} \quad \text{or} \quad [\, \Phi \,] = \begin{bmatrix} 1.00 & 1.00 & 1.00 \\ 1.83 & 0.00 & -3.08 \\ 2.50 & 2.00 & 2.50 \end{bmatrix}$$

To reduce the model to a single degree of freedom system, we assume a suitable displacement shape $\{\psi\}$, where in this case $\{\psi\}$ is a vector with three elements. That is,

$$\{u\}(t) = a(t)\{\psi\}$$

in which $a(t)$ is the unknown generalized coordinate. The strain energy of the system is

$$U = \tfrac{1}{2}\{u\}^T[\, K \,]\{u\} = \tfrac{1}{2}a(t)^2\{\psi\}^T[\, K \,]\{\psi\}$$

and the kinetic energy is

$$T = \tfrac{1}{2}\{\dot{u}\}^T[\, M \,]\{\dot{u}\} = \tfrac{1}{2}\dot{a}(t)^2\{\psi\}^T[\, M \,]\{\psi\}$$

Let us now assume that the vibration shape is

$$\{\psi\} = \left\{ \begin{array}{c} 1 \\ 2 \\ 3 \end{array} \right\}$$

This is a reasonable assumption for the first mode of cantilever-type structures. The reduced stiffness and mass are computed to be

$$\tilde{K} = \{\psi\}^T[\, K \,]\{\psi\} = 6k, \qquad \tilde{M} = \{\psi\}^T[\, M \,]\{\psi\} = 22m$$

Assuming that the motion is simple harmonic, we get the following value of the fundamental frequency

$$\omega^2 = \frac{\tilde{K}}{\tilde{M}} = \frac{6k}{22m}, \qquad \omega = 0.522\sqrt{\frac{k}{m}}$$

This is very close to the actual value of 0.519.

Often the shape of the first mode is not obvious; in this case it is usual to assume the guess as a fully populated matrix. That is, assume

$$\{\psi\} = \left\{ \begin{array}{c} 1 \\ 1 \\ 1 \end{array} \right\}$$

This gives the reduced stiffness and mass as

$$\tilde{K} = 3k, \qquad \tilde{M} = 8m$$

These are significantly different from the previous values but the frequency is determined to be

$$\omega^2 = \frac{\tilde{K}}{\tilde{M}} = \frac{3k}{8m}, \qquad \omega = 0.612\sqrt{\frac{k}{m}}$$

This value is higher than the previous estimate and the exact value.

As a final example, suppose a random vector of the form

$$\{\psi\} = \left\{ \begin{array}{c} 1 \\ -1 \\ 1 \end{array} \right\}$$

is assumed. This gives the reduced stiffness and mass as

$$\tilde{K} = 15k, \qquad \tilde{M} = 8m$$

and the frequency is determined to be

$$\omega^2 = \frac{\tilde{K}}{\tilde{M}} = \frac{15k}{8m}, \qquad \omega = 1.37\sqrt{\frac{k}{m}}$$

This value of frequency is actually very close to the third resonance value of 1.49.

It is apparent from these examples that the choice of Ritz vector can have a significant effect on the value of the results.

**Example 12.9:**   For the same three story building frame of Figure 12.7, reduce the modeling to a two degree of freedom system.

Although we know the first choice of the last example was good, we will use the second and third choices instead to illustrate what happens as the number of degrees of freedom are increased. That is, we choose for the Ritz vectors

$$\{\psi\}_1 = \left\{ \begin{array}{c} 1 \\ 1 \\ 1 \end{array} \right\}, \qquad \{\psi\}_2 = \left\{ \begin{array}{c} 1 \\ -1 \\ 1 \end{array} \right\} \qquad \text{or} \qquad [\,\Psi\,] = \left[ \begin{array}{cc} 1 & 1 \\ 1 & -1 \\ 1 & 1 \end{array} \right]$$

The transformed stiffness and mass matrices are now obtained as

$$[\,\tilde{K}\,] = k \left[ \begin{array}{cc} 3 & 3 \\ 3 & 15 \end{array} \right], \qquad [\,\tilde{M}\,] = m \left[ \begin{array}{cc} 8 & 4 \\ 4 & 8 \end{array} \right]$$

This system admits two resonances which are

$$\omega^2 = \{0.349, 2.151\} \times \frac{k}{m} \qquad \text{or} \qquad \omega = \{0.590, 1.467\} \times \sqrt{\frac{k}{m}}$$

We see that we get an improved value for the first mode although still not as good as when $\{\psi\}_1 = \{1, 2, 3\}^T$ is used. Notice that the estimate of the second frequency is now very close to that of the third exact frequency.

## 12.7   Rayleigh Quotient

We conclude this chapter by discussing a very interesting concept in the theory of
vibrations, namely, Rayleigh's quotient. This also has practical value as a means
of estimating the fundamental frequency of a system and as a tool in speeding up
convergence to the solution of the eigenvalue problem in matrix iteration.

The free vibration of a system is given by

$$[K]\{u\} - \lambda[M]\{u\} = 0, \qquad \lambda = \omega^2$$

If the system size is $N$, then there are $N$ eigenpairs $(\lambda_m, \{\phi\}_m)$. For a particular pair,
premultiply both sides by $\{\phi\}^T$, divide the resulting equation through by the mass
term to obtain

$$\lambda = \omega^2 = \frac{\{\phi\}^T[K]\{\phi\}}{\{\phi\}^T[M]\{\phi\}}$$

Suppose we do not know the eigenvector $\{\phi\}$, but only a guess for it $\{\psi\}$, we can still
form the above ratio. That is,

$$\rho(\{\psi\}) = \frac{\{\psi\}^T[K]\{\psi\}}{\{\psi\}^T[M]\{\psi\}}$$

where $\rho(\{\psi\})$ is a scalar whose value depends not only on the matrices $[M]$ and $[K]$
but also on the guess $\{\psi\}$. Whereas matrices $[M]$ and $[K]$ reflect the system characteristics, the vector $\{\psi\}$ is arbitrary, so that for a given system $\rho(\{\psi\})$ depends on
the vector $\{\psi\}$ alone. The scalar $\rho(\{\psi\})$ is called *Rayleigh's quotient* and it possesses
some very interesting properties. Clearly, if $\{\psi\}$ coincides with one of the system
eigenvectors, then the quotient reduces to the associated eigenvalue. Moreover, the
quotient has stationary values in the neighborhood of the system eigenvectors. To
show this, let us use the expansion theorem and represent the arbitrary vector $\{\psi\}$
as a linear combination of the system eigenvectors in the form

$$\{\psi\} = \sum_{m=1}^{N} c_m\{\phi\}_m = [\phi]\{c\}$$

where $[\phi]$ is the modal matrix, and $\{c\}$ a vector with its elements consisting of the
coefficients $c_m$. Let the eigenvectors be normalized so that the modal matrix satisfies
the orthogonality conditions

$$[\phi]^T[K][\phi] = \lceil \Lambda \rfloor, \qquad [\phi]^T[M][\phi] = \lceil I \rfloor$$

Introducing the expansion into the Rayleigh quotient, we obtain

$$\rho(\{\psi\}) = \frac{\{c\}^T[\phi]^T[K][\phi]\{c\}}{\{c\}^T[\phi]^T[M][\phi]\{c\}} = \frac{\{c\}^T\lceil \Lambda \rfloor\{c\}}{\{c\}^T\lceil I \rfloor\{c\}} = \frac{\sum_{i=1}^{n}\lambda_i c_i^2}{\sum_{i=1}^{n} c_i^2}$$

Suppose the guess is close to a mode $\{\phi\}_m$, then all the coefficients except $c_m$ are of order $\epsilon$. Thus the Rayleigh quotient becomes

$$\rho(\{\psi\}) \approx \frac{\lambda_m c_m^2 + O(\epsilon^2)}{c_m^2 + O(\epsilon^2)} \approx \lambda_m + O(\epsilon^2)$$

Hence, if the trial vector $\{\psi\}$ differs from the eigenvector $\{\phi\}_m$ by a small quantity of first order, then the quotient $\rho(\{\psi\})$ differs from the eigenvalue $\lambda_m$ by a small quantity of second order. The implication is that Rayleigh's quotient has a stationary value in the neighborhood of an eigenvector, where the stationary value is the corresponding eigenvalue.

In the neighborhood of the fundamental mode, Rayleigh's quotient actually has a minimum value. Hence, Rayleigh's quotient is never lower than the first eigenvalue, and the minimum value it can take is that of the first eigenvalue itself. We therefore conclude that a practical application of Rayleigh's quotient is to obtain estimates for the fundamental frequency of the system. To this end, a very good estimate can be obtained by using as a trial vector $\{\psi\}$ the vector of static displacements obtained by subjecting the masses to forces proportional to their weights.

**Example 12.10:**   The stiffness matrix and the mass matrix of a system are given by

$$[K] = \begin{bmatrix} 3 & -1 \\ -1 & 1 \end{bmatrix}, \qquad \lceil M \rfloor = \begin{bmatrix} 2 & 0 \\ 0 & 1 \end{bmatrix}$$

Obtain the fundamental frequency of the system by minimizing the Rayleigh quotient. The exact solution is given by the eigenpairs

$$(\tfrac{1}{2}, \left\{ \begin{matrix} 1 \\ 2 \end{matrix} \right\}) \qquad (2, \left\{ \begin{matrix} 1 \\ -1 \end{matrix} \right\})$$

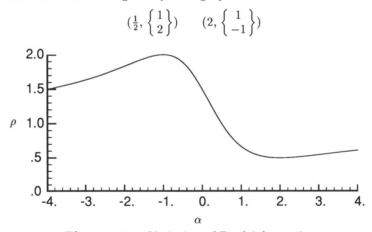

**Figure 12.8:** Variation of Rayleigh quotient.

To illustrate the use of the Rayleigh quotient, we will assume the displacement vector as

$$\{\psi\} = \left\{ \begin{matrix} 1 \\ \alpha \end{matrix} \right\}$$

where $\alpha$ is a parameter that is yet to be determined. We now form the Rayleigh quotient

$$\rho = \frac{\{\psi\}^T[\,K\,]\{\psi\}}{\{\psi\}^T[\,M\,]\{\psi\}} = \frac{(3 - 2\alpha + \alpha^2)}{(2 + \alpha^2)}$$

The complete behavior of the quotient over a range of values of $\alpha$ is shown in Figure 12.8. To obtain the value of $\alpha$ for which $\rho$ will be a stationary value, we set $d\rho/d\alpha = 0$. This gives

$$\frac{d\rho}{d\alpha} = \frac{-4 + 2\alpha^2 - 2\alpha}{(2 + \alpha^2)^2} = 0 \qquad \text{or} \qquad \alpha^2 - \alpha - 2 = 0$$

The solution of this equation gives $\alpha = 2$ or $\alpha = -1$. One of these two values will give a minimum for $\rho$, while the other will give a maximum. For $\alpha = 2$, we get $\rho = k/2m$, while for $\alpha = -1, \rho = 2k/m$. Thus $\alpha = 2$ minimizes $\rho$ and the fundamental frequency $\omega_1$ is given by

$$\omega_1 = \sqrt{\frac{1}{2}}, \qquad \{\psi\}_1 = \left\{ \begin{matrix} 1 \\ 2 \end{matrix} \right\}$$

The maximum value of $\rho$ leads to the second frequency

$$\omega_2 = \sqrt{\frac{2}{1}} = \sqrt{2}, \qquad \{\psi\}_2 = \left\{ \begin{matrix} 1 \\ -1 \end{matrix} \right\}$$

These, of course, are the exact values.

## Problems

**12.1** The operators $\delta$, $d/dt$, and $\partial/\partial x$ are commutative, that is,

$$\delta \dot{u} = \delta\left(\frac{du}{dt}\right) = \frac{d}{dt}(\delta u), \qquad \delta\frac{\partial u}{\partial x} = \frac{\partial}{\partial x}(\delta u)$$

Generate the following identities

$$\delta \ddot{u} = \frac{d^2}{dt^2}(\delta u), \qquad \delta\frac{\partial^2 u}{\partial x^2} = \frac{\partial^2}{\partial x^2}(\delta u)$$

**12.2** Use the Love theory of rods to show that the resonant frequencies for a fixed-free rod are

$$\omega = \frac{(n + \frac{1}{2})\pi}{L}\sqrt{\frac{EA}{\rho A}} \Big/ \sqrt{1 + \left(\frac{n\pi}{L}\right)^2 \frac{\rho J}{\rho A}}$$

**12.3** Show that the Timoshenko equations can be rewritten as (with $q(x) = 0$)

$$EA\frac{\partial^4 v}{\partial x^4} - \left(\frac{\rho A}{GA}EI + \rho I\right)\frac{\partial^4 v}{\partial x^2 \partial t^2} + \rho A\frac{\partial^2 v}{\partial t^2} + \frac{\rho A}{GA}\rho I\frac{\partial^4 v}{\partial t^4} = 0$$

[Reference [43], pp. 205]

**12.4** Assume a solution of the Timoshenko equations in the form

$$v(x,t) = Ce^{-ikx}e^{i\omega t}$$

Show that for low frequencies the wavenumber $k$ is the same as for the Bernoulli-Euler beam. That is

$$k \approx \pm\sqrt{\omega}[\frac{\rho A}{EI}]^{\frac{1}{4}} \, , \pm i\sqrt{\omega}[\frac{\rho A}{EI}]^{\frac{1}{4}}$$

[Reference [14], pp. 121]

**12.5** The stiffness and mass matrices for a system are given by

$$[K] = \begin{bmatrix} 2 & -1 & 0 \\ -1 & 4 & -1 \\ 0 & -1 & 2 \end{bmatrix}, \qquad \lceil M \rfloor = \frac{1}{2}\begin{bmatrix} 1 & & \\ & 2 & \\ & & 1 \end{bmatrix}$$

The exact eigenvalues are $\lambda_1 = 2, \lambda_2 = 4, \lambda_3 = 6$. Show that if the following Ritz vectors

$$[\Psi] = \frac{1}{12}\begin{bmatrix} 7 & 1 \\ 2 & 2 \\ 1 & 7 \end{bmatrix}$$

are used as approximate solutions to the eigenproblem then the reduced stiffness and mass matrices are

$$[\tilde{K}] = \frac{1}{12}\begin{bmatrix} 7 & 1 \\ 1 & 7 \end{bmatrix}, \qquad [\tilde{M}] = \frac{1}{144}\begin{bmatrix} 29 & 11 \\ 11 & 29 \end{bmatrix}$$

and that the solution of the reduced eigenproblem gives the eigenpairs

$$(\rho_1, x_1) = \left(2.4004, \begin{Bmatrix} 1.3418 \\ 1.3418 \end{Bmatrix}\right), \qquad (\rho_2, x_2) = \left(4.0032, \begin{Bmatrix} 2.008 \\ -2.008 \end{Bmatrix}\right)$$

[Reference [5], pp. 590]

**12.6** Continue the previous exercise by taking the initial Ritz vectors as

$$[\Psi] = \frac{1}{6}\begin{bmatrix} 5 & 1 \\ 4 & 2 \\ 5 & 1 \end{bmatrix}$$

Show that the reduced stiffness and mass matrices are

$$[\tilde{K}] = \frac{1}{3}\begin{bmatrix} 7 & 4 \\ 4 & 1 \end{bmatrix}, \qquad [\tilde{M}] = \begin{bmatrix} 41 & 13 \\ 13 & 5 \end{bmatrix}$$

and that the solution of the associated eigenproblem gives the eigenpairs

$$(\rho_1, x_1) = \left(2.0000, \begin{Bmatrix} 0.70711 \\ 0.70711 \end{Bmatrix}\right), \qquad (\rho_2, x_2) = \left(6.0000, \begin{Bmatrix} -2.1213 \\ 6.3640 \end{Bmatrix}\right)$$

[Reference [5], pp. 591]

**12.7** The lumped representation of a distributed load is given by

$$P_i = \int_0^L q(x) f_i(x)\, dx$$

where $f_i$ are the shape functions. Investigate the use of this formula (with $q$ replaced with $\rho$) as a means to determine a lumped mass representation for the rod and beam.

**12.8** Consider a system for which the Lagrangian $L \equiv T - (U - V)$ does not contain time explicitly and on which only conservative forces are acting, i.e., $L = L(u_i, \dot{u}_i)$. Show that a first integral of Lagrange's equation leads to

$$T + (U + V) = \mathcal{E} = constant$$

where E is a constant of integration.

[Reference [44], pp. 92]

**12.9** Show that the expressions for the strain energy and the kinetic energy in principal coordinates take on the following particularly simple form

$$2U = \omega_1^2 \eta_1^2 + \omega_2^2 \eta_2^2 + \cdots, \qquad 2T = \dot{\eta}_1^2 + \dot{\eta}_2^2 + \cdots$$

[Reference [7], pp. 67]

## Exercises

**12.1** A particle describes the path $x = t$, $y = t^2$, $z = t^3$, in which $t$ denotes time. The motion is resisted by a force that is opposite to the velocity and that is proportional to the speed. Calculate the work that the resisting force perform during the interval $t = 1$ to $t = 2$.          $[W = -c7084/15]$

**12.2** A uniform simple beam of length $L$ and stiffness $EI$ carries a mass $m_c$ at the center. Neglecting gravity and the mass of the beam, use Lagrange's equation to derive the differential equation for free vibrations of the mass.
$$[m_c \ddot{v} + (48 EI/L^2) v = 0]$$

**12.3** A rigid bar of length $L_2$ is attached to a light string of length $L_1$. The string is fixed at the other end so that the combination forms a pendulum. Use Lagrange's equation to derive the differential equation for free vibrations.
$$[2L_1 \ddot{\theta}_1 + L_2 \ddot{\theta}_2 + 2g\theta_1 = 0 \quad 3L_1 \ddot{\theta}_1 + 2L_2 \ddot{\theta}_2 + 3g\theta_1 = 0]$$

**12.4** Estimate the fundamental frequency of a cantilever beam using as the Ritz function guess $\psi(x) = x^2$.          $[\omega = 4.47\sqrt{EI/\rho AL^4}]$

**12.5** Estimate the fundamental frequency of a cantilever beam using as the Ritz function guess the static deflection under its own weight. $[\omega = 3.53\sqrt{EI/\rho AL^4}]$

**12.6** A steel shaft $4\,ft$ long is carried in bearings at each end. The middle $1\,ft$ is $2\,in$ in diameter with the remainder being $1\,in$ in diameter. Use an approximate method to estimate the lowest natural frequency.          $[\omega_{exact} = 195\,r/s]$

# Chapter 13

# Computer Methods II

Determining the dynamic response of a structure is one of the most demanding challenges for implementing matrix methods on a computer. This divides into two distinct but highly related problems: direct integration in time of the dynamic equilibrium equations, and performing a modal analysis. Reference [5] is an excellent source of additional material.

Direct integration will be conceived as a sequence of pseudo-static problems (one at each time step) with a time varying load that also depends on the inertia properties. Crucial considerations in this type of incremental solution are the questions of accuracy and numerical stability. We first develop a few basic tools that aid in answering these questions, and then analyze two methods of time integration.

The power of modal analysis is that it shows the way for replacing a large dynamic system by one of a much smaller size. Indeed, for many structural dynamics problems, it is usually only the first ten or twenty modes that are of interest. Therefore, emphasis will be given to solving the *partial* eigenvalue problem; that is, only the lower eigenvalues of a large system will be solved for. We first introduce the concept of repeated orthogonal transformations as a means to reduce a matrix to diagonal form. This is implemented in the Jacobi rotation method. This method, however, is very inefficient for large systems, but the ideas it embodies are incorporated with the vector iteration method to produce the subspace iteration scheme. In this scheme, we can iterate simultaneously on many eigenvectors to give a robust partial solution.

## 13.1 Finite Differences

When it is inconvenient to integrate a differential equation by analytical means, we usually resort to a step-by-step integration procedure. In this approach, the response is evaluated for a series of short time increments $\Delta t$. The condition of dynamic equilibrium is established at the beginning and end of each interval, and the motion of the system during the time increment is evaluated approximately. The complete response is obtained by using the velocity and displacement computed at the end of one interval as the initial conditions for the next interval.

In the interval, the functions are replaced by simple polynomial representations in terms of just the end values. Such approximations are called *finite differences*. While the application of this process seems straight-forward, several numerical difficulties arise which can profoundly impact on the quality of the approximate solutions. This section addresses some of these issues before we proceed to the application to our structural systems.

When a differential equation is approximated by a difference equation, there is introduced an error called *truncation error*; to find the conditions under which the truncation error can be satisfactorily controlled is the problem of *convergence*. When the difference equation is solved numerically, additional errors due to round-offs are introduced. To find the conditions under which the round-off errors remain sufficiently small as many steps are executed is the problem of *stability*. Unless the numerical representation of the differential equations is both convergent and stable, the results derived from its use are generally not good approximations to the true values.

## Difference Equations

With reference to Figure 13.1, suppose we want an approximation to the derivative of the function at time $t$. Assume we know the function $f(t)$ at discrete times

$$\ldots, \quad f(t-h), \quad f(t), \quad f(t+h), \quad \ldots$$

where $h$ is the spacing between the known values. These discrete values are related to each other through the following Taylor series approximations

$$f(t+h) \approx f(t) + \frac{df(t)}{dt}h + \ldots$$

$$f(t-h) \approx f(t) - \frac{df(t)}{dt}h + \ldots$$

We can now use these expansions to get various approximations for the derivative at time $t$. From the first we get the forward approximation, the backward approximation from the second, and the combination of the two will give the central difference approximation. That is, respectively,

$$\text{forward:} \qquad \frac{df(t)}{dt} \approx \frac{1}{h}[f(t+h) - f(t)]$$

$$\text{backward:} \qquad \frac{df(t)}{dt} \approx \frac{1}{h}[f(t) - f(t-h)]$$

$$\text{central:} \qquad \frac{df(t)}{dt} \approx \frac{1}{2h}[f(t+h) - f(t-h)]$$

Similarly, if we view the function $f(t)$ as being the derivative of another function $f(t) = dg(t)/dt$, then we can construct formulas for the second derivatives as

$$\text{forward:} \qquad \frac{d^2g}{dt^2} \approx \frac{1}{h}[g'(t+h) - g'(t)] = \frac{1}{h^2}[g(t+2h) - 2g(t+h) + g(t)]$$

backward:   $\dfrac{d^2g}{dt^2} \approx \dfrac{1}{h}[g'(t) - g'(t-h)] = \dfrac{1}{h^2}[g(t) - 2g(t-h) + g(t-2h)]$

central:   $\dfrac{d^2g}{dt^2} \approx \dfrac{1}{2h}[g'(t+h) - g'(t-h)] = \dfrac{1}{h^2}[g(t+h) - 2g(t) + g(t-h)]$

(In this last relation, the step was changed as $2h \rightarrow h$). These formulas give us a scheme for replacing derivatives by the discrete solution values.

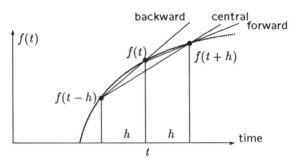

**Figure 13.1**: Three difference schemes for the derivative.

We will now show an application of these formulas to finding numerical solutions to differential equations. Suppose we have a differential equation

$$\frac{du}{dt} + au = 0, \qquad u(0) = u_o$$

where $a$ is a positive constant and we want to find $u(t)$. In the approximate solution, we seek the function at equi-spaced time values such that $u_n \equiv u(nh)$ and replace derivatives by their difference approximations in terms of these values. For example, the forward difference approximation evaluated at time $t = t_n = nh$ gives

$$\frac{u_{n+1} - u_n}{h} + au_n = 0 \qquad \text{or} \qquad u_{n+1} = (1 - ah)u_n$$

The second form is called a *recursion relation* because it allows the next value of the solution to be determined from the previous ones by recursively using the same formula. If we let $A = (1 - ah)$, then we have the sequence

$$\begin{aligned} u_1 &= Au_o \\ u_2 &= Au_1 = A^2 u_o \\ u_3 &= Au_2 = A^3 u_o \end{aligned}$$

and so on. Similarly, the backward difference formula (by expanding about the time $t = (n+1)h$) gives

$$\frac{u_{n+1} - u_n}{h} + au_{n+1} = 0 \qquad \text{or} \qquad u_{n+1} = \frac{1}{(1 + ah)}u_n$$

| t | forward h = 1.0 | h= .5 | h=.25 | backward h = 1.0 | h=.5 | h= .25 | exact |
|---|---|---|---|---|---|---|---|
| 0 | 1.000 | 1.000 | 1.000 | 1.000 | 1.000 | 1.000 | 1.000 |
| 1 | .5000 | .5625 | .5861 | .6666 | .6400 | .6243 | .6065 |
| 2 | .2500 | .3164 | .3436 | .4444 | .4096 | .3897 | .3678 |
| 3 | .1250 | .1779 | .2014 | .2963 | .2621 | .2433 | .2231 |
| 4 | .0625 | .1001 | .1180 | .1975 | .1677 | .1519 | .1353 |
| 5 | .0312 | .0563 | .0692 | .1316 | .1073 | .0948 | .0820 |
| steps | 5 | 10 | 20 | 5 | 10 | 20 | |

Table 13.1: Convergence for small step size.

This is evaluated in precisely the same manner as for the forward difference method.

The magnitude of error involved in replacing a derivative by a difference can be determined by means of a Taylor series expansion. To illustrate, consider the central difference approximation for the second derivative; use the Taylor expansion to get

$$f(t \pm h) \approx f(t) \pm \frac{df}{dt}h + \frac{1}{2}\frac{d^2f}{dt^2}h^2 \pm \frac{1}{6}\frac{d^3f}{dt^3}h^3 + \frac{1}{24}\frac{d^4f}{dt^4}h^4 + \cdots$$

Substitute this into the difference equation and get

$$\begin{aligned}
\frac{d^2f}{dt^2} &\approx \frac{1}{h^2}\left[\left(f(t) + \frac{df}{dt}h + \frac{1}{2}\frac{d^2f}{dt^2}h^2 + \frac{1}{6}\frac{d^3f}{dt^3}h^3 + \frac{1}{24}\frac{d^4f}{dt^4}h^4\right) - 2f(t)\right. \\
&\quad \left. + \left(f(t) - \frac{df}{dt}h + \frac{1}{2}\frac{d^2f}{dt^2}h^2 - \frac{1}{6}\frac{d^3f}{dt^3}h^3 + \frac{1}{24}\frac{d^4f}{dt^4}h^4\right)\right] \\
&\approx \frac{d^2f}{dt^2} + \frac{d^4f}{dt^4}\frac{h^2}{12} + O(h^4)
\end{aligned}$$

This indicates that the *truncation* error involved in the difference relation is of the order of magnitude $h^2$, hence as the step size is halved, the error is quartered. A similar error analysis of both the forward and backward difference approximations for the first derivatives shows that in both cases

$$\frac{df}{dt} \approx \frac{df}{dt} \pm \frac{d^2f}{dt^2}\frac{1}{2}h + O(h^3)$$

The error is of order $h$ and this indicates that as the step is halved, so is the error.

**Example 13.1:**   Let the initial value be $u_0 = 1$ and $a = 0.5$ for the differential equation $\dot{u} + au = 0$. Compute the results for different time steps and compare with the exact solution given as $u(t) = u_o e^{-at}$.

Table 13.1 shows the results as the time step is decreased; it is obvious that both solutions get better. This illustrates the idea of convergence as the time step is decreased.

The table of results also bear out that, approximately, the error is halved as the step size is halved.

**Example 13.2:**   Estimate the time step needed in the central difference approximation for the second order system

$$\frac{d^2 u}{dt^2} + a^2 u = 0, \qquad u(0) = 0, \quad \dot{u}(0) = C$$

if the error is to be less than 1%.
   The exact solution is

$$u(t) = C \sin at$$

Hence the error in acceleration is

$$error = \frac{h^2}{12} \frac{d^4 u}{dt^4} = \frac{h^2}{12} a^4 C \sin at$$

We want 1% error, so make the comparison at the maximum value, that is,

$$\left| \frac{\Delta \ddot{u}}{\ddot{u}} \right| = \frac{Ch^2 a^4 / 12}{Ca^2} = \frac{1}{12} h^2 a^2 \le 0.01$$

This gives, approximately,

$$h \le \frac{\sqrt{12}}{10a} \approx \frac{1}{20} T$$

where $T = 2\pi/a$ is the period of oscillation of the solution. A reasonable rule of thumb is that the time step should be less than one-tenth of the period of oscillation.

## Numerical Stability

It must be kept in mind that even though it may be possible to have a difference equation where the truncation error is arbitrarily small, it is also possible that round-off errors in the use of this equation can accumulate at each integration step yielding useless results. This is the problem of numerical stability and is quite separate from the issue of accuracy.
   Consider the first order case of the last section; as shown, we have the sequence of solution values

$$
\begin{aligned}
u_1 &= Au_o \\
u_2 &= Au_1 = A^2 u_o \\
u_3 &= Au_2 = A^3 u_o \\
&\ \ \vdots \\
u_n &= Au_{n-1} = A^n u_o
\end{aligned}
$$

where $A = (1 - ah)$ for the forward difference scheme. If $|A| < 1$ then we have $A^n$ tending to zero as $n$ increases. Suppose the step size is $h = 1/a$, then $A = 0$ and after the first step the solution is zero. Obviously, the accuracy is not very good but it is an "approximation" to the long term value of zero. Contrast this with what happens when we increase the step to $h > 2/a$ giving $-A > 1$. Now the solution blows up as $n$ increases. This is an example of instability. The lack of accuracy in this situation is quite different from the previous case because the errors grow irrespective of the initial value of $u_o$. To emphasize this point, if $u_o$ is just some initial condition due to round-off error, then nonetheless an unstable solution ensues. Clearly, the method is unstable if $h > 2/a$, for then $u_n \to \infty$ as $n \to \infty$. A similar analysis of the backward difference scheme with $A = 1/(1 + ah)$ shows the solution to be always stable and hence no such restrictions apply to the step size. Therefore, in order to use difference equations with confidence, we must be aware of their stability conditions.

Before we analyze some difference relations, we first need to develop some tools. Recall that a technique in solving differential equations with constant coefficients is to assume a solution of the form $Ce^{\lambda t}$ and determine $\lambda$ from a characteristic equation. For example, the second order differential equation would give

$$\ddot{u} + a^2 u = 0 \quad \Rightarrow \quad [\lambda^2 + a^2]Ce^{\lambda t} = 0 \quad \text{or} \quad \lambda = \pm ia$$

We can use the same technique for difference equations; simply rewrite

$$Ce^{\lambda t} \to Ce^{\lambda n h} = C[e^{\lambda h}]^n = C\rho^n$$

For example, the first order system using forward differences gives

$$u_{n+1} = (1 - ah)u_n \quad \Rightarrow \quad C\rho^{n+1} = (1 - ah)C\rho^n \quad \text{or} \quad \rho = (1 - ah)$$

Our criterion for stability is that the magnitude of $\rho$ be less than unity. That is,

$$|\rho| = |(1 - ah)| < 1 \quad \text{or} \quad h < 2/a$$

This is in agreement with what we had already concluded.

Now consider the forward approximation of the second order system

$$\frac{u_{n+2} - 2u_{n+1} + u_n}{h^2} + a^2 u_n = 0 \quad \text{or} \quad u_{n+2} - 2u_{n+1} + (1 + a^2 h^2)u_n = 0$$

Assume a trial solution of the form $u_n = C\rho^n$ giving the characteristic equation

$$\rho^2 - 2\rho + (1 + a^2 h^2) = 0$$

There are two possible solutions to this given by

$$\rho = 1 \pm \sqrt{1 - (1 + a^2 h^2)} = 1 \pm iah$$

There are two things to note about this solution. First, the magnitude is greater than unity and therefore, irrespective of the step size $h$, the solution is always unstable !

Second, $\rho$ is complex and therefore the solution will exhibit oscillatory behavior. We can see this from

$$1 \pm iah = Ae^{\pm i\phi}, \qquad A = \sqrt{1 + a^2 h^2}, \qquad \tan \phi = ah$$

The solution of the difference equation is

$$u_n = C\rho^n = CA^n e^{\pm i\phi n} = C\,A^n[\cos \phi n \pm i \sin \phi n]$$

If we now associate increasing $n$ with increasing time such that $t = nh$, then we can write, for example,

$$\cos \phi n = \cos \frac{\phi}{h} nh = \cos \omega^* t$$

The effective frequency of oscillation is given by

$$\omega^* = \frac{1}{h}\phi = \frac{1}{h}\tan^{-1} ah \approx a[1 - \frac{1}{3}a^2 h^2 + \ldots]$$

Therefore, in the consideration of the difference schemes, we will look at the amplitude and the frequency shift. In this case, we see that the frequency decreases resulting in a period elongation.

**Example 13.3:**   Analyze the second order system $u'' + a^2 u = 0$ for stability when approximated by the backward difference formulas.

   The backward difference approximation of the second order system gives

$$\frac{u_n - 2u_{n-1} + u_{n-2}}{h^2} + a^2 u_n = 0 \qquad \text{or} \qquad (1 + a^2 h^2)u_n - 2u_{n-1} + u_{n-2} = 0$$

Assume a trial solution of the form $u_n = C\rho^n$ giving the characteristic equation

$$(1 + a^2 h^2)\rho^2 - 2\rho + 1 = 0 \qquad \text{or} \qquad \rho = \frac{2 \pm \sqrt{4 - 4(1 + a^2 h^2)}}{2(1 + a^2 h^2)} = \frac{1 \pm iah}{(1 + a^2 h^2)}$$

This gives the amplitude and phase as

$$A = \frac{1}{\sqrt{1 + a^2 h^2}}, \qquad \tan \phi = ah$$

We therefore conclude that the scheme is unconditionally stable. However, it will exhibit an amplitude decay over time (called *artificial damping*) as well as a period elongation.

# 13.2   Direct Integration Methods

This section introduces the *direct integration* methods for finding the dynamic response of structures; that is, the dynamic equilibrium equations are integrated directly in a step-by-step fashion. This is done without the modal decoupling used in the modal superposition approach of Chapter 11, and therefore we operate with the full structural matrices in the general form

$$[K]\{u\} + [C]\{\dot{u}\} + [M]\{\ddot{u}\} = \{P\}$$

The equations for direct integration are either *explicit* or *implicit*. In the explicit equations, the equations of motion are written at the current time and as a result neither $\{u\}$ nor $\{\dot{u}\}$ at the current time is a function of the acceleration $\{\ddot{u}\}$ at the next time $t + \Delta t$. In the implicit equations, on the other hand, the equations of motion are used at the next time, $t + \Delta t$. An example of both will be given.

## Central Difference and Average Acceleration Methods

To construct the central difference algorithm, we begin with difference expressions for the nodal velocities and accelerations at the current time $t$

$$\{\dot{u}\}_t = \frac{1}{2\Delta t}\{u_{t+\Delta t} - u_{t-\Delta t}\}$$

$$\{\ddot{u}\}_t = \frac{1}{\Delta t^2}\{u_{t+\Delta t} - 2u_t + u_{t-\Delta t}\} \tag{13.1}$$

Substitute these into the equations of motion written at time $t$ to get

$$[K]\{u\}_t + [C]\frac{1}{2\Delta t}\{u_{t+\Delta t} - u_{t-\Delta t}\} + [M]\frac{1}{\Delta t^2}\{u_{t+\Delta t} - 2u_t + u_{t-\Delta t}\} = \{P\}_t$$

This scheme is therefore explicit. Rearrange this equation so that only quantities evaluated at time $t + \Delta t$ are on the left hand side

$$\left[\frac{1}{2\Delta t}C + \frac{1}{\Delta t^2}M\right]\{u\}_{t+\Delta t} = \{P\}_t \tag{13.2}$$

$$- \left[K - \frac{2}{\Delta t^2}M\right]\{u\}_t + \left[\frac{1}{2\Delta t}C - \frac{1}{\Delta t^2}M\right]\{u\}_{t-\Delta t}$$

Note that the stiffness is on the right hand side in the effective load vector; therefore, these equations cannot recover the static solution in the limit of very slow loading. Further, if the mass matrix is not positive definite (that is, if it has some zeros on the diagonal) then the scheme does not work because the square matrix on the left hand side is not invertible.

The algorithm for the step-by-step solution operates as follows: We start at $t = 0$, initial conditions prescribe $\{u\}_0$ and $\{\dot{u}\}_0$, from these and the equation of motion, we find the acceleration $\{\ddot{u}\}_0$ if it is not prescribed. These equations also yield the displacements $\{u\}_{-\Delta t}$ needed to start the computations; that is, from the differences for the velocity and acceleration we get

$$\{u\}_{-\Delta t} = \{u\}_0 - (\Delta t)\{\dot{u}\}_0 + \tfrac{1}{2}(\Delta t)^2\{\ddot{u}\}_0$$

The set of Equations(13.1) and (13.2) are then used repeatedly; the equation of motion gives $\{u\}_{\Delta t}$, then the difference equations gives $\{\ddot{u}\}_{\Delta t}$ and $\{\dot{u}\}_{\Delta t}$, and then the process is repeated. The solution is simple if the mass matrix is diagonal and the damping is zero or diagonal. This is a significant advantage. The computational cost, in general, is approximately

$$cost = \tfrac{1}{2}NB_m^2 + [2N(2B_m - 1) + N2B]q$$

where $q$ is the number of time increments and $B_m$ is the semi-bandwidth of the mass (and damping) matrix. When the mass matrix is diagonal this reduces to

$$cost = 2NBq$$

The cost is linear in the number of time steps.

We will now derive an entirely different integration scheme. Assume that the acceleration is constant over the small time step $\Delta t$ and given by its average value. That is,

$$\ddot{u}(t) = \tfrac{1}{2}(\ddot{u}_t + \ddot{u}_{t+\Delta t}) = constant$$

Integrate this to give the velocity and displacement as

$$\dot{u}(t) = \dot{u}_t + \tfrac{1}{2}(\ddot{u}_t + \ddot{u}_{t+\Delta t})t$$
$$u(t) = u_t + \dot{u}_t t + \tfrac{1}{4}(\ddot{u}_t + \ddot{u}_{t+\Delta t})t^2$$

These can be rearranged to give difference formulas for the new acceleration and velocity (at time $t + \Delta t$) in terms of the new displacement as

$$\{\ddot{u}\}_{t+\Delta t} = \frac{4}{\Delta t^2}\{u_{t+\Delta t} - u_t\} - \frac{4}{\Delta t}\{\dot{u}\}_t - \{\ddot{u}\}_t$$
$$\{\dot{u}\}_{t+\Delta t} = \frac{2}{\Delta t}\{u_{t+\Delta t} - u_t\} - \{\dot{u}\}_t \qquad (13.3)$$

Substitute these into the equations of motion at the new time $t + \Delta t$ to obtain the implicit scheme

$$[K]\{u\}_{t+\Delta t} + [C]\left\{\frac{2}{\Delta t}\{u_{t+\Delta t} - u_t\} - \{\dot{u}\}_t\right\}$$
$$+ [M]\left\{\frac{4}{\Delta t^2}\{u_{t+\Delta t} - u_t\} - \frac{4}{\Delta t}\{\dot{u}\}_t - \{\ddot{u}\}_t\right\} = \{P\}_{t+\Delta t}$$

All terms that have been evaluated at time $t$ are now shifted to the right hand side. The rearranged equations of motion are then

$$\left[K + \frac{2}{\Delta t}C + \frac{4}{\Delta t^2}M\right]\{u\}_{t+\Delta t} = \{P\}_{t+\Delta t} \qquad (13.4)$$
$$+ [C]\left\{\frac{2}{\Delta t}u + \dot{u}\right\}_t + [M]\left\{\frac{4}{\Delta t^2}u + \frac{4}{\Delta t}\dot{u} + \ddot{u}\right\}_t$$

The new displacements are obtained by solving this system of equations, then the acceleration and velocity are updated from Equations(13.3).

The algorithm operates as follows: We start at $t = 0$, initial conditions prescribe $\{u\}_0$ and $\{\dot{u}\}_0$. From these and the equations of motion (written at time $t = 0$) we find $\{\ddot{u}\}_0$ if it is not prescribed. Then the above system of equations are solved for the displacement $\{u\}_{\Delta t}$, from which estimates of the accelerations $\{\ddot{u}\}_{\Delta t}$ and the velocities $\{\dot{u}\}_{\Delta t}$ can also be obtained. These are used to obtain current values of the right hand side. Then solving the equation of motion again yields $\{u\}_{2\Delta t}$, and so on. The solution procedure for $\{u\}_{t+\Delta t}$ is not trivial, but the coefficient matrix need be reduced to $[\,U\,]^T\lceil\,D\,\rfloor[\,U\,]$ form only once if $\Delta t$ or any of the system matrices do not change during the integration. Note that in the limit of large $\Delta T$ we recover the static solution.

The computational cost is approximately

$$cost = \tfrac{1}{2}NB^2 + [2NB + 2N(2B_m - 1)]q$$

When the mass matrix is diagonal this reduces to

$$cost = \tfrac{1}{2}NB^2 + 2NBq$$

Except for the cost of the initial decomposition the total cost is the same as for the central difference method. When the mass matrix is banded the cost is

$$cost = \tfrac{1}{2}NB^2 + 6NBq$$

which is the same for both methods. As an idea of the total computational cost, consider running this on our benchmark machine of 1 MFlops. The time to solve a medium sized system of order 1000 with 10% bandwidth for 1000 time steps is approximately

$$time = [\tfrac{1}{2} \times 1000 \times 100^2 + 6 \times 1000 \times 100 \times 1000] \div (1.0 \times 10^6) = 10\,minutes$$

In this case, the initial decomposition cost is only about 1% of the total cost. For a large sized system of 10,000 the time is about $16\,hours$.

## Stability

A point to note is that if all the equations are integrated directly then the same results are obtained if a modal transformation is first performed, the integration done numerically, and the physical responses reconstructed. Therefore, to study the accuracy of direct integration we may focus attention only on integrating a single modal equation. In this way the only variables to be considered are $\Delta t$, $\omega_m$, and $\zeta_m$. Furthermore, because all equations are similar we need only study the integration of a typical one given by

$$\ddot{u} + 2\zeta\omega\,\dot{u} + \omega^2 u = p$$

where $\omega$ is the modal frequency and $\zeta$ is the modal damping.

To investigate the central difference algorithm, we begin with finite difference expressions in time for nodal velocities and accelerations at the current time $t$ as given in Equation(13.1) and substitute these into the equations of motion, written at time $t$, to get

$$(1 + \zeta\omega\Delta t)u_{n+1} - (2 - \omega^2\Delta t^2)u_n + (1 - \zeta\omega\Delta t)u_{n-1} = p_n\Delta t^2$$

The characteristic equation associated with this difference equation is

$$\rho^2 - \rho\frac{(2 - \omega^2\Delta t^2)}{(1 + \zeta\omega\Delta t)} + \frac{(1 - \zeta\omega\Delta t)}{(1 + \zeta\omega\Delta t)} = 0$$

Hence the roots are

$$\rho = \left[(2 - \omega^2\Delta t^2) \pm \sqrt{\omega^4\Delta t^4 - 4\omega^2\Delta t^2(1 - \zeta^2)}\right]/(2 + 2\zeta\omega\Delta t)$$

We need to have oscillating solutions (because of our second order system), hence the radical must be negative. Thus, we require that

$$\omega^4\Delta t^4 - 4\omega^2\Delta t^2(1 - \zeta^2) < 0 \qquad \text{or} \qquad \omega\Delta t < 2\sqrt{1 - \zeta^2}$$

Hence the method is only conditionally stable since it is possible for this criterion not to be satisfied in some circumstances. The amplitude and phase are given by

$$A = \sqrt{\frac{1 - \zeta\omega\Delta t}{1 + \zeta\omega\Delta t}}, \qquad \tan\phi = \frac{\omega\Delta t\sqrt{4(1 - \zeta^2) - \omega^2\Delta t^2}}{(2 - \omega^2\Delta t^2)}$$

When damping is negligible, these reduce to

$$A = 1, \qquad \tan\phi = \frac{\omega\Delta t\sqrt{1 - \omega^2\Delta t^2/4}}{(1 - \omega^2\Delta t^2/2)}$$

Hence there are no amplitude errors. This requires that

$$\omega\Delta t < 2 \qquad \text{or} \qquad \Delta t < \frac{2}{\omega} = \frac{T}{\pi}$$

where $T$ is the period associated with the frequency $\omega$. Thus for stability the step size must be less than one-third the period. This seems easily achieved since we have already established that the step size should be less than one-tenth the period for an accurate solution. As we will see for multiple degree of freedom systems, this is not so straight-forward.

A similar analysis gives for the average acceleration method

$$\left\{\begin{array}{c} u \\ \dot{u} \\ \ddot{u} \end{array}\right\}_{n+1} = \frac{1}{D}\left[\begin{array}{ccc} 1 + \zeta\omega\Delta t & 1 + \frac{1}{2}\zeta\omega\Delta t & \frac{1}{4} \\ -\frac{1}{2}\omega^2\Delta t^2 & 1 - \frac{1}{4}\omega^2\Delta t^2 & \frac{1}{2} \\ -\omega^2\Delta t^2 & -(2\zeta\omega\Delta t + \omega^2\Delta t^2) & -\zeta\omega\Delta t - \frac{1}{4}\omega^2\Delta t^2 \end{array}\right]\left\{\begin{array}{c} u \\ \dot{u} \\ \ddot{u} \end{array}\right\}_n$$

where $D = 1 + \zeta\omega\Delta t + \frac{1}{4}\omega^2\Delta t^2$. We can obtain a characteristic equation for this system by assuming a solution of the form

$$\{u\} = \{C\}\rho^n$$

After substitution, this gives rise to an eigenvalue problem, the roots of which are given as

$$\rho = \frac{1 - \frac{1}{4}\omega^2\Delta t^2 \pm i\omega\Delta t}{1 + \frac{1}{4}\omega^2\Delta t}$$

for the no damping case. We first notice that there is automatically an imaginary part, hence the solution will exhibit the desired oscillations. The magnitude and phase are given by

$$A = 1, \qquad \tan\phi = \frac{\omega\Delta t}{1 - \frac{1}{4}\omega^2\Delta t^2}$$

Hence we conclude that the system is *unconditionally stable*, gives no amplitude decay, but will exhibit a frequency shift.

## Explicit Versus Implicit Schemes

The foregoing analysis shows that the computational cost for the explicit central difference scheme can be less than the implicit average acceleration method. Further, the stability analysis shows that the criterion

$$\Delta t < \frac{2}{\omega} = \frac{T}{\pi}$$

for the explicit scheme will automatically be satisfied because of accuracy considerations. Hence we could conclude that the explicit scheme is preferable.

A very important factor was overlooked in the above; a multiple degree of freedom system will have many modes and when direct integration is used, this is equivalent to integrating each mode with the same time step $\Delta t$. Therefore, the above stability criterion must be applied to the highest modal frequency of the system even if our interest is in the low frequency response. In other words, if we energize the system in such a way as to excite only the lower frequencies, we must nonetheless choose an integration step corresponding to the highest possible mode. The significance of this is that the matrix analyses of structures produce so-called *stiff equations*. In the present context, stiff equations characterize a structure whose highest natural vibration frequencies are much greater than the lowest. Especially stiff structures therefore include those with a very fine mesh, and a structure with near-rigid support members. If the conditionally stable algorithm is used for these structures, $\Delta t$ must be very small, possibly orders of magnitude smaller than for the implicit scheme.

In summary, explicit methods are conditionally stable and therefore require a small $\Delta t$ but produce equations that are inexpensive to solve. The implicit methods are (generally) unconditionally stable and therefore allow a large $\Delta t$ but produce equations that are more expensive to solve. The size of $\Delta t$ is governed by considerations of accuracy rather than stability; that is, we can adjust the step size appropriate to the excitation force or the number of modes actually excited. The difference factor can be orders of magnitude and will invariably outweigh any disadvantage in having to decompose the system matrices. Based on these considerations, we therefore prefer the implicit scheme and will develop the algorithm fully in the next section.

## 13.3   Newmark's Method

The foregoing indicates that there is a variety of methods available for solving transient problems by direct integration. The algorithm used in **STADYN** and given in module **TRANSient**, is of the implicit type and chosen primarily because of its stability and accuracy under a wide range of element size and time step variations. It is based on the average acceleration scheme but given the name *Newmark integration*.

### General Form

We will motivate the development of the Newmark formulas by reconsidering the average acceleration method. First rewrite the formulas in the form

$$\ddot{u}(t) = \ddot{u}_t + \tfrac{1}{2}(\ddot{u}_{t+\Delta t} - \ddot{u}_t) = constant$$
$$\dot{u}(t) = \dot{u}_t + \ddot{u}_t t + \tfrac{1}{2}(\ddot{u}_{t+\Delta t} - \ddot{u}_t)t$$
$$u(t) = u_t + \dot{u}_t t + \tfrac{1}{2}\ddot{u}_t t^2 + \tfrac{1}{4}(\ddot{u}_{t+\Delta t} - \ddot{u}_t)t^2$$

The term $(\ddot{u}_{t+\Delta t} - \ddot{u}_t)$ is the core of the approximation; by adjusting its coefficient the approximation for the acceleration can be made to have the acceleration value at $t$ or $t + \Delta t$. Hence, rather than use the factor $\tfrac{1}{2}$ we use the parameter $\gamma$. A similar argument shows that we can replace the factor $\tfrac{1}{4}$ in the displacement equation with a second parameter $\beta$. Thus, we rewrite the integration formulas as

$$\dot{u}(t) = \dot{u}_t + \ddot{u}_t t + \gamma(\ddot{u}_{t+\Delta t} - \ddot{u}_t)t$$
$$u(t) = u_t + \dot{u}_t t + \tfrac{1}{2}\ddot{u}_t t^2 + \beta(\ddot{u}_{t+\Delta t} - \ddot{u}_t)t^2$$

where $\gamma$ and $\beta$ are selected to give the best accuracy and stability performance. The equations of motion can now be written as

$$\left[K + \frac{\gamma}{\beta\Delta t}C + \frac{1}{\beta\Delta t^2}M\right]\{u\}_{t+\Delta t} = \{P\}_{t+\Delta t} \tag{13.5}$$

$$+ \ [C]\left\{\frac{\gamma}{\beta\Delta t}u + \left(\frac{\gamma}{\beta} - 1\right)\dot{u} + \Delta t\left(\frac{\gamma}{2\beta} - 1\right)\ddot{u}\right\}_t$$

$$+ \ [M]\left\{\frac{1}{\beta\Delta t^2}u + \frac{1}{\beta\Delta t}\dot{u} + \left(\frac{1}{2\beta} - 1\right)\ddot{u}\right\}_t$$

An advantage of the Newmark method is that it reduces to the static approach when the mass and damping effects are neglected or when the time step is very large. Once the new displacements are obtained, then estimates for the new accelerations and velocities are obtained from

$$\{\ddot{u}\}_{t+\Delta t} \;=\; \frac{1}{\beta \Delta t^2}\left(\{u\}_{t+\Delta t} - \{u\}_t\right) - \frac{1}{\beta \Delta t}\{\dot{u}\}_t - \left(\frac{1}{2\beta} - 1\right)\{\ddot{u}\}_t$$

$$\{\dot{u}\}_{t+\Delta t} \;=\; \{\dot{u}\}_t + \Delta t\,\{(1-\gamma)\ddot{u}_t + \gamma \ddot{u}_{t+\Delta t}\} \tag{13.6}$$

The method is implicit and has many of the characteristics of the average acceleration method.

It can be shown that Newmark's method is unconditionally stable if $\gamma \geq 0.5$ and $\beta \geq (2\gamma + 1)^2/16$. A good choice of parameters are the ones used for the average acceleration method, that is, $\gamma = 0.5$ and $\beta = 0.25$; these are the default values used by STADYN. With the above choice of parameters there are no amplitude errors in the response, but the frequencies of the response are underestimated.

There are other choices for the Newmark parameters. For example, $\gamma = \frac{1}{2}$, $\beta = \frac{1}{6}$ gives the *linear acceleration method*. This is only conditionally stable with the criterion being $\omega \Delta t < \sqrt{12}$. When this is satisfied the method gives no amplitude decay and a minimum of period elongation. The *constant acceleration method* with $\gamma = 0$ and $\beta = 0$, is also conditionally stable but, in addition, gives amplitude decay and period elongation.

## Computer Algorithm

The basic scheme for the Newmark Method can be stated as:

**Step 1:** Specify $\Delta t$, $\beta$, and $\gamma$. Usually take $\beta = 0.25$ and $\gamma = 0.5$.

**Step 2:** Read the stiffness, damping, and mass matrices.

**Step 3:** Form the effective stiffness matrix as

$$[K_{eff}] \equiv \left[ K + \frac{\gamma}{\beta \Delta t} C + \frac{1}{\beta \Delta t^2} M \right]$$

**Step 4:** Decompose the effective stiffness to

$$[K_{eff}] = [\,U\,]^T \lceil\, D\,\rfloor [\,U\,]$$

**Step 5:** Specify the initial conditions for $\{u\}_0$ and $\{\dot{u}\}_0$. Obtain $\{\ddot{u}\}_0$ from the equations of motion.

**Step 6:** Read the load vector $\{P\}_{t+\Delta t}$. It may be necessary to interpolate this from non-equispaced values.

**Step 7:** Form the effective load vector

$$\{P_{eff}\}_{t+\Delta t} \equiv \{P\}_{t+\Delta t}$$
$$+ \frac{1}{\beta\Delta t}[C]\left\{\gamma u + (\gamma - \beta)\Delta t\,\dot{u} + (\gamma - 2\beta)\tfrac{1}{2}\Delta t^2\ddot{u}\right\}_t$$
$$+ \frac{1}{\beta\Delta t^2}[M]\left\{u + \Delta t\,\dot{u} + (1 - 2\beta)\tfrac{1}{2}\Delta t^2\ddot{u}\right\}_t$$

**Step 8:** Solve for the new displacements from

$$[U]^T[D][U]\{u\}_{t+\Delta t} = \{P_{eff}\}_{t+\Delta t}$$

**Step 9:** Update the acceleration and velocity

$$\{\ddot{u}\}_{t+\Delta t} = \frac{1}{\beta\Delta t^2}\{u_{t+\Delta t} - u_t\} - \frac{1}{\beta\Delta t}\{\dot{u}\}_t - (\frac{1}{2\beta} - 1)\{\ddot{u}\}_t$$
$$\{\dot{u}\}_{t+\Delta t} = \{\dot{u}\}_t + \Delta t\left\{(1 - \gamma)\ddot{u}_t + \gamma\ddot{u}_{t+\Delta t}\right\}$$

**Step 10:** Repeat from Step 6 for each time increment.

The computational cost approximately is

$$cost = \tfrac{1}{2}NB^2 + [2NB + 2NB(2B_m - 1)]q$$

where $q$ is the number of time steps. Notice that at each time step, the cost is basically that of forward and back substitution (which is of order $2NB$) plus what it takes to form the effective load vector ($4NB$ for banded arrays). If the mass and damping are diagonal then this cost is simply $4N$ and this can be a substantial saving.

The right hand side is a load vector that is updated at each time step. It is calculated as a product of mass and damping matrices with the displacements and its derivatives. Thus, the mass and damping matrices must be kept in-core. The significant storage requirements are therefore

$$memory = 3NB$$

This can be reduced to $NB$ if the mass and damping matrices are diagonal.

**Example 13.4:** A three story building has a water tower on top as shown in Figure 13.2. Model this as a shear building and obtain the equations of motion. Assume the flexural stiffness of all walls are the same, the mass of each floor is $m$, and the mass of the water tower is $0.01m$. Neglect damping.

Following Chapter 4, we neglect the joint rotations and axial deformation; therefore, each segment of the building is modeled as

$$\left\{\begin{matrix} V_1 \\ V_2 \end{matrix}\right\} = \frac{12EI}{L^3}\begin{bmatrix} 1 & -1 \\ -1 & 1 \end{bmatrix}\left\{\begin{matrix} u_1 \\ u_2 \end{matrix}\right\}$$

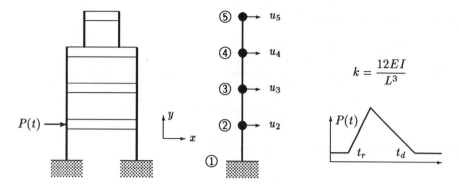

**Figure 13.2**: Building with water tower.

Let

$$k \equiv \frac{12EI}{L^3}$$

then the equations of motion for the reduced system are

$$\begin{bmatrix} k & -k & 0 & 0 \\ -k & 2k & -k & 0 \\ 0 & -k & 2k & -k \\ 0 & 0 & -k & k \end{bmatrix} \begin{Bmatrix} u_2 \\ u_3 \\ u_4 \\ u_5 \end{Bmatrix} + \begin{bmatrix} m & 0 & 0 & 0 \\ 0 & m & 0 & 0 \\ 0 & 0 & m & 0 \\ 0 & 0 & 0 & .01m \end{bmatrix} \begin{Bmatrix} \ddot{u}_2 \\ \ddot{u}_3 \\ \ddot{u}_4 \\ \ddot{u}_5 \end{Bmatrix} = \begin{Bmatrix} P_2 = P \\ P_3 = 0 \\ P_4 = 0 \\ P_5 = 0 \end{Bmatrix}$$

The force is applied only at the second node.

**Example 13.5:** Obtain the dynamic responses for the shear building of the previous example, when subjected to the following different force histories:

$$\begin{array}{lll}
\text{(a)} & t_r = 10.0\,s & t_d = 20.0\,s \\
\text{(b)} & t_r = 1.00\,s & t_d = 2.00\,s \\
\text{(c)} & t_r = 0.10\,s & t_d = 0.20\,s
\end{array}$$

where $t_r$ is the rise time, and $t_d$ is the decay time as shown in Figure 13.2. In each case take the maximum force as $1000\,N$ and the properties of the building as $k = 2.4\,GN/m$, $m = 200\,Mg$.

The results for the response of the second floor are shown in Figure 13.3. It is apparent that the sharper loading pulse generates a higher frequency response dominated by the resonance of the fundamental mode. Note that a different time step was used for each loading case; this could not be done with an explicit integration scheme. It is significant that loading history (a) essentially causes a static response.

Figure 13.4 shows the response at each floor due to loading history (b). The magnitudes are little different, but the character of the response of each floor is the same.

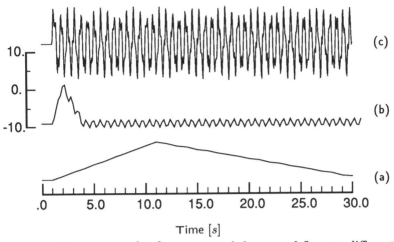

Time [$s$]

**Figure 13.3**: Displacement [$\mu m$] responses of the second floor to different forces.

## 13.4   Complete Solution of Eigensystems

As shown in Chapter 7, solving an eigenvalue problem can be a complicated and time consuming process. That chapter introduced the vector iteration method for finding the lowest eigenvalues, and by use of shifting the other eigenvalues could be obtained also. If multiple roots, eigenclusters, or rigid body modes are present then the iteration algorithm needs further refinement. In this section, we consider the Jacobi transformation method because of its relative simplicity, and its stability. Unlike the vector iteration method, it obtains all the eigenvalues simultaneously and thereby can cope with repeated roots. It can also be used to calculate negative, zero, or positive eigenvalues.

### Transformation Methods

The grand strategy of the transformation methods is to find a transformation matrix $[\,\Phi\,]$ by iteration, so that the system matrices tend to diagonal form. That is, we want

$$[\,\Phi\,]^T[\,K\,][\,\Phi\,] = \lceil\,\tilde{K}\,\rfloor, \qquad [\,\Phi\,]^T[\,M\,][\,\Phi\,] = \lceil\,\tilde{M}\,\rfloor$$

where $\lceil\,\tilde{K}\,\rfloor$ and $\lceil\,\tilde{M}\,\rfloor$ are diagonal. Under this circumstance, the matrix $[\,\Phi\,]$ is the *modal matrix* and its columns are the collection of eigenvectors. The $N$ eigenvalues can then be obtained from

$$\lambda_m = \frac{\tilde{K}_{mm}}{\tilde{M}_{mm}}$$

where $\tilde{K}_{mm}$ and $\tilde{M}_{mm}$ are the diagonal elements of the transformed matrices. The scheme is to reduce the stiffness and mass matrices into diagonal form using successive

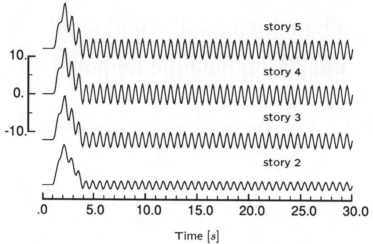

**Figure 13.4**: Displacement $[\mu m]$ responses of the different floors to the second force.

pre- and post-multiplication by matrices $[Q^k]^T$ and $[Q^k]$ , respectively, where $k = 1, 2 \ldots$ is the iteration counter. Initially, we define $[K^1] = [\,K\,]$ and $[M^1] = [\,M\,]$, and we form the sequences

$$
\begin{aligned}
[K^2] &= [Q^1]^T[K^1][Q^1] & [M^2] &= [Q^1]^T[M^1][Q^1] \\
[K^3] &= [Q^2]^T[K^2][Q^2] & [M^3] &= [Q^2]^T[M^2][Q^2] \\
&\;\;\vdots & &\;\;\vdots \\
[K^{k+1}] &= [Q^k]^T[K^k][Q^k] & [M^{k+1}] &= [Q^k]^T[M^k][Q^k]
\end{aligned}
$$

where the matrices $[Q_k]$ are selected to bring $[K_k]$ and $[M_k]$ closer to diagonal form. That is, we are trying to achieve the limits

$$
[K^{k+1}] \to \lceil \tilde{K} \rfloor \quad \text{and} \quad [M^{k+1}] \to \lceil \tilde{M} \rfloor \qquad \text{as} \quad k \to \infty
$$

We have, at any stage, an estimate for the modal matrix and eigenvalue as

$$
[\Phi^k] = [Q^1][Q^2]\cdots[Q^k] = [\Phi^{k-1}][Q^k], \qquad \lambda_m^k = K_{mm}^k / M_{mm}^k
$$

A number of different iteration methods have been proposed that use the basic idea described above; we shall discuss in the next sections only the Jacobi method because of its relative simplicity and robustness. In this case, we specifically choose matrices $[\,Q\,]$ that are orthogonal.

## The Jacobi Method

For simplicity, we first develop the Jacobi method for solving the standard eigenproblem when the mass matrix $[\,M\,]$ is the identity matrix. The general case is left until the next section. For the standard eigenproblem

$$[K]\{\phi\} = \lambda\{\phi\}$$

the $k^{th}$ iteration step reduces to

$$[K^{k+1}] = [Q^k]^T[K^k][Q^k], \qquad [Q^k]^T[\ I\ ][Q^k] = \lceil\ I\ \rfloor$$

where $[Q^k]$ is to be an orthogonal matrix. Specifically, we choose $[Q^k]$ to be a plane rotation matrix selected in such a way that an off-diagonal element in $[K_{ij}]$ is zeroed. That is,

$$[Q^k(i,j)] \equiv \begin{bmatrix} 1 & & & & & & & \\ & \ddots & & & & & & \\ & & \cos\Theta & & -\sin\Theta & & i & row \\ & & & \ddots & & & & \\ & & \sin\Theta & & \cos\Theta & & j & row \\ & & & & & \ddots & & \\ & & & & & & 1 & \end{bmatrix}$$

where all diagonal elements (except the two indicated) are unity. Multiply the $k^{th}$ transformation out, and get for the affected elements (using as shorthand the definitions $c \equiv \cos\Theta$ and $s \equiv \sin\Theta$)

$$\begin{aligned}
K_{ri}^{(k+1)} &= K_{ri}^{(k)}c + K_{rj}^{(k)}s & r \neq i, \quad r \neq j \\
K_{rj}^{(k+1)} &= K_{rj}^{(k)}c - K_{ri}^{(k)}s & r \neq i, \quad r \neq j \\
K_{ii}^{(k+1)} &= K_{ii}^{(k)}c^2 + K_{jj}^{(k)}s^2 + K_{ij}^{(k)}2cs \\
K_{jj}^{(k+1)} &= K_{ii}^{(k)}s^2 + K_{jj}^{(k)}c^2 - K_{ij}^{(k)}2cs \\
K_{ij}^{(k+1)} &= K_{ij}^{(k)}(c^2 - s^2) - (K_{ii}^{(k)} - K_{jj}^{(k)})cs
\end{aligned}$$

where only elements in the $i$ and $j$ rows and columns are affected. We want the term $K_{ij}^{(k+1)}$ to be zero, hence choose

$$\tan 2\Theta = \frac{2cs}{c^2 - s^2} \equiv \frac{2K_{ij}^{(k)}}{K_{ii}^{(k)} - K_{jj}^{(k)}}$$

Should the diagonal terms be equal, then choose $\Theta = \pi/4$. Let $t \equiv \tan\Theta$ and $T \equiv \tan 2\Theta$, then we find the rotation angle from the following quadratic expression

$$T = 2t/(1 - t^2) \qquad \text{or} \qquad t^2T - 2t - T = 0$$

This gives the rotation parameters as

$$t = \frac{1 \pm \sqrt{1 + T^2}}{T} = \frac{T}{1 + \sqrt{1 + T^2}}, \qquad c = \frac{1}{\sqrt{1 + t^2}}, \qquad s = tc$$

The transformation equations are now rearranged so that the new quantity is equal to the old quantity plus a correction. That is,

$$
\begin{aligned}
K_{ri}^{(k+1)} &= K_{ri}^{(k)} + \left[ K_{rj}^{(k)} s - K_{ri}^{(k)} (1-c) \right] \\
K_{rj}^{(k+1)} &= K_{rj}^{(k)} - \left[ K_{ri}^{(k)} s + K_{rj}^{(k)} (1-c) \right] \\
K_{ii}^{(k+1)} &= K_{ii}^{(k)} + K_{ij}^{(k)} t \\
K_{jj}^{(k+1)} &= K_{jj}^{(k)} - K_{ij}^{(k)} t \\
K_{ij}^{(k+1)} &= 0
\end{aligned}
\tag{13.7}
$$

It should be noted that the numerical evaluation of $[K^{(k+1)}]$ requires only the linear combination of two rows and columns. Another important point to keep in mind is that although the above transformation reduces an off-diagonal element to zero, this element will again become non-zero during the transformations that follow. As a result, during the transformations the initial bandedness of the arrays could be temporarily increased even though it will eventually be diagonal. Therefore it is necessary that the working array be a full $[N \times N]$ matrix.

For the design of an actual algorithm, we have to decide which element to reduce to zero. One choice is to zero the largest off-diagonal element. However, the search for this largest element is time consuming, and so a preferable procedure is simply to carry out the Jacobi transformations systematically, row-by-row, running once over every off-diagonal element in one *sweep*. The disadvantage of this procedure is that, regardless of its size, an off-diagonal element is always zeroed. That is, the element may already be nearly zero and a rotation is still applied. A refinement on the method is to do a *threshold check* to see if a rotation is required. To define an appropriate threshold note that, physically, in the diagonalization of $[K]$ we want to reduce the coupling between the degrees of freedom $i$ and $j$. A measure of this coupling is given by $(K_{ij}^2 / K_{ii} K_{jj})$, and it is this factor that can be used effectively in deciding whether to apply a rotation. That is, we check

$$
\left[ \frac{K_{ij}^{(k+1)} K_{ij}^{(k+1)}}{K_{ii}^{(k+1)} K_{jj}^{(k+1)}} \right] \le 10^{-2s}, \qquad \text{all } i,j \quad i < j
$$

where $2s$ is chosen to be about 10.

At the end of each sweep, it is also necessary to measure convergence. We have in the limit that

$$
[K^{(k+1)}] \approx \lceil \Lambda \rfloor
$$

and hence we say that convergence to a tolerance $2s$ has been achieved during the sweep provided that

$$
\frac{|\lambda_m^{(k+1)} - \lambda_m^{(k)}|}{|\lambda_m^{(k+1)}|} \le 10^{-2s}, \qquad \text{all } i = 1 \ldots, N
$$

This states that the current and last approximations to the eigenvalues did not change in the first $2s$ digits.

At the end of each rotation, the eigenvectors are updated as

$$\Phi_{ri}^{(k+1)} = \Phi_{ri}^{(k)}c + \Phi_{rj}^{(k)}s$$
$$\Phi_{rj}^{(k+1)} = \Phi_{rj}^{(k)}c - \Phi_{ri}^{(k)}s \tag{13.8}$$

Note that only the $i$ and $j$ columns are affected.

## Convergence and Computational Cost

An explicit algorithm for the Jacobi method will be given in the next section; of interest now are estimates of the convergence rate and the computational cost.

We begin by noting that a transformation of the form $[\,Q\,]^{-1}[\,A\,][\,Q\,]$ is called a *similarity* transformation. It has the useful property that the eigenvalues are unchanged for any non-null square matrix $[\,Q\,]$ (providing its inverse exists). This can be seen from

$$\det([\,Q\,]^{-1}[\,A\,][\,Q\,] - \lambda[\,I\,]) = \det([\,Q\,]^{-1}[A - \lambda I][\,Q\,])$$
$$= (\det[\,Q\,]^{-1})\det([\,A\,] - \lambda[\,I\,])(\det[[\,Q\,])$$
$$= \det([\,A\,] - \lambda[\,I\,])$$

We also have the result that the norm is invariant under a similarity transformation. Specifically, using the Euclidean norm, we can say that the sum of the squares of the matrix elements is a constant. From Equations(13.7) we see that for the affected diagonal terms

$$|K_{ii}^{(k+1)}|^2 + |K_{jj}^{(k+1)}|^2 = |K_{ii}^{(k)} + K_{ij}^{(k)}t|^2 + |K_{jj}^{(k)} - K_{ij}^{(k)}t|^2$$
$$= |K_{ii}^{(k)}|^2 + |K_{jj}^{(k)}|^2 + 2|K_{ij}^{(k)}|^2t^2$$

Since the diagonal terms are increased then the off-diagonal terms must have decreased. Therefore we get convergence. A more rigorous proof of convergence can be found References [18, 48].

The following is another aspect of convergence; it shows the *rate* of convergence. Suppose at one stage in the iteration that the diagonal terms are of order $K$ and the off-diagonal terms of order $\epsilon K$ where $\epsilon$ is a small parameter. When the next element is zeroed then

$$\tan 2\Theta = \frac{2\epsilon K}{O(K)} = O(\epsilon)$$

That is, $\Theta \approx \epsilon$, consequently $s \approx \epsilon$, and $c \approx 1$. In this rotation, the other off-diagonal terms are changed by

$$K_{ri}^{(k+1)} = K_{ri}^{(k)} - K_{rj}^{(k)}\epsilon = K_{ri}^{(k)} - \epsilon^2 K$$

In other words, an element that had already been zeroed, say, will change by only $\epsilon^2 K$. Thus we say the convergence is *quadratic*, which means that once the method has begun to converge then the rate of convergence is very rapid.

In estimating the total computational cost, we have that one sweep uses about $\frac{1}{2}N(N-1)$ Jacobi rotations. Typical large matrices require about 10 sweeps to achieve convergence, or about $5N^2$ Jacobi rotations. Each rotation requires of order $4N$ operations, and so the total labor is at least $20N^3$ operations. Calculation of the eigenvectors as well as the eigenvalues changes the operation count from $4N$ to $6N$ per rotation. The total cost is about

$$cost = 30N^3$$

As an idea of the meaning of this total computational cost, consider running this on our benchmark machine of 1 MFlops. The time to solve a small system of order 100 is approximately

$$time = 30 \times (100)^3 \div (1.0 \times 10^6) = 30 \, secs$$

Now suppose the system size is 1000, then the time is increased to 8 hours. A large system of 10000 degrees of freedom would take nearly a full year !

It is obvious that the Jacobi scheme is not to be used for solving large systems, i.e., those larger than about 100. This rules it out from use on structural systems, unless some scheme is developed for reducing the size of the structural system. In fact, it is developed here precisely because it plays a fundamental role in the subspace iteration scheme to be developed later.

**Example 13.6:**   Use Jacobi rotations to find the complete eigensolution to the following matrix

$$\begin{bmatrix} 5 & 2 & -1 \\ 2 & 2 & 2 \\ -1 & 2 & 5 \end{bmatrix}$$

This matrix was chosen because its solution will exhibit some of the robust features of the Jacobi rotation method. It is straight forward to show that the characteristic equation is

$$\lambda(\lambda - 6)(\lambda - 6) = 0$$

This system has one root that is zero, and two roots that are the same (repeated roots). Both of these situations are often challenging to many eigensolver algorithms. The modal matrix is

$$[\,\Phi\,] = \begin{bmatrix} 1 & 1 & 1 \\ -2 & x & y \\ 1 & -1+2x & -1+2y \end{bmatrix}, \qquad y = \frac{2(x-1)}{(5x-2)}$$

Any independent choice of $x$ and $y$ will give the eigenvectors associated with the repeated roots as orthogonal to the first vector. The relation between $x$ and $y$ forces these two vectors to be orthogonal to each other. Even so, there are many values of $x$ that will make all three vectors orthogonal.

We begin the sweep by zeroing the first off-diagonal term. Thus, for $i = 1$, $j = 2$, get $c = .089442$, $s = .044721$ giving for the matrices $[\,Q\,]^T[\,K\,][\,Q\,]$ and $[\,\Phi\,][\,Q\,]$, respectively,

$$
\begin{bmatrix}
6.00000 & 0.00000 & 0.00000 \\
0.00000 & 1.00000 & 2.23606 \\
0.00000 & 2.23606 & 5.00000
\end{bmatrix},
\qquad
\begin{bmatrix}
.089442 & -.044721 & 0.00000 \\
.044721 & .089442 & 0.00000 \\
0.00000 & 0.00000 & 1.00000
\end{bmatrix}
$$

Note that initially $[\,\Phi\,] = [\,I\,]$. The next non-zero off-diagonal term at $i = 2$, $j = 3$, for which we get $c = .091287$, $s = -.040824$ giving for the two matrices

$$
\begin{bmatrix}
6.00000 & 0.00000 & 0.00000 \\
0.00000 & 0.00000 & 0.00000 \\
0.00000 & 0.00000 & 6.00000
\end{bmatrix},
\qquad
\begin{bmatrix}
.089442 & -.040824 & -.018257 \\
.044721 & .081649 & .036514 \\
0.00000 & -.040824 & .091287
\end{bmatrix}
$$

This completes the first sweep. It turns out that the process has converged in one sweep; this is usually not the case and more sweeps must be performed. The eigenvalues are the same as the exact values, and the eigenvectors satisfy the orthogonality condition.

It is important to realize that no special precautions were used in order to ensure the orthogonality of the repeated roots; this is a natural consequence of using orthogonal transformations.

# 13.5   Generalized Jacobi Method

The natural eigenproblem in structural analysis is the non-standard case; we now present a generalized Jacobi method which operates directly on $[\,K\,]$ and $[\,M\,]$. The algorithm is a natural extension of the standard Jacobi solution scheme and reduces to it when $[\,M\,]$ is the identity matrix.

In the generalized Jacobi iteration, we use the following transformation matrix:

$$
[Q^k] =
\begin{bmatrix}
1 \\
 & \ddots \\
 & & 1 & & \alpha \\
 & & & \ddots \\
 & & \gamma & & 1 \\
 & & & & & \ddots \\
 & & & & & & 1
\end{bmatrix}
$$

where the constants $\alpha$ and $\gamma$ are selected in such a way as to reduce the off-diagonal elements $K_{ij}$ and $M_{ij}$ to zero, *simultaneously*. Therefore, the values of $\alpha$ and $\gamma$ are a function of the stiffness elements $K_{ij}^{(k)}$, $K_{ii}^{(k)}$, $K_{jj}^{(k)}$ and the mass elements $M_{ij}^{(k)}$, $M_{ii}^{(k)}$,

$M_{jj}^{(k)}$. The rotation multiplications gives for the stiffness terms

$$
\begin{aligned}
K_{ri}^{(k+1)} &= K_{ri}^{(k)} + \gamma K_{rj}^{(k)} \\
K_{rj}^{(k+1)} &= K_{rj}^{(k)} + \alpha K_{ri}^{(k)} \\
K_{ii}^{(k+1)} &= K_{ii}^{(k)} + \gamma^2 K_{jj}^{(k)} + 2\gamma K_{ij}^{(k)} \\
K_{jj}^{(k+1)} &= K_{jj}^{(k)} + \alpha^2 K_{ii}^{(k)} + 2\alpha K_{ij}^{(k)} \\
K_{ij}^{(k+1)} &= K_{ij}^{(k)}(1 + \alpha\gamma) + \gamma K_{ii}^{(k)} + \alpha K_{jj}^{(k)}
\end{aligned}
\tag{13.9}
$$

The corresponding relations for the mass terms are similar. It is interesting to note that these relations are already in the form of the original term plus a correction; this occurred since the rotation matrix has unity on all diagonal elements. We now use the condition that the off-diagonal element be zero, hence obtain from the last equation

$$
(1 + \alpha\gamma)K_{ij}^{(k)} + \alpha K_{ii}^{(k)} + \gamma K_{jj}^{(k)} = 0, \qquad (1 + \alpha\gamma)M_{ij}^{(k)} + \alpha M_{ii}^{(k)} + \gamma M_{jj}^{(k)} = 0
$$

These equations are of the same form as each other, and are sufficient to determine the coefficients. To solve for them, first note that

$$
\alpha = -\frac{M_{ij}^{(k)} + \gamma M_{jj}^{(k)}}{M_{ii}^{(k)} + \gamma M_{ij}^{(k)}}
$$

and by defining the products

$$
A \equiv K_{jj}^{(k)} M_{ij}^{(k)} - M_{jj}^{(k)} K_{ij}^{(k)}, \qquad B \equiv K_{ii}^{(k)} M_{jj}^{(k)} - K_{jj}^{(k)} M_{ii}^{(k)}, \qquad C \equiv K_{ii}^{(k)} M_{ij}^{(k)} - M_{ii}^{(k)} K_{ij}^{(k)}
$$

we see that the $\gamma$ coefficient can be obtained from the following quadratic equation

$$
A\gamma^2 - B\gamma - C = 0
$$

Solve this to get

$$
\gamma = \frac{-\beta}{A} = -\frac{C}{\beta}, \qquad \alpha = \frac{A}{\beta}
$$

where $\beta$ is determined using

$$
\beta \equiv \tfrac{1}{2}B + \tfrac{1}{2}\mathrm{sign}(B)\sqrt{B^2 + 4AC}
$$

The above values for $\alpha$ and $\gamma$ have been chosen so that $\det[Q^k] \neq 0$.

If it should happen that the submatrices considered are scalar multiples of each other, that is,

$$
\frac{K_{ii}^{(k)}}{M_{ii}^{(k)}} = \frac{K_{jj}^{(k)}}{M_{jj}^{(k)}} = \frac{K_{ij}^{(k)}}{M_{ij}^{(k)}}
$$

then $A = B = C = 0$, and the above formulas break down. However, this is a trivial case and by choosing $\alpha = 0$ and $\gamma = -K_{ij}^{(k)}/K_{jj}^{(k)}$ we can proceed. Note that if the mass is already diagonalized and in unit form, the transformation relations yield $\alpha = -\gamma$, and we recognize that $[Q^k]$ is a multiple of the rotation matrix used in the standard Jacobi method.

The complete solution process is analogous to the standard Jacobi iteration method, the differences lie in that now a mass coupling factor must also be calculated, and the transformation is applied to both $[K^k]$ and $[M^k]$ simultaneously. Convergence is measured by comparing successive eigenvalue approximations and by testing if all off-diagonal elements are small enough. That is, at the end of each sweep, we test for convergence by

$$\lambda_m^{(k+1)} = \frac{K_{mm}^{(k+1)}}{M_{mm}^{(k+1)}}, \qquad \frac{|\lambda_m^{(k+1)} - \lambda_m^{(k)}|}{\lambda_m^{(k+1)}} \leq 10^{-2s}, \qquad \text{all } m = 1, \ldots, N$$

and

$$\left[\frac{K_{ij}^{(k+1)} K_{ij}^{(k+1)}}{K_{ii}^{(k+1)} K_{jj}^{(k+1)}}\right] \leq 10^{-2s}, \qquad \left[\frac{M_{ij}^{(k+1)} M_{ij}^{(k+1)}}{M_{ii}^{(k+1)} M_{jj}^{(k+1)}}\right] \leq 10^{-2s}, \qquad \text{all } i, j \quad i < j, \quad j = 1, N$$

where again $10^{-2s}$ is the convergence tolerance. Note that we test the off-diagonal terms of both the stiffness *and* mass matrices.

## Computer Algorithm

The steps of the algorithm can now be stated as:

**Step 1:** Initialize the eigenvalues and eigenvectors as

$$\lambda_m = K_{mm}/M_{mm}, \qquad [\,\Phi\,] = [\,I\,]$$

**Step 2:** Begin a sweep. Use a sliding threshold based on the current number of sweeps $q$

$$\epsilon_{threshold} = 10^{-4q}$$

Hence by the fourth or so sweep, this is effectively zero.

**Step 3:** Sweep over all $(i, j)$ off-diagonal elements of $[K^k]$ and $[M^k]$ in the upper triangle and evaluate

$$\frac{K_{ij}^{(k)} K_{ij}^{(k)}}{K_{ii}^{(k)} K_{jj}^{(k)}} \leq \epsilon_{threshold}, \qquad \frac{M_{ij}^{(k)} M_{ij}^{(k)}}{M_{ii}^{(k)} M_{jj}^{(k)}} \leq \epsilon_{threshold}$$

If both inequalities are satisfied, do not do a rotation and look at the next element. If either criterion is not satisfied then proceed.

**Step 4:** Evaluate the rotation parameters $\alpha$ and $\gamma$ from

$$A = K_{jj}^{(k)} M_{ij}^{(k)} - M_{jj}^{(k)} K_{ij}^{(k)}, \quad B = K_{ii}^{(k)} M_{jj}^{(k)} - K_{jj}^{(k)} M_{ii}^{(k)}, \quad C = K_{ii}^{(k)} M_{ij}^{(k)} - M_{ii}^{(k)} K_{ij}^{(k)}$$

$$\beta = \tfrac{1}{2} B + \tfrac{1}{2} \text{sign}(B) \sqrt{B^2 + 4AC}, \qquad \gamma = -\frac{C}{\beta}, \qquad \alpha = \frac{A}{\beta}$$

**Step 5:** Perform a generalized rotation in place

$$[K^{(k+1)}] = [Q^{(k)}]^T [K^{(k)}][Q^{(k)}], \qquad [M^{(k+1)}] = [Q^{(k)}]^T [M^{(k)}][Q^{(k)}]$$

Only rows $(i, j)$ and columns $(i, j)$ are affected.

**Step 6:** Update the eigenvectors in place

$$[\Phi^{k+1}] = [\Phi^k][Q^k]$$

Only columns $(i, j)$ are affected.

**Step 7:** If off-diagonal terms are still remaining, then go to Step 3. Otherwise, this completes a sweep.

**Step 8:** Update the eigenvalues

$$\lambda_m^{(k+1)} = K_{mm}^{(k+1)} / M_{mm}^{(k+1)}$$

**Step 9:** Check for convergence on the eigenvalues

$$\frac{|\lambda_m^{(k+1)} - \lambda_m^{(k)}|}{\lambda_m^{(k+1)}} \leq 10^{-2s}$$

where generally $2s = 12$. If this is not satisfied for all eigenvalues, then go to Step 2 to begin another complete sweep. Otherwise proceed.

**Step 10:** Check convergence to diagonal matrices. This is similar to the threshold check on the off-diagonal terms.

$$\frac{K_{ij}^{(k+1)} K_{ij}^{(k+1)}}{K_{ii}^{(k+1)} K_{jj}^{(k+1)}} \leq 10^{-2s}, \qquad \frac{M_{ij}^{(k+1)} M_{ij}^{(k+1)}}{M_{ii}^{(k+1)} M_{jj}^{(k+1)}} \leq 10^{-2s}$$

where generally $2s = 12$. If neither of these are satisfied for any off-diagonal term, then go to Step 2 to begin another complete sweep. Otherwise convergence has been achieved.

**Step 11:** Once convergence and diagonal matrices have been obtained, then scale the eigenvectors by the diagonal mass matrix

$$\{\phi\}_m = \{\phi\}_m / \sqrt{M_{mm}}$$

**Step 12:** Sort the eigenvalues in ascending order.

The convergence of this algorithm is quadratic (Reference [48]) once the off-diagonal elements are small, and therefore little extra cost is involved in solving the eigensystem to high accuracy once an approximate solution has been obtained. Typically, about six to twelve sweeps are required for solution of the eigensystem to high accuracy. Although convergence is usually rapid, a maximum limit is set at 15 or 20 sweeps in case the system is ill-conditioned.

The computational cost of this algorithm is about

$$cost = q[6 + 12 + 4N + 2N]\tfrac{1}{2}N^2$$

where $q$ is the number of sweeps. Note that the computation of the eigenvectors is only a 50 percent overhead so is automatically performed. For ten sweeps, this approximates to

$$cost = 30N^3$$

The total number of operations in one sweep, as given here, are an upper bound because it is assumed that both matrices are full and that all off-diagonal elements are zeroed (i.e., the threshold tolerance is never passed). In practice, and particularly as part of the subspace iteration scheme to be presented next, the matrices are nearly diagonal and the computational cost is significantly less than the above.

One of the weaker parts of this algorithm, especially from the microcomputer perspective, is its memory requirements. The modal matrix is a full $[N \times N]$ matrix irrespective of the bandedness of the structural matrices. Further, the working array must also be of order $[N \times N]$ even though it eventually becomes diagonal. Thus, in the general case, the memory requirements are

$$Memory = 3[N \times N]$$

From this it is apparent that the largest system size (if about 1 MByte of RAM is available) can be no bigger than 160 degrees of freedom ($160 \times 160 \times 3 \approx 80000$) since this corresponds (using double precision variables) to about $600kB$ of data and the rest for code. As shown above, however, it would be unwise to attempt to solve problems any bigger than this because of the time involved.

In the next section we develop another scheme that can solve the partial eigenproblem and therefore requires less storage.

**Example 13.7:**   Use the generalized Jacobi rotation scheme to obtain the eigenvalues and eigenvectors of the shear building problem of Figure 13.2.

After 6 sweeps involving 20 rotations the process converges. The rotations per sweep were 6, 6, 3, 3, 1, 1. The following are the computed eigenpairs given in the

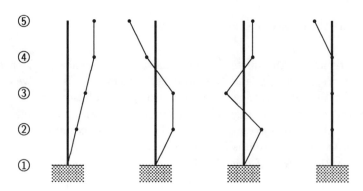

**Figure 13.5**: Mode shapes for shear building.

form $(\lambda, \{\phi\})$.

$$\left(98.4332, \begin{Bmatrix} .07305 \\ .13172 \\ .16444 \\ .16476 \end{Bmatrix} \times 10^{-2}\right), \qquad \left(774.261, \begin{Bmatrix} -.16463 \\ -.07416 \\ .13122 \\ .13328 \end{Bmatrix} \times 10^{-2}\right),$$

$$\left(1620.70, \begin{Bmatrix} .13251 \\ -.16477 \\ .07236 \\ .07478 \end{Bmatrix} \times 10^{-2}\right), \qquad \left(50473.5, \begin{Bmatrix} -.00000 \\ .00022 \\ -.02247 \\ 2.2247 \end{Bmatrix} \times 10^{-2}\right)$$

These are shown schematically in Figure 13.5. Note that the first three exhibit a cantilever beam type behavior in that each higher mode has an additional zero. The fourth mode is particularly interesting because all of the deformation is in the water tower.

**Example 13.8:**  Use modal superposition to get the response for the building of Figure 13.2.

The modal parameters are calculated to be

| mode | $\omega_m$ | $\tilde{K}_{mm}$ | $\tilde{M}_{mm}$ | $\zeta_m$ | $\tilde{P}_m$ |
|------|-----------|------------------|------------------|-----------|---------------|
| 1 | 9.92135 | 98.4332 | 1.00000 | .00000 | +.730568E-03 |
| 2 | 27.8255 | 774.260 | 1.00000 | .00000 | -.164631E-02 |
| 3 | 40.2578 | 1620.69 | 1.00000 | .00000 | +.132511E-02 |
| 4 | 224.663 | 50473.4 | 1.00000 | .00000 | -.229261E-07 |

Note that because the eigenvectors are orthonormalized, the modal masses are unity and that, numerically, $\tilde{K}_{mm} = \omega_m^2$. Also, although a single force is applied there are significant non-zero modal forces for the first three modes.

When all the modes are used then the direct method and the modal superposition give the exact same results as can be seen from Table 13.2. Actually, the inclusion of only three modes also gives the exact response.

| time | direct | 4 modes | 3 modes | 2 modes | 1 mode |
|------|--------|---------|---------|---------|--------|
| 1.00 | .50639E-12 | .50639E-12 | .50639E-12 | .40280E-12 | .12765E-12 |
| 2.00 | .98823E-05 | .98823E-05 | .98823E-05 | .87935E-05 | .53057E-05 |
| 3.00 | .53805E-05 | .53805E-05 | .53805E-05 | .48576E-05 | .31136E-05 |
| 4.00 | -.63817E-06 | -.63817E-06 | -.63817E-06 | -.66953E-06 | -.66986E-06 |
| 5.00 | .80883E-06 | .80883E-06 | .80883E-06 | .84876E-06 | .84823E-06 |

Table 13.2: Modal responses for Node 2 at selected times.

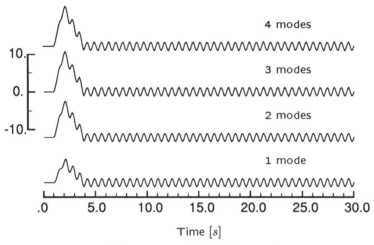

Figure 13.6: Modal Comparison.

The extended response is shown in Figure 13.6. We conclude that for the force histories given only three modes are of interest. Even two modes could be sufficient. This example shows that it is desirable to have a scheme that can evaluate only the lower modes.

It is instructive to correlate these results with the frequency spectrums of the forces given in Figure B.3. There it is shown that all significant frequency components are less than $10\,Hz$ or $60\,r/s$.

## 13.6 Subspace Iteration

The number of modes of a structure is equal to the number of degree of freedoms. In vibration studies, however, it is usually the lower order resonances which are of interest, so it would be useful if the number of degrees of freedom could be reduced in such a way to be comparable to the number of modes of interest. This can be done by using a technique known as *static condensation*; the algorithm is usually called *Guyan reduction* in the context of a vibration problem. The basic idea is that if the mass associated with a particular degree of freedom is small, then the contribution that this degree of freedom makes to the overall dynamics is also small. Hence we can write

$$u_i = -\sum_{j \neq i} K_{ij} u_j$$

and condense the system by removing $u_i$. The literature is large on this topic and Reference [12] can be used as a beginning.

The number of reduced degrees of freedom may be as high as 90% of the total number. But this is purely a question of how well the degrees of freedom were selected. In fact, the main objection to static condensation is that the accuracy obtainable depends precisely on this choice, and this is a difficult task to automate as part of an algorithm. Another objection to this method is that the bandwidth of the system matrices increase as the degrees of freedom are reduced; if a sufficient number are reduced we are left with nearly fully populated matrices that are expensive to solve. A final objection is that the criterion for elimination of a degree of freedom is based on physical quantities (size of mass) and does not utilize dynamical information. For example, consider modeling a rod with many elements of the same size; this gives rise to stiffness and mass matrices where most of the diagonal elements are equal. Thus we could not decide which degrees of freedom to reduce even though we know the modal frequencies are quite distinct. We therefore prefer the dynamical approach to reduction that is presented in this section.

Recall that the Jacobi method is very stable, but all the eigenvalues must be obtained at once. Conversely, inverse iteration can obtain as few as a single eigenvalue but can become unstable when used to determine a sequence of closely spaced eigenvalues. The *subspace iteration* scheme attempts to combine the best of these two methods in that a reduced eigensystem is first established (by iteration) and all of its eigenvectors found. Care will be taken to insure that the subspace spans the range of eigenvalues of interest, and therefore it is not likely that any eigenvalues are missed. The point of the subspace approach (in comparison to vector iteration) is that it is much easier to establish an $m$-dimensional subspace which spans the actual vector space than to find $m$ vectors that individually are close to the required eigenvectors. That is, because iteration is performed with a subspace, convergence of the subspace is easier to achieve. The meaning of this will become clearer later.

## Basic Approach

The basic objective in the subspace iteration method is to solve for the lowest $m$ eigenvalues and corresponding eigenvectors satisfying the system

$$[\,K\,][\,\Phi\,] = \lceil\,\Lambda\,\rfloor[\,M\,][\,\Phi\,]$$

where $\lceil\,\Lambda\,\rfloor = \text{diag}\,(\lambda_m)$ and $[\,\Phi\,] = [\{\phi\}_1, \ldots, \{\phi\}_m]$ is the modal matrix of size $[N \times m]$. This notation emphasizes that multiple eigenpairs will be considered simultaneously. The algorithm attempts to find an orthogonal basis of vectors so as to preserve stability in the iteration process. In this way, the trial vectors will converge to different modes shapes and not all to lowest one.

Borrowing ideas from inverse iteration, suppose we set up the following iteration scheme. For the iteration steps $k = 1, 2 \ldots$, solve

$$[K][\Psi]_{k+1} = [\tilde{R}]_k = [M][\Psi]_k$$

starting with an initial guess for $[\Psi]_1$. If this system is solved iteratively for the individual vectors in $[\Psi]$, then they will all eventually converge to the same fundamental mode. This follows from our discussion of the inverse iteration method of Chapter 7. Therefore, before proceeding with the next iteration step, the vectors in $[\Psi]$ must be orthogonalized so that they will converge to different eigenvectors and not all to the lowest one. Further, the eigenvectors should be normalized in some way so that the numbers remain within reasonable bounds. The two requirements are satisfied by solving a new eigenvalue problem and using its eigenvectors (which we know to be orthogonal) as the basis for forming our new guesses.

The approach will be motivated by way of a Ritz analysis. Indeed, we may think of the subspace iteration as a repeated application of the Ritz method in which the eigenvector approximations calculated in the previous iteration are used to form the right-hand side load vectors in the current iteration. Consider a typical vector taken from $[\Psi]$ and expanded in terms of a second set of vectors

$$\{\psi\}_i = c_{1i}\{\tilde{\psi}\}_1 + c_{2i}\{\tilde{\psi}\}_2 + \ldots = \sum_j c_{ji}\{\tilde{\psi}\}_j = [\tilde{\Psi}]\{c\}_i$$

where $\{\tilde{\psi}\}_j$ are known guesses for the eigenvectors. We now wish to find that set of coefficients $c_{ji}$ which give the combination of $\{\tilde{\psi}\}_j$ that is a "best" solution for the problem. Using this expansion in the potential energy expression gives for each mode

$$
\begin{aligned}
\Pi_i &= \{\psi\}_i^T[K]\{\psi\}_i - \lambda_i\{\psi\}_i^T[M]\{\psi\}_i \\
&= \sum_j \sum_k c_{ji}c_{ki}\{\tilde{\psi}\}_j^T[K]\{\tilde{\psi}\}_k - \lambda_i \sum_j \sum_k c_{ji}c_{ki}\{\tilde{\psi}\}_j^T[M]\{\tilde{\psi}\}_k
\end{aligned}
$$

To make the representation more compact, we can introduce new stiffness and mass matrices whose elements are defined as

$$\tilde{K}_{jk} \equiv \{\tilde{\psi}\}_j^T[K]\{\tilde{\psi}\}_k, \qquad \tilde{M}_{jk} \equiv \{\tilde{\psi}\}_j^T[M]\{\tilde{\psi}\}_k$$

If the vectors $\{\tilde{\psi}\}_j$ are orthogonal (with respect to $[K]$ and $[M]$) then the reduced matrices $[\tilde{K}]$ and $[\tilde{M}]$ would be diagonal. At present, since $\{\tilde{\psi}\}_j$ are just arbitrary guesses these matrices are generally populated. We are now in a position to write the total potential as

$$\Pi_i = \sum_j \sum_k c_{ji}c_{ki}\tilde{K}_{jk} - \lambda_i \sum_j \sum_k c_{ji}c_{ki}\tilde{M}_{jk} = \{c\}_i^T[\tilde{K}]\{c\}_i - \lambda_i\{c\}_i^T[\tilde{M}]\{c\}_i$$

Realizing that the only parameters that can be varied in this relationship are the coefficients $c_{ni}$, then the necessary condition for $\Pi_i$ to be a stationary value is that

$$\frac{\partial \Pi_i}{\partial c_{ni}} = 0$$

Carrying out this partial differentiation, and regrouping terms gives

$$\sum_k \left[ \tilde{K}_{nk} - \lambda_i \tilde{M}_{nk} \right] c_{ki} = 0$$

This, of course, is an eigenvalue problem. Indeed, each vector term gives rise to the same equation, thus we can write in matrix form

$$[\tilde{K}]\{c\}_i = \lambda_i [\tilde{M}]\{c\}_i$$

where the stiffness and mass matrices are of order $[m \times m]$. The collection of vectors $[C]$ (which are the eigenvector solutions of this reduced eigenvalue problem) can be thought of as Ritz solutions. There are $m$ of them, and each is of length $m$. This eigenvalue problem is of a comparatively small size, being of the order $[m \times m]$.

Returning now to our original problem, we find that the improved guesses for the eigenvectors

$$\{\psi\}_i = c_{1i}\{\tilde{\psi}\}_1 + c_{2i}\{\tilde{\psi}\}_2 + \ldots = \sum_j c_{ji}\{\tilde{\psi}\}_j = [\tilde{\Psi}]\{c\}_i$$

are now automatically orthogonal with respect to the original stiffness and mass matrices. This is easily shown by

$$[\Psi]^T [K][\Psi] = [C]^T [\tilde{\Psi}]^T [K][\tilde{\Psi}][C] = [C]^T [\tilde{K}][C] = \lceil \Lambda \rfloor$$

A similar result is obtained for the mass. The key point is that the coefficients are orthogonal and this was achieved by identifying them as the eigenvectors of a reduced eigenvalue problem. In using this scheme to orthogonalize the trial vectors, we are assured that the trial vectors will therefore converge to different mode shapes. That is, provided the vectors in $[\Psi]_1$ are not orthogonal to one of the required eigenvectors, we have in the limit for large iterations

$$\lceil \Lambda^{k+1} \rfloor \to \lceil \Lambda \rfloor \quad \text{and} \quad [\Psi^{k+1}] \to [\Phi] \quad \text{as} \quad k \to \infty$$

Before this outline is stated formally as an algorithm, a few other issues must be addressed first. These concern the starting guesses and the conververgence criterion; both are considered in the next few sections. An observation worth making now is that the reduced arrays $[\tilde{K}^{k+1}]$ and $[\tilde{M}^{k+1}]$ tend toward diagonal form as the number of iterations increase. Before that stage, however, they have no special character (except symmetry); hence, it follows from our earlier discussion that the generalized Jacobi method is an effective choice for the solution of this reduced eigenproblem.

## Starting Iteration Vectors

As in any iteration scheme, good starting values can help the rate of convergence. To motivate the selection of the starting eigenvectors, we will first consider a couple of simple special cases.

Suppose there are only $m$ non-zero masses in a diagonal mass matrix then there are only $m$ finite resonances. Choosing the starting vectors as unit vectors with the entries +1 corresponding to the mass degrees of freedom, then the iteration converges in one step. (Keep in mind, of course, that an $[m \times m]$ eigenvalue problem must still be solved.) In this case, we have effectively done a static condensation analysis (or a Guyan reduction). This follows because a subspace iteration embodies a Ritz analysis.

As a second case, consider when both $[K]$ and $[M]$ are diagonal. Again, if the iteration vectors are unit vectors with the entries +1 corresponding to those degrees of freedom that have the largest ratios $M_{ii}/K_{ii}$, the iteration also converges in one step. These vectors are, in effect, the eigenvectors corresponding to the smallest eigenvalues, and this is why convergence is achieved in one step.

These results concerning the construction of the starting iteration vectors indicate how an effective set of starting iteration vectors may be selected. The first column in $[R]_1$ is the diagonal of the mass matrix $[M]$. This assures that all mass degree of freedoms are excited. The other columns are unit vectors with entries +1 at the coordinates with the largest ratios $M_{ii}/K_{ii}$. Just in case this set is orthogonal to one of the desired vectors, the last column is composed of a random vector. In fact, during each iteration, the last column is reset to being random.

**Example 13.9:**  Consider a system described by the following diagonal matrices:

$$[K] = \begin{bmatrix} 4 & & & \\ & 3 & & \\ & & 2 & \\ & & & 1 \end{bmatrix}, \qquad [M] = \begin{bmatrix} 3 & & & \\ & .1 & & \\ & & 1 & \\ & & & 2 \end{bmatrix}$$

Obtain the starting vectors.

The ratios $M_{ii}/K_{ii}$ are $\{0.75, 0.03, 0.5, 2.0\}$. The most effective starting vectors to use would therefore be

$$[\Psi]_1 = \left[ \begin{Bmatrix} 0 \\ 0 \\ 0 \\ 1 \end{Bmatrix} \begin{Bmatrix} 1 \\ 0 \\ 0 \\ 0 \end{Bmatrix} \begin{Bmatrix} 0 \\ 0 \\ 1 \\ 0 \end{Bmatrix} \begin{Bmatrix} 0 \\ 1 \\ 0 \\ 0 \end{Bmatrix} \right]$$

The subspace stiffness and mass matrices formed are now

$$[\tilde{K}] = \begin{bmatrix} 1 & & & \\ & 4 & & \\ & & 2 & \\ & & & 3 \end{bmatrix}, \qquad [\tilde{M}] = \begin{bmatrix} 2 & & & \\ & 3 & & \\ & & 1 & \\ & & & 0 \end{bmatrix}$$

These are also diagonal.

The general scheme, however, would give

$$[\Psi]_1 = \begin{bmatrix} 3 & 0 & 1 & 1 \\ .1 & 0 & 0 & -1 \\ 1 & 0 & 0 & 1 \\ 2 & 1 & 0 & -1 \end{bmatrix}$$

The last column would normally be a properly generated random vector. Also, each column should be normalized in some way. The reduced matrices are

$$[\tilde{K}] = \begin{bmatrix} 42.03 & 2 & 12 & 11.7 \\ 2 & 1 & 0 & -1 \\ 12 & 0 & 4 & 4 \\ 11.7 & -1 & 4 & 10 \end{bmatrix}, \qquad [\tilde{M}] = \begin{bmatrix} 36.001 & 4 & 9 & 5.99 \\ 4 & 2 & 0 & -2 \\ 9 & 0 & 3 & 3 \\ 5.99 & -2 & 3 & 6.1 \end{bmatrix}$$

Note that these are fully populated. The eigenvalue solution gives the coefficients matrix as

$$[C] = \begin{bmatrix} 0 & 0 & .91 & -2.88 \\ .71 & 0 & -1.73 & 8.62 \\ 0 & .58 & -2.82 & 5.75 \\ 0 & 0 & .10 & 2.88 \end{bmatrix}$$

Finally, new estimates for the starting vectors are

$$[\tilde{\Psi}] = \begin{bmatrix} 0 & .58 & 0 & 0 \\ 0 & 0 & 0 & 2.6 \\ 0 & 0 & 1 & 0 \\ .71 & 0 & 0 & 0 \end{bmatrix}$$

Aside from a scaling factor, we see that this is the same as first obtained. Therefore on the next iteration the reduced matrices will be diagonalized.

## Convergence Criterion

Assume that between two consecutive iterations, the eigenvalue approximations $\lambda^k$ and $\lambda^{k+1}$, have been calculated. Convergence is measured by requiring that

$$\frac{|\lambda^{k+1} - \lambda^k|}{|\lambda^{k+1}|} \leq 10^{-2s}$$

where the eigenvalues are to be accurate to about $2s$ digits. This criterion is applied to all $m$ eigenvalues until it is satisfied. For example, if we iterate until all ratios are smaller that $10^{-8}$, we very likely have that $\lambda_m$ has been approximated to about eight-digit accuracy, and the lower eigenvalues have been evaluated more accurately. The eigenvector approximations are accurate to only about $s$ digits.

In practice, the iteration is performed with $r$ vectors where

$$r = \min(2m, m + 10)$$

(as recommended in References [5, 38]) but convergence is measured only on the approximations obtained for the $m$ smallest eigenvalues. The number of iterations required for convergence depends on the system matrices and on the accuracy required in the eigenvalues and eigenvectors. Generally, with the starting vectors described above on the order of ten iterations are needed. The typical value $2s = 10$ may seem a bit low; however, because the procedure is basically a Ritz method convergence is quite rapid in the lower eigenvalues. Thus by the time the higher modes have converged, the lower modes have achieved an even greater accuracy.

Another important aspect when using the subspace iteration technique is in verifying that all the required eigenvalues and vectors have indeed been calculated. Once the convergence limit is satisfied, we can do this by making use of the Sturm sequence property of the system matrices. This property yields that in the factorization of a system shifted by $\mu$, that is,

$$[K - \mu M] \rightarrow [\, U \,]^T \lceil \, D \, \rfloor [\, U \,]$$

the number of negative elements in $\lceil \, D \, \rfloor$ is equal to the number of eigenvalues smaller the $\mu$. Hence, for a shift corresponding to the largest eigenvalue $\lambda_m$, we should have $m$ negative elements in $\lceil \, D \, \rfloor$. In this counting only the smallest eigenvalues that converged to a tolerance of $10^{2s}$ should be included. This is applied to the last cluster, and only if that indicates a discrepancy is the full Sturm sequence check performed.

## Computer Algorithm

All the pieces are now gathered together in order to state the algorithm. The basic outline is similar to that of inverse iteration but with two differences. First, a number of vectors are iterated on simultaneously and this will be signified by the arrays being rectangular. Second, the vector guesses are kept orthogonal by solving a reduced eigenvalue problem.

The steps of the algorithm may now be stated as:

**Step 1:** In preparation for solving a system of equations (the pseudo-static problem), decompose the stiffness matrix as:

$$[\, K \,] = [\, U \,]^T \lceil \, D \, \rfloor [\, U \,]$$

At this stage the system is checked for rigid body modes. If some are detected, a shift is performed and the system decomposed again.

**Step 2:** Pick a set of $r$ initial trial vectors $[\, R \,]_1$ with the last being random.

**Step 3:** Estimate improved vectors $[\tilde{\Psi}^{k+1}]$ by solving the system

$$[\, K \,][\tilde{\Psi}^{k+1}] = [R^k]$$

**Step 4:** Form the reduced stiffness matrix as

$$[\tilde{K}^{k+1}] = [\tilde{\Psi}^{k+1}]^T[K][\tilde{\Psi}^{k+1}] = [\tilde{\Psi}^{k+1}]^T[R^{k+1}]$$

**Step 5:** Improved load vectors are obtained from

$$[Q^{k+1}] = [M][\tilde{\Psi}^{k+1}]$$

**Step 6:** Now form the reduced mass matrix as

$$[\tilde{M}^{k+1}] = [\tilde{\Psi}^{k+1}]^T[M][\tilde{\Psi}^{k+1}] = [\tilde{\Psi}^{k+1}]^T[Q^{k+1}]$$

This sequencing of Steps 4, 5 and 6 saves computations as well as memory.

**Step 7:** Use the Jacobi method to solve for the eigenpairs of the reduced system

$$[\tilde{K}^{k+1}]\{c\} = \lambda[\tilde{M}^{k+1}]\{c\}$$

**Step 8:** Form the coefficients matrix as

$$[C^{k+1}] = [\{c\}_1\{c\}_2 \ldots]$$

**Step 9:** Obtain new estimates of the load vector

$$[R^{k+1}] = [Q^{k+1}][C^{k+1}]$$

**Step 10:** Check for convergence. Convergence can be measured by comparing two successive values of $\lambda$, thus

$$\frac{\left|\lambda^{k+1} - \lambda^k\right|}{\lambda^{k+1}} \leq 10^{-2s}$$

This is done for each mode up to $m$. If convergence is not achieved in any one of the modes then go back to Step 3.

**Step 11:** When convergence has occurred (or the maximum number of iterations exceeded), find the final values for the eigenvectors from

$$[\Psi^{k+1}] = [\tilde{\Psi}^{k+1}][C^{k+1}]$$

Steps 3, 4, 5, 6, 7 and 8 are used to construct the right hand side.

**Step 12:** Check the quality of the eigenvectors by using the error norm

$$\|Z\| = \frac{\|[K]\{\psi^{k+1}\} - \lambda^{k+1}[M]\{\psi^{k+1}\}\|}{\|[K]\{\psi^{k+1}\}\|}$$

for each eigenpair. Ignore any eigenpairs whose error norm exceeds $10^{-2}$.

**Step 13:** Do a Sturm sequence check on the shifted system

$$[K - \mu M] \rightarrow [\,U\,]^T [\,D\,][\,U\,]$$

using the shift $\mu = \lambda_m$. If there is a discrepancy between this and the number actually found, then do a Sturm sequence check on each eigencluster to locate the problem area.

Note that normalization is not required (as was done with inverse iteration) because this occurs as part of the Jacobi rotations.

The subspace iteration method, as presented above, is most suitable for the solution of the smallest eigenvalues and corresponding eigenvectors (say, $m < 50$). If $q$ iterations are performed, then the computational cost is about

$$cost = \tfrac{1}{2}NB^2 + q\left[2NBr + \tfrac{1}{2}Nr^2 + NBr + \tfrac{1}{2}Nr^2 + 20r^2 + Nr^2\right] + 4Nm + \tfrac{1}{2}NB^2$$

Typically, the number of iterations is about 10 and since $m \ll N$, this cost may be approximated as

$$cost \approx NB^2 + 30NBm$$

We also used the assumption that for large systems $B \gg m$. From this cost formula we see that on our benchmark machine of 1 MFlops, the time to solve 20 modes of a medium system of order 1000 with 10% bandwidth is approximately

$$time = [1000 \times 100^2 + 30 \times 1000 \times 100 \times 20] \div (1.0 \times 10^6) \approx 1\,minute$$

A large problem of size 10000 would take about four hours. In comparison to the Jacobi method, the significant achievements of this algorithm is that advantage can be taken of the banded nature of the arrays, and that a partial solution can be found. Note that the cost per eigenvalue is approximately the same as for vector iterations with shifts. What is achieved here, however, is that multiple vectors are iterated on simultaneously and therefore multiple roots and eigenclusters are handled in a robust manner.

The computational cost for the solution of problems for a larger number of eigenpairs (say, $m > 50$) rises rapidly. This comes about because with $r = (m + 10)$ a relatively large number of iteration vectors is used throughout all subspace iterations, even though convergence of the smallest eigenvalues is generally already achieved in the first few iterations. As the number of iteration vectors $r$ increase, the storage requirements also increase significantly. Finally, we should note that the number of numerical operations required in the solution of the reduced eigenproblem becomes significant when $r$ is large, and these numerical operations cannot be neglected in the operations count. Reference [4] discusses a number of schemes for accelerating convergence of the subspace scheme.

**Example 13.10:** Use subspace iteration to find the first two eigenvalues and vectors for the problem of Example 7.5 in Chapter 7.

We will iterate on three vectors, making the third a random vector. In Step 2, we choose the initial matrix

$$[\,R\,]_1 = \begin{bmatrix} .18257 & .00000 & -.77284 \\ .36514 & .00000 & .02439 \\ .54772 & 1.0000 & -.62119 \\ .73029 & .00000 & -.12744 \end{bmatrix}$$

The last vector will always be made random. In Step 3, we begin the iteration by using the above matrix as a collection of load vectors and solving to get

$$[\,\tilde{\Psi}\,] = \begin{bmatrix} 3.1037 & 1.5000 & -3.7031 \\ 2.9211 & 1.5000 & -2.9303 \\ 2.3734 & 1.5000 & -2.1818 \\ 1.8257 & 1.0000 & -1.4970 \end{bmatrix}$$

The reduced stiffness and mass matrices are obtained in Steps 4,5, and 6

$$[\,\tilde{K}\,] = \begin{bmatrix} 4.2666 & 2.3734 & -4.0344 \\ 2.3734 & 1.5000 & -2.1818 \\ -4.0344 & -2.1818 & 4.3366 \end{bmatrix}, \quad [\,\tilde{M}\,] = \begin{bmatrix} 56.933 & 31.402 & -55.082 \\ 31.402 & 17.500 & -30.152 \\ -55.082 & -30.152 & 54.134 \end{bmatrix}$$

In Step 7, the reduced eigenvalue problem is solved using the Jacobi method. Convergence was achieved after 6 sweeps. The eigenvalues are

$$\lambda = .07459, .55897, 1.6464$$

The coefficients matrix is obtained from the eigenvectors in Step 8

$$[\,C\,] = \begin{bmatrix} .11221 & .44815 & -2.5733 \\ -.00885 & .70908 & 2.8420 \\ -.02599 & .86389 & -1.0391 \end{bmatrix}$$

We now form improved estimates of the loads in Step 9

$$[\,R\,] = \begin{bmatrix} .43128 & -.74456 & -.75005 \\ .78140 & -.31746 & .09559 \\ .92935 & .72714 & -.46123 \\ .93977 & .93593 & .46426 \end{bmatrix}$$

This completes the first iteration. At this stage we need to test convergence; we do not have any previous estimates, hence Step 10 becomes

$$1.000000 \le 1.0000 \times 10^{-8}$$

which is obviously not satisfied.

We will now just state the major results for the next few iterations.

$$ITERATION:2 \quad [\,\tilde{\Psi}\,] \;=\; \begin{bmatrix} 5.7968 & -1.3729 & -2.6137 \\ 5.3655 & -.62840 & -1.8637 \\ 4.1528 & .43361 & -1.2092 \\ 3.0818 & .60105 & -.65142 \end{bmatrix}$$

$$[\tilde{K}] = \begin{bmatrix} 13.448 & -.11534 & -4.3196 \\ -.11534 & 2.0996 & 1.0487 \\ -4.3196 & 1.0487 & 2.0376 \end{bmatrix}, \quad [\tilde{M}] = \begin{bmatrix} 180.91 & -1.8907 & -58.247 \\ -1.8907 & 4.6839 & 2.7917 \\ -58.247 & 2.7917 & 19.863 \end{bmatrix}$$

$$\lambda = .07433, .44844, 1.8595$$

$$[C] = \begin{bmatrix} .07462 & -.03004 & 1.0706 \\ -.00150 & .51452 & -1.5650 \\ .00090 & -.11056 & 3.3762 \end{bmatrix}, \quad [R] = \begin{bmatrix} .43228 & -.59161 & -.73462 \\ .79931 & -.55697 & .57694 \\ .92445 & .69610 & .35132 \\ .91391 & 1.1547 & .06344 \end{bmatrix}$$

$$CONV: \quad 3.4982 \times 10^{-3} \le 1.0000 \times 10^{-8}$$

$$ITERATION: 3 \quad [\tilde{\Psi}] = \begin{bmatrix} 5.8118 & -1.2641 & -.53839 \\ 5.3795 & -.67255 & .19623 \\ 4.1479 & .47602 & .35391 \\ 3.0699 & .70227 & .25709 \end{bmatrix}$$

$$[\tilde{K}] = \begin{bmatrix} 13.452 & -.00217 & .48625 \\ -.00217 & 2.2648 & .75246 \\ .48625 & .75246 & .64938 \end{bmatrix}, \quad [\tilde{M}] = \begin{bmatrix} 180.97 & -.03585 & 6.5433 \\ -.03585 & 5.1553 & 1.6442 \\ 6.5433 & 1.6442 & 1.0070 \end{bmatrix}$$

$$\lambda = .07433, .43869, 1.5592$$

$$[C] = \begin{bmatrix} .07433 & .00181 & -.07312 \\ -.00004 & .45546 & -.63409 \\ .00006 & -.04749 & 2.0189 \end{bmatrix}, \quad [R] = \begin{bmatrix} .43202 & -.53968 & -.76606 \\ .79983 & -.61180 & -.43086 \\ .92500 & .62254 & -.33247 \\ .91274 & 1.2528 & -.34199 \end{bmatrix}$$

$$CONV: \quad 1.0729 \times 10^{-5} \le 1.0000 \times 10^{-8}$$

Notice that to the number of digits shown the first eigenvalue has already converged. Complete convergence occurred after 8 iterations. The final round gives

$$[\tilde{\Psi}] = \begin{bmatrix} 5.8119 & -1.2302 & -1.2486 \\ 5.3799 & -.69106 & -.69856 \\ 4.1480 & .45385 & -.54574 \\ 3.0696 & .72794 & -.19845 \end{bmatrix}$$

$$[\tilde{K}] = \begin{bmatrix} 13.452 & .00000 & -1.7841 \\ .00000 & 2.2817 & .51746 \\ -1.7841 & .51746 & .60652 \end{bmatrix}, \quad [\tilde{M}] = \begin{bmatrix} 180.97 & .00000 & -24.001 \\ .00000 & 5.2062 & 1.1807 \\ -24.001 & 1.1807 & 3.5860 \end{bmatrix}$$

$$\lambda = .07433, .43826, 1.8681$$

$$[C] = \begin{bmatrix} .07433 & .00000 & .36070 \\ .00000 & .43826 & -.61680 \\ -.00000 & .00000 & 2.7197 \end{bmatrix}, \quad [\Psi] = \begin{bmatrix} .43202 & -.53917 & -.80549 \\ .39991 & -.30287 & -.50518 \\ .30834 & .19890 & .08841 \\ .22817 & .31903 & .29690 \end{bmatrix}$$

Because this example problem is so small, re-seeding the random vector on each iteration actually caused a slower rate of convergence.

# 13.7 Selecting a Dynamic Solver

Generally, modal and implicit direct methods are economical in transient vibration type problems where only a few modes are needed to describe the response. For example, an earthquake may excite only the lowest mode of a building, and a vibrating machine may excite only a narrow range of frequencies in its support structure. The direct integration methods are more economical in wave propagation and shock loading problems because many modes are excited. Under some circumstances, an explicit direct method might be favored if the mass is lumped.

The form of the matrices influence the choice of algorithm. The damping matrix $[C]$ must be orthogonal for the modal methods to work efficiently. It must be diagonal or null for an efficient explicit algorithm but need have no special form for an implicit algorithm. If $[M]$ is not diagonal, the cost of an explicit algorithm increases by a much greater factor than the cost of an implicit algorithm.

The discussions in this chapter and Chapter 7 show that several alternative methods are available for the solution of an eigenproblem. Which method is most effective in a particular case depends on the type of solution desired and the characteristics of the matrices involved. A transformation method is most useful when the matrices are of comparatively small order and are fully populated or have a large bandwidth, or all eigenvalues are required. It is an excellent scheme when good quality eigenvectors are required for all the modes. The Jacobi transformation method (and its generalized variant) is simple and stable, and one of the better methods in this category. The stiffness matrix need not be positive definite and the eigenvalues being sought may be negative, zero, or positive. The Jacobi method is not appropriate when the matrices involved are of a large order and only a few eigenvalues need to be determined.

Iteration methods are most effective when the matrices involved are of a large order and only a few eigenvalues are desired. They may be applied directly to linearized eigenvalue problems without the need of transforming the latter to a standard form. Inverse iteration leads to the lowest eigenvalue. When an eigenvalue other than the most dominant or the least dominant is to be determined, the eigenvectors already determined must be swept off from the trial vector by using Gram-Schmidt orthogonalization. The accuracy of subsequent eigenvectors depends on the precision with which the previous eigenvectors have been determined. Therefore, the method is most suitable for determining only the first few eigenpairs. When more than a few lower-order eigenpairs are required, inverse iteration with shifts must be done. The shifting procedure, however, can be computationally costly.

For problems of large size where a limited number of eigenpairs are required (say, $m < 50$), the subspace iteration method is very effective. The scheme presented here of subspace iteration combined with Jacobi transformation and a Sturm sequence check, has proven to be quite effective and robust. There are other schemes of solving the partial eigenvalue problem; of particular note is the Lanczos method. This is treated in great detail in Sehmi's book [38] where he makes a direct comparison with the subspace iteration scheme. Generally, the Lanczos method can be anywhere from two to ten times faster than subspace iteration. However, additional programming is required to ensure that eigenvalues are not missed, as well as to ensure that the vectors are orthogonal.

In addition to the dynamic methods discussed here, it is becoming more popular to analyze structures under forced loading by the frequency domain approach. Fourier series expansions eliminate the time variable, and the fast Fourier transform (FFT) algorithm can yield computational efficiency. These are treated in some detail in References [14, 15] which show specific applications to wave propagation problems. The efficiency comes from being able to use the exact dynamic stiffness matrix and thereby have a relatively small system size.

In summary, modern computing environments should provide a variety of tools and algorithms for doing structural dynamic analyses. Each tool has its proponents (and its detractors) and it is often difficult to decide which is "best"; the truth is probably that none of these methods are best but some are better than others is certain circumstances. Thus a wise approach for the analyst is to be familiar with many tools and combine them as the problem requires. For example, a Fourier analysis of the input force history can tell immediately the number of modes needed in a modal superposition method. This, in turn, will suggest the level of discretization needed in the approximate structural model. A Fourier analysis would also indicate an appropriate time step to be used with an implicit integration scheme.

## Problems

**13.1** Show that to have a truncation error of 1% when using the forward difference approximation for a first order differential equation $\dot{u} + au = 0$, that we must have

$$h < .02/a$$

**13.2** Show that in the decision between using modal superposition and direct integration that the rule of thumb that the ratio $q/m > 10$ favors the modal method. $q$ is the number of time steps and $m$ is the number of modes.

**13.3** Show that the effect of the damping term $2ca\,\dot{u}$ when added to the second order system is to give a spectral radius of

$$\rho = (1 - cah) \pm iah\sqrt{1 - c^2}$$

Under what circumstance does the solution not "blow-up"? Is the solution still "good"?

## Exercises

**13.1** A freely suspended aluminum rod of area $1\,in^2$ and length $40\,in$, is struck at one end with a triangular force that reaches a peak of $1000\,lb$ in $50\,\mu s$. Using

elements 1 *in* long, determine the maximum particle velocity at the impact site.
$$[19.28\,in/s \text{ at } 51\,\mu s;\ 34.69\,in/s \text{ at } 461\,\mu s]$$

**13.2** Continuing the previous problem, determine the maximum velocity and the time at which it occurs, for a position 20 *in* from the impact site.
$$[18.30\,in/s \text{ at } 151\,\mu s]$$

**13.3** A freely suspended aluminum beam of cross-section 1 *in* × 1 *in* and length 80 *in*, is struck transversely at the center with a triangular force that reaches a peak of 1000 *lb* in 50 *μs*. Using elements 1 *in* long, determine the maximum particle velocity at the impact site.
$$[136.3\,in/s \text{ at } 66\,\mu s]$$

**13.4** Continuing the previous problem, determine the maximum velocity and the time at which it occurs, for a position 20 *in* from the impact site.
$$[33.95\,in/s \text{ at } 346\,\mu s;\ 41.33\,in/s \text{ at } 559\,\mu s]$$

**13.5** Consider a structure in the form of a cross with the four extremities fixed. Each member is of length 10 *in*, area 0.25 *in²*, moment of inertia 0.0012 *in⁴* and made of aluminum. Model the structure as a 2-D frame and divide each member into ten equal sized elements for a total of 40 elements. Show that the total system size may be reduced to [111 × 15].

**13.6** Continuing the previous problem, test various eigensolver methods and compare with the following results. Note that some roots are repeated (2,3), (6,7), and that some roots are very close (3,4), (7,8).

| Eigenvalue | Jacobi | Subspace | Eigenvalue | Jacobi |
|:---:|:---:|:---:|:---:|:---:|
| 1 | 2052.0 | 2052.00 | 21 | 40731. |
| 2 | 2971.5 | 2971.47 | 22 | 40731. |
| 3 | 2971.5 | 2971.47 | 23 | 51785. |
| 4 | 2977.7 | 2977.70 | 24 | 56091. |
| 5 | 6650.8 | 6650.82 | 25 | 56263. |
| 6 | 8157.4 | 8157.43 | 26 | 56263. |
| 7 | 8157.4 | 8157.43 | 27 | 60597. |
| 8 | 8210.0 | 8210.00 | 28 | 60597. |
| 9 | 13884. | 13884.2 | 29 | 70146. |
| 10 | 15847. | 15847.2 | 30 | 74199. |
| 11 | 15847. | 15847.2 | 31 | 74199. |
| 12 | 16107. | 16106.6 | 32 | 75200. |
| 13 | 23775. | 23775.4 | 33 | 87376. |
| 14 | 24923. | 24923.0 | 34 | 87376. |
| 15 | 24923. | 24923.0 | 35 | 91615. |
| 16 | 26669. | 26668.9 | 36 | 97339. |
| 17 | 29976. | 29976.5 | 37 | 99508. |
| 18 | 29976. | 29976.5 | 38 | 99508. |
| 19 | 36377. | 36376.7 | 39 | |
| 20 | 39961. | 39961.2 | 40 | |

# Matrices and Linear Algebra

The computer methods described in this book deal with ordered set of entities which may be either numbers or functions. In particular, they deal with an ordinary sequence of the form

$$x_1, x_2, \ldots, x_n$$

or with a two-dimensional array such as the following rectangular arrangement

$$
\begin{array}{cccc}
a_{11}, & a_{12}, & \cdots, & a_{1n} \\
a_{21}, & a_{22}, & \cdots, & a_{2n} \\
\vdots & \vdots & \ddots & \vdots \\
a_{m1}, & a_{m2}, & \cdots, & a_{mn}
\end{array}
$$

consisting of $m$ rows and $n$ columns. When these sets obey certain laws of equality, addition, subtraction, and multiplication, then these arrays are called *matrices*. This appendix summarizes such laws. The discussion is modest and its main function is to collect together those mathematical concepts from linear algebra explicitly needed in the text and that would be too cumbersome to discuss there. Fuller discussions of the topics can be found in books such as References [7, 19].

## A.1  Matrix Notation

To make descriptions easier, special names are given to matrices that have special properties. There is also a notation to describe these special properties. Unfortunately, a wide variety of notations has evolved, therefore this section will concentrate only on the special terms and notations introduced in the text.

### Brackets and Braces

The rectangular array of the coefficients $a_{ij}$ is enclosed in square brackets and denoted by

$$
[\,a\,] \equiv [a_{ij}] \equiv
\begin{bmatrix}
a_{11} & a_{12} & \cdots & a_{1n} \\
a_{21} & a_{22} & \cdots & a_{2n} \\
\vdots & \vdots & \ddots & \vdots \\
a_{m1} & a_{m2} & \cdots & a_{mn}
\end{bmatrix}
$$

This is called an $[m \times n]$ *matrix*. In the symbol $a_{ij}$, representing a typical component, the first subscript (here $i$) denotes the row and the second subscript (here $j$) the column occupied by the component.

The sets of quantities $x_i (i = 1, 2, \cdots, n)$ are represented as matrices of one column each and are called *vectors*. In order to emphasize the fact that this matrix consists of only one column, it is enclosed in braces (rather than brackets), and written as

$$\{x\} \equiv \{x_i\} \equiv \left\{ \begin{array}{c} x_1 \\ x_2 \\ \vdots \\ x_n \end{array} \right\} \quad or \quad \{x\} = \{x_i\} = \{x_1, x_2, \cdots, x_n\}$$

Other symbols, such as single and double vertical lines, are also used to enclose matrix arrays. The meaning of these are explained as they are introduced in the text.

## Symmetric, Anti-symmetric Matrices

The components of a symmetric array $[\ S\ ]$ and anti-symmetric array $[\ A\ ]$ have the properties

$$S_{ij} = S_{ji}, \qquad A_{ij} = -A_{ji}$$

Every array can be decomposed into its symmetric and anti-symmetric parts since

$$[B_{ij}] = [\tfrac{1}{2}(B_{ij} + B_{ji}) + \tfrac{1}{2}(B_{ij} - B_{ji})] = [S_{ij}] + [A_{ij}]$$

For example, consider the decomposition of the following matrix

$$\begin{bmatrix} 1 & 2 & 3 \\ 4 & 5 & 6 \\ 7 & 8 & 9 \end{bmatrix} = \begin{bmatrix} 1 & 3 & 5 \\ 3 & 5 & 7 \\ 5 & 7 & 9 \end{bmatrix} + \begin{bmatrix} 0 & -1 & -2 \\ 1 & 0 & -1 \\ 2 & 1 & 0 \end{bmatrix}$$

Note the zeros on the diagonal of the anti-symmetric matrix.

## Diagonal Matrix

A square matrix having components $a_{ii}$ along the diagonal with all other components equal to zero is a diagonal matrix. Its form is

$$\begin{bmatrix} a_{11} & 0 & 0 \\ 0 & a_{22} & 0 \\ 0 & 0 & a_{33} \end{bmatrix}, \qquad \lceil\ I\ \rfloor \equiv \begin{bmatrix} 1 & 0 & 0 \\ 0 & 1 & 0 \\ 0 & 0 & 1 \end{bmatrix}$$

The unit matrix, $\lceil\ I\ \rfloor$, is a square matrix in which the diagonal components are unity with all other components equal to zero.

## Banded Matrix

A matrix whose non-zero components are clustered near the main diagonal is call a *banded* matrix. The largest spread is called the *bandwidth*.

$$\begin{bmatrix} 1 & 2 & 0 & 0 & 0 \\ 3 & 4 & 0 & 0 & 0 \\ 0 & 5 & 0 & 6 & 0 \\ 0 & 0 & 7 & 8 & 0 \\ 0 & 0 & 0 & 9 & 1 \end{bmatrix}$$

This is of bandwidth three. For symmetric arrays, the bandwidth is usually measured off the diagonal, and the number of columns (including the diagonal) is called the semi-bandwidth.

## A.2 Matrix Operations

There is an algebra for manipulating matrices. That is, just as for common numbers, matrices can be added, multiplied, and so on. The rules for doing these operations, however, differ slightly from the usual way of accomplishing them. Since all rectangular matrices can be augmented to be square, the following descriptions generally will be restricted to square matrices.

## Addition and Subtraction

The simplest matrix operations are those of addition and subtraction. These, of course, can be applied only to matrices of the same size. (Note, however, that this can always be achieved by suitably augmenting the matrix.) Thus

$$[\,a\,]+[\,b\,]=[\,c\,]$$

where the individual components must be added as

$$c_{ij} = a_{ij} + b_{ij}$$

It is this requirement that the addition (or subtraction) be done on a term by term basis that requires the two matrices to be of the same size.

## Multiplication

Consider the set of equations

$$\sum_{k=1}^{n} a_{ik}x_k = y_i \qquad (i = 1, 2, \cdots, m)$$

which leads to the matrix equation

$$[\,a\,]\{x\} = [a_{ik}]\{x_k\} \equiv \{\sum_{k=1}^{n} a_{ik}x_k\} = \{y_i\} = \{y\}$$

Formally, we merely replace the *column* subscript in the general term of the first factor by a *dummy index* $k$, replace the *row* subscript in the general term of the second factor by the same dummy index, and sum over that index. The definition clearly is applicable only when the number of *columns* in the first factor is equal to the number of *rows* (elements) in the second factor. Unless this condition is satisfied, the product is undefined.

Now suppose that the $n$ variables $x_1, \cdots, x_n$ are expressed as linear combinations of $s$ new variables $z_1, \cdots, z_s$, that is, a set of relations holds of the form

$$x_i = \sum_{k=1}^{s} b_{ik} z_k \qquad (i = 1, 2, \cdots, n)$$

The result of the substitution of this into the first form gives

$$\sum_{l=1}^{s} \left( \sum_{k=1}^{n} a_{ik} b_{kl} \right) z_i = y_i \qquad (i = 1, 2, \cdots, m)$$

In matrix notation, this transformation is expressed as

$$[\ a\ ][\ b\ ]\{z\} = [\ c\ ]\{z\} = \{y\}$$

Thus it follows that the result of operating on $\{y\}$ by $[\ b\ ]$, and then by $[\ a\ ]$, is the same as the result of operating on $\{y\}$ directly by the matrix $[\ c\ ]$. We accordingly define the matrix to be the product

$$[\ c\ ] \equiv [\ a\ ][\ b\ ] = [a_{ik}][b_{kj}] \equiv \left[ \sum_{k=1}^{n} a_{ik} b_{kj} \right]$$

The relation

$$[\ a\ ]([\ b\ ]\{z\}) = ([\ a\ ][\ b\ ])\{z\}$$

then is a consequence of this definition.

Recalling that the first subscript in each case is the row index and the second the column index, we see that if the first factor has $m$ rows and $n$ columns, and the second $n$ rows and $s$ columns, the index $i$ in the right-hand member may vary from 1 to $m$ while the index $j$ in that member may vary from 1 to $s$. Hence, the product of an $[m \times n]$ matrix by an $[n \times s]$ matrix is an $[m \times s]$ matrix. The component $c_{ij}$ in the $i^{th}$ row and the $j^{th}$ column of the product is formed by multiplying together corresponding components of the $i^{th}$ row of the first factor and the $j^{th}$ column of the second factor, and adding the results algebraically.

Multiplication, in general, is not commutative. That is

$$[\ a\ ][\ b\ ] \neq [\ b\ ][\ a\ ] \qquad \text{or} \qquad \sum_{k} a_{ik} b_{kj} \neq \sum b_{ik} a_{kj}$$

as shown in the second case below.

## Examples

An array and a vector are multiplied as

$$\begin{bmatrix} 1 & 2 \\ 3 & 4 \\ 5 & 6 \end{bmatrix} \left\{ \begin{matrix} 7 \\ 8 \end{matrix} \right\} = \left\{ \begin{matrix} 1 \times 7 + 2 \times 8 \\ 3 \times 7 + 4 \times 8 \\ 5 \times 7 + 6 \times 8 \end{matrix} \right\} = \left\{ \begin{matrix} 23 \\ 53 \\ 83 \end{matrix} \right\}$$

However, the same objects arranged in the following order

$$\left\{ \begin{array}{c} 7 \\ 8 \end{array} \right\} \left[ \begin{array}{cc} 1 & 2 \\ 3 & 4 \\ 5 & 6 \end{array} \right] = \quad \text{not defined}$$

is not defined because of incompatible sizes. Two arrays are multiplied as

$$\left[ \begin{array}{cc} 1 & 2 \\ 3 & 4 \end{array} \right] \left[ \begin{array}{ccc} 3 & 4 & 5 \\ 6 & 7 & 8 \end{array} \right] = \left[ \begin{array}{ccc} 3+12 & 4+14 & 5+16 \\ 9+24 & 12+28 & 15+32 \end{array} \right] = \left[ \begin{array}{ccc} 15 & 18 & 21 \\ 33 & 40 & 47 \end{array} \right]$$

## Transpose

The transpose $[\ a\ ]^T$ of a matrix $[\ a\ ]$ is one in which the rows and columns are interchanged. For example

$$[\ a\ ] = \left[ \begin{array}{ccc} a_{11} & a_{12} & a_{13} \\ a_{21} & a_{22} & a_{23} \\ 0 & 0 & 0 \end{array} \right], \quad [\ a\ ]^T = \left[ \begin{array}{ccc} a_{11} & a_{21} & 0 \\ a_{12} & a_{22} & 0 \\ a_{13} & a_{23} & 0 \end{array} \right]$$

The transpose of a column matrix is a row matrix

$$\{x\} = \left\{ \begin{array}{c} x_1 \\ x_2 \\ x_3 \end{array} \right\}, \quad \{x\}^T = \{x_1, x_2, x_3\}$$

A generalization is that if

$$[\ B\ ] = [\ A\ ]_1 [\ A\ ]_2 \cdots [\ A\ ]_n \quad \text{then} \quad [\ B\ ]^T = [\ A\ ]_n^T [\ A\ ]_{n-1}^T \cdots [\ A\ ]_1^T$$

## Partitioning and Augmenting

The idea of augmenting (or padding with zeros) to make the matrix bigger or of a certain size, is a fundamental concept associated with sets. This is part of a more general idea known as partitioning. For example, a square matrix $[\ a\ ]$ may be partitioned as

$$\left[ \begin{array}{ccc} 2 & 4 & \vdots & -1 \\ 0 & -3 & \vdots & 4 \\ 1 & 2 & \vdots & 2 \\ \cdots & \cdots & \vdots & \cdots \\ 3 & -1 & \vdots & -5 \end{array} \right] = \left[ \begin{array}{ccc} [\ A\ ] & \vdots & [\ B\ ] \\ \cdots & \vdots & \cdots \\ [\ C\ ] & \vdots & [\ D\ ] \end{array} \right]$$

where the submatrices are

$$[\ A\ ] = \left[ \begin{array}{cc} 2 & 4 \\ 0 & -3 \\ 1 & 2 \end{array} \right], \quad [\ B\ ] = \left\{ \begin{array}{c} -1 \\ 4 \\ 2 \end{array} \right\}, \quad [\ C\ ] = \{3 \ -1\}, \quad [\ D\ ] = [-5]$$

Partitioned matrices obey the normal rules of matrix algebra and can be added, subtracted, and multiplied as though the submatrices were ordinary matrix components. Thus

$$\begin{bmatrix} A & \vdots & B \\ \cdots & \cdots & \cdots \\ C & \vdots & D \end{bmatrix} \left\{ \begin{matrix} x \\ \cdots \\ y \end{matrix} \right\} = \left\{ \begin{matrix} [\,A\,]\{x\} & + & [\,B\,]\{y\} \\ [\,C\,]\{x\} & + & [\,D\,]\{y\} \end{matrix} \right\}$$

$$\begin{bmatrix} [\,A\,] & \vdots & [\,B\,] \\ \cdots & \cdots & \cdots \\ [\,C\,] & \vdots & [\,D\,] \end{bmatrix} \begin{bmatrix} \{E\} & \vdots & \{F\} \\ \cdots & \cdots & \cdots \\ \{G\} & \vdots & \{H\} \end{bmatrix} = \begin{bmatrix} \{AE + BG\} & \{AF + BH\} \\ \{CE + DG\} & \{CF + DH\} \end{bmatrix}$$

The original array can be made square by augmenting with a column of zeros as

$$\begin{bmatrix} 2 & 4 & -1 & 0 \\ 0 & -3 & 4 & 0 \\ 1 & 2 & 2 & 0 \\ 3 & -1 & -5 & 0 \end{bmatrix}$$

Note that this could have been done in a variety of ways.

## Inverse and Orthogonal Matrix

There is no concept of division in matrix analysis; the operation that achieves essentially the same thing is that of inversion. Thus, for example, if

$$[\,A\,]\{x\} = \{y\} \qquad \text{then} \qquad \{x\} = [\,A\,]^{-1}\{y\}$$

where $[\,A\,]^{-1}$ is the inverse of $[\,A\,]$ and satisfies the relationships

$$[\,A\,]^{-1}[\,A\,] = [\,A\,][\,A\,]^{-1} = [\,I\,]$$

A generalization is that if

$$[\,B\,] = [\,A\,]_1[\,A\,]_2 \cdots [\,A\,]_n \qquad \text{then} \qquad [\,B\,]^{-1} = [\,A\,]_n^{-1}[\,A\,]_{n-1}^{-1} \cdots [\,A\,]_1^{-1}$$

Later we will discuss various schemes for computing the inverse of a matrix.

Consider

$$[\,A\,] = \begin{bmatrix} 1 & 2 \\ 3 & 4 \end{bmatrix}, \qquad [\,A\,]^{-1} = \tfrac{1}{2}\begin{bmatrix} -4 & 2 \\ 3 & -1 \end{bmatrix}$$

Check the result by multiplying with the original matrix and see if the unit matrix is obtained

$$\tfrac{1}{2}\begin{bmatrix} -4 & 2 \\ 3 & -1 \end{bmatrix}\begin{bmatrix} 1 & 2 \\ 3 & 4 \end{bmatrix} = \begin{bmatrix} 1 & 0 \\ 0 & 1 \end{bmatrix}$$

An orthogonal matrix $[\,A\,]$ satisfies the relationship

$$[\,A\,]^T[\,A\,] = [\,A\,][\,A\,]^T = [\,I\,]$$

From the definition of an inverse matrix it is evident that for an orthogonal matrix

$$[\,A\,]^T = [\,A\,]^{-1}$$

From this, it is obvious that orthogonal matrices can be inverted easier than ordinary ones. It can be easily verified, for example, that the following is an orthogonal matrix

$$\frac{1}{\sqrt{6}} \begin{bmatrix} \sqrt{2} & 1 & -\sqrt{3} \\ \sqrt{2} & -2 & 0 \\ \sqrt{2} & 1 & \sqrt{3} \end{bmatrix}$$

## Triangular Decomposition

A matrix that has only zeros below the diagonal is called an upper triangular matrix. Similarly, if there are only zeros above the diagonal it is called a lower triangular matrix. Every matrix can be decomposed into the product of two triangular matrices. That is

$$[\,A\,] = [\,L\,][\,U\,] = \begin{bmatrix} 1 & 0 & 0 \\ L_{21} & 1 & 0 \\ L_{31} & L_{32} & 1 \end{bmatrix} \begin{bmatrix} U_{11} & U_{12} & U_{13} \\ 0 & U_{22} & U_{23} \\ 0 & 0 & U_{33} \end{bmatrix}$$

for a [3 × 3] matrix. The components of $[\,L\,]$ and $[\,U\,]$ are obtained by multiplying them out and equating to the components of $[\,A\,]$. Thus, for this [3 × 3] matrix the first row of terms becomes

$$\begin{aligned} A_{11} &= U_{11} & \Rightarrow & & U_{11} &= A_{11} \\ A_{12} &= U_{12} & \Rightarrow & & U_{12} &= A_{12} \\ A_{13} &= U_{13} & \Rightarrow & & U_{13} &= A_{13} \end{aligned}$$

The second row of products gives

$$\begin{aligned} A_{21} &= L_{21}U_{11} & \Rightarrow & & L_{21} &= A_{21}/U_{11} \\ A_{22} &= L_{21}U_{12} + U_{22} & \Rightarrow & & U_{22} &= A_{22} - L_{21}U_{12} \\ A_{23} &= L_{21}U_{13} + U_{23} & \Rightarrow & & U_{23} &= A_{23} - L_{21}U_{13} \end{aligned}$$

And the third row gives

$$\begin{aligned} A_{31} &= L_{31}U_{11} & \Rightarrow & & L_{31} &= A_{31}/U_{11} \\ A_{32} &= L_{31}U_{12} + L_{32}U_{22} & \Rightarrow & & L_{32} &= (A_{32} - L_{31}U_{12})/U_{22} \\ A_{33} &= L_{31}U_{13} + L_{32}U_{23} + U_{33} & \Rightarrow & & U_{33} &= A_{33} - L_{31}U_{13} - L_{32}U_{23} \end{aligned}$$

Thus, they are obtained as a sequence of algebraic operations. The above equations can be summarized for a general array as

$$U_{ij} = A_{ij} - \sum_{k=1}^{i-1} L_{ik}U_{kj} \qquad i < j$$

$$L_{ij} = [A_{ij} - \sum_{k=1}^{j-1} L_{ik}U_{kj}]/L_{jj} \qquad i > j$$

$$U_{ii} = 1 \qquad i = j$$

If the matrix $[\,A\,]$ is symmetric then

$$[\,A\,] = [\,U\,]^T[\,U\,] = [\,L\,]^T[\,L\,]$$

For example, consider the following decomposition of an arbitrary matrix

$$\begin{bmatrix} 1 & 2 & 3 \\ 4 & 5 & 6 \\ 7 & 8 & 9 \end{bmatrix} = \begin{bmatrix} 1 & 0 & 0 \\ 4 & 1 & 0 \\ 7 & 2 & 1 \end{bmatrix} \begin{bmatrix} 1 & 2 & 3 \\ 0 & -3 & -6 \\ 0 & 0 & 0 \end{bmatrix}$$

# A.3   Vector and Matrix Norms

Vectors and matrices are functions of many components, but we often need to measure their 'magnitude' as a single entity. For example, in iterative solution processes using vectors and matrices we need a measure of convergence — a single number representing the whole array by which we can judge if the measure is getting smaller. This number is the *norm*.

A norm is a single number which depends on the magnitude of all components in the vector or matrix and is written as $||A||$ for the matrix $[\,A\,]$. Norms have the following properties

1. $||A|| \geq 0$

2. $||cA|| = |c|\,||A||$

3. $||A + B|| \leq ||A|| + ||B||$

4. $||AB|| \leq ||A|| \cdot ||B||$

The third relation is the *triangle inequality*. The last condition must be satisfied in order to be able to use matrix norms when matrix products occur.

The following three vector norms are commonly used, and are called the infinity, one and two vector norms, respectively,

$$||v||_\infty = \max |v_i|\,, \qquad ||v||_1 = \sum_{i=1}^{n} |v_i|\,, \qquad ||v||_2 = \left( \sum_{i=1}^{n} |v_i|^2 \right)^{1/2}$$

The last is also known as the Euclidean vector norm. Geometrically, this norm is equal to the length of the vector. It can be shown that these three norms are related by

$$||v||_\infty \leq ||v||_1 \leq n||v||_\infty\,, \qquad ||v||_\infty \leq ||v||_2 \leq \sqrt{n}||v||_\infty$$

Note that although the actual vector length is calculated using the Euclidean norm, the 1 and $\infty$ norms also provide some measure of the length of the vector. A vector whose norm is unity is called a *unit vector*. Any non-zero vector can be normalized so as to form a unit vector by simply dividing the vector by its norm. That is,

$$\frac{\{v\}}{||v||}$$

The following are frequently used matrix norms:

$$||A||_\infty = \max_i \sum_{j=1}^n |a_{ij}|, \quad ||A||_1 = \max_j \sum_{i=1}^n |a_{ij}|, \quad ||A||_2 = \sqrt{\tilde{\lambda}_n}, \quad ||A||_E = \sqrt{\sum_{i,j} |a_{ij}|^2}$$

For a symmetric matrix we have $||A||_\infty = ||A||_1$. The norm $||A||_2$ is called the spectral norm of $[\ A\ ]$ and $\tilde{\lambda}_n$ is the maximum eigenvalue of $[\ A\ ]^T[\ A\ ]$. The $||A||_E$ is the Euclidean norm; this is the 'sum of squares' norm and is compatible with the $||v||_2$ norm for vectors. In order to obtain useful information by applying norms to vector and matrix products, we need to employ compatible norms. Thus we have for a matrix $[\ A\ ]$ and vector $\{v\}$

$$||Av|| \leq ||A|| \cdot ||v||$$

where $||Av||$ and $||v||$ are evaluated using the vector norm and $||A||$ is evaluated using the matrix norm.

We can now measure the convergence of a sequence of matrices $[\ A\ ]_1, [\ A\ ]_2, [\ A\ ]_3, \ldots, [\ A\ ]_k$ to a matrix $[\ A\ ]$. For the sequence to converge, it is necessary and sufficient that

$$\lim_{k \to \infty} ||A_k - A|| = 0$$

for any one of the given matrix norms.

## Example

Calculate the 1, 2, and $\infty$ norms of the given vector $\{v\}$, and verify the relations among the norms.

$$\{v\} = \{2, -6, 1\}$$

We have

$$||v||_\infty = 6, \quad ||v||_1 = 2 + 6 + 1 = 9, \quad ||v||_2 = \sqrt{4 + 36 + 1} = \sqrt{41} = 6.4$$

The relationships among the norms are

$$6 \leq 9 \leq (3)(6), \quad 6 \leq \sqrt{41} \leq (\sqrt{3})(6)$$

## Example

Calculate the $\infty$ and $E$ norms of the following matrix

$$[\ A\ ] = \begin{bmatrix} 5 & -4 & -7 \\ -4 & 2 & -4 \\ -7 & -4 & 5 \end{bmatrix}$$

Using the definitions we have

$$||A||_\infty = \max \left\{ \begin{array}{rcl} 1+2+3 & = & 6 \\ 2+4+5 & = & 11 \\ 3+5+6 & = & 14 \end{array} \right\} = 14, \quad ||A||_E = \sqrt{216} = 14.7$$

# A.4    Determinants

The determinant of a matrix $[\ a\ ]$, denoted $\det[\ a\ ]$, is a number with a unique value. It can be expanded in the form

$$\det[\ a\ ] = \sum_{k=1}^{n} a_{ik}C_{ik} \qquad \text{or} \qquad \det[\ a\ ] = \sum_{k=1}^{n} a_{kj}C_{kj}$$

where $C_{ij}$ are the cofactors of the determinant. This is called of the *Laplace expansion* of the determinant.

A *minor (determinant)* $M_{ij}$ of the component $a_{ij}$ is a determinant formed by deleting the $i^{th}$ row and the $j^{th}$ column from the original determinant. The *cofactor* $C_{ij}$ of the component $a_{ij}$ is related to the minor and is defined by the equation

$$C_{ij} = (-1)^{i+j} M_{ij}$$

Thus the above definition for the determinant is really a recursive relation. A beginning point can be taken as a $[2 \times 2]$ matrix for which

$$\det[\ A\ ] = a_{11}a_{22} - a_{21}a_{12}$$

A matrix is said to be *singular* if its determinant is zero. A matrix is *positive definite* if

$$\{u\}^{T}[\ a\ ]\{u\} > 0$$

for all possible non-null vectors $\{u\}$. A necessary and sufficient condition for a symmetric matrix to be positive definite is that all determinants formed from it be positive. In other words, a positive definite matrix is non-singular.

The following are some of the significant properties of determinants:

- The interchange of any two columns or rows changes the sign of the determinant but not its value.

- If two rows or two columns are identical, or the components are in a constant ratio, the determinant is zero.

- If every component of any row or column is zero, its determinant is zero.

- If all components of one row or column are multiplied by a number $k$, the determinant is multiplied by $k$.

- If to the components of any row (column) are added $k$ times the corresponding components of any other row (column), the determinant is unchanged.

From these properties others may be deduced. For example, it follows that to any row may be added any linear combination of the other rows without changing the determinant of the matrix. By combining this result with the first property, we deduce that if any row of a square matrix is a linear combination of the other rows, then the determinant of that matrix is zero.

A generalization is that if

$$[\ B\ ] = [\ A\ ]_1[\ A\ ]_2 \cdots [\ A\ ]_n \qquad \text{then} \qquad \det[\ B\ ] = \det[\ A\ ]_n \det[\ A\ ]_{n-1} \cdots \det[\ A\ ]_1$$

**Example**

Given the third order matrix
$$\begin{bmatrix} 2 & 1 & 5 \\ 4 & 2 & 1 \\ 2 & 0 & 3 \end{bmatrix}$$

The minor of the term $a_{21} = 4$ is

$$M_{21} \text{ of } \begin{bmatrix} 2 & 1 & 5 \\ 4 & 2 & 1 \\ 2 & 0 & 3 \end{bmatrix} = \det \begin{bmatrix} 1 & 5 \\ 0 & 3 \end{bmatrix} = 1 \times 3 - 0 \times 5 = 3$$

and its cofactor is
$$C_{21} = (-1)^{2+1} 3 = -3$$

The determinant, expanded in terms of the first row, is

$$\det \begin{bmatrix} 2 & 1 & 5 \\ 4 & 2 & 1 \\ 2 & 0 & 3 \end{bmatrix} = 2(-1)^{1+1} \det \begin{bmatrix} 2 & 1 \\ 0 & 3 \end{bmatrix} + 1(-1)^{1+2} \det \begin{bmatrix} 4 & 1 \\ 2 & 3 \end{bmatrix} + 5(-1)^{1+3} \det \begin{bmatrix} 4 & 2 \\ 2 & 0 \end{bmatrix}$$
$$= 12 - 10 - 20 = -18$$

The determinant, actually, can be expanded in terms of any row or column. For example, in terms of the second column, it is

$$\det \begin{bmatrix} 2 & 1 & 5 \\ 4 & 2 & 1 \\ 2 & 0 & 3 \end{bmatrix} = 1(-1)^{1+2} \det \begin{bmatrix} 4 & 1 \\ 2 & 3 \end{bmatrix} + 2(-1)^{2+2} \det \begin{bmatrix} 2 & 5 \\ 2 & 3 \end{bmatrix} + 0(-1)^{3+2} \det \begin{bmatrix} 2 & 5 \\ 4 & 1 \end{bmatrix}$$
$$= -10 - 8 = -18$$

Of course, the same value is obtained.

# A.5   Solution of Simultaneous Equations

Systems of simultaneous equations are at the core of the computerized structural methods. They generally appear in the following form

$$\begin{aligned} a_{11}x_1 + a_{12}x_2 + \cdots + a_{1n}x_n &= y_1 \\ a_{21}x_1 + a_{22}x_2 + \cdots + a_{2n}x_n &= y_2 \\ \vdots\; &=\; \vdots \\ a_{n1}x_1 + a_{n2}x_2 + \cdots + a_{nn}x_n &= y_n \end{aligned}$$

where the vectors $\{x\}, \{y\}$ are the vectors of the unknowns and knowns, respectively. If the matrix $[\,a\,]$ is non-singular then its inverse exists and we can solve the above system by multiplying both sides by the inverse. That is

$$[\,a\,]^{-1}[\,a\,]\{x\} = [\,a\,]^{-1}\{y\} \qquad \text{or} \qquad \{x\} = [\,a\,]^{-1}\{y\}$$

In the special case when $\{y\}$ is zero (that is, all of its components are individually zero) then $\{x\}$ is zero also. Another interesting case is when $[\,a\,]$ is singular; the question arises as to whether there are still solutions. Multiply both sides by the solution vector to get

$$\{x\}^T[\,a\,]\{x\} = \{x\}^T\{y\}$$

Since the matrix is singular, it is possible to find a non-null $\{x\}$ so that the triple product is zero. Hence the solutions are those for which

$$\{x\}^T[\,a\,]\{x\} = 0, \qquad \{x\}^T\{y\} = 0$$

There are, in fact, an infinity of such solutions.

Efficient schemes of solving systems of equations are essential because for structural problems they can be of the order of 10,000 or more. We will concern ourselves now only with some simple schemes for solving them. We will concentrate on non-singular systems.

## Direct Method

The common method of reduction and substitution can be illustrated by the following example

$$
\begin{aligned}
2x_1 + x_2 + 5x_3 &= 5 \\
x_1 + x_2 - 3x_3 &= -1 \\
3x_1 + 6x_2 - 2x_3 &= 8
\end{aligned}
$$

Multiply the second equation by 2 and subtract it from the first equation. Multiply the third equation by $\frac{2}{3}$ and subtract it from the first equation also. This removes $x_1$ leaving two equations

$$
\begin{aligned}
-x_2 + 11x_3 &= 7 \\
-3x_2 + 6\tfrac{1}{3}x_3 &= -\tfrac{1}{3}
\end{aligned}
$$

Multiply the second of these equation by $\frac{1}{3}$ and subtract it from the first. This removes $x_2$ leaving

$$(11 - 2\tfrac{1}{9})x_3 = 7 + \frac{1}{9}$$

This leaves one equation with one unknown $x_3$, which can be easily solved to give

$$x_3 = \frac{8}{10}$$

Working backwards, we can then obtain the remainder of the solution as

$$x_2 = \frac{18}{10}, \qquad x_1 = -\frac{4}{10}$$

Variations of this method are *Gaussian elimination* and *Jordan elimination*.

## Solution by Cramer's Rule

Consider the system of equations $[A]\{x\} = \{y\}$. Denote $[A_i]$ $(i = 1, 2, \cdots, n)$ as the matrix obtained from $[A]$ by replacing its $i^{th}$ column with $\{y\}$. Then the system has the unique solution:

$$x_1 = \frac{\det[A_1]}{\det[A]}, \qquad x_2 = \frac{\det[A_2]}{\det[A]}, \qquad \cdots, \qquad x_n = \frac{\det[A_n]}{\det[A]}$$

This is known as Cramer's rule. From this it is apparent that a system of $n$ nonhomogeneous equations in $n$ unknowns has a unique solution provided that $\det[A] \neq 0$.

Consider the following system of equations

$$\begin{aligned} 2x_1 + x_2 + 5x_3 &= 5 \\ x_1 + x_2 - 3x_3 &= -1 \\ 3x_1 + 6x_2 - 2x_3 &= 8 \end{aligned}$$

The various determinants are

$$\det[A] = \det \begin{bmatrix} 2 & 1 & 5 \\ 1 & 1 & -3 \\ 3 & 6 & -2 \end{bmatrix} = 40$$

$$\det[A_1] = \det \begin{bmatrix} 5 & 1 & 5 \\ -1 & 1 & -3 \\ 8 & 6 & -2 \end{bmatrix} = -16$$

$$\det[A_2] = \det \begin{bmatrix} 2 & 5 & 5 \\ 1 & -1 & -3 \\ 3 & 8 & -2 \end{bmatrix} = 72$$

$$\det[A_3] = \det \begin{bmatrix} 2 & 1 & 5 \\ 1 & 1 & -1 \\ 3 & 6 & 8 \end{bmatrix} = 32$$

The solution is then

$$x_1 = \frac{\det[A_1]}{\det[A]} = -\frac{4}{10}, \qquad x_2 = \frac{18}{10}, \qquad x_3 = \frac{8}{10}$$

This is a convenient method for solving very small systems of equations, but is highly inefficient for large systems.

## Gauss-Seidel Iterative Method

The system of equations can be rewritten in the following form with the diagonal terms on the left hand side

$$\begin{aligned} x_1 &= \frac{1}{a_{11}}(y_1 - a_{12}x_2 - a_{13}x_3 - \cdots - a_{1n}x_n) \\ x_2 &= \frac{1}{a_{22}}(y_2 - a_{21}x_1 - a_{23}x_3 - \cdots - a_{2n}x_n) \\ &\vdots \\ x_n &= \frac{1}{a_{nn}}(y_n - a_{n1}x_1 - a_{n2}x_2 - \cdots - a_{n,n-1}x_{n-1}) \end{aligned}$$

It is noticed that the unknowns appear on both sides, hence iteration is used to solve the system. Start by setting guesses for $\{x\}$, i.e., $x_2 = x_3 \cdots x_n = 0$, and obtain

$$x_1 = \frac{1}{a_{11}} y_1$$

Substitute this value back into the equations to obtain new values for the remaining terms $x_2, x_3, \cdots, x_n$ as

$$x_2 = \frac{1}{a_{22}}(y_2 - \frac{a_{21}}{a_{11}} y_1)$$

$$x_3 = \frac{1}{a_{33}}(y_3 - \frac{a_{31}}{a_{11}} y_1)$$

$$\vdots =$$

$$x_n = \frac{1}{a_{nn}}(y_n - \frac{a_{n1}}{a_{11}} y_1)$$

Using the new values of $x_2, x_3 \cdots, x_n$, then in turn obtain a modified $x_1$ from the first equation. This process can be continued until the desired accuracy of solution is achieved.

### Example

Use iteration to find a solution to the system of equations

$$3x_1 + 2x_2 = 1$$
$$x_1 + x_2 = 2$$

Rewrite the above equations as

$$x_1 = \tfrac{1}{3} - \tfrac{2}{3}x_2$$
$$x_2 = 2 - x_1$$

The iterative procedure is shown in the following table starting with $x_2 = 0$

$$x_1: \quad \frac{1}{3} \quad \frac{-7}{9} \quad \frac{-41}{27} \quad \cdots \quad -3$$

$$x_2: \quad 0 \quad \frac{5}{3} \quad \frac{25}{9} \quad \frac{95}{27} \quad \cdots \quad 5$$

Note that sometimes convergence can be slow. It is also essential that the diagonal terms be dominant.

## A.6   Eigenvectors and Eigenvalues

A frequently encountered problem is that of determining those values of a constant $\lambda$ for which non-trivial solutions exist to a homogeneous set of equations of the form

$$a_{11}x_1 + a_{12}x_2 + \cdots + a_{1n}x_n = \lambda x_1$$
$$a_{21}x_1 + a_{22}x_2 + \cdots + a_{2n}x_n = \lambda x_2$$
$$\vdots = \vdots$$
$$a_{n1}x_1 + a_{n2}x_2 + \cdots + a_{nn}x_n = \lambda x_n$$

Such a problem is known as an *eigenvalue* problem; values of $\lambda$ for which non-trivial solutions exist are called the *eigenvalues* of the matrix $[\ a\ ]$, and corresponding vector solutions $\{x\}$ are known as *eigenvectors* of the matrix. An array made up of the eigenvectors is called a *modal matrix*. In matrix notation, the above equations take the form

$$[\ a\ ]\{x\} = \lambda\{x\} \qquad \text{or} \qquad [[\ a\ ] - \lambda[\ I\ ]]\{x\} = 0$$

and is called the *standard* eigenvalue problem. When the system is such that $\lambda[\ I\ ]$ is replaced with a general array $\lambda[\ b\ ]$ this is called the *generalized* eigenvalue problem.

The present discussion is restricted to real symmetric matrices. As is apparent by an application of Cramer's rule, this homogeneous problem possesses non-trivial solutions if and only if the determinant of the coefficient matrix $[[\ a\ ] - \lambda[\ I\ ]]$ vanishes:

$$\det[[\ a\ ] - \lambda[\ I\ ]] \equiv \det \begin{bmatrix} a_{11} - \lambda & a_{12} & \cdots & a_{1n} \\ a_{21} & a_{22} - \lambda & \cdots & a_{2n} \\ \vdots & \vdots & \ddots & \vdots \\ a_{n1} & a_{n2} & \cdots & a_{nn} - \lambda \end{bmatrix} = 0$$

This condition requires that $\lambda$ be a root of an algebraic equation of degree $n$, known as the characteristic equation. The $n$ solutions $\lambda_1, \lambda_2, \cdots, \lambda_n$, (which need not all be distinct) are the characteristic numbers or roots of the matrix. Corresponding to each such value $\lambda_m$, there exists at least one vector solution (the eigenvector) which is determined within an arbitrary multiplicative constant. For a real symmetric matrix, we may draw the following important conclusions:

- The characteristic numbers of such a matrix are always real.

- Two characteristic vectors corresponding to different characteristic numbers are orthogonal.

## Example

Consider solving the following system of equations

$$\begin{bmatrix} 1 & 2 & 0 \\ 2 & 3 & 0 \\ 0 & 0 & 4 \end{bmatrix} \begin{Bmatrix} x_1 \\ x_2 \\ x_3 \end{Bmatrix} - \lambda \begin{bmatrix} 1 & 0 & 0 \\ 0 & 1 & 0 \\ 0 & 0 & 1 \end{bmatrix} \begin{Bmatrix} x_1 \\ x_2 \\ x_3 \end{Bmatrix} = 0$$

First combine as

$$\begin{bmatrix} 1 - \lambda & 2 & 0 \\ 2 & 3 - \lambda & 0 \\ 0 & 0 & 4 - \lambda \end{bmatrix} \begin{Bmatrix} x_1 \\ x_2 \\ x_3 \end{Bmatrix} = \begin{Bmatrix} 0 \\ 0 \\ 0 \end{Bmatrix}$$

It is possible to have a non-trivial solution provided the determinant of the system is zero. We will now see if indeed this is possible. Multiplying the determinant out gives

$$\det = (1 - \lambda)(3 - \lambda)(4 - \lambda) - (2)(2)(4 - \lambda) = 0$$

This can be rearranged into the following product form

$$(\lambda - 2 - \sqrt{5})(\lambda - 4)(\lambda - 2 + \sqrt{5}) = 0$$

There are three values of $\lambda$ which make the determinant zero. Specifically, these are

$$\lambda_1 = 2 - \sqrt{5}, \qquad \lambda_2 = 4, \qquad \lambda_3 = 2 + \sqrt{5}$$

These are called the eigenvalues. Corresponding to each of these are the eigenvectors. For example, for the $\lambda_1$ eigenvalue we have

$$\begin{bmatrix} -1 + \sqrt{5} & 2 & 0 \\ 2 & 1 + \sqrt{5} & 0 \\ 0 & 0 & 2 + \sqrt{5} \end{bmatrix} \begin{Bmatrix} x_1 \\ x_2 \\ x_3 \end{Bmatrix} = \begin{Bmatrix} 0 \\ 0 \\ 0 \end{Bmatrix}$$

giving the solution

$$x_1 = x_1, \qquad x_2 = (\frac{1 - \sqrt{5}}{2})x_1, \qquad x_3 = 0$$

Notice that only two of the equations are independent and hence the results are obtained only in terms of the first component. For the second mode we have

$$\begin{bmatrix} -3 & 2 & 0 \\ 2 & -1 & 0 \\ 0 & 0 & 0 \end{bmatrix} \begin{Bmatrix} x_1 \\ x_2 \\ x_3 \end{Bmatrix} = \begin{Bmatrix} 0 \\ 0 \\ 0 \end{Bmatrix}$$

giving the solution

$$x_1 = 0, \qquad x_2 = 0, \qquad x_3 = x_3$$

Notice that in this case $x_3$ is arbitrary because it is multiplied by a zero. The third eigenvector is

$$x_1 = x_1, \qquad x_2 = (\frac{1 + \sqrt{5}}{2})x_1, \qquad x_3 = 0$$

The modal matrix formed from the eigenvectors is

$$\begin{bmatrix} 1 & 0 & 1 \\ (\frac{1 - \sqrt{5}}{2}) & 0 & (\frac{1 + \sqrt{5}}{2}) \\ 0 & 1 & 0 \end{bmatrix}$$

It is easy to show that this is orthogonal.

## A.7   Vector Spaces

Consider a set of vectors $\{v\}_1, \{v\}_2, ...., \{v\}_n$. Let us form another vector given by

$$\{w\} = c_1\{v\}_1 + c_2\{v\}_2 + \cdots + c_n\{v\}_n$$

with $c_i$ being scalar constants. This new vector is said to be a linear combination of the original vectors.

If no combination of these vectors can be found such that $\{w\} = 0$ (when at least one of the $c_i$ is not zero) then the vectors $\{v\}_n$ are said to be *linearly independent*. They are also said to form the *basis* of an n-dimensional vector space.

A vector $\{u\}$ formed from only a subset of $\{v\}_n$ is said to *span* a *subspace* of $\{v\}_n$. Conversely, a vector $\{u\}$ formed from a set of vectors $\{v\}_m$ of which only $n$ are independent, is said to span the entire space of $\{v\}_n$.

This underlies the importance of the basis vectors, since they are the smallest number of vectors that span the space considered.

# Appendix B

# Spectral Analysis

The use of Fourier analysis as a means of describing motion is essential to the study of structural dynamics. The basic idea is that an arbitrary time signal can be thought of as the superposition of many sinusoidal components. That is, it has a *spectrum* of frequency components. Working in terms of the spectrum is called *spectral analysis* or sometimes *frequency domain analysis*.

The time domain for a motion or response is from minus infinity to plus infinity. Such a signal is represented by a continuous distribution of components which is known as the continuous Fourier transform (CFT). However, periodic functions have spectrums that are discrete and these are called Fourier series transforms. We review both of these transforms, but greater depth can be found in References [9, 14].

## B.1    Continuous Fourier Transform

The continuous Fourier transform pair of a function $F(t)$, defined on the time domain from $-\infty$ to $+\infty$, is given as:

$$2\pi F(t) = \int_{-\infty}^{\infty} C(\omega)e^{+i\omega t}d\omega, \qquad C(\omega) = \int_{-\infty}^{\infty} F(t)e^{-i\omega t}dt$$

where $C(\omega)$ is the continuous Fourier transform, $\omega$ is the angular frequency, and $i$ is the complex $\sqrt{-1}$. The first form is called the *inverse* transform while the second is the *forward* transform — this arbitrary convention arises from the fact that the signal to be transformed usually originates in the time domain.

The process of obtaining the Fourier transform of a signal separates the waveform into its constituent sinusoids (or spectrum) and thus a plot of $C(\omega)$ against frequency represents a diagram displaying the amplitude of each of the constituent sinusoids. The spectrum $C(\omega)$ usually has non-zero real and imaginary parts.

> **Example B.1:** Determine the continuous Fourier transform of a rectangular pulse of duration $a$. Investigate the phase behavior by positioning the pulse at different times.
>
> First consider when the pulse is symmetric with respect to $t = 0$. The function is given by
> $$F(t) = F_o \qquad -a/2 \leq t \leq a/2$$

and zero otherwise. Substituting into the forward transform and integrating gives

$$C(\omega) = \int_0^T F_o e^{-i\omega t} dt = 2F_o \left\{ \frac{\sin (\omega a/2)}{\omega} \right\} = F_o a \left\{ \frac{\sin (\omega a/2)}{\omega a/2} \right\}$$

In this particular case the transform is real only and symmetric about $\omega = 0$ as shown in Figure B.1. The term inside the braces is called a *sinc* function, and has the characteristic behavior of starting at unity magnitude and oscillating with decreasing amplitude. It is noted that the value of the transform at $\omega = 0$ is just the area under the time function. This, in fact, is a general result as seen from

$$C(0) = \int_{-\infty}^{\infty} F(t) e^{-i0t} dt = \int_{-\infty}^{\infty} F(t) dt$$

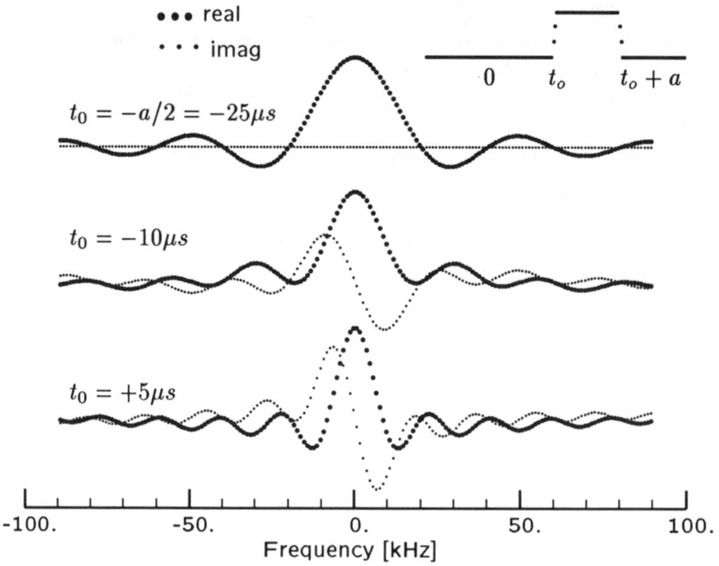

**Figure B.1**: Continuous transform of a rectangular pulse with $a = 50\mu s$.

When the pulse is displaced along the time axis such that the function is given by

$$F(t) = F_o \qquad t_o \le t \le t_o + a$$

and zero otherwise, the transform is then

$$C(\omega) = F_o a \left\{ \frac{\sin(\omega a/2)}{\omega a/2} \right\} e^{-i\omega(t_o + a/2)} = C_o(\omega) e^{-i\omega(t_o + a/2)}$$

which has both real and imaginary parts and is not symmetric with respect to $\omega = 0$. On closer inspection, however, we see that the magnitudes of the two transforms are the same; it is just that the latter is given an extra phase shift of amount $\omega(t_o + a/2)$. Figure B.1 shows the transform for different amounts of shift. We therefore associate phase shifts with shifts of the signal along the time axis.

# B.2   Periodic Functions: Fourier series

Let the time function $F(t)$ we are interested in be periodic on a period $T$. We can view this either as separate functions of duration $T$ placed one after the other; or as separate functions of infinite duration but with non-zero behavior only over the period $T$. We adopt this latter view as shown in Figure B.2.

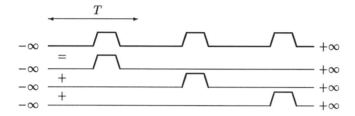

**Figure B.2**: Periodic signal as a superposition of infinite signals.

We saw from the previous example that if $C_o(\omega)$ is the transform of a pulse, the transform of the same pulse displaced an amount $T$ is $C_o(\omega)e^{-i\omega T}$. It is obvious therefore that the transform of the periodic signal can be represented as

$$C(\omega) = C_o(\omega)[1 + e^{-i\omega T} + e^{-i\omega 2T} + e^{-i\omega 3T} + \ldots]$$

This transform will show infinite peaks whenever the frequency is one of the following discrete values $\omega_n = 2\pi n/T$. Under this circumstance, each of the exponential terms is unity and there is an infinity of them. (This result should not be surprising since the $C(0)$ component is the area under the curve, and for the periodic rectangle of Figure B.2 we see that it is infinite.) At other frequencies, the exponentials are as likely to be positive as negative and hence their sum will be relatively small. Therefore our first conclusion about the transform of a periodic signal is that it will show very sharp spectral peaks. We can go further and say that the transform is zero everywhere except at the discrete frequency values $\omega_n = 2\pi n/T$ where it has an infinite value. We can represent this behavior by use of the *delta function*, $\delta(x)$; this special function is zero everywhere except at $x = 0$ where it is infinite. It has the very important additional property that its integral over the whole domain is unity. Thus

$$
\begin{aligned}
C(\omega) &= C_o(\omega)A[\delta(\omega) + \delta(\omega - 2\pi/T) + \delta(\omega - 4\pi/T) + \ldots] \\
&= C_o(\omega)A\sum_n \delta(\omega - 2\pi n/T) = C_o(\omega)A\sum_n \delta(\omega - \omega_n)
\end{aligned}
$$

$A$ is a proportionality constant which we determine next. We reiterate that although the transform $C(\omega)$ is a continuous function of frequency it effectively evaluates to discrete non-zero values.

The remainder of the transform pair is given by

$$2\pi F(t) = \int_{-\infty}^{\infty} C(\omega)e^{+i\omega t}d\omega = \int_{-\infty}^{\infty} C_o(\omega)A\sum_n \delta(\omega - \omega_n)e^{+i\omega t}d\omega$$

Interchanging the summation and the integration, and using the properties of the delta function gives

$$2\pi F(t) = A \sum_n C_o(\omega_n)e^{+i\omega_n t}$$

Integrate both sides of this equation over a time period $T$, and realizing that all terms on the right hand side are zero except the first, gives that

$$2\pi \int_0^T F(t)\,dt = AC_o(0)T$$

But $C_o(0)$ is the area under the pulse, hence we conclude that $A = 2\pi/T$.

We now have the representation of a periodic function $F(t)$ in terms of its transform over a single period. That is,

$$F(t) = \frac{1}{T} \sum_n C_o(\omega_n)e^{+i\omega_n t}, \qquad C_o(\omega_n) = \int_0^T F(t)e^{-i\omega_n t}dt$$

Except for a normalizing constant, this is a complex Fourier series representation of a periodic signal. We now recover the more usual form. Introduce the real coefficients $a_n$ and $b_n$ such that

$$(a_n - ib_n) \equiv \frac{2}{T}C_n$$

From this, it is apparent that

$$a_{-n} = a_n, \qquad b_{-n} = -b_n, \qquad n = 1, 2, 3, \ldots$$

which is the Fourier series result requiring that the reconstructed signal be real only. Taking off the real and imaginary parts of the transform, we get the usual Fourier series representation

$$F(t) = \frac{1}{2}a_o + \sum_{n=1}^{\infty} [a_n \cos(2\pi n \frac{t}{T}) + b_n \sin(2\pi n \frac{t}{T})]$$

where the Fourier coefficients $a_n$ and $b_n$ are obtained from

$$a_n = \frac{2}{T} \int_0^T F(t)\cos(2\pi n \frac{t}{T})dt, \qquad n = 0, 1, 2, \ldots$$

$$b_n = \frac{2}{T} \int_0^T F(t)\sin(2\pi n \frac{t}{T})dt, \qquad n = 1, 2, \ldots$$

and we have used $\omega_n = 2\pi n/T$.

**Example B.2:** Determine the Fourier series coefficients of a signal composed of repeated rectangles.

We use the specification of the rectangle of the last example. The $a_n$ coefficients are

$$a_n = \frac{2}{T} \int_{t_o}^{t_o+a} F_o \cos(2\pi nt/T)\,dt = \frac{F_o a}{T} \left\{ \frac{\sin \omega_n a/2}{\omega_n a/2} \right\} \cos \omega_n (t_o + a/2)$$

where $\omega_n = 2\pi n/T$. Similarly for the $b_n$ coefficients

$$b_n = \frac{F_o a}{T} \left\{ \frac{\sin \omega_n a/2}{\omega_n a/2} \right\} \sin \omega_n(t_o + a/2)$$

It is seen that the Fourier series transform is very similar to that obtained from the continuous transform.

# B.3   Discrete Fourier Transform

The discrete coefficients in the Fourier series are obtained by performing continuous integrations over the time period. We now replace these integrations by summations as a further step in the numerical implementation of the continuous transform.

The essential starting point for the discrete transform is the series transform pair rearranged as

$$f(t) = \frac{1}{T} \sum_{-\infty}^{\infty} C_n e^{i\omega_n t}, \qquad C_n(\omega) = \int_0^T f(t) e^{-i\omega_n t} \, dt$$

Let the time function $f(t)$ be divided into $M$ piecewise constant segments whose heights are $f_m$ and base $\Delta T = T/M$. The coefficients are now obtained from

$$
\begin{aligned}
C_n \approx D_n &= \sum_{m=0}^{M-1} f_m \int_{t_m - \Delta T/2}^{t_m + \Delta T/2} e^{-i\omega_n t} \, dt = \Delta T \sum_m f_m \left\{ \frac{\sin \omega_n \Delta T/2}{\omega_n \Delta T/2} \right\} e^{-i\omega_n t_m} \\
&= \Delta T \left\{ \frac{\sin \omega_n \Delta T/2}{\omega_n \Delta T/2} \right\} \sum_m f_m e^{-i\omega_n t_m}
\end{aligned}
$$

We see that this is the sum of the transforms of a series of rectangles each shifted in time by $t = t_m + \Delta T/2$. The contribution of each of these will now be evaluated more closely.

First look at the summation term. If $n > M$, that is if $n = M + n^*$, then the exponential term evaluates the same irrespective of $M$ since

$$i\omega_n t_m = i 2\pi \frac{n}{T} \frac{mT}{M} = i2\pi \frac{(M + n^*)m}{M} = i2\pi m + i2\pi \frac{n^* m}{M}$$

and the exponential of the first term is always unity. More specifically, if $M = 8$ say, then $n = 9, 11, 17$ evaluates the same as $n = 1, 3, 1$, respectively. The discretization process has forced a periodicity into the frequency description. Now look at the other contribution; we see that the sinc function term does depend on the value of $n$ and is given by

$$\text{sinc}(x) \equiv \frac{\sin(x)}{x}, \qquad x = \omega_n \Delta T/2 = \pi \frac{n}{T} \frac{T}{M} = \pi \frac{n}{M}$$

The sinc function is such that it decreases rapidly with increasing argument and is negligible beyond its first zero. The first zero occurs where $n = M$; if $M$ is made very large, that is, the integration segments are made very small, then it will be the higher order coefficients (i.e., large $n$) that are in the vicinity of the first zero. Let it be further assumed that the magnitude of these higher order coefficients are negligibly small. Then an approximation for the coefficients is

$$D_n \approx \Delta T \{1\} \sum_m^M f_m e^{-i\omega_n t_m}$$

on the assumption that it is good for $n < M$ and that $C_n \approx 0$ for $n \geq M$.

Since there is no point in evaluating the coefficients for $n > M - 1$, the approximation for the Fourier series coefficients is now taken as

$$f_m = f(t_m) \approx \frac{1}{T} \sum_{n=0}^{N-1} D_n e^{+i\omega_n t_m} = \frac{1}{T} \sum_{n=0}^{N-1} D_n e^{+i2\pi nm/N}$$

$$D_n = D(\omega_n) \approx \Delta T \sum_{m=0}^{N-1} f_m e^{-i\omega_n t_m} = \Delta T \sum_{m=0}^{N-1} f_m e^{-i2\pi nm/N}$$

where both $m$ and $n$ range from 0 to $N - 1$. These are the definition of what is called the *discrete Fourier transform*. It is interesting to note that the exponentials do not contain dimensional quantities; only the integers $n, m, N$ appear. In this transform, both the time and frequency domains are discretized, and as a consequence, the transform behaves periodically in both domains. The dimensional scale factors $\Delta T, 1/T$ have been retained so that the discrete transform gives the same numerical values as the continuous transform. There are other possibilities for these scales found in the literature.

The discrete transform enjoys all the same properties of the continuous transform; the only significant difference is that both the time domain and frequency domain functions are now periodic. To put this point into perspective consider the following: A discrete Fourier transform seeks to represent a signal (known over a finite time $T$) by a finite number of frequencies. Thus it is the continuous Fourier transform of a periodic signal. Alternatively, the continuous Fourier transform itself can be viewed as a discrete Fourier transform of a signal with an infinite period. The lesson is that by choosing a large signal sample length, the effect due to the periodicity assumption can be minimized and the discrete Fourier transform approaches the continuous Fourier transform.

**Example B.3:**   A real only function is given by the following sampled values in time

$$f_1 = f_2 = 1, \qquad f_0 = f_3 = f_4 = f_5 = f_6 = f_7 = 0$$

Determine its discrete Fourier transform.

This function has the shape of a rectangular pulse. Eight points are given, thus it is implicit that the function repeats itself beyond that. That is, the next few values are 0, 1, 1, 0, etc. The transform becomes (since $\Delta T = 1, N = 8$)

$$D_n = \sum_{m=0}^{7} f_m e^{-i2\pi nm/8} = f_1 e^{-i\pi n/4} + f_2 e^{-i\pi n/2}$$

The first ten transform points, in explicit form, are

$$
\begin{aligned}
D_0 &= 2.0 \\
D_1 &= 0.707 - 1.707i \\
D_2 &= -1.0 - 1.0i \\
D_3 &= -0.707 + 0.293i \\
D_4 &= 0.0 \\
D_5 &= -0.707 - 0.293i
\end{aligned}
$$

$$D_6 = -1.0 + 1.0i$$
$$D_7 = 0.707 + 1.707i$$
$$D_8 = 2.0$$
$$D_9 = 0.707 - 1.707i$$

The obvious features of the transform is that it is complex and that it begins to repeat itself beyond $n = 7$. Note also that $D_4$ (the $(\frac{1}{2}N + 1)^{th}$ value) is the *Nyquist* value. The real part of the transform is symmetric about the Nyquist frequency, while the imaginary part is anti-symmetric. It follows from this that the sum $\frac{1}{2}[D_n + D_{N-n}]$ gives only the real part, i.e.,

$$2.0, \quad 0.707, \quad -1.0, \quad -0.707, \quad 0.0, \quad -0.707, \quad \cdots$$

while the difference $\frac{1}{2}[D_n - D_{N-n}]$ gives the imaginary part

$$0, \quad -1.707i, \quad -i, \quad +0.293i, \quad 0, \quad -0.293i, \quad \cdots$$

These two functions are the even and odd decompositions, respectively, of the transform. Also note that $D_0$ is the area under the function.

# B.4   Fast Fourier Transform Algorithm

The final step in the numerical implementation is the development of the fast Fourier transform (FFT) algorithm for performing the summations of the discrete Fourier transform in a highly efficient manner. This is not a different transform; the numbers obtained from the FFT are exactly the same in every respect as those obtained from the DFT. The intention of this section is to just survey the major features of the FFT algorithm and to point out how the great speed is achieved. More detailed accounts can be found in the References [9, 14].

Consider the generic forward transform written as

$$S_n = \sum_{m=0}^{N-1} f_m e^{-i2\pi \frac{nm}{N}}, \qquad n = 0, 1, \ldots, N-1$$

We write this in the expanded form

$$S_0 = \{f_0 + f_1 + f_2 + \cdots\}$$
$$S_1 = \{f_0 + f_1 e^{-i2\pi \frac{1}{N}} + f_2 e^{-i2\pi \frac{2}{N}} + \cdots\}$$
$$S_2 = \{f_0 + f_1 e^{-i2\pi \frac{2}{N}} + f_2 e^{-i2\pi \frac{4}{N}} + \cdots\}$$
$$\vdots$$
$$S_n = \{f_0 + f_1 e^{-i2\pi \frac{n}{N}} + f_2 e^{-i2\pi \frac{n2}{N}} + \cdots\}$$

and so on. For each $S_n$, there are $(N - 1)$ complex products and $(N - 1)$ complex sums. Consequently, the total number of computations (in round terms) is on the order of $2N^2$. The purpose of the FFT is to take advantage of the special form of the exponential terms to reduce the number of computations to less than $N^2$.

The key to understanding the FFT algorithm lies in seeing the repeated forms of numbers. This will be motivated by considering the special case of $N$ being 8. First consider the matrix of the exponents $-i2\pi\left(\frac{mn}{N}\right)$

$$\frac{-i2\pi}{N}\begin{bmatrix} 0 & 0 & 0 & 0 & \cdots & 0 \\ 0 & 1 & 2 & 3 & \cdots & (N-1) \\ 0 & 2 & 4 & 6 & \cdots & 2(N-1) \\ 0 & 3 & 6 & 9 & \cdots & 3(N-1) \\ \vdots & \vdots & \vdots & \vdots & \vdots & \vdots \\ 0 & (N-1) & 2(N-1) & 3(N-1) & \cdots & (N-1)(N-1) \end{bmatrix}$$

It is apparent that for an arbitrary value of $N$, $2\pi$ will be multiplied by non-integer numbers (in general). These exponents, however, can be made quite regular if $N$ is highly composite. For example, if $N$ is one of the following

$$N = 2^\gamma = 2, 4, 8, 16, 32, 64, 112, 256, 512, 1024, \ldots$$

then the effective number of different integers in the matrix is decreased. Thus if $N = 8$ we get (with $\lambda = -i2\pi/8$)

$$\lambda\begin{bmatrix} 0 & 0 & 0 & 0 & 0 & 0 & 0 & 0 \\ 0 & 1 & 2 & 3 & 4 & 5 & 6 & 7 \\ 0 & 2 & 4 & 6 & 0 & 8+2 & 8+4 & 8+6 \\ 0 & 3 & 6 & 8+1 & 8+4 & 8+7 & 16+2 & 16+5 \\ 0 & 4 & 8+0 & 8+4 & 16+0 & 16+4 & 24+0 & 24+0 \\ 0 & 5 & 8+2 & 8+7 & 16+4 & 24+1 & 24+6 & 32+3 \\ 0 & 6 & 8+4 & 16+2 & 24+0 & 24+6 & 32+4 & 40+2 \\ 0 & 7 & 8+6 & 16+5 & 24+4 & 32+3 & 40+2 & 48+1 \end{bmatrix} = \lambda\begin{bmatrix} 0 & 0 & 0 & 0 & 0 & 0 & 0 & 0 \\ 0 & 1 & 2 & 3 & 4 & 5 & 6 & 7 \\ 0 & 2 & 4 & 6 & 0 & 2 & 4 & 6 \\ 0 & 3 & 6 & 1 & 4 & 7 & 2 & 5 \\ 0 & 4 & 0 & 4 & 0 & 4 & 0 & 0 \\ 0 & 5 & 2 & 7 & 4 & 1 & 6 & 3 \\ 0 & 6 & 4 & 2 & 0 & 6 & 4 & 2 \\ 0 & 7 & 6 & 5 & 4 & 3 & 2 & 1 \end{bmatrix}$$

This comes about because the exponentials take on the following simple forms

$$e^{-i2\pi[0]} = e^{-i2\pi[1]} = e^{-i2\pi[2]} = e^{-i2\pi[3]} = \ldots = 1$$

The regularity is enhanced even more if $(N/2 = 4)$ is added to the latter part of the odd rows. That is, if it is written as

$$\frac{-i2\pi}{8}\begin{bmatrix} 0 & 0 & 0 & 0 & (0 & 0 & 0 & 0) + 0 \\ 0 & 1 & 2 & 3 & (0 & 1 & 2 & 3) + 4 \\ 0 & 2 & 4 & 6 & (0 & 2 & 4 & 6) + 0 \\ 0 & 3 & 6 & 1 & (0 & 3 & 6 & 1) + 4 \\ 0 & 4 & 0 & 4 & (0 & 4 & 0 & 4) + 0 \\ 0 & 5 & 2 & 7 & (0 & 5 & 2 & 7) + 4 \\ 0 & 6 & 4 & 2 & (0 & 6 & 4 & 2) + 0 \\ 0 & 7 & 6 & 5 & (0 & 7 & 6 & 5) + 4 \end{bmatrix}$$

We see that many of the computations used in forming one of the summations is also used in the others. For example, $S_0, S_2, S_4, S_6$ all use the sum $(f_0 + f_4)$. Realizing that $e^{-i2\pi 4/8} = -1$, then we also see that all the odd summations contain common terms such as $(f_0 - f_4)$. This re-use of the same computations is the reason a great reduction of computational effort is afforded by the FFT. The algorithm sets up the book-keeping so that this is done in a systematic way.

The number of computations with and without the FFT algorithm is given by

$$\tfrac{3}{2}N \log_2 N \qquad \text{versus} \qquad 2N^2$$

When $N = 8$, this gives a speed factor of only 3.5:1, but when $N = 1024$, this jumps to over 100:1. It is this excellent performance at large N that makes the application of Fourier analysis feasible for practical problems. On our benchmark machine of $1MFlops$, a 1024 point transform takes significantly less than one second.

There are many codes available for performing the FFT as can be found in Reference [9] and well-documented FORTRAN routines are described in Reference [34]. The computer program CFFTCOMP used in the example to follow has its source code listed in Reference [14].

**Example B.4:**   Use the fast Fourier transform to estimate the amplitude spectrum of the following two force histories.

|  | $t$ | $P(t)$ |  | $t$ | $P(t)$ |
|---|---|---|---|---|---|
|  | 0 | 0 |  | 0 | 0 |
|  | 1 | 0 |  | 1 | 0 |
| force 1: | 2 | 1000 | force 2: | 1.1 | 10000 |
|  | 4 | 0 |  | 1.3 | 0 |
|  | 10 | 0 |  | 10 | 0 |

Note that the peak forces are different but that the impulses (the area under the curve) are the same.

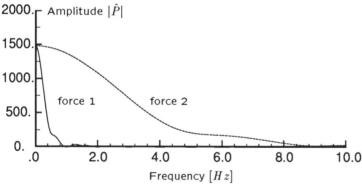

**Figure B.3**: Amplitude spectrum of input force history.

The results are shown in Figure B.3. A $\Delta T$ of .04 s and an $N$ of 512 was used; the intermediate values are obtained by linear interpolation. Note that the narrower pulse in the time domain gives the broader frequency response. Also note that significant amplitudes extend only to about 10 Hz for the narrower pulse; this will be of significance in Chapter 13 when we consider the structural response caused by this pulse.

The fast Fourier transform algorithm is so efficient that it has revolutionized the whole area of spectral analysis. It can be shown quite simply that it enjoys all the same properties of the continuous transform. Therefore, in the subsequent analysis, we will assume that any time input or response can be represented in the spectral form

$$F(t) = \sum_{n=o}^{N-1} C_n e^{+i\omega_n t}$$

and the tasks of forward and inverse transforms are accomplished with a computer program.

# Appendix C

# Computer Source Code

This appendix contains the complete source code listing for the program STADYN: a program capable of doing the static and dynamic analysis of 3-D frame structures. It implements most of the topics covered in the text; this includes static analysis, stability analysis, transient analysis, and some modal analysis. Unfortunately, because of space restrictions, we have not been able to include any of the pre- and post-processing capabilities such as automatic node renumbering and graphics. However, we have included an abbreviated manual with a set of tutorials for running STADYN.

An electronic form of the source code plus the *Makefiles* is available from *ikayex* Software Tools, 615 Elston Road, Lafayette, Indiana 47905, USA.

## C.1 Compiling the Source Code

All the source code is written in FORTRAN-77, and every attempt has been made to make it as portable as possible. The actual dialect used is that of Microsoft V5.0 for the microcomputer, hence there are a number of slight differences in comparison to Unix installations. The Unix alternatives are indicated in the code. The coding has some practices that older compilers may complain about. Examples are the use of Real*8 instead of Double Precision, and the Do - EndDo construct. In general, the complaints can be ignored; it is expected that compilers that are compliant with the release of the FORTRAN-90 standard should have no difficulties at all with the source code.

STADYN uses Jacobi rotations for solving the eigenvalue problem and therefore does not have an implementation of the vector iteration and subspace iteration routines. The program MODDYN which is designed specifically to do modal analysis of large systems does implement these. Therefore the source codes for these modules are taken from MODDYN; however, both MODDYN and STADYN are completely compatible hence there should be no difficulty in combining the modules with STADYN.

The style of coding used is basically as recommended in Reference [34]; this book describes an excellent no nonsense approach to programming.

409

# C.2   Manual and Tutorial

This is a collection of short tutorials to get you acquainted with running STADYN as quickly as possible. It is advisable that the first time through, you follow the instructions strictly. This is because some of the later tutorials make reference back to the earlier ones.

The program STADYN is capable of doing the static and dynamic analysis of 3-D structures. Its main capabilities involve determining the member displacements and loads, and structural reactions for:

- Static loading
- Applied transient load history
- Modal analysis
- Stability analysis

The tutorials cover examples from the first and last of these. The program is menu driven in such a way that its functioning is apparent.

As STADYN proceeds, it creates a number of files among which are

|           |           |           |
|-----------|-----------|-----------|
| stadyn.log | stadyn.stf | stadyn.dis |
| stadyn.out | stadyn.mas | stadyn.snp |
| stadyn.dyn | stadyn.geo |           |

The LOG file echoes all the input responses as well as having some extra information that might prove useful during post analysis of the results. The OUT file is the usual location for STADYN output. The second column of files are associated with the stiffness matrix and mass matrix — these are left on disk in case they may be of value for some other purpose. They are in binary form. The last column of files are output files. They too are in binary form for compactness but can be read for further post-processing.

## Data File Format

The input data file can be in free format with blanks or commas used as separators. You need an editor or word processor to create your own data files, and when you do this make sure you store them as ASCII files without any word-processing hidden symbols.

Note that the data input is arranged in groups, and that each group must have the word END as its last line. This acts as an additional data checker.

## HEADER GROUP

```
TITLE
IPROB
IFLAG1 IFLAG2 IFLAG3 IFLAG4
end
```

TITLE : Short title (up to 50 characters) describing the problem.

IPROB : General problem type.[integer]
  1-D: 11=rods, 12=beams, 13=shafts
  2-D: 21=truss 22=frames, 22=grids
  3-D: 31=truss 32=frames

IFLAG : Echo flags for the input
  1=connectivity, 2=material, 3=numbering, 4=loads

## ELEMENT GROUP

The number of lines are equal to NEL. Every element must have an input for it, although they do not have to be input in exact sequential order. It is allowable to mix truss and frame elements in the same structure.

```
NEL
ELM  TYP  NPI  NPJ
 :    :    :    :
 :    :    :    :
end
```

NEL : Number of elements
ELM : The unique number assigned to each element.
TYP : The type of element, either frame or truss.[integer]
  1: element is a rod or truss.
  2: element is a beam or frame.
NPI : The number of the #1 node on the element.
NPJ : The number of the #2 node on the element.

## Material Lines

The number of lines are equal to NMAT. Every element must have a material, however, it is possible to specify overlaying numbers. For example, to make one member different from the rest then specify 1 - 10 followed by 7 - 7, say.

```
NMAT
EL1  EL2   E    G    A    Ix   Iy   Iz   RHO  GAMA
 :    :    :    :    :    :    :    :    :    :
 :    :    :    :    :    :    :    :    :    :
end
```

NMAT : Number of element materials [integer]
EL1  : First element of this material [integer]
EL2  : Last element of this material [integer]
E    : Young's modulus of the element.[real]
G    : Shear modulus of the element.[real]
A    : Cross sectional area of the element.[real]
Ix   : The polar moment of inertia about x-axis.[real]
Iy   : The moment of inertia about y-axis. [real]
Iz   : The moment of inertia about z-axis. [real]
RHO  : The density of the element. [real]
GAMA : Orientation of principal axes. [real]

## NODE GROUP

The number of lines input must equal NNP. The lines do not have to be input in order from 1 to NNP, but every node must have a line.

```
NNP
NODE   XORD   YORD   ZORD
 :      :      :      :
 :      :      :      :
end
```

NNP  : Number of nodal points
NODE : The unique number assigned to each node.
XORD : x-coordinate of the node point.
YORD : y-coordinate of the node point.
ZORD : z-coordinate of the node point.

## Boundary Conditions

The number of lines input must equal NBC. Again the sequence can be random. The default status for each degree of freedom is free, so the boundary conditions need be imposed only for those nodes that are different.

```
NBC
NODE   XDOF   YDOF   ZDOF   XROT   YROT   ZROT
 :      :      :      :      :      :      :
 :      :      :      :      :      :      :
end
```

NBC   :  The number of a node at which at least one degree
         of freedom is being fixed.

NODE :  The number of a node at which at least one degree
         of freedom is being fixed.

XDOF :  Motion in the x-direction. [integer]

YDOF :  Motion in the y-direction. [integer]

ZDOF :  Motion in the y-direction. [integer]

XROT :  Motion about the x-axis. [integer]

YROT :  Motion about the y-axis. [integer]

ZROT :  Motion about the z-axis. [integer]

In each case: 0=fixed, 1=free

## Nodal Loads

The number of lines of input must equal NLOAD. The lines do not have
to be input in order from 1 to NNP. Each node is assumed to have zero
applied load and zero concentrated mass unless imposed otherwise. Note
that the actual applied load history for transient problems is input from the
menu within STADYN — the entries here essentially say where the loads are
located, as well as their relative scaling.

```
NLOAD
NODE    XLOAD    YLOAD    ZLOAD    XMOM    YMOM    ZMOM    CMASS
 :        :        :        :        :       :       :       :
 :        :        :        :        :       :       :       :
end
```

NLOAD : Number of nodal loads or mass points.

NODE  : The unique number assigned to each node.

XLOAD : Force applied in the x-direction at the node.

YLOAD : Force applied in the y-direction at the node.

ZLOAD : Force applied in the z-direction at the node.

XMOM : Moment applied about the x-axis at the node.

YMOM : Moment applied about the y-axis at the node.

ZMOM : Moment applied about the z-axis at the node.

CMASS : Concentrated mass at the node.

## SPECIALS GROUP

This section is a mechanism to allow the input of special global properties.
For STADYN, the only recognized code is for damping.

The number of lines input must equal NSP. The lines do not have to
be input in order.

```
NSP
CODE    C1    C2    C3
 :       :     :     :
 :       :     :     :
end
```

NSP   : Number of lines

CODE : The unique code number for each special attribute.[integer]
       2110=damping, $[ C ] = c_1 + c_2[ K ] + c_3[ M ]$

C1    : First constant [real]

C2    : Second constant [real]

C3    : Third constant [real]

## Example

The following example is for a simple truss structure.

```
exs.1 casf pp121                                ::HEADER GROUP
21
1 1 1 1
end
3                                               ::ELEMENT GROUP
1    1    1    2
2    1    2    3
3    1    3    1
end
1                                               ::material props
1    3  200e9  70e6  3000e-6  1.0 1.0   1.0   0.0  0.0
end
3                                               ::NODE GROUP
1    0.0    0.0    0.0
2    3.0    3.0    0.0
3    6.0    0.0    0.0
end
2                                               ::boundary condns
1    0    0    0    0    0    0
3    1    0    0    0    0    0
end
1                                               ::applied loads
2    100.0e3   200.0e3   0.0    0.0    0.0   0.0  0.0
end
1                                               ::SPECIALS GROUP
2110    0.0 0.0 0.0
end
```

This is the input file for a simple truss problem. Note that this data is
inputted in free format with white spaces used only as separators. In general,
the file can be documented by adding comments on the remainder of a line.
For this particular example, the units are in metric — STADYN will handle
any set of units as long as they are consistent. Notice how the boundary
conditions are imposed: Node 1 is fixed in all directions, but Node 3 is free
to move in the x-direction and not in any of the others. That is, Node 3 is
on rollers.

## Checking Input

This tutorial shows how you can use some of the built-in diagnostics of
STADYN to help ensure that the structure data file is correct.

To run the program simply type

```
C> sdcomp
```

Note that sdcomp is the name of the executable version. The menu will
appear and you then just respond to the questions. Since you want to input
the structural data respond

```
2
```

and you are asked for the data filename. Type

```
exs.1
```

If STADYN thinks it has read the data correctly, it will acknowledge so and
then present the main menu again. To quit choose

```
0
```

If you look in your directory you will find a file called STADYN.LOG. It is an
ASCII file so peruse it. There appears to be a big jumble of numbers and
symbols such as

```
@@ STADYN version 2.05
@@ DATE 7-21-1990  TIME 18:43
                   2   ::MAIN
exs.1                  ::filename
@@ HEADER GROUP
@@ Title      = exs.1 casf pp121
@@ Problem type = 21
```

This file is actually a record, or log, of the session you just did above. There are two types of lines in this file. The ones that have double colons : : are the responses you gave to STADYN. What follows the colons are brief descriptions of what you typed or where you typed it. The other lines that begin with double ats @@ are STADYN's information or answers to you. It gives you extra insight into its workings that can be very useful when trying to backtrack to nail a bug. In this particular case it can be used to judge if all the structural data was read as intended. If STADYN detects inconsistencies in the input data it will flag them and by looking through this log file you can determine, approximately, where the inconsistency occurred.

Look further through the file and see the manner in which the structural data is echoed. The single most common source of errors for a finite element analysis is the data input, so you should attempt to become familiar with this file because it can be an immensely useful tool for error checking.

## Static Analysis

Since we know the data in the above file to be in order we will now use it to do a simple static analysis. Run the program and read the data as before

```
C> sdcomp
2
exs.1
```

Before we can obtain the static solution we must first form the stiffness matrix

```
3
```

Since this is your first example, echo the stiffness matrix to STADYN.OUT just to see what it looks like

```
1
```

You probably noticed some activity with your hard disk, that is because the stiffness matrix (and the mass matrix also) are not kept in RAM unless they are immediately needed. Since the loads were already read from EXS.1 we can now obtain the static solution by typing

```
6
```

You are now asked to choose the form of output

```
CHOOSE:  0=return,  -1=all members,  # of single node
```

Since we want everything choose

```
-1
```

Choosing a positive number outputs the information for all the elements connected to that node. As you can see, it is a recurring menu so multiple, separated nodes can be chosen. To quit and get out of STADYN type

```
0
0
```

That is all there is to it!

The output can now be surveyed by viewing the ASCII file STADYN.OUT. This file is fairly well documented so there is little difficulty in interpreting it. Reference [3] gives for the displacement of Node 2

```
x: 0.457mm   y: 1.664mm   z: 0.0mm
```

whereas the program gives

```
x: 4.571E-4  y: 1.664E-3  z: 0.0
```

These two sets agree.

## Using Driver Files

Actually, the main reason for doing this example is to introduce you to a unique feature of STADYN that underlies its design philosophy. If you look in your directory you will find a new version of the file STADYN.LOG and as pointed out before, this gives extra insight into the workings of the program.

While this is useful enough, its main purpose is to allow you to create a driver or response file. If you are not familiar with this idea let me explain. Make a copy of the log file by

```
C> copy stadyn.log instatic
```

Now use your editor to edit instatic. The only editing you need do is remove blank lines and lines beginning with @@. That is simple enough (although tedious on large files) and what you are left with is the same collection of responses as on the previous page but now it is documented. After you exit your editor all you need do to run the example of this section is

```
C> sdcomp < instatic
```

Go ahead, do it. The screen will scroll and everything is done automatically for you.

After a while you will find yourself doing the same operations repeatedly. Then you can make driver files for each of them. More importantly, the driver file is a record of how you processed the data. In a week, a month or six months, if you need to reconstruct what you did, it is all there laid out for you. This emphasis on driver files explains the willingness to allow the user interface to be somewhat awkward and cryptic. The idea is that the driver file approach allows the user a greater variety of inputs and gives greater control of the program. And quite complicated sequences of instructions can be pieced together to give very fine control over your data manipulations. It also allows the program to be used in batch mode and thereby blend in with your other programs more productively.

## Stability Analysis

STADYN is capable of obtaining the buckling loads and buckled shapes for a 3-D structure. Keep in mind that the number of modes obtained is equal to the number of free degrees of freedom in the system. Further, it is usually the lower loads of most interest. Hence, it is prudent to keep the system size small consistent with the desired usage of the output. A complicating factor is that to get accurate lower loads it is necessary to introduce more degrees of freedom (and thus be required to evaluate the higher loads also.) As an approximate rule of thumb, a system size $N$ gives about $N/2$ good values of buckling load. This must be treated with caution since it depends on the location of the nodes, and also if the mode shapes themselves are of interest.

The following structure file is for a two member truss structure.

```
exb.1 tmsa pp397                      ::HEADER GROUP
21
1 1 1 1
end
2                                     ::ELEMENT GROUP
1 1 1 2
2 1 1 3
end
1                                     ::material props
1 2    10e6   4e6   1.0    1.0 1.0 1.0   0.0   0.0
end
3                                     ::NODE GROUP
1  10.0 10.0  0.0
3  0.0   0.0  0.0
2  10.0  0.0  0.0
end
2                                     ::boundary condns
2 0   0   0   0   0   0
3 0   0   0   0   0   0
end
1                                     ::applied loads
1 0.0 -1.0    0.0 0.0 0.0 0.0 0.0
end
0                                     ::SPECIALS GROUP
end
```

The material is nominally aluminum and its properties are given in customary units.

Both the stiffness and geometric matrices must be assembled before the stability analysis itself can be performed. Do this by typing:

```
C> sdcomp
   2
   exb.1
   3
   1
```

It is now necessary to form the geometric matrix

```
   5
   1
```

Note that in order to establish this matrix, it is first necessary to solve the corresponding static problem. The material that is echoed to the screen reflects this. With both the stiffness and geometric matrices formed, the eigenvalue problem can be solved

```
   9
```

STADYN does two types of eigenvalue problems; vibration and stability. In response to

```
      CHOOSE: 0=return  1=stability  2=vibration
```

choose

```
   1
```

The program now performs the complete analysis. An interesting quantity to keep your eye on is the DIFF NORM. Finding eigenvalues is an iterative process and this quantity gives you an idea of the rate of convergence. The next response expected of you is to determine the form of output storage

```
   CHOOSE Storage:
      0=return  1=mode shapes  2=modal vectors  3=full <bin>
```

The difference between the mode shapes and the modal vectors is that the former has the boundary conditions already factored in. This is a recurring prompt so, if desired, all the information can be stored. The third option would be used if the results are to be piped into a modal analysis post-processor. To illustrate the difference between the first two options, choose

```
   1
```

This option then wants the number of modes to be reported. It can be from one to the system size. If you give a number outside this range, then the program takes the appropriate allowable limit. Choose

```
   6
```

Note that there are only two eigenvalues. Go through the process again and this time store the modal vector

```
   2
   2
```

Now exit the program by typing

```
   0
   0
```

Scan through the file STADYN.OUT to see the results. The first portion of it contains the stiffness and geometric matrices written in band storage form. Reference [35] shows that the first buckling load should be

$$P_{crit} = \frac{2\sqrt{2}-1}{7}EA = 0.2612\,EA$$

The output from STADYN gives the critical loads as

```
   0.26106E07        0.16252E14
```

The complete physical interpretation of the modal vector is given in a format identical to that of the static output. The Modal Vector option gives the same information but in compacted form (i.e., without the zero degree of freedoms).

Experiment with reversing the sign of the loads and note that STADYN still obtains the correct buckling load. This is important in complex structures where it is not obvious which members are in compression.

## C.3   Source Code for STADYN

Following is a complete listing of the FORTRAN-77 source code for the program STADYN. The common statements referred to in the include statement are cccc

```
      common/files/  iout,idyn , isnp, ilog , idat,imat,idrv
      common/datfil/ istf, imss, idis, ilod,icms,igeo
      common/params/ neq , nnp, nel, iband,ibandm
      common/material/eta,dampkk,dampmm,dampcc,gx,gy,gz
cccc
```

### Main Control

```
cccc
      block data
         include 'scommons'
         data isnp/60/,ilog /67/,iout/65/,idyn/63/,idrv/68/
         data idat/3/, imss/71/,istf/70/,idis/73/
         data ilod/74/,icms/75/,imat/76/,igeo/77/
      end
c
c     MAIN
      include 'scommons'
c
      parameter( maxnode =1000, maxelem = 1005)
      parameter( maxdof = maxnode*6)
      parameter( isize= 30000)
c
      character*40 filena
      integer npi(maxelem),npj(maxelem),nelt(maxelem),jbc(maxdof)
      real   xyz(maxnode*3)
      real*8 respce(isize)/isize*0.0/, ekmg1212(12,12)/144*0.0/
c
      write(*,'(//)')
      write(*,*)' STADYNSTADYNSTADYNSTADYNSTADYNSTADYNSTADYN'
      write(*,*)'   a program for STAtic & DYNamic  analysis'
      write(*,*)'          of 3-D framed structures '
      write(*,*)'        mmPC version 2.05, January 1991  '
      write(*,*)'            (c) ikayex SOFTWARE TOOLS     '
      write(*,*)' STADYNSTADYNSTADYNSTADYNSTADYNSTADYNSTADYN'
      write(*,*)' '
c
c     open files ready for work
      open( istf,file='stadyn.stf', form='unformatted')
      open( imss,file='stadyn.mas', form='unformatted')
      open( igeo,file='stadyn.geo', form='unformatted')
      open( idis,file='stadyn.dis', form='unformatted')
      open( ilod,file='stadyn.lod')
      open( icms,file='stadyn.tmp')
      open( ilog,file='stadyn.log')
      open( iout,file='stadyn.out')
      open( idyn,file='stadyn.dyn')
      open( isnp,file='stadyn.snp', form='unformatted')
      open( imat,file='stadyn.mat', form='formatted',
     >             access='direct',recl=95)
      rewind(iout)
      rewind(ilog)
      rewind(idyn)
      write(ilog,*)'@@ STADYN version 2.05'
      call timday(ilog)
c
1     continue
      write(*,*)' '
      write(*,*)'     MAIN menu: '
      write(*,*)'          0:  Quit '
      write(*,*)' '
      write(*,*)'          2:  Read in structure data'
      write(*,*)'          3:  Form stiffness matrix'
      write(*,*)'          4:  Form mass matrix'
      write(*,*)'          5:  Form geometric matrix'
      write(*,*)' '
      write(*,*)'          6:  Static loading      '
      write(*,*)'          8:  Transient loading    '
```

```
      write(*,*)'          9:  Eigenvalue problems'
      write(*,*)' '
      write(*,*)'          110:  System services '
      write(*,'(a\)' ) ' SELECT one-->'
      read(*,*) imain
      write(ilog,*) imain,' ::MAIN'
      write(*,*)' '
c
      if (imain .eq. 0) then
          close ( istf )
          close ( imss )
          close ( idis )
          close ( ilod )
          close ( icms )
          close ( ilog )
          close ( iout )
          close ( idyn )
          close ( isnp )
          close ( imat )
          stop'@@STADYN: stopped from menu'
      endif
      if (imain .eq. 2) goto 200
      if (imain .eq. 3) goto 300
      if (imain .eq. 4) goto 400
      if (imain .eq. 5) goto 500
      if (imain .eq. 6) goto 600
      if (imain .eq. 8) goto 800
      if (imain .eq. 9) goto 900
      if (imain .eq. 110) then
          pause' Type: system command'
cUWiX     use  call system( )
      endif
      go to 1
c
c READ structure's data, and bcs
 200  continue
      write(*,'(a\)')' TYPE:  input datafilename--> '
      read(*   ,'(a40)') filena
      write(ilog,*) filena,'      ::filename'
      open( idat, file = filena)
      rewind(idat)
c
c     storage pointers
      ir1 = 1
      ir2 = ir1 + maxnode
      ir3 = ir2 + maxnode
      ir4 = ir3 + maxnode
      ir5 = ir4 + maxnode
      ir6 = ir5 + maxnode
      ir7 = ir6 + maxnode
      ir8 = ir7 + maxnode
      ir9 = ir8 + maxnode*6
      call ckspace(ilog,isize,'DATAin',ir9)
c
      call datain(xyz(1),xyz(maxnode+1),xyz(maxnode*2+1),
     >            respce(ir1),respce(ir2),respce(ir3),
     >            respce(ir4),respce(ir5),respce(ir6),
     >            respce(ir7),
     >            jbc, maxdof, maxelem, maxnode,
     >            npi, npj, nelt)
c
      write(*,*)'@@ now converting'
      call datacnv(respce(ir1),respce(ir2),respce(ir3),
     >            respce(ir4),respce(ir5),respce(ir6),
     >            respce(ir7),respce(ir8),
     >            jbc, maxdof, maxelem, maxnode,
     >            npi, npj, nelt, iprob)
c
      close(idat)
      write(*,212)'@@ system size = [',neq,' X ',iband,']'
 212  format(1x,a,i5,a,i5,a)
      goto 1
c
c     STIFFNESS  MATRIX
 300  continue
c     storage pointers
```

```
      ir1 = 1
      ir2 = ir1 + neq*iband
      call ckspace(ilog,isize,'FORMstff',ir2)
c
      call formstff(respce(ir1), npi,npj,
     >            xyz(1),xyz(maxnode+1),xyz(maxnode*2+1),
     >            jbc, nelt, ekmg1212)
      goto 1
c
c     MASS MATRIX
 400  continue
c     storage pointers
      ir1 = 1
      ir2 = ir1 + neq*iband
      call ckspace(ilog,isize,'FORMmass',ir2)
c
      call formmass(respce(ir1), npi, npj,
     >            xyz(1),xyz(maxnode+1),xyz(maxnode*2+1),
     >            jbc, nelt, ekmg1212)
      write(ilog,*)'@@ mass half bandwidth =',ibandm
      write(*  ,*)'@@ mass half bandwidth =',ibandm
      goto 1
c
c     GEOMETRIC MATRIX
 500  continue
c     FIRST solve the static problem
c     storage pointers
      ir1=1
      ir2=ir1+neq*iband
      ir3=ir2+neq
      ir4=ir3+neq
      ir5 = ir4 + nnp*6
      call ckspace(ilog,isize,'FORMgeom',ir5)
c
      write(*,*)'@@ solve STATIC problem'
      call   static( respce(ir1),respce(ir2),respce(ir3))
c
c     SECOND assemble the geometric stiffness
      call formgeom(respce(ir1), npi, npj,
     >            xyz(1),xyz(maxnode+1),xyz(maxnode*2+1),
     >            jbc, nelt, ekmg1212,
     >            respce(ir3),respce(ir4))
      goto 1
c
c     STATIC DISPLACEMENTS
 600  continue
c     storage pointers
      ir1=1
      ir2=ir1+neq*iband
      ir3=ir2+neq
      ir4=ir3+neq
      ir5 = ir4 + nnp*6
      call ckspace(ilog,isize,'STATic',ir5)
c
      call static( respce(ir1),respce(ir2),respce(ir3))
c
c     STATIC OUTPUT
      call statout(respce(ir3), respce(ir4), npi,npj,
     >            xyz(1),xyz(maxnode+1),xyz(maxnode*2+1),
     >            jbc, nelt, ekmg1212)
      goto 1
c
c     TRANSIENT  LOADING
 800  continue
c     storage pointers
      ir1=1
      ir2=ir1+neq*iband
      ir3=ir2+neq*ibandm
      ir4=ir3+neq
      ir5=ir4+neq
      ir6=ir5+neq
      ir7=ir6+neq
      ir8=ir7+neq
      ir9=ir8+neq
      ir10=ir9+neq
      ir11=ir10+neq
```

```
        call ckspace(ilog,isize,'TRNSient',ir11)
c
        write(*,*)' '
        call trnsient(respce(ir1),respce(ir2),respce(ir3),
     >              respce(ir4),respce(ir5),respce(ir6),
     >              respce(ir7),respce(ir8),respce(ir9),
     >              respce(ir10), jbc, xyz,maxnode)
        goto 1
c
c
c       MODE SHAPES & FREQUENCIES
900     continue
        write(*,*)'CHOOSE: 0=return 1=stability 2=vibration'
        write(*,'(a\)')' --> '
        read(* ,* ) ivib
        write(ilog,*) ivib ,' ::2=vibn '
c
c       storage pointers
        if (ivib .eq. 0) then
             goto 1
        elseif (ivib .eq. 2) then
            if (ibandm .eq. 1) then
                  neqmas=1
            else
                  neqmas=neq
            endif
        elseif (ivib .eq. 1) then
              neqmas=neq
        endif
c
c       memory pointers
        ir1=1
        ir2=ir1+neq*neq
        ir3=ir2+neq
        ir4=ir3+neq*neq
        ir5=ir4+neq*neqmas
        ir6=ir5+neq
        call ckspace(ilog,isize,'EIGEN',ir6)
c
        call eigen(respce(ir1),respce(ir2),respce(ir3),
     >            respce(ir4),respce(ir5),neqmas,ivib)
c
        call eigout(respce(ir1),respce(ir2),
     >            respce(ir3), jbc,ivib)
        goto 1
c
c
999     continue
c
      end
c
c
      SUBROUTINE RDSTFF(stf,iwidth)
        include 'scommons'
        real*8 stf(neq,iwidth)
c
        write(*,*)'@@ reading <<stadyn.stf>>'
        rewind (istf)
        read(istf) neq,iband
        do 8 i=1,neq
          read(istf) (stf(i,j), j=1,iband)
8       continue
      return
      end
c
      SUBROUTINE RDMASS(b,iwidth)
        include 'scommons'
        real*8 b(neq,iwidth)
c
        write(*,*)'@@ reading <<stadyn.mas>>'
        rewind(imss)
        read(imss) n,ibandm
        if (ibandm.eq.1) then
          read(imss) (b(i,1), i=1,n)
        else
          do 3 i=1,n
```

```
          read(imss) (b(i,j), j=1,ibandm)
3         continue
        endif
      return
      end
c
      SUBROUTINE RDLOAD(load)
        include 'scommons'
        real*8 load(neq)
        write(*,*)'@@ reading <<stadyn.lod>>'
        rewind (ilod)
        read(ilod,*) (load(i), i=1,neq )
      return
      end
c
      SUBROUTINE RDGEOM(a,iwidth)
        include 'scommons'
        real*8 a(neq,iwidth)
c
        write(*,*)'@@ reading <<stadyn.geo>>'
        rewind(igeo)
        read(igeo) n,ibandg
        write(*,*)'@@ neq ibandg ',n,ibandg
        do 4 i=1,neq
          read(igeo) (a(i,j), j=1,ibandg)
4       continue
      return
      end
c
c       SUBROUTINE CHECKSPACE
c       This subroutine checks the space requirements
c
      subroutine ckspace(ilog,isize,subname,iend)
        character*(*) subname
        write(ilog,*)'@@ ',subname,': memory used ',
     >              iend,' of',isize
        write(*    ,*)'@@ ',subname,': memory used ',iend,
     >              ' of',isize
        if (isize .le. iend) then
          write(*,*)'ERROR: not enough reserved memory'
          stop
        else
          return
        endif
      return
      end
c
c
cccc
```

## DataIn

```
cccc

      SUBROUTINE DATAIN(xord,yord,zord,xload,yload,zload,
     >                  xmom,ymom,zmom,cmass,
     >                  jbc, maxdof, maxelem, maxnode,
     >                  npi, npj, nelt)
c     Inputs the structural data
        include 'scommons'
c
        integer xdof,ydof,zdof, xrot,yrot,zrot, jbc(maxdof)
        integer npi(maxelem), npj(maxelem), nelt(maxelem)
        real    xord(maxnode), yord(maxnode),zord(maxnode)
        real*8 xload(maxnode),yload(maxnode),zload(maxnode)
        real*8 cmass(maxnode)
        real*8 xmom(maxnode), ymom(maxnode),zmom(maxnode)
        character*50 title,endin
c
c       define function for testing ranges
        logical range
        range(ii,jj,kk) = (ii .gt. jj) .or. (jj .gt. kk)
c
c CHECK all GROUP sizes first
        write(*,*)'@@ reading GROUP sizes '
```

```
        call grpchk(idat,ilog)                              if (nnp .gt. maxnode) goto 9015
        write(*,*)'@@ GROUP sizes OK '                      do 231 j= 1, nel
        rewind(idat)                                           if (range(1,npi(j),nnp) .or. range(1,npj(j),nnp))
c                                                     >            goto 9030
c READ header  data                              231       continue
        write(ilog,*)'@@ reading HEADER lines'              do 230 i= 1, nnp
        read(idat,'(1a50)') title                              read(idat,*) j,xord(j),yord(j),zord(j)
        write(*  ,*) title                                     if (range(1,j,nnp)) goto 9040
        read(idat,*) iprob                          230       continue
        read(idat,*) iflag1,iflag2,iflag3,iflag4            read(idat,'(1a50)') endin
        read(idat,'(1a50)') endin                           call chkend(endin,ichk)
        call chkend(endin,ichk)                             if (ichk.eq.0) then
        if (ichk.eq.0) then                                   write(ilog,*)'@@ Incorrect Coordinate data length'
          write(ilog,*)'@@ Incorrect HEADER group length'    write(*  ,*)'@@ Incorrect Coordinate data length'
          write(*  ,*)'@@ Incorrect HEADER group length'     stop'!! ERROR'
          stop'!! ERROR'                                   endif
        endif                                               ichk=1
        ichk=1                                       c
        write(ilog,1004)title,iprob,iflag1,iflag2,iflag3,iflag4  c READ bcs: set default = 1    i.e. free
c                                                           do 241 i= 1, nnp*6
c READ elem data: el# eltype node#1 node#2                    jbc(i) = 1
        write(ilog,*)'@@ reading connectivities lines'  241   continue
        read(idat,*) nel                                    write(ilog,*)'@@ reading bc lines'
        if (nel .lt. 1) goto 9000                           read(idat,*)  nbc
        if (nel .gt. maxelem) goto 9005                     if (nbc .gt. nnp) goto 9020
        do 210 i= 1, nel                                    do 240 i=1,nbc
          read(idat,*)j,nelt(j),npi(j),npj(j)                  read(idat,*) node, xdof,ydof,zdof, xrot,yrot,zrot
          if (range( 1, j, nel)) goto 9025                     if (range( 1, node, nnp)) goto 9045
 210      continue                                             if (range( 0, xdof, 1)) goto 9050
        read(idat,'(1a50)') endin                              if (range( 0, ydof, 1)) goto 9055
        call chkend(endin,ichk)                                if (range( 0, zdof, 1)) goto 9060
        if (ichk.eq.0) then                                    if (range( 0, xrot, 1)) goto 9050
          write(ilog,*)'@@ Incorrect Connectivities data size' if (range( 0, yrot, 1)) goto 9055
          write(*  ,*)'@@ Incorrect Connectivities data size'  if (range( 0, zrot, 1)) goto 9060
          stop'!! ERROR'                                       ind = node*6
        endif                                                  jbc(ind-5) = xdof
        ichk=1                                                 jbc(ind-4) = ydof
c       check the data and count the number of beams and rods  jbc(ind-3) = zdof
        nbeam = 0                                              jbc(ind-2) = xrot
        ntruss = 0                                             jbc(ind-1) = yrot
        do 222 j= 1, nel                                      jbc(ind)   = zrot
          if (range(1,nelt(j),2)) goto 9035           240   continue
c              rod=1  beam=2                                  read(idat,'(1a50)') endin
          if (nelt(j).eq.1) ntruss=ntruss+1                  call chkend(endin,ichk)
          if (nelt(j).eq.2) nbeam=nbeam+1                    if (ichk.eq.0) then
 222    continue                                               write(ilog,*)'@@ Incorrect BCs data length'
c                                                              write(*  ,*)'@@ Incorrect BCs data length'
c READ elem material  properties                              stop'!! ERROR'
        write(ilog,*)'@@ reading material lines'             endif
        read(idat,*) nmat                                    ichk=1
        if (nmat .lt. 1) goto 9007                    c
        do 212 i= 1, nmat                             c READ  nodal loads: node# xyz forces, xyz moments, cmass
          read(idat,*)nm1,nm2,e0,g0,a0,zix0,ziy0,zi0,rho0,beta0  write(ilog,*)'@@ reading load lines'
          if (range(1,nm1,nel)) goto 9027                    read(idat,*) nload
          if (range(1,nm2,nel)) goto 9027                    if (nload .gt. nnp) goto 9022
          do 214 j=nm1,nm2                                   do 232 i= 1, nload
            write(imat,922,rec=j),e0,g0,a0,zix0,ziy0,zi0,      read(idat,*) j,xload(j),yload(j),zload(j),
     >                           rho0,beta0            >              xmom(j),ymom(j),zmom(j),cmass(j)
 922        format(1x,i4,1x,8(g10.4,1x))                      if (range(1,j,nnp)) goto 9047
 214      continue                                    232   continue
 212    continue                                            read(idat,'(1a50)') endin
        read(idat,'(1a50)') endin                           call chkend(endin,ichk)
        call chkend(endin,ichk)                             if (ichk.eq.0) then
        if (ichk.eq.0) then                                   write(ilog,*)'@@ Incorrect Loads data length'
          write(ilog,*)'@@ Incorrect Material data length'    write(*  ,*)'@@ Incorrect Loads data length'
          write(*  ,*)'@@ Incorrect Material data length'     stop'!! ERROR'
          stop'!! ERROR'                                   endif
        endif                                               ichk=1
        ichk=1                                        c
c                                                           write(ilog ,1006) nnp, nbc,nload
        write(ilog ,1005) nel, nbeam, ntruss,nmat     c
c                                                     c READ  special materials properties
c READ  nodal coordinate data: node# x    y   z            write(ilog,*)'@@ reading special material lines'
        write(ilog,*)'@@ reading coordinate lines'          read(idat,*) nspmat
        read(idat,*)  nnp                                   write(ilog ,1007)
        if (nnp .lt. 2) goto 9010                               dampkk=0.0
```

```
          dampmm=0.0                                    write(*,*)'data at',j
          dampcc=0.0                                    stop
            gx=0.0                            9035 write(*,*)'Element type not set to 1 or 2 !!!'
            gy=0.0                                    write(*,*)'data at',j
            gz=0.0                                    stop
       do 242 i= 1, nspmat                    9040 write(*,*)'Node number out of range !!!'
          read(idat,*) icode, c1,c2,c3                 write(*,*)   j,' > ',nnp
          if (icode .eq. 2110) then                    stop
             dampkk=c2                        9045 write(*,*)'Boundary node number out of range !!!'
             dampmm=c3                                 write(*,*)node,' > ',nnp
             dampcc=c1                                 stop
             write(ilog,1008)' Proportional Damping'  9047 write(*,*)'Load node number out of range !!!'
          endif                                        write(*,*)   j,' > ',nnp
242    continue                                        stop
       read(idat,'(1a50)') endin                9050 write(*,*)'X degree of freedom not set to 0 or 1 !!!'
       call chkend(endin,ichk)                        write(*,*)' dof =',xdof
       if (ichk.eq.0) then                             stop
          write(ilog,*)'@@ Incorrect SPecials data length'  9055 write(*,*)'Y degree of freedom not set to 0 or 1 !!!'
          write(*   ,*)'@@ Incorrect SPecials data length'       write(*,*)' dof =',ydof
          stop'!! ERROR'                               stop
       endif                                    9060 write(*,*)'Theta degree of freedom not set 0 or 1 !!!'
       ichk=1                                          write(*,*)' dof =',zdof
c                                                      stop
                                               c
       goto 999                                999  continue
c                                                    write(*,*) '@@ DATAin:  Input data read  OK'
c              FORMATS                               return
c                                                    end
1004 format (' ',/,'@@ HEADER GROUP          ',/  c
     >      ,'@@',3x,'Title        =',3x,a50/        SUBROUTINE CHKEND(endin,ichk)
     >      ,'@@',3x,'Problem type =',i4/            character*50 endin
     >      ,'@@',3x,'element flag =',i4/            character*3 end1,end2
     >      ,'@@',3x,'element flag =',i4/            end1='end'
     >      ,'@@',3x,'node    flag =',i4/            end2='END'
     >      ,'@@',3x,'node    flag =',i4/)           i1=index(endin,end1)
1005 format(' ',/,'@@ ELEMENT GROUP          ',/     i2=index(endin,end2)
     >      ,'@@', 3x,'number of elements    =',i4/  if (i1.eq.0 .and. i2.eq.0) then
     >      ,'@@', 3x,'number of beam elements   =',i4/     ichk = 0
     >      ,'@@', 3x,'number of truss elements  =',i4/  else
     >      ,'@@', 3x,'number of materials   =',i4/ )    ichk = 1
1006 format(' ',/,'@@ NODE GROUP             ',/     endif
     >      ,'@@', 3x,'number of nodal points    =',i4/  return
     >      ,'@@', 3x,'number of boundary nodes  =',i4/  end
     >      ,'@@', 3x,'number of loaded nodes    =',i4/ )  c
1007 format(' ',/,'@@ SPECIALS GROUP       ')         SUBROUTINE GRPCHK(idat,ilog)
1008 format('@@', 3x,a )                        c    Check sizes (length) of each data group
c                                                    character*50 endin
c              ERROR MESSAGES                        character*1 ch
c                                               c
9000 write(*,*)'Number of elements are less than 1 !!!'  rewind idat
       write(*,*)'nel =',nel                    c READ header  data
       stop                                          write(ilog,*)'@@ reading HEADER group'
9005 write(*,*)'Number of elements greater than maximum!!!'  read(idat,'(1a1)') ch
       write(*,*)'nel =',nel                         read(idat,'(1a1)') ch
       stop                                          read(idat,'(1a1)') ch
9007 write(*,*)'Number of element matls is less than 1 !!!'  read(idat,'(1a50)') endin
       write(*,*)'nmat =',nmat                        call chkend(endin,ichk)
       stop                                          if (ichk.eq.0) then
9010 write(*,*)'Number of nodes less than 2 !!!'         write(ilog,*)'@@ Incorrect HEADER group length'
       write(*,*)'nnp =',nnp                          write(*   ,*)'@@ Incorrect HEADER group length'
       stop                                          stop'!! ERROR'
9015 write(*,*)'Number of nodes greater than maximum !!!'  endif
       write(*,*)'nnp =',nnp                         ichk=1
       stop                                     c
9020 write(*,*)'Number boundary nodes > number nodes !!!'  c READ elem GROUP size
       write(*,*) nbc,' > ',nnp                       write(ilog,*)'@@ reading connectivities GROUP'
       stop                                          read(idat,*) nel
9022 write(*,*)'Number load nodes > number nodes !!!'   if (nel .lt. 1) goto 9000
       write(*,*) nload,' > ',nnp                     do 210 i= 1, nel
       stop                                             read(idat,'(1a1)') ch
9025 write(*,*)'Element number out of range !!!'   210   continue
       write(*,*)'data at',i                         read(idat,'(1a50)') endin
       stop                                          call chkend(endin,ichk)
9027 write(*,*)'Material element number out of range !!!'  if (ichk.eq.0) then
       write(*,*)'data at',i                            write(ilog,*)'@@ Incorrect Connect. GROUP size'
       stop                                             write(*   ,*)'@@ Incorrect Connect. GROUP size'
9030 write(*,*)'Node number on element out of range !!!'
```

```
          stop'!! ERROR'
        endif
        ichk=1
c
c READ elem material GROUP size
        write(ilog,*)'@@ reading material GROUP'
        read(idat,*) nmat
        if (nmat .lt. 1) goto 9007
        do 212 i= 1, nmat
          read(idat,'(1a1)') ch
 212    continue
        read(idat,'(1a50)') endin
        call chkend(endin,ichk)
        if (ichk.eq.0) then
          write(ilog,*)'@@ Incorrect Material GROUP size'
          write(*   ,*)'@@ Incorrect Material GROUP size'
          stop'!! ERROR'
        endif
        ichk=1
c
c READ nodal coordinate  GROUP size
        write(ilog,*)'@@ reading coordinate GROUP'
        read(idat,*) nnp
        if (nnp .lt. 2) goto 9010
        do 230 i= 1, nnp
          read(idat,'(1a1)') ch
 230    continue
        read(idat,'(1a50)') endin
        call chkend(endin,ichk)
        if (ichk.eq.0) then
          write(ilog,*)'@@ Incorrect Coordinate GROUP size'
          write(*   ,*)'@@ Incorrect Coordinate GROUP size'
          stop'!! ERROR'
        endif
        ichk=1
c
c READ bcs GROUP size
        write(ilog,*)'@@ reading bc GROUP'
        read(idat,*) nbc
        do 240 i=1,nbc
          read(idat,'(1a1)') ch
 240    continue
        read(idat,'(1a50)') endin
        call chkend(endin,ichk)
        if (ichk.eq.0) then
          write(ilog,*)'@@ Incorrect BCs GROUP size'
          write(*   ,*)'@@ Incorrect BCs GROUP size'
          stop'!! ERROR'
        endif
        ichk=1
c
c READ nodal load GROUP size
        write(ilog,*)'@@ reading load GROUP'
        read(idat,*) nload
        do 232 i= 1, nload
          read(idat,'(1a1)') ch
 232    continue
        read(idat,'(1a50)') endin
        call chkend(endin,ichk)
        if (ichk.eq.0) then
          write(ilog,*)'@@ Incorrect Loads GROUP size'
          write(*   ,*)'@@ Incorrect Loads GROUP size'
          stop'!! ERROR'
        endif
        ichk=1
c
c
c READ SPecials GROUP size
        write(ilog,*)'@@ reading SPecial materials GROUP'
        read(idat,*) nspmat
        do 242 i= 1, nspmat
          read(idat,'(1a1)') ch
 242    continue
        read(idat,'(1a50)') endin
        call chkend(endin,ichk)
        if (ichk.eq.0) then
          write(ilog,*)'@@ Incorrect SPecials GROUP size'
          write(*   ,*)'@@ Incorrect SPecials GROUP size'
          stop'!! ERROR'
        endif
        ichk=1
        write(ilog,*)'@@ finished checking group sizes'
c
        goto 999
c
c        ERROR MESSAGES
c
 9000 write(*,*)'Number of elements are less than 1 !!!'
      write(*,*)'nel =',nel
      stop
 9007 write(*,*)'Number of element matls are less than 1 !!!'
      write(*,*)'nmat =',nmat
      stop
 9010 write(*,*)'Number of nodes less than 2 !!!'
      write(*,*)'nnp =',nnp
      stop
c
 999  continue
      write(*,*)'@@ GROUP sizes  OK'
      return
      end
cccc
c
      SUBROUTINE DATACNV(xload,yload,zload,
     >                   xmom,ymom,zmom,cmass,loads,
     >                   jbc, maxdof, maxelem, maxnode,
     >                   npi, npj, nelt,iprob)
c   Convert the structural data input
      include 'scommons'
c
      integer jbc(maxdof)
      integer npi(maxelem), npj(maxelem), nelt(maxelem)
      real*8 xload(maxnode),yload(maxnode),zload(maxnode)
      real*8 xmom(maxnode), ymom(maxnode),zmom(maxnode)
      real*8 loads(maxnode*6), cmass(maxnode)
      character*50 title
      character*5 eltype
c
c READ header  data again to get flags
      rewind (idat)
      read(idat,'(1a50)') title
      write(*   ,*) title
      read(idat,*) iprob
      read(idat,*) iflag1,iflag2,iflag3,iflag4
c
c CHECKS
      if (iflag1.gt.0) then
c       echo connectivities
        write(ilog ,10021)
        do 260 i = 1, nel
          if (nelt(i) .eq. 2) then
            eltype = ' beam'
          else
            eltype = 'truss'
          endif
          write(ilog,10031) i, npi(i), npj(i), eltype
 260    continue
      endif
c
c MINIMUM allowable values for area, modulus and inertia are set
      call chkmat(maxelem,nelt   )
c
c       echo the input if needed
      if (iflag2.gt.0) then
c       echo materials
        write(ilog ,1002)
        rewind (imat)
        do 261 i = 1, nel
          read(imat,922,rec=i)m,e0,g0,a0,zix0,ziy0,zi0,
     >                        rho0,beta0
 922      format(1x,i4,1x,8(g10.4,1x))
          write(ilog ,1003) i,e0,g0,a0,zix0,ziy0,zi0,
```

```
      >                       rho0,beta0
261           continue
            endif
c
c   Impose conditions for special GLOBAL shapes
            call spclbc(maxdof,maxelem,jbc,npi,npj,nelt,iprob)
c
c           Number the equations in JBC from 1 up to the order.
c           Start assigning equation numbers for non-zero dof's
c           from 1 up; only non-zero given a number
            neq   = 0
            do 2430 i= 1, nnp*6
              if (jbc(i) .gt. 0) then
                  neq   = neq   + 1
                  jbc(i) = neq
              else
                  jbc(i) = 0
              endif
2430        continue
c
            if (iflag3.gt.0) then
c             echo equation numbering
              write(ilog ,1008)
              do 280 i=1,nnp
                ind=(i-1)*6
                write(ilog ,1007)i,(jbc(j+ind),j=1,6)
280           continue
            endif
c
c           JBC() contains the eqn # sequence 1 2 3 4 .....
c           first the masses
            do 250 i= 1, nnp
              ind = (i-1)*6
              if (jbc(ind+1).ne.0) loads(jbc(ind+1))=cmass(i)
              if (jbc(ind+2).ne.0) loads(jbc(ind+2))=cmass(i)
              if (jbc(ind+3).ne.0) loads(jbc(ind+3))=cmass(i)
              if (jbc(ind+4).ne.0) loads(jbc(ind+4))=0
              if (jbc(ind+5).ne.0) loads(jbc(ind+5))=0
              if (jbc(ind+6).ne.0) loads(jbc(ind+6))=0
250         continue
            rewind(icms)
            do 38 i=1,neq
              write(icms,*) loads(i)
38          continue
c
c           then the forces
            do 252 i=1,neq
252           loads(i)=0.0
            do 254 i= 1, nnp
              ind = (i-1)*6
              if (jbc(ind+1).ne.0) loads(jbc(ind+1))=xload(i)
              if (jbc(ind+2).ne.0) loads(jbc(ind+2))=yload(i)
              if (jbc(ind+3).ne.0) loads(jbc(ind+3))=zload(i)
              if (jbc(ind+4).ne.0) loads(jbc(ind+4))=xmom(i)
              if (jbc(ind+5).ne.0) loads(jbc(ind+5))=ymom(i)
              if (jbc(ind+6).ne.0) loads(jbc(ind+6))=zmom(i)
254         continue
            rewind(ilod)
            write(ilod,*) (loads(i), i= 1, neq  )
c
            if (iflag4.gt.0) then
c             echo loads and conc masses
              write(ilog ,1018)
              do 284 i=1,nnp
                ind=(i-1)*6
                write(ilog,1017)i,(loads(jbc(j+ind)),j=1,6),
      >                           cmass(i)
284           continue
            endif
c
            call maxbnd(npi,npj,jbc,iband ,maxdof,maxelem,nel)
            write(*   ,*)'@@ from DATAck:  half band =',iband
            write(ilog,1022) neq  ,iband ,ixxxx
c
c           STORE results for future reference
            open(unit=33,file='stadyn.jbc')
```

```
            rewind 33
            write(33,33) neq  ,nnp,nel,iband
            write(33,33) (jbc(n), n=1,nnp*6)
            write(33,33) (npi(n), n=1,nel)
            write(33,33) (npj(n), n=1,nel)
            close (33)
33          format(1x,12(i5,1x))
c
            goto 999
c
c
c               FORMATS
10021   format(' ',/,'@@ Connectivities',/,
      >         '@@ el.   i    j   type ' )
10031   format('@@',3i4,1x,a5)
1002    format(' ',/,'@@ Materials',/,
      >         '@@ el.      e           g       ',
      >         '  a      ix      iy      iz    density beta')
1003    format('@@',i4,2(1x,g9.3),4(1x,g9.3),2(1x,g9.3))
1004    format (' ',/,'@@ HEADER GROUP                ',/
      >          ,'@@',3x,'Title        =',3x,a50/
      >          ,'@@',3x,'Problem type =',i4/
      >          ,'@@',3x,'element flag =',i4/
      >          ,'@@',3x,'element flag =',i4/
      >          ,'@@',3x,'node    flag =',i4/
      >          ,'@@',3x,'node    flag =',i4)
1005    format(' ',/,'@@ ELEMENT GROUP              ',/
      >          ,'@@', 3x,'number of elements      ='i4/
      >          ,'@@', 3x,'number of beam elements  ='i4/
      >          ,'@@', 3x,'number of truss elements ='i4/
      >          ,'@@', 3x,'number of materials     ='i4 )
1006    format(' ',/,'@@ NODE GROUP                 ',/
      >          ,'@@', 3x,'number of nodal points   ='i4/
      >          ,'@@', 3x,'number of boundary nodes ='i4/
      >          ,'@@', 3x,'number of loaded nodes   ='i4 )
1007    format('@@',i5,6i8)
1008    format( /,'@@  numbering of equations'/,
      >         '@@  node     x       y       z       x',
      >         '      y       z' )
1017    format('@@',i4,6f10.2,1f10.4)
1018    format( /,'@@  applied nodal loads and conc masses'/,
      >         '@@  node     Fx      Fy      Fz      Mx',
      >         '      My      Mz      mass')
1022    format(' ',/,'@@ SYSTEM GROUP               ',/
      >          ,'@@', 3x,'system size (order)     ='i4/
      >          ,'@@', 3x,'half bandwidth          ='i4/
      >          ,'@@', 3x,'xxxxxxxxxxxxxxx         ='i4/)
c
c
999     continue
        return
        end
c
        SUBROUTINE MAXBND(npi,npj,jbc,iband,maxdof,maxelem,nel)
c       calculate maxbandwidth
        integer jbc(maxdof),npi(maxelem), npj(maxelem)
        integer ipv(12)
        ibndm1=0
        do 2108 n=1,nel
          idof = (npi(n)-1)*6
          jdof = (npj(n)-1)*6
          do 2102 i=1, 6
            ipv(i) = idof + i
            ipv(i+6) = jdof + i
2102      continue
c
          do 2104 i= 1, 12
            ieqn1 = jbc(ipv(i))
            if (ieqn1 .gt. 0) then
              do 2106 j= i,12
                ieqn2 = jbc(ipv(j))
                if (ieqn2 .gt. 0) then
                  ibndm1=max0(ibndm1,iabs(ieqn1-ieqn2))
                endif
2106          continue
            endif
```

```
2104        continue                                    c        3-D truss
2108      continue                                              do 2531 i=1,nnp
        iband=ibndm1+1                                            jbc(i*6-0) = 0
      return                                                     jbc(i*6-1) = 0
      end                                                        jbc(i*6-2) = 0
c                                                      2531      continue
      SUBROUTINE SPCLBC(maxdof,maxelem,jbc,npi,npj,nelt,iprob)   endif
c     Impose conditions for special GLOBAL shapes         return
      include 'scommons'                                  end
c                                                    c
      integer jbc(maxdof)                                 SUBROUTINE CHKMAT(maxelem,nelt)
      integer npi(maxelem), npj(maxelem), nelt(maxelem)   include 'scommons'
c                                                    c
c     Make sure all nodes which are only parts of trusses  integer nelt(maxelem)
c     have 0 dof for phi rotation.                   c
      do 2410 i= 1, nnp                              c     set the MINIMUM allowable values for area, etc
        do 2420 j= 1, nel                                  atemp = 1.0e-10
          if (nelt(j) .eq. 2) then                         etemp = 1.0e-10
            if (npi(j).eq.i .or. npj(j).eq.i) goto 2410    ztemp = 1.0e-10
          endif                                            ytemp = 1.0e-10
2420      continue                                         rewind (imat)
        jbc(i*6-0) = 0                                     do 220 j= 1, nel
        jbc(i*6-1) = 0                                       read(imat,922,rec=j)mat,e0,g0,a0,zix0,ziy0,ziz0,
        jbc(i*6-2) = 0                                  >                             rho0,beta0
2410    continue                                            if (a0.le.0.0) a0=atemp
c                                                           if (e0.le.0.0) e0=etemp
c                                                           if (g0.le.0.0) g0=etemp
c     if (iprob.eq.11) then
c        1-D rods                                    c     if it is a rod element force the bending & torsion
        do 2511 i=1,nnp                              c     stiffnesses to zero
          jbc(i*6-0) = 0                                   if (nelt(j) .eq. 1)then
          jbc(i*6-1) = 0                                     zix0=1.0e-10
          jbc(i*6-2) = 0                                     ziy0=1.0e-10
          jbc(i*6-3) = 0                                     ziz0=1.0e-10
          jbc(i*6-4) = 0                                   endif
2511      continue                                         if (ziz0 .le. 0.0) ziz0 =ztemp
      elseif (iprob.eq.12) then                            if (zix0 .le. 0.0) zix0 =ztemp
c        1-D beams                                         if (ziy0 .le. 0.0) ziy0 =ztemp
        do 2512 i=1,nnp                              c
          jbc(i*6-1) = 0                                   write(imat,922,rec=j)j,e0,g0,a0,zix0,ziy0,ziz0,
          jbc(i*6-2) = 0                                >                           rho0,beta0
          jbc(i*6-3) = 0                             922          format(1x,i4,1x,8(g10.4,1x))
          jbc(i*6-5) = 0                             220     continue
2512      continue                                   c
      elseif (iprob.eq.13) then                         return
c        1-D shafts                                     end
        do 2513 i=1,nnp                              cccc
          jbc(i*6-0) = 0
          jbc(i*6-1) = 0
          jbc(i*6-3) = 0
          jbc(i*6-4) = 0
          jbc(i*6-5) = 0
2513      continue
      elseif (iprob.eq.21) then
c        2-D truss
        do 2521 i=1,nnp
          jbc(i*6-0) = 0
          jbc(i*6-1) = 0
          jbc(i*6-2) = 0
          jbc(i*6-3) = 0
2521      continue
      elseif (iprob.eq.22) then
c        plane frames
        do 2522 i=1,nnp
          jbc(i*6-1) = 0
          jbc(i*6-2) = 0
          jbc(i*6-3) = 0
2522      continue
      elseif (iprob.eq.23) then
c        plane grillage
        do 2523 i=1,nnp
          jbc(i*6-0) = 0
          jbc(i*6-4) = 0
          jbc(i*6-5) = 0
2523      continue
      elseif (iprob.eq.31) then
```

## ASSEMble

```
cccc

c
      SUBROUTINE ASSEMB( aa, a, jbc, i1, j1)
c     Assemble the element matrices in upper symm band form;
c     called separately from FORMSTIF, FORMMASS, FORMGEOM
      include 'scommons'
c
      integer jbc(nnp*6), ipv(12)
      real*8  a(12,12), aa(neq,iband )
c
c     Set ipv to the positions in the array for the nodes.
      idof = (i1-1)*6
      jdof = (j1-1)*6
      do 10 i= 1, 6
        ipv(i)   = idof + i
        ipv(i+6) = jdof + i
10    continue
c
c     Store the values for invidual array in global array
      do 20 i= 1, 12
        ieqn1 = jbc(ipv(i))
        if (ieqn1 .gt. 0) then
          do 30 j= i, 12
            ieqn2 = jbc(ipv(j))
```

```
          if (ieqn2 .gt. 0) then
            if (ieqn1 .gt. ieqn2) then
               jband=( ieqn1- ieqn2)+1
               aa(ieqn2,jband)=aa(ieqn2,jband)+a(i,j)    c
            else
               jband=(ieqn2- ieqn1)+1
               aa(ieqn1,jband)=aa(ieqn1,jband)+a(i,j)    c
            endif
          endif
30        continue                                       c
        endif                                            c
20    continue
    return
    end
c
c
      SUBROUTINE FORMSTFF( stf,npi, npj, xord, yord,zord,
     >                     jbc,nelt,ek)
c  Form the global stiffness matrix [K]
      include 'scommons'
c
      integer npi(nel), npj(nel), jbc(nnp*6), nelt(nel)
      real xord(nnp), yord(nnp),zord(nnp), l,m,n
      real*8 stf(neq,iband), ek(12,12)
c
      write(*,*)'CHOOSE: 0=no echo,  1=echo   ',
     >                   'to <<stadyn.out>>'                 c
      write(*,'(a\)')'  --> '
      read(*,*) iecho
      write(ilog,*) iecho,'  ::echo'
c
c       initialize [K] to zero
      do 30 i=1, neq
        do 32 j=1,iband
          stf(i,j) = 0.0e0
32      continue
30    continue
c
c       form each element matrix, and assemble
      do 50 i=1,nel
        i1=npi(i)
        j1=npj(i)
        dx= xord(j1) - xord(i1)
        dy= yord(j1) - yord(i1)
        dz= zord(j1) - zord(i1)
        xl=sqrt(dx*dx+dy*dy+dz*dz)
        l=dx/xl
        m=dy/xl
        n=dz/xl
c
c         Calculate the stiffness matrix and assemble     c
        read(imat,922,rec=i)mat,e0,g0,a0,zix0,ziy0,ziz0,   c
     >                      rho0,beta0                      c
922     format(1x,1i4,1x,8(g10.4,1x))
        call elmstf(xl,zix0,ziy0,ziz0,a0,e0,g0,ek      )
c
c         use trans3d to transform from local to global
        call trans3d(l,m,n,ek,beta0)
        call assemb( stf, ek, jbc, i1, j1)
c
50    continue
c
c       STORE stiffness  matrix on disk in case          c
      if (iecho .eq. 1) then
        write(iout,*)'STIFFNESS: upper banded'
        do 11 i=1,neq
          write(iout,22) (stf(i,j), j= 1, iband)
22        format(1x,6(g13.6))
11      continue
      endif
      rewind(istf)
      write(istf) neq ,iband
      do 12 i=1,neq
12      write(istf)  (stf(i,j), j= 1, iband)
c
      write(*    ,*) '@@ FORMSTFF:  Formed  [K]   OK'
```

```
      write(ilog ,*) '@@ FORMSTFF:  Formed  [K]   OK'
    return
    end
c
      SUBROUTINE ELMSTF(length, ix,iy,iz,area, emod,gmod,ek)
c  calculate the element stiffness matrix.
c
      real  area, length, ix, iy, iz, emod, gmod
      real*8 ek(12,12)
c
c       initialize all ek elements to zero
      do 90 i=1,12
        do 90 j=1,12
90        ek(i,j)=0.0
c
c       STIFFNESS matrix in local coordinates
      emlen  = emod/length
      emlen2 = emlen/length
      emlen3 = emlen2/length
c
      ek(1,1)   =    area*emlen
      ek(2,2)   =    12.0*emlen3*iz
      ek(3,3)   =    12.0*emlen3*iy
      ek(4,4)   =    gmod*ix/length
      ek(5,5)   =    4.0*emlen*iy
      ek(6,6)   =    4.0*emlen*iz
c
      ek(2,6)   =    6.0*emlen2*iz
      ek(3,5)   =   -6.0*emlen2*iy
c
      ek(7,7)   =    ek(1,1)
      ek(8,8)   =    ek(2,2)
      ek(9,9)   =    ek(3,3)
      ek(10,10) =    ek(4,4)
      ek(11,11) =    ek(5,5)
      ek(12,12) =    ek(6,6)
c
      ek(1,7)   =   -ek(1,1)
      ek(2,8)   =   -ek(2,2)
      ek(2,12)  =    ek(2,6)
      ek(3,9)   =   -ek(3,3)
      ek(3,11)  =    ek(3,5)
      ek(4,10)  =   -ek(4,4)
      ek(5,9)   =   -ek(3,5)
      ek(5,11)  =    ek(5,5)/2.0
      ek(6,8)   =   -ek(2,6)
      ek(6,12)  =    ek(6,6)/2.0
c
      ek(8,12)  =   -ek(2,6)
      ek(9,11)  =   -ek(3,5)
c
c       impose the symmetry
      do 10 i= 1, 12
        do 10 j= i, 12
10        ek(j,i) = ek(i,j)
c
    return
    end
c
c
      SUBROUTINE TRANS3D (l,m,n,ek,beta)
c  Makes 3-D coordinate transformations.
      real*8  ek(12,12),rt(3,3),r(3,3),ktemp(12,12)
      real m,n,l,beta
      pi=4.0*atan(1.0)
c
      sb=sin(beta*pi/180)
      cb=cos(beta*pi/180)
      d=sqrt(1-n**2)
      if (abs(n).gt.0.995) then
        r(1,1)  =  0.0
        r(1,2)  =  0.0
        r(1,3)  =  n
        r(2,1)  = -n*sb
        r(2,2)  =  cb
```

```
      r(2,3)  =  0.0
      r(3,1)  = -n*cb
      r(3,2)  = -sb
      r(3,3)  =  0.0
      else
      r(1,1)  =  1
      r(1,2)  =  m
      r(1,3)  =  n
      if (abs(beta) .le. .01) then
        r(2,1)  = -m/d
        r(2,2)  =  1/d
        r(2,3)  =  0.0
        r(3,1)  = -l*n/d
        r(3,2)  = -m*n/d
        r(3,3)  =  d
      else
        r(2,1)  = -(m*cb+l*n*sb)/d
        r(2,2)  =  (l*cb-m*n*sb)/d
        r(2,3)  =  d*sb
        r(3,1)  =  (m*sb-l*n*cb)/d
        r(3,2)  = -(l*sb+m*n*cb)/d
        r(3,3)  =  d*cb
      endif
      endif
c
      do 7 in=1,3
       do 7 jn=1,3
7        rt(jn,in)=r(in,jn)
c
c take [Rtrans][K][R] using the nature of [R] for speed.
c k is sectioned off into 3x3s then multiplied [rtrans][k][r]
c
      do 22 i=0,3
       do 22 j=0,3
        do 23 k=1,3
         do 23 ii=1,3
          j1=i*3
          j2=j*3
          ktemp(j1+k,j2+ii)=0.0
          do 23 jj=1,3
           ktemp(j1+k,j2+ii) = ktemp(j1+k,j2+ii)
     >                       + ek(j1+k,j2+jj)*r(jj,ii)
23       continue
        do 24 k=1,3
         do 24 ii=1,3
          ek(j1+k,j2+ii)=0.0
          do 24 jj=1,3
           ek(j1+k,j2+ii) = ek(j1+k,j2+ii)
     >                    + rt(k,jj)*ktemp(j1+jj,j2+ii)
24       continue
22     continue
      return
      end
c
c
      SUBROUTINE FORMMASS(mss, npi, npj, xord, yord, zord,
     >                    jbc, nelt, em)
c     Form the global mass matrix [M]
      include 'scommons'
c
      integer npi(nel), npj(nel), jbc(nnp*6), nelt(nel)
      real xord(nnp), yord(nnp), zord(nnp), m,n,l
      real*8 mss(neq,iband), em(12,12)
c
      write(*,*)'CHOOSE mass: 1=lumped 2=consist. | ',
     >          ' 1=echo to <<stadyn.out>>'
      write(*,'(a\)')' --> '
      read(*,*) ilump,iecho
      write(ilog,*) ilump,iecho,' ::1=lumped 1=echo'
c
c     initialize [M] to zero
      do 30 i=1, neq
       do 32 j=1,iband
        mss(i,j) = 0.0
32     continue
30    continue

c
c     form each element matrix, and assemble
      do 50 i=1,nel
       i1=npi(i)
       j1=npj(i)
       dx= xord(j1) - xord(i1)
       dy= yord(j1) - yord(i1)
       dz= zord(j1) - zord(i1)
       xl=sqrt(dx*dx+dy*dy+dz*dz)
       l=dx/xl
       m=dy/xl
       n=dz/xl
c
c      Calculate the mass matrix and assemble
       read(imat,922,rec=i)mat,e0,g0,a0,zix0,ziy0,ziz0,
     >                     rho0,beta0
922    format(1x,1i4,1x,8(g10.4,1x))
       call elmmas( rho0, a0, xl, zix0, em,ilump)
c      use trans3d to transform from local to global
       call trans3d(l,m,n,em,beta0)
       call assemb( mss, em, jbc, i1, j1)
50    continue
c
c     modify mass matrix for concentrated masses
      rewind(icms)
      do 60 n=1,neq
       read(icms,*) concms
       mss(n,1)=mss(n,1)+concms
60    continue
c
c     Store mass matrices on disk
      if (iecho .eq. 1) then
       if (ilump.eq.1) then
        write(iout,*)'MASS: diagonal'
        write(iout,22) (mss(i,1), i= 1, neq )
       else
        write(iout,*)'MASS: upper banded'
        do 11 i=1,neq
         write(iout,22) (mss(i,j), j=1,iband)
11      continue
       endif
22     format(1x,6(g13.6))
      endif
c
      if (ilump .eq. 1) then
       ibandm=1
      else
       ibandm=iband
      endif
      rewind(imss)
      write(imss) neq ,ibandm
      if (ilump.eq.1) then
       write(imss)    (mss(i,1), i= 1, neq )
      else
       do 12 i=1,neq
        write(imss)    (mss(i,j), j=1,ibandm)
12     continue
      endif
c
      write(*  ,*) '@@ FORMMASS:  Formed  [M]  OK'
      write(ilog,*) '@@ FORMMASS:  Formed  [M]  OK'
      return
      end
c
      SUBROUTINE ELMMAS( rho, area, length, zix, em,ilump)
c     Calculate the element mass matrix
c
      real   rho, area, length,zix
      real*8 em(12,12)
c
      do 10 i = 1,12
       do 12 j = 1,12
        em(i,j) = 0.0
12     continue
10    continue
c
```

```
      if (ilump.ne.1) goto 30
c
c     contibutions to lumped mass matrix
      roal = rho*area*length/2.0
      alfa = 1.0e-6
c
      em(1,1) = roal
      em(2,2) = roal
      em(3,3) = roal
      em(4,4) = roal*zix/area
      em(5,5) = roal*length*length*alfa
      em(6,6) = roal*length*length*alfa
      em(7,7)   =   em(1,1)
      em(8,8)   =   em(2,2)
      em(9,9)   =   em(3,3)
      em(10,10) =   em(4,4)
      em(11,11) =   em(5,5)
      em(12,12) =   em(6,6)
      return
c
 30   continue
c     contributions to consistent  mass matrix
      roala = rho*area*length/6.0
      roalb = rho*area*length/420.0
      roalc = rho*area*length/12.0
c
      em(1,1) =  roala*2.0
      em(1,7) =  roala
c
      em(2,2) =  roalb*156.0
      em(2,6) =                + roalb*22.0*length
      em(2,8) =   roalb*54.0
      em(2,12) =              - roalb*13.0*length
c
      em(3,3) =  roalb*156.0
      em(3,5) =               - roalb*22.0*length
      em(3,9) =   roalb*54.0
      em(3,11) =              + roalb*13.0*length
c
      em(4,4) =                roala*2.0*zix/area
      em(4,10) =              roala*zix/area
c
      em(5,5) =                roalb*4.0*length*length
      em(5,9) =              - roalb*13.0*length
      em(5,11) =             - roalb*3.0*length*length
c
      em(6,6) =                roalb*4.0*length*length
      em(6,8) =              roalb*13.0*length
      em(6,12) =            - roalb*3.0*length*length
c
      em(7,7)   =   em(1,1)
      em(8,8)   =   em(2,2)
      em(9,9)   =   em(3,3)
      em(10,10) =   em(4,4)
      em(11,11) =   em(5,5)
      em(12,12) =   em(6,6)
      em(8,12)  =  -em(2,6)
      em(9,11)  =  -em(3,5)
c
c     impose  symmetry
      do 20 i= 2, 12
         do 22 j= 1, i-1
            em(i,j) = em(j,i)
 22      continue
 20   continue
c
      return
      end
cccc

      SUBROUTINE FORMGEOM( geo,npi, npj, xord, yord,zord,
     >                     jbc, nelt, eg, disp, dispp)
c     Form the global geometric matrix [G]
         include 'mcommons'
c
      integer npi(nel), npj(nel), jbc(nnp*6), nelt(nel)
```

```
      real    xord(nnp), yord(nnp),zord(nnp), l,m,n
      real*8 geo(neq,iband),eg(12,12),disp(neq),dispp(nnp*6)
c
      write(*,*)'CHOOSE: 0=no echo,  1=echo   ',
     >                   'to <<stadyn.out>>'
      write(*,'(a\)')'   -->  '
      read(*,*) iecho
      write(ilog,*) iecho,'   ::echo'
c
c     initialize [Kg]  to zero
      do 30 i=1, neq
         do 32 j=1,iband
            geo(i,j) = 0.0e0
 32      continue
 30   continue
c
c     Assign displacements to each node
      do 70 idof = 1, nnp*6
         ieqnum = jbc(idof)
         if(ieqnum .gt. 0) then
            dispp(idof) = disp(ieqnum)
         else
            dispp(idof) = 0.0e0
         endif
 70   continue
c
c     form each element matrix, and assemble
      do 50 i=1,nel
         i1=npi(i)
         j1=npj(i)
         dx= xord(j1) - xord(i1)
         dy= yord(j1) - yord(i1)
         dz= zord(j1) - zord(i1)
         xl=sqrt(dx*dx+dy*dy+dz*dz)
         l=dx/xl
         m=dy/xl
         n=dz/xl
c
c        Calculate the stiffness matrix and assemble
         read(imat,922,rec=i)mat,e0,g0,a0,zix0,ziy0,ziz0,
     >                        rho0,beta0
 922     format(1x,1i4,1x,8(g10.4,1x))
c
c        Find axial force (s) on each element
         iu=npi(i)*6-5
         iv=npi(i)*6-4
         iw=npi(i)*6-3
         ju=npj(i)*6-5
         jv=npj(i)*6-4
         jw=npj(i)*6-3
         s=-(e0*a0/xl)*( l*(dispp(ju)-dispp(iu))+
     >                   m*(dispp(jv)-dispp(iv))+
     >                   n*(dispp(jw)-dispp(iw)))
c
         if (nelt(i) .eq. 1) s=s*30.0/36.0
         call elmgeom( xl,eg ,s  )
c
c        use trans3d to transform from local to global
         call trans3d(l,m,n,eg,beta0)
         call assemb( geo, eg, jbc, i1, j1)
 50   continue
c
c     CHECK for zero diag; replace with fraction of norm
      geonorm=0.0
      do i=1,neq
         geonorm=geonorm + abs(geo(i,1))
      enddo
      geonorm=geonorm/neq
      do i=1,neq
         if (abs(geo(i,1)) .eq. 0.0) then
            geo(i,1) = geonorm*1.0e-6
         endif
      enddo
c
c     STORE stiffness matrix on disk in case
```

```
      if (iecho .eq. 1) then
        write(iout,*)'GEOMETRIC: upper banded'
        do 11 i=1,neq
          write(iout,22) (geo(i,j), j= 1, iband)
22        format(1x,6(g13.6))
11      continue
      endif
c
      ibandg=iband
      rewind(igeo)
      write(igeo) neq  ,ibandg
      do 12 i=1,neq
12      write(igeo)  (geo(i,j), j= 1, ibandg)
c
      write(*   ,*) '@@ FORMGEOM:  Formed [G]  OK'
      write(ilog ,*) '@@ FORMGEOM:  Formed [G]  OK'
      return
      end
c
      SUBROUTINE ELMGEOM( length, eg,s   )
c     calculates the element geometric matrix
      real     length
      real*8 eg(12,12)
c
c     initialize all eg elements to zero
      do 90 i=1,12
        do 90 j=1,12
90        eg(i,j)=0.0
c
c
      emlenz = s/(30.0*length)
      alpha=(s/length)*1.0e-6
      if (abs(alpha) .lt. 1.0e-10) alpha=1.0e-10
c
      eg(1,1)   =   alpha
      eg(2,2)   =   36*emlenz
      eg(3,3)   =   36*emlenz
      eg(4,4)   =   alpha
      eg(5,5)   =   4.0*emlenz*length*length
      eg(6,6)   =   4.0*emlenz*length*length
c
      eg(2,6)   =   3.0*emlenz*length
      eg(3,5)   =  -3.0*emlenz*length
c
      eg(7,7)   =   eg(1,1)
      eg(8,8)   =   eg(2,2)
      eg(9,9)   =   eg(3,3)
      eg(10,10) =   eg(4,4)
      eg(11,11) =   eg(5,5)
      eg(12,12) =   eg(6,6)
c
      eg(1,7)   =  -eg(1,1)
      eg(2,8)   =  -eg(2,2)
      eg(2,12)  =   eg(2,6)
      eg(3,9)   =  -eg(3,3)
      eg(3,11)  =   eg(3,5)
      eg(4,10)  =  -eg(4,4)
      eg(5,9)   =  -eg(3,5)
      eg(5,11)  =  -eg(5,5)/4.0
      eg(6,8)   =  -eg(2,6)
      eg(6,12)  =  -eg(6,6)/4.0
c
      eg(8,12)  =  -eg(2,6)
      eg(9,11)  =  -eg(3,5)
c
c     impose the symmetry
      do 10 i= 1, 12
        do 10 j= i, 12
10        eg(j,i) = eg(i,j)
c
      return
      end
cccc
```

## SOLVEr

```
      SUBROUTINE STATIC( stf,  load,  wk )
c     Solves the static system
      include 'scommons'
c
      real*8 stf(neq,iband), load(neq), wk(neq)
c
      iwidth=iband
      call rdstff(stf,iwidth)
      call rdload(load)
      write(*   ,*)'@@ STADYN  Reloaded  [K] & {P}  '
      write(ilog,*)'@@ STADYN  Reloaded  [K] & {P}  '
c
c     Decompose effective stiffness matrix
      ier1=0
      call  udu(stf,neq,iband,ier1)
      if (ier1.eq.0) then
          write(*,*)'ERROR: zero diagonal term'
          return
      endif
      write(*   ,*)'@@ UDU: mults & divs',ier1
      write(ilog,*)'@@ UDU: mults & divs',ier1
c
c     Solve for new displacements and save in case ...
      ier1=0
      call  bak(stf,load,neq,iband,wk,ier1)
      rewind(idis)
      write(idis) (wk(i), i= 1, neq  )
      return
      end
c
c
      SUBROUTINE UDU(a,neq,iband,imult)
c     solution of banded system using UDU decomposition
c     based on Weaver & Gere pp 469-472.
      real*8 a(neq,iband), temp,sum
c
      if (a(1,1).eq.0.0d0) then
        imult=0
        return
      endif
      if (neq .eq. 1) then
        imult=1
        return
      endif
c
      do 10 j=2,neq
        jm1=j-1
        j2=j-iband+1
        if (j2.lt.1) then
          j2=1
        endif
c
c       off-diagonal terms
        if (jm1.eq.1) then
          sum=a(j,1)
        else
          do 20 i=j2+1,jm1
            im1=i-1
            sum=a(i,j-i+1)
            do 21 k=j2,im1
              sum=sum-a(k,i-k+1)*a(k,j-k+1)
              imult=imult+1
21          continue
            a(i,j-i+1)=sum
20        continue
          sum=a(j,1)
        endif
c
c       diagonal terms
        do 30 k=j2,jm1
          temp=a(k,j-k+1)/a(k,1)
          sum=sum-temp*a(k,j-k+1)
          a(k,j-k+1)=temp
```

```
                  imult=imult+2                         real    xord(nnp), yord(nnp),zord(nnp)
30        continue                            c
          if (sum.eq.0.0d0) then              c         reload nodal displacements
              imult=0                                   rewind(idis)
              return                                    read(idis) (disp(i), i=1,neq  )
          endif                                         write(*,*)'@@ reloaded displacements'
          a(j,1)=sum                          c
10        continue                            c         fill in full displacement vector
c                                                       do 670 idof=1,nnp*6
      return                                               ieqnum = jbc(idof)
      end                                                  if (ieqnum .gt. 0) then
c                                                             dispp (idof) = disp (ieqnum)
      SUBROUTINE BAK(a,b,neq,iband,wk,imult)               else
          real*8 a(neq,iband),b(neq),wk(neq), sum             dispp (idof) = 0.0
c                                                          endif
c         forward substitutions              670        continue
          do 10 i=1,neq                                 write(*,*)'@@ reassigned displacements'
          j=i-iband+1                                   call memload (dispp,npi,npj,xord,yord,zord,nelt,ek)
          if (i.le.iband) then                c
              j=1                                 return
          endif                                 end
          sum=b(i)                            c
          km1=i-1                                 SUBROUTINE OUTDIS( dispp, str1)
          if (j.gt.km1) then                  c   Report the relative displacements of the structure
              wk(i)=sum                             include 'scommons'
          else                                c
              do 11 k=j,km1                         character*50 str1
                  sum=sum-a(k,i-k+1)*wk(k)           real*8    dispp(nnp*6)
                  imult=imult+1               c
11            continue                            xmax=0.0
              wk(i)=sum                           ymax=0.0
          endif                                   zmax=0.0
10        continue                                xrmax=0.0
c                                                 yrmax=0.0
c         middle terms                            zrmax=0.0
          do 30 i=1,neq                       c
          wk(i)=wk(i)/a(i,1)                  c         calculate maximum values for easy reference
              imult=imult+1                         write(iout ,1100) str1
30        continue                                  do 10 i=1, nnp*6, 6
c                                                       node = 1 + (i/6)
c         backward substitution                       val = dispp(i)
          do 50 i1=1,neq                              if (abs(val) .gt. abs(xmax)) then
          i=neq-i1+1                                     xmax= val
          j=i+iband-1                                    nxmax= node
          if (j.gt.neq) then                          endif
              j=neq                                   val = dispp(i+1)
          endif                                       if (abs(val) .gt. abs(ymax)) then
          sum=wk(i)                                      ymax= val
          k2=i+1                                         nymax= node
          if (k2.gt.j) then                           endif
              wk(i)=sum                               val = dispp(i+2)
          else                                        if (abs(val) .gt. abs(zmax)) then
              do 40 k=k2,j                               zmax= val
                  sum=sum-a(i,k-i+1)*wk(k)                nzmax= node
                  imult=imult+1                       endif
40            continue                                val = dispp(i+3)
              wk(i)=sum                               if (abs(val) .gt. abs(xrmax)) then
          endif                                          xrmax= val
50        continue                                       nxrmax= node
c                                                     endif
      return                                          val = dispp(i+4)
      end                                             if (abs(val) .gt. abs(yrmax)) then
cccc                                                     yrmax= val
                                                        nyrmax= node
                                                      endif
                                                      val = dispp(i+5)
OutPut                                                if (abs(val) .gt. abs(zrmax)) then
                                                         zrmax= val
cccc                                                     nzrmax= node
                                                      endif
      SUBROUTINE STATOUT (disp,dispp,npi,npj,          write(iout ,1110) node, (dispp(i+j),j=0,5)
     >                    xord,yord,zord, jbc,nelt,ek)  10    continue
c   Report the static results                   c
      include 'scommons'                        write(iout,1120)
c                                               write(iout,1130) nxmax,nymax,nzmax ,nxrmax,nyrmax,nzrmax
      integer npi(nel), npj(nel), nelt(nel), jbc(nnp*6)   write(iout,1140) xmax,ymax,zmax,xrmax,yrmax,zrmax
      real*8 disp(neq), dispp(nnp*6), ek(12,12)
```

```
c
c          FORMATS
1100 format(1x,a50,/,
    >' node     x-disp     y-disp     z-disp ',
    >'          x-rot      y-rot      z-rot' )
1110 format(1x,i4,2x,6(1pg11.3,1x))
1120 format(1x, ' maximum displacements')
1130 format(1x,'node',6x,6(i4,8x))
1140 format(1x,'value',2x,6(1pg11.3,1x))
c
     return
     end
c
     SUBROUTINE MEMLOAD(dispp,npi,npj,xord,yord,zord,nelt,ek)
c    Determine member loads referred to local S of M convent.
     include 'scommons'
c
     integer npi(nel), npj(nel), nelt(nel)
     real*8   dispp(nnp*6), delt(12), ek(12,12), nload(12)
     real xord(nnp), yord(nnp),zord(nnp), dx,dy,xl, l,m,n
     character*50 str1
c
     str1='Static displacements'
     call outdis(dispp,str1)
c
80   continue
     write(*,*)'CHOOSE: 0=return, -1=all members,',
    >                 ' # of single node'
     write(*,'(a\)')' --> '
     read(*,*) iprnode
     write(ilog,*) iprnode,'  ::print node -1=all'
     if (iprnode .eq. 0) return
c
c    determine nodal loads by {F}=[k]{u}
     do 90 i=1,nel
       i1=npi(i)
       j1=npj(i)
c      check if single node
       if ( iprnode .ne. -1) then
         if (iprnode .ne. i1) .and. (iprnode .ne. j1))
    >        goto 90
       endif
c
       dx = xord(j1) - xord(i1)
       dy = yord(j1) - yord(i1)
       dz = zord(j1) - zord(i1)
       xl=sqrt(dx*dx+dy*dy+dz*dz)
         l=dx/xl
         m=dy/xl
         n=dz/xl
c
       read(imat,922,rec=i)mat,e0,g0,a0,zix0,ziy0,ziz0,
    >                     rho0,beta0
922         format(1x,i4,1x,8(g10.4,1x))
       call elmstf(xl, zix0,ziy0,ziz0, a0,e0,g0, ek)
c      use trans3d to transform from local to global stifness
       call trans3d(l,m,n,ek,beta0)
c
       in1 = (i1-1)*6
       in2 = (j1-1)*6
       do 70 k=1,6
         delt( k)=dispp(in1+k)
         delt(6+k)=dispp(in2+k)
70     continue
c
c      multiply [ek]{delt}={nload}
       call matAbd( ek, delt, nload, 12)
       call trns3dv(nload,l,m,n,delt,beta0)
c
c      Write output to storage file
       if (nelt(i) .eq. 1) then
         write(iout ,1901) i, 'truss'
       else
         write(iout ,1901) i, 'beam '
       endif
       write(iout ,1902)
```

```
       write(iout ,1903) npi(i),(-delt(iii),iii=1,6)
       write(iout ,1903) npj(i),( delt(iii),iii=7,12)
       write(iout,*)' '
       write(iout,*)' loads in global coords'
       write(iout ,1909) npi(i),(nload(iii),iii=1,6)
       write(iout ,1909) npj(i),(nload(iii),iii=7,12)
90   continue
     goto 80
c
c          FORMATS
1901 format(/,' element number:',i4,' type: ',a5)
1902 format('        |           Member Loads          |',
    >       '        |           Member Moments         |',/,
    >       '   node |  axial     shearY     shearZ  |',
    >       '          ' torque    momentY    momentZ |')
1903 format(2x,i4,3x, 1pg10.3,1x, 1pg10.3,1x, 1pg10.3,3x,
    >                 1pg10.3,1x, 1pg10.3,1x, 1pg10.3)
1909 format(2x,i4,3x, 1pg10.3,1x, 1pg10.3,1x, 1pg10.3,3x,
    >                 1pg10.3,1x, 1pg10.3,1x, 1pg10.3)
c
     return
cccc end
c
cccc SUBROUTINE MATABD( a, b, c, n)
c    Multiply [A]{b}={c}
     real*8   a(n,n), b(n), c(n)
c
     do 10 i = 1,n
       c(i) = 0.0e0
       do 20 j = 1,n
         c(i) = c(i) + a(i,j) * b(j)
20     continue
10   continue
     return
     end
c
     SUBROUTINE TRNS3DV (gloads,l,m,n,lloads,beta)
c    Makes 3-D vector transformations.
     real*8 gloads(12),r(3,3),lloads(12),lload3(3),
    >       gload3(3)
     real m,n,l,d
c
     pi=4.0*atan(1.0)
     cb=cos(beta*pi/180)
     sb=sin(beta*pi/180)
     d=sqrt(1-n**2)
     if (d.lt.1e-3)then
       r(1,1)  = 0.0
       r(1,2)  = 0.0
       r(1,3)  = n
       r(2,1)  = -n*sb
       r(2,2)  = cb
       r(2,3)  = 0.0
       r(3,1)  = -n*cb
       r(3,2)  = -sb
       r(3,3)  = 0.0
     else
       r(1,1)  = l
       r(1,2)  = m
       r(1,3)  = n
       r(2,1)  = -(m*cb+l*n*sb)/d
       r(2,2)  = (l*cb-m*n*sb)/d
       r(2,3)  = d*sb
       r(3,1)  = (m*sb-l*n*cb)/d
       r(3,2)  = -(l*sb+m*n*cb)/d
       r(3,3)  = d*cb
     endif
c
     do 11 i=1,12,3
       gload3(1)=gloads(i)
       gload3(2)=gloads(i+1)
       gload3(3)=gloads(i+2)
       call matABd(r,gload3,lload3,3)
       lloads(i)=lload3(1)
       lloads(i+1)=lload3(2)
       lloads(i+2)=lload3(3)
```

```
11        continue                                      elseif (istore.eq.3) then
c                                                           do 84 nn=1,neq
      return                                      84        write(isnp) eigv(nn), (a(i,nn), i=1,neq)
      end                                             endif
c                                                     goto 82
c                                                  return
      SUBROUTINE EIGOUT(a, eigv, disp,jbc ,ivib )  end
c     Report the mode shapes for eigenvalue problems  cccc
      include 'scommons'
c                                          EIGENvalue Solver
      integer jbc(nnp*6)                       cccc
      character*50 str1
      real*8  a(neq,neq), eigv(neq), disp(nnp*6)     SUBROUTINE EIGEN(x,eigv,a,b,d,neqmas,ivib)
c                                                  c   Solve eigenvalue problem using JACOBI rotations
82        continue                                     include 'scommons'
      write(*,*)'CHOOSE modal storage:'           c
      write(*,*)'   0=return 1=mode shapes ',           real*8 a(neq,neq),b(neq,neqmas),eigv(neq),d(neq),
     >          '2=modal vectors 3=full <binary>'      >     x(neq,neq), rtol
      write(*,'(a\)')'  --> '                      c
      read(*,*) istore                                 rtol=1e-12
      write(*,*) istore                            c
      write(ilog,*)istore,'  ::istore 1=shape 2=vector',   ividth=neq
     >           '  3=matrix<bin>'                    call rdstff(x,ividth)
      if (istore.eq.0) then                        c   reassign to full form
          return                                       do 2 i=1,neq
      elseif (istore.eq.1) then                          jmax=min(iband,(neq  -i+1))
          write(*,*)'INPUT: # of modes to report'        do 21 j=1,jmax
          write(*,'(a\)')'  --> '                            a(i,i+j-1)=x(i,j)
c         neigen=intget(0,neq  )                   21          a(i+j-1,i)=x(i,j)
          read(*,*) neigen                         2    continue
          if (neigen .gt. neq) neigen=neq          c
          write(ilog,*) neigen,'  ::# of modes'        if (ivib .eq. 2) then
          write(*  ,*) neigen,'  ::# of modes '          call rdmass(x,neqmas)
          do 20 nn=1,neigen                        c      reassign to [b] matrix
c             Fill in  the displacement vector            if (ibandm.eq.1) then
          do 70 idof = 1, nnp*6                             do 31 i=1,neq
              ieqnum = jbc(idof)                               b(i,1) = x(i,1)
              if (ieqnum .gt. 0) then             31           continue
                  disp(idof) = a(ieqnum,nn)               else
              else                                          do 32 i=1,neq
                  disp(idof) = 0.0e0                            jmax=min(ibandm,(neq  -i+1))
              endif                                            do 33 j=1,jmax
70            continue                                             b(i,i+j-1)=x(i,j)
c                                              33                   b(i+j-1,i)=x(i,j)
c             Print the displacements.         32           continue
          if (ivib .eq. 2) then                           endif
              freq=abs(eigv(nn))                          write(*,  *)'@@ EIGN: reloaded [K] [M]   OK '
              freq=sqrt(freq)                             write(ilog,*)'@@ EIGN: reloaded [K] [M]   OK '
              write(*,23)'RESONANT FREQ:',freq,'rad/s'  c
              write(iout,*)' '                         elseif (ivib .eq. 1) then
              write(iout,23)'RESONANT FREQ:',freq,'rad/s'    ividth=neq
              str1= 'Mode shapes     '                       call rdgeom(x,ividth)
              call outdis( disp, str1)             c      reassign to [b]
          elseif (ivib .eq. 1) then                        do 4 i=1,neq
              freq=eigv(nn)                                    jmax=min(iband,(neq  -i+1))
              write(*,23)'BUCKLING LOAD:',freq                 do 42 j=1,jmax
              write(iout,*)' '                                     b(i,i+j-1)=x(i,j)
              write(iout,23)'BUCKLING LOAD:',freq   42               b(i+j-1,i)=x(i,j)
              str1= 'Mode shapes     '             4    continue
              call outdis( disp, str1)                    write(*,  *)'@@ EIGN: reloaded [K] [G]   OK '
          endif                                           write(ilog,*)'@@ EIGN: reloaded [K] [G]   OK '
23        format(1x,a,3x,1g13.5,1x,a)                     endif
20        continue                                 c
      elseif (istore.eq.2) then                        ibandm=neqmas
          write(*,*)'INPUT: # of vectors to report'     nsmax=15
          write(*,'(a\)')'  --> '                       call jacobi(a,b,eigv,d,neq,rtol,nsmax,ibandm,x,neqmas)
cc        neigen=intget(0,neq  )                        write(ilog,*)'@@ # of sweeps = ',nsmax
          read(*,*) neigen                              write(*  ,*)'@@ # of sweeps = ',nsmax
          if (neigen .gt. neq) neigen=neq               call eigsrt(eigv,x,neq)
          write(ilog,*) neigen,'  ::# of vectors'       rewind isnp
          write(*  ,*) neigen,'  ::# of vectors '       do 84 i=1,neq
          write(iout,*)' '                                  write(isnp) eigv(i),(x(j,i),j=1,neq)
          write(iout,*)'MODAL VECTORS:'            84       continue
          do 80 nn=1,neigen                        c
80            write(iout,81) eigv(nn), (a(i,nn), i=1,neq)
81            format(1x,7(g12.5,1x))
```

```
      return
      end
c
      SUBROUTINE EIGSRT(eigv,x,neq)
c     Sort the eigenvalues in ascending order
      implicit real*8(a-h,o-z)
      dimension eigv(neq),x(neq,neq)
c
      do 13 i=1,neq-1
      k=i
      p=eigv(i)
c     search for lowest value
      do 11 j=i+1,neq
      if (abs(eigv(j)) .lt. abs(p)) then
      k=j
      p=eigv(j)
      endif
11    continue
c
c     re-arrange vectors
      if (k.ne.i) then
      eigv(k)=eigv(i)
      eigv(i)=p
      do 12 j=1,neq
      p=x(j,i)
      x(j,i)=x(j,k)
      x(j,k)=p
12    continue
      endif
13    continue
c
      return
cccc  end
c
cccc
      SUBROUTINE JACOBI(a,b,eigv,d,n,rtol,nsmax,ibandm,
     >                  x,neqmas)
c     Based on Bathe pp 643-645.
      implicit real*8(a-h,o-z)
      dimension a(n,n),b(n,neqmas),eigv(n),d(n),x(n,n)
c
c     check for zero diagonal terms
      do 10 i=1,n
      if (ibandm.eq.1) then
      bii=b(i,1)
      else
      bii=b(i,i)
      endif
      biiabs=abs(bii)
      if (a(i,i).gt.0.0 .and. biiabs .gt. 0.0) goto 4
      write(*,2020)
      stop'ERROR: 1'
4     continue
      d(i)=a(i,i)/bii
      eigv(i)=d(i)
10    continue
c
c     initialize the modal matrix to the unit matrix
      do 30 i=1,n
      do 20 j=1,n
      x(i,j)=0.0
20    continue
      x(i,i)=1.0
30    continue
c
      if (n.eq.1) return
c
c     set sweep counter
      nsweep=0
      nr=n-1
40    continue
      nsweep=nsweep+1
c
c     check off-diags
      eps=(0.01**nsweep)**2
      iknt=0
      do 210 j=1,nr

      jj=j+1
      do 210 k=jj,n
c     check that off-diag term exceeds the threshold
      if (ibandm.eq.1) then
      eptola=(a(j,k)*a(j,k))/(a(j,j)*a(k,k))
      if (eptola.lt.eps) then
      goto 210
      endif
      akk=-b(k,1)*a(j,k)
      ajj=-b(j,1)*a(j,k)
      ab=a(j,j)*b(k,1)-a(k,k)*b(j,1)
      iknt=iknt+1
      else
      eptola=(a(j,k)*a(j,k))/(a(j,j)*a(k,k))
      eptolb=(b(j,k)*b(j,k))/(b(j,j)*b(k,k))
      if ((eptola.lt.eps) .and.
     >              (eptolb.lt.eps)) then
      goto 210
      endif
      akk=a(k,k)*b(j,k)-b(k,k)*a(j,k)
      ajj=a(j,j)*b(j,k)-b(j,j)*a(j,k)
      ab=a(j,j)*b(k,k)-a(k,k)*b(j,k)
      iknt=iknt+1
      endif
      radicl=(ab*ab+4.0*akk*ajj)/4.0
      if (radicl .lt. 0.0) then
      write(*,*)'case 2'
      write(*,2020)
      stop'ERROR 2'
      endif
c
      sqch=sqrt(radicl)
      d1=ab/2.0+sqch
      d2=ab/2.0-sqch
      den=d1
      if (abs(d2).gt.abs(d1)) den=d2
      if (den .eq. 0.0) then
      ca=0.0
      cg=-a(j,k)/a(k,k)
      else
      ca=akk/den
      cg=-ajj/den
      endif
c
c     do generalized rotation
      if (n.eq.2) goto 190
      jp1=j+1
      jm1=j-1
      kp1=k+1
      km1=k-1
      if (jm1.lt.0) goto 130
c     columns
      do 120 i=1,jm1
      aj=a(i,j)
      ak=a(i,k)
      a(i,j)=aj+cg*ak
      a(i,k)=ak+ca*aj
      if (ibandm.ne.1) then
      bj=b(i,j)
      bk=b(i,k)
      b(i,j)=bj+cg*bk
      b(i,k)=bk+ca*bj
      endif
120   continue
130   continue
      if (kp1.gt.n) goto 160
c     rows
      do 150 i=kp1,n
      aj=a(j,i)
      ak=a(k,i)
      a(j,i)=aj+cg*ak
      a(k,i)=ak+ca*aj
      if (ibandm.ne.1) then
      bj=b(j,i)
      bk=b(k,i)
      b(j,i)=bj+cg*bk
```

```
              b(k,i)=bk+ca*bj
            endif
150     continue
160     continue
        if (jp1.gt.km1) goto 190
c       mixture
        do 180 i=jp1,km1
          aj=a(j,i)
          ak=a(i,k)
          a(j,i)=aj+cg*ak
          a(i,k)=ak+ca*aj
          if (ibandm.ne.1) then
            bj=b(j,i)
            bk=b(i,k)
            b(j,i)=bj+cg*bk
            b(i,k)=bk+ca*bj
          endif
180     continue
190     continue
c       do diagonal terms
        ak=a(k,k)
        a(k,k)=ak+2.0*ca*a(j,k)+ca*ca*a(j,j)
        a(j,j)=a(j,j)+2.0*cg*a(j,k)+cg*cg*ak
c       force off-diagonal term to exact zero
        a(j,k)=0.0
        if (ibandm.eq.1) then
          bk=b(k,1)
          b(k,1)=bk+ca*ca*b(j,1)
          b(j,1)=b(j,1)+cg*cg*bk
        else
          bk=b(k,k)
          b(k,k)=bk+2.0*ca*b(j,k)+ca*ca*b(j,j)
          b(j,j)=b(j,j)+2.0*cg*b(j,k)+cg*cg*bk
c         force off-diagonal term to exact zero
          b(j,k)=0.0
        endif
c       update eigenvectors
        do 200 i=1,n
          xj=x(i,j)
          xk=x(i,k)
          x(i,j)=xj+cg*xk
          x(i,k)=xk+ca*xj
200     continue
210   continue
c     end sweep loop
c
c     update after each sweep
      do 220 i=1,n
        if (abs(a(i,i)).le.1e-12) a(i,i)=1e-12
        if (ibandm.eq.1) then
          bii=b(i,1)
        else
          bii=b(i,i)
        endif
        if (abs(bii).le.1e-20) bii  =1e-20
        eigv(i)=a(i,i)/bii
220   continue
c
c     check convergence
      do 240 i=1,n
ccccccccccccccc tol=rtol*d(i)+rtol
        tol=rtol*d(i)
        dif=abs(eigv(i)-d(i))
c                         tol=abs(tol)
        if (dif.gt.tol) goto 280
240   continue
c     check off-diagonals
      eps=rtol**2
      do 250 j=1,nr
        jj=j+1
        do 252 k=jj,n
          epsa=(a(j,k)*a(j,k))/(a(j,j)*a(k,k))
          if (ibandm.eq.1) then
            epsb=0.0
          else
            epsb=(b(j,k)*b(j,k))/(b(j,j)*b(k,k))
```

```
            endif
            epsa=abs(epsa)
            epsb=abs(epsb)
            if ((epsa.lt.eps) .and. (epsb.lt.eps))goto 252
c             need more iterating
              goto 280
252       continue
250     continue
c
c       scale eigenvectors
        do 270 j=1,n
          if (ibandm .eq. 1) then
            bbjj=  b(j,1)
          else
            bbjj=  b(j,j)
          endif
          bbjj=abs(bbjj)
          bb=sqrt(bbjj)
          do 272 k=1,n
            x(k,j)=x(k,j)/bb
272       continue
270     continue
c
c       converged  return
        nsmax=nsweep
        return
c
c       update d matrix for another round
280     continue
        znorm=0.0
        do 290 i=1,n
          diff = d(i)-eigv(i)
          d(i)=eigv(i)
c         znorm=amax1(znorm,abs(diff))
          znorm= max (znorm,abs(diff))
290     continue
        write(*,*)'@@ DIFF NORM:', znorm,' rotns =',iknt
        if (nsweep.lt.nsmax) goto 40
        return
c
2020    format( ' matrices not positive definite')
c
      return
      end
cccc
```

## TRANSient

```
cccc

        SUBROUTINE TRANSIENT(stf,mass,load,disp,vel,acc,fmag,
     >                  olddis,wk,damp,jbc,loadin,maxnode)
c     Transient analysis by NEWMARK time integration
        include 'scommons'
c
        integer inod(111,2), jbc(nnp*6)
        real*8 stf(neq,iband), mass(neq,ibandm), load(neq)
        real*8 vel(neq), acc(neq), fmag(neq), olddis(neq)
        real*8 disp(neq), wk(neq), damp(neq), aterm
c
        parameter( sigma=0.5e0, alpha=0.25e0)
        real   loadin(maxnode*3)
        ildin=maxnode*3
c
c       Get things ready for time integration loop
        write(*,*)' '
        write(*,*)'TYPE: time inc | # of incs | print count |'
        write(*,'(a\)')' --> '
        read (*,*) deltat,npt,iprcnt
        if (npt .gt. ildin) then
          write(*,*)'@@ time steps npt > load size ',
     >                      npt,' > ',ildin
          write(ilog,*)'@@ time steps npt > load size ',
     >                      npt,' > ',ildin
          npt=ildin-1
```

```fortran
        endif
        write(ilog,*) deltat,npt,iprcnt,' ::dt #pts print'
        startt=0.
        endt=real(npt)*deltat
c
c       Set integration constants for Newmark method
        a0 = 1.0/(alpha*deltat*deltat)
        a1 = sigma/(alpha*deltat)
        a2 = 1.0/(alpha*deltat)
        a3 = 1.0/(alpha*2.0) - 1.0
        a4 = sigma/alpha - 1.0
        a5 = (sigma/alpha - 2.0)*0.5*deltat
        a6 = (1.0 - sigma)*deltat
        a7 = sigma*deltat
c
c    READ  STIFFNESS, MASS & LOAD
        iwidth=iband
        call rdstff(stf,iwidth)
        iwidth=ibandm
        call rdmass(mass,iwidth)
        call rdload(fmag)
        write(*    ,*) '@@ Reloaded  [K]  [M] {P}   OK '
        write(ilog,*) '@@ Reloaded  [K]  [M] {P}   OK '
c
        write(ilog,*)'@@ damping coeffs: ', dampkk,dampmm
        if (dampkk .gt. 0.0 .or. dampmm .gt. 0.0) then
            idamp=1
        else
            idamp=0
        endif
c
        if (idamp.eq.1) then
            do 24 i= 1, neq
            damp(i) = dampkk*stf(i,1)+dampmm*mass(i,1)
     >                          + dampcc
24          continue
        endif
c
c    Form effective stiffness matrix
        if (ibandm .eq. 1) then
            do 20 i= 1, neq
20          stf(i, 1) = stf(i, 1) + a0*mass(i,1)
        else
            do 21 i= 1, neq
            do 22 j=1,iband
22              stf(i,j) = stf(i,j) + a0*mass(i,j)
21          continue
        endif
        if (idamp.eq.1) then
            do 25 i= 1, neq
            stf(i, 1) = stf(i, 1) + a1*damp(i)
25          continue
        endif
c
c    Decompose effective stiffness matrix
        ier1=0
        call udu(stf,neq,iband,ier1)
        if (ier1.eq.0) then
            write(*,*)'ERROR: zero diagonal term'
            return
        endif
        write(*    ,*)'@@ UDU: mults & divs',ier1
        write(ilog,*)'@@ UDU: mults & divs',ier1
c
c    Input load history from a file and interpolate
        call getload( npt,deltat, loadin,ildin,ilog)
c
        write(*,*)' '
        write(*,*)'CHOOSE nodal output:'
        nout=0
27      continue
        nout=nout+1
        write(*,*)' TYPE: node # | dof | rate <0 0 0 to end>'
        write(*,'(a\)')'   --> '
        read(*,*) inode,idof,irate
        jdof = (inode-1)*6 + idof

        inod(nout,1) = jbc(jdof)
        inod(nout,2) = irate
        write(ilog,*) inode,idof,irate,' :: node  dof rate'
        write(ilog,*)'@@ eqn # ',inod(nout,1)
        if (inode.ne.0) go to 27
        nout=nout-1
c
c  START TIME INTEGRATION LOOP
        write(*    ,*)'@@ Beginning transient analysis'
        write(ilog ,*)'@@ Beginning transient analysis'
c
c       Initialize   times, displacement, velocity, accln
        write(*,*)'@@ Initial disps, vels and accels set=0'
        do 620 i= 1, neq
            disp(i) = 0.0e0
            vel(i) = 0.0e0
            acc(i) = 0.0e0
            read(*,*) acc(i)
620     continue
        forc=loadin(1)
        call nodeout(time,forc,disp,vel,acc,neq,nout,inod,idyn)
c
        kount = 0
        write(*,*)' '
c  BIG TIME LOOP
        do 30 itime= 2,npt
            if (mod((itime-1),10) .eq. 0) then
                write(*,'('' + step: '',i4)') itime-1
            endif
            time = real(itime-1)*deltat
            kount = kount + 1
c
c       Save displacements
        do 40 i= 1, neq
            olddis(i) = disp(i)
40      continue
c
c       Fmag says where the load is applied
c       Done this way in case distributed load applied
        do 421 i= 1, neq
            load(i) = loadin(itime)*fmag(i)
421     continue
c
c       Form effective load vector
        if (idamp.eq.1) then
            do 26 i= 1, neq
            atermc = a1*disp(i)+a4*vel(i)+a5*acc(i)
            load(i) = load(i) + atermc*damp(i)
26          continue
        endif
c       the effective acceleration
        do 60 i= 1, neq
            aterm  = a0*disp(i) + a2*vel(i) + a3*acc(i)
            disp(i) = aterm
60      continue
        if (ibandm .eq. 1) then
            do 50 i= 1, neq
            load(i) = load(i) + disp(i)*mass( i,1)
50          continue
        else
            call abband( mass, disp, load,neq,ibandm )
        endif
c
c  SOLVE for new displacements: UDU already obtained
c                             : do back-substitution
        ier1=0
        call bak(stf,load,neq,iband,wk,ier1)
c
c       Obtain new velocities, accelerations
        do 90 i= 1, neq
            oldvel = vel(i)
            oldacc = acc(i)
            disp(i)= wk(i)
            acc(i) = a0*(disp(i) - olddis(i))
     >               -a2*oldvel   - a3*oldacc
            vel(i) = oldvel + a6*oldacc + a7*acc(i)
90      continue
```

```
c             Print out nodal results
c             if (iprcnt .eq. kount) then
                  kount = 0
                  forc=loadin(itime)
                  call nodeout(time,forc,disp,vel,acc,neq,
     >                              nout,inod,idyn)
               endif
c
 30         continue
c           BOTTOM of time integration loop
c
 122        format(1x,8(g12.5,1x))
 123        format(1x,i3,1x,6(g12.5,1x))
c
      return
      end
c
      SUBROUTINE NODEOUT(time,forc,disp,vel,acc,neq,nout,
     >                        inod,idyn)
         integer  inod(111,2)
         real*8   disp(neq), vel(neq), acc(neq)
         real     velout(111)
c
         do 72 n=1,nout
            iprnode=inod(n,1)
            if (inod(n,2) .eq. 0) then
               velout(n)=disp(iprnode)
            elseif (inod(n,2) .eq. 1) then
               velout(n)= vel(iprnode)
            elseif (inod(n,2) .eq. 2) then
               velout(n)= acc(iprnode)
            endif
 72         continue
         write(idyn,122)time,forc,(velout(n), n=1,nout)
 122        format(1x,8(g12.5,1x))
      return
      end
c
      SUBROUTINE ABBAND( matrix, vecin, vecout,neq,iband)
c     Multiplies  [ banded ]{vector} = {vector}
c
         real*8 vecout(neq),matrix(neq,iband),vecin(neq),val
c
         do 10 i= 1,   neq
            jlim=max(1,(i-iband+1))
            do 20 j= jlim,i
               val = vecin(j)
               vecout(i)=vecout(i)+val*matrix(j,(i-j+1))
 20            continue
            jlim=min(iband,(neq-i+1))
            do 30 j=2,jlim
               val = vecin(i+j-1)
               vecout(i) = vecout(i) + val*matrix(i,j)
 30            continue
 10         continue
      return
      end
c
      SUBROUTINE GETLOAD( npt,dt,loadin,nmax,ilog)
c     Gets applied load history by interpolation from input
         real     loadin(nmax)
         character*40 fyl2
         data iload/24/
c
c        read data file
         write(*,*)' '
         write(*,'(a\)')' TYPE: force filename --> '
         read(*,'(1a40)') fyl2
         write(ilog,*) fyl2,' ::filename'
c
         open(unit=iload,file=fyl2)
         rewind iload
c
c        adjust times and interpolate
c
```

```
         time = 0.0
         read(iload,*,end=240) t1,f1
         tzero=t1
         t1=t1-tzero
         loadin(1)=f1
c
         k = 1
         time = time+dt
         do 210 i = 2,4100
            t0=t1
            f0=f1
            read(iload,*,end=220) t1,f1
            t1=t1-tzero
 214        continue
            if (k.eq.npt) go to 220
            if (t1.gt.time) then
               xk4=(f1-f0)/(t1-t0)*(time-t0)+f0
               k = k+1
               time = time+dt
               loadin(k)=xk4
               go to 214
            endif
 210        continue
c
 220        continue
         write(ilog,*)'@@ # of force points read ', i-1
         write(*  ,*)'@@ # of force points read ', i-1
c        pad remainder with zeroes
         do 230 i = (k+1),npt
            loadin(i)=0.0
 230        continue
         close (iload)
         return
c
 240        continue
         write(ilog,*)'@@ NO DATA in load file'
c
      return
      end
cccc
```

# C.4 Source Code from MODDYN

Following is some FORTRAN-77 source code for the program MODDYN. The modules are for inverse vector iteration and subspace iteration.

## Inverse Vector Iteration

```
cccc
c
      SUBROUTINE HIGHLOW( stf, mass, load, lod0,wk, wka)
c     Finds highest and lowest resonances by vector itern
         implicit real*8 (a-h,o-z)
         include 'commons'
         integer  bmss,bstf,fdis,flod
         real*8 stf(neq,iband),mass(neq,ibandm),load(neq)
         real*8 lod0(neq),wk(neq),wka(neq,iband)
c
         write(*,*)'@@ reading <<stadyn.stf>>'
         rewind (bstf)
         read(bstf) neq,iband
         write(*  ,*)'@@ neq iband',neq,iband
         write(ilog,*)'@@ neq iband',neq,iband
         do 8 i=1,neq
            read(bstf) (stf(i,j), j=1,iband)
 8          continue
         write(*,*)'@@ reading <<stadyn.mas>>'
         rewind (bmss)
         read(bmss) neqm,ibandm
         write(*   ,*)'@@ neqm ibandm',neqm,ibandm
         write(ilog,*)'@@ neqm ibandm',neqm,ibandm
         if (ibandm.eq.1) then
            read(bmss) (mass(i,1), i=1,neqm)
```

```fortran
      else
        do 81 i=1,neqm                                    c
          read(bmss) (mass(i,j), j=1,ibandm)              c
81      continue                                     200
      endif                                               c
      write(*,*)'@@ [K] [M] reread   OK '                 c
c
c     make copy of stiffness
      do 82 i=1,neq
        do 82 j=1,iband                              221
82        wka(i,j)=stf(i,j)                               c
c
c     establish norm                                 690
      sumk=0.0
      summ=0.0
      do 70 i=1,neq
        sumk=sumk+abs(stf(i,1))
        summ=summ+abs(mass(i,1))
70    continue                                            c
      znorm=sumk/summ
      write(*,*)' '
      write(*   ,*)'@@ zlam norm: ',znorm                 c
      write(ilog,*)'@@ zlam norm: ',znorm                 c
c
10    continue
      write(*,*)'CHOOSE: 0=return 1=low 2=high 4=single'
      write(*,'(a\)')' --> '
      read(*,*) iopt
      write(*,*)' '
      write(ilog,*) iopt,'   ::1=low'                     c
c
      if (iopt .eq. 0) return
      if (iopt .eq. 1) goto 100
      if (iopt .eq. 2) goto 200
      if (iopt .eq. 4) goto 400
        goto 10
c
100   continue
c            LOWEST
      write(*,*)'LOWEST'
c     Update stiffness
      zlam=0.0
      call shift( stf, mass, wka, neq,iband,ibandm,zlam)
c
c     Decompose effective stiffness matrix
      ier1=0
      call udu(stf,neq,iband,ier1)
      if (ier1.eq.0) then
        write(*,*)'@@ ZERO diagonal term: try (-) shift'
        zlam=zlam-1.0e-1
        call shift( stf, mass, wka, neq,iband,ibandm,zlam) c
c       Decompose effective stiffness matrix              c
        ier2=0                                            c
        call udu(stf,neq,iband,ier2)
        if (ier2.eq.0) then
          write(*,*)'@@ ZERO diag term AGAIN: give up'  22
          return                                          c
        endif
      endif                                               c
c
c     ITERATE      INVERSE iteration                      c
c     initial vector
      do 20 i=1,neq                                       c
        lod0(i)=1.0
20    continue                                            c
c
      call veciter( stf, mass, load, lod0,wk, neq,iband,
     >               ibandm, rho)                         c
      zmin=rho+zlam
c     Print out results of interest                       c
      write(*  ,*)' LAMBDA= ', zmin,' OMEGA= ', sqrt(abs(zmin))
      write(*   ,622) ( wk(i), i=1,neq )             400
      write(iout,622) zmin,( wk(i), i=1,neq )            c
      write(*,*)'@@ '
622   format(1x,8(g13.6,1x))
c

        goto 10                                           c
c
200   continue
c            HIGHEST
c     first get crude estimat
      write(*,*)'HIGHEST'
      zlammax=znorm*2.0
      write(*,221)'@@ CRUDE estimate: ',zlammax
221   format(1x,a,3(1x,g14.6))                            c
c
      iultimate=0                                         c
690   continue
      iultimate=iultimate+1
      if (iultimate .gt. 100) then
        write(*,*)'DONE this too many times !!!'
        return
      endif                                               c
c     INVERSE iteration  Shift by eigenvalue estimate
      zlam=zlammax
      call shift( stf, mass, wka, neq,iband,ibandm,zlam)
c
c     Decompose effective stiffness matrix
      ier1=0
      call udu(stf,neq,iband,ier1)
      if (ier1.eq.0) then
        write(*,*)'@@ ZERO diagonal term: try (-) shift'
        zlam=zlam-1.0e-1
        call shift( stf, mass, wka, neq,iband,ibandm,zlam)
c       Decompose effective stiffness matrix
        ier2=0
        call udu(stf,neq,iband,ier2)
        if (ier2.eq.0) then
          write(*,*)'@@ ZERO diag term AGAIN: give up'
          return
        endif
      endif                                               c
c
      call countd(stf,neq, iband,iless1)
      if (iless1 .lt. neq-1) then                         c
        try higher value
        zlammax=zlammax*(neq-iless1+1)
        write(*,*)'@@N-1: NEW estimate of LAM: ',zlammax
        goto 690
      elseif (iless1 .lt. neq) then                       c
        try higher value
        zlammax=zlammax*2.0
        write(*,*)'@@N-0: NEW estimate of LAM: ',zlammax
        goto 690
      endif
c
c     ITERATE
c     initial vector
      do 22 i=1,neq
        lod0(i)=(-1.0)**i
22    continue                                            c
c
      call veciter( stf, mass, load, lod0,wk, neq,iband,
     >               ibandm, rho)
c
      zmax=zlammax+rho
      write(*,*)'@@ FINAL lam ',zmax
c     Print out results of interest
      write(*,*)' LAMBDA= ',zmax,' OMEGA= ',sqrt(abs(zmax))
      write(*   ,622)      (wk(i), i=1,neq)
      write(iout,622) zmax, (wk(i), i=1,neq)
      write(*,*)'@@ '
c
      goto 10
c
400   continue
c            SINGLE
      write(*,*)'INPUT: zlam'
      read(*,*) zlam
      call shift( stf, mass, wka, neq,iband,ibandm,zlam)
```

```
c
c        Decompose effective stiffness matrix
         ier1=0
         call  udu(stf,neq,iband,ier1)
         if (ier1.eq.0) then
             write(*,*)'@@ ZERO diagonal term: try (-) shift'
             zlam=zlam-1.0e-1
             call shift( stf,mass,wka, neq,iband,ibandm,zlam)
c            Decompose effective stiffness matrix
             ier2=0
             call udu(stf,neq,iband,ier2)
             if (ier2.eq.0) then
                 write(*,*)'@@ ZERO diag term AGAIN: give up'
                 return
             endif
         endif
c
         call countd(stf,neq   ,iband,iless1)
         write(*,*)'@@ lams < 0 ',iless1
c
c        ITERATE
c        initial vector
         do 420 i=1,neq
             lod0(i)=1.0
420      continue
c
         call veciter( stf, mass, load, lod0,wk, neq,iband,
     >                             ibandm, rho)
c
         zlam=zlam+rho
c        Print out results of interest
         write(*,*)' LAMBDA=',zlam,' OMEGA= ',sqrt(abs(zlam))
         write(*   ,622)      (wk(i), i=1,neq)
         write(iout,622) zlam, (wk(i), i=1,neq)
         write(*,*)'@@ '
c
         goto 10
c
      return
      end
c
      SUBROUTINE VECITER( stf, mass, load, lod0,wk,
     >                    neq,iband,ibandm,rho)
c     Uses vector iteration
c
         implicit real*8 (a-h,o-z)
         real*8 stf(neq,iband), mass(neq,ibandm) ,load(neq)
         real*8 lod0(neq ),wk(neq )
c
c        s is the number of significant digits in frequency
         s=5
         rtol=10**(-2*s)
c
         do 20 iter=1,20
c            Solve new vector by substitution. Put in wk(i):
             ier1=0
             call bak(stf,lod0,neq,iband,wk,ier1)
c
c            Form new load matrix by adding inertia terms
             if (ibandm.eq.1) then
                 do 30 i = 1, neq
                     load(i) = mass(i,1)*wk(i)
30               continue
             else
                 call bandAxd(mass, wk, load ,neq,iband )
             endif
c
             sum0=0.0
             sumz=0.0
             do 40 k=1,neq
                 sum0=sum0+wk(k)*lod0(k)
                 sumz=sumz+wk(k)*load(k)
40           continue
             rho0=rho
             rho=sum0/sumz
             sqrz=sqrt(sumz)
```

```
c
             do 42 i=1,neq
                 lod0(i)=load(i)/sqrz
42           continue
c
c            check convergence
             rtoltest=abs((rho-rho0)/rho)
             if (rtoltest .le. rtol) then
                 do 44 i=1,neq
                     wk(i)=wk(i)/sqrz
44               continue
                 return
             endif
20       continue
c
         write(*,*)'@@ WARNING !!!   did max iterations'
      return
      end
c
      SUBROUTINE SHIFT(stf,mass,wka,neq,iband,ibandm,zlam)
c     Shift the stiffness matrix by zlam
c
         implicit real*8 (a-h,o-z)
         real*8 stf(neq,iband),mass(neq,ibandm),wka(neq,iband)
c
c        Shift matrix by eigenvalue estimate
         if (ibandm.eq.1) then
             do 20 i = 1, neq
                 stf(i,1) = wka(i,1) - zlam*mass(i,1)
                 do 22 j=2,iband
                     stf(i,j) = wka(i,j)
22               continue
20           continue
         else
             do 24 i = 1, neq
                 do 26 j=1,iband
                     stf(i,j) = wka(i,j) - zlam*mass(i,j)
26               continue
24           continue
         endif
c
      return
      end
cccc
```

## Subspace Iteration

```
cccc
c
      SUBROUTINE SUBSPCE(stf,mass,load,wk,modemax,maxroot,
     >            ar,br,eigval,eigvec,dj,r,d)
c     Finds first m eigenvalues and vectors using Subspace
c     iteration, based on Bathe pp 685-689.
         implicit real*8 (a-h,o-z)
         include 'commons'
c
         integer  bmss,bstf,fdis,flod
         integer  icluster(100)
         real*8 stf(neq,iband), mass(neq,ibandm) ,load(neq)
         real*8 wk(neq),r(neq,maxroot),dj(maxroot),d(maxroot)
         real*8 ar(maxroot,maxroot),br(maxroot,maxroot)
         real*8 eigval(maxroot),eigvec(maxroot,maxroot)
c
         zlam=0.0
         iconv=0
         rtol=1.0D-8
         rtolj=1.0D-12
         itermax=16
         do i=1,maxroot
             d(i)=0.0
         enddo
c
c        STIFFNESS and MASS
         write(*   ,*)'@@ reading <<stadyn.stf>>'
         write(ilog,*)'@@ reading <<stadyn.stf>>'
```

```
        rewind (bstf)                                              write(ilog,*)'@@  test for rigid modes', wnorm
        read(bstf) neq,iband                                       write(*,*)' '
        write(*   ,*)'@@ neq iband',neq,iband                      write(*   ,*)'@@ !!! possible RIGID modes !!!'
        write(ilog,*)'@@ neq iband',neq,iband                      write(ilog,*)'@@ !!! possible RIGID modes !!!'
        do 8 i=1,neq                                               write(*,*)' '
           read(bstf) (stf(i,  j), j=1,iband)                      write(*,*)'CHOOSE: 0=ignore 1=shift'
8       continue                                                   write(*,'(a\)')' --> '
        write(*   ,*)'@@ reading <<stadyn.mas>>'                   read(*,*) ignore
        write(ilog,*)'@@ reading <<stadyn.mas>>'                   write(ilog,*) ignore,'  ::0=ignore'
        rewind (bmss)                                              if (ignore .gt. 0) then
        read(bmss) neqm,ibandm                                        rewind 99
        write(*   ,*)'@@ neqm ibandm',neqm,ibandm                     read(99) stf
        write(ilog,*)'@@ neqm ibandm',neqm,ibandm                     call zerodiag(stf,mass,neq,iband,ibandm,
        if (ibandm .eq.1) then                          >                      zlam,dzz,ilog,ier2)
           read(bmss) (mass(i,1), i=1,neqm)                           if (ier2.eq.0) return
        else                                                      endif
           do 81 i=1,neqm                                         goto 19
              read(bmss) (mass(i,  j), j=1,ibandm)             endif
81         continue                                        19  continue
        endif                                           c
        write(*,*)'@@ [K] [M] reloaded  OK'             c
c                                                       c   INITIAL load vector: first column is mass diag
c                                                           sum=0.0
c       establish norm                                     do 20 i=1,neq
        sumk=0.0                                              sum=sum+mass(i,1)*mass(i,1)
        summ=0.0                                     20     continue
        do 70 i=1,neq                                       sqmas=sqrt(sum)
           sumk=sumk+abs(stf(i,1))                          do 201 i=1,neq
           summ=summ+abs(mass(i,1))                            r(i,1)=mass(i,1)/sqmas
           load(i)=mass(i,1)/stf(i,1)               201    continue
70      continue                                            do 22 i=1,neq
        sumk=sumk/neq                                          do 24 j=2,maxroot
        summ=summ/neq                                             r(i,j)=0.0
        znorm=sumk/summ                              24        continue
        write(*   ,*)'@@ zlam norm: ',znorm          22     continue
        write(ilog,*)'@@ zlam norm: ',znorm          c
           dzz= (znorm/neq)*1.0e-6                   c       find largest of M/K
cX                                                           nd=neq/maxroot
c          NORMALIZE stiff and mass                         l=neq
           do i=1,neq                               c       helps distribution for equal values of M/K
              do j=1,iband                                  do 30 j=2,maxroot
                 stf(i,j)=stf(i,j)/sumk                        l=l-nd
              enddo                                           rt=0.0
           enddo                                              do 32 i=1,l
           do i=1,neq                                            if (load(i) .ge. rt) then
              if (ibandm .eq. 1) then                              rt=load(i)
                 mass(i,1)=mass(i,1)/summ                           iloc=i
              else                                                endif
                 do j=1,iband                         32        continue
                    mass(i,j)=mass(i,j)/summ                     do 34 i=1,neq
                 enddo                                              if (load(i) .gt. rt) then
              endif                                                   rt=load(i)
           enddo                                                      iloc=i
c                                                                  endif
c       make copy of stiffness                        34        continue
        open(unit=99,form='unformatted',status='scratch')          r(iloc,j)=1.0
        rewind 99                                                  load(iloc)=0.0
        write(99) stf                                 30     continue
c                                                       c
c       DECOMPOSE stiffness matrix                      c   RANDOM vector
        ier1=0                                              call ranvec(wk,neq,10)
        call  udu(stf,neq,iband,ier1)                       do 37 i=1,neq
        if (ier1.eq.0) then                                    r(i,maxroot)=wk(i)
           write(*,*)' '                               37  continue
           write(*,*)'@@ ZERO diagonal term: try shift'  c
           write(ilog,*)'@@ ZERO diagonal term: try shift'  c   ITERATIONs begin
           rewind 99                                        iter=0
           read(99) stf                              100  continue
           call zerodiag(stf,mass,neq,iband,ibandm,zlam,    iter=iter+1
>                       dzz,ilog,ier2)                      write(*,*)'ITERATION: ',iter
           if (ier2.eq.0) return                      c
        endif                                          c   SOLVE {u}k+1  &   simultaneously reduce K
c                                                           do 110 j=1,maxroot
c       CHECK if near RIGID body modes                         call  bak(stf,r(1,j) ,neq,iband,wk,ier1)
        do n=1,neq                                             do 130 i=j,maxroot
           wnorm=stf(n,1)/sumk                                    sum=0.0
           if (wnorm  .le. 10e-10) then
```

```
c       since [K]{u}k+1 = [r]k then only pre-mult
        do 140 k=1,neq
           sum=sum+wk(k)*r(k,i)
140     continue
        ar(i,j)=sum
        ar(j,i)=sum
130     continue
c
c       store the solution vectors for later use with mass
        do 150 i=1,neq
           r(i,j)=wk(i)
150     continue
110     continue
c
c       MULTIPLY {u}[M]{u} for reduced M
        do 160 j=1,maxroot
           call bandAxd(mass, r(1,j) ,wk,neq,ibandm)
           do 180 i=j,maxroot
              sum=0.0
              do 190 k=1,neq
                 sum=sum+r(k,i)*wk(k)
190           continue
              br(i,j)=sum
              br(j,i)=sum
180        continue
           if (iconv .gt. 0) goto 160
c
c          store [M]{u} as new R load vectors
           do 200 i=1,neq
              r(i,j)=wk(i)
200        continue
160     continue
c
c       SOLVE reduced eigenvalue prob by JACOBI rotations
        nsmax=15
        n=maxroot
        rtolj=1e-12
        call jacobi(ar,br,eigval,dj,n,rtolj,nsmax,n,eigvec,n)
        write(*    ,*)'@@ # of sweeps = ',nsmax
        call eigsrt(eigval,eigvec,n)
c
c       calculate R times eigenvectors for new loads
        do  420 i=1,neq
           do 422 j=1,maxroot
              dj(j)=r(i,j)
422        continue
           do 424 j=1,maxroot
              sum=0.0
              do 430 k=1,maxroot
                 sum=sum+dj(k)*eigvec(k,j)
430           continue
              r(i,j)=sum
424        continue
420     continue
c       enforce random last vector
        call ranvec(wk,neq,iter)
        do 1137 i=1,neq
           r(i,maxroot)=wk(i)
1137    continue
c
        if (iconv .gt. 0) goto 500
c
c       check for convergence
        do 380 i=1,modemax
           dif=abs(eigval(i)-d(i))
           rtolv=dif/eigval(i)
           if (rtolv .gt. rtol) then
              write(*,*)'TRIGGER rtolv: ',i
              write(*,*)'            ',rtolv,' vs ',rtol
              goto 400
           endif
380     continue
        write(ilog,*)'@@ NORMAL SUBspace convergence'
        write(*,*)'@@ NORMAL SUBspace convergence'
        write(*,*)'@@ another round'
        iconv=1
```

```
        goto 100
c
400     continue
        write(ilog,*)'@@ # of sweeps = ',nsmax,
     >              '      TRIGGER rtolv: ',i
c       see if max iterations have occurred
        if (iter .eq. itermax) then
           write(*,*)'@@ NO convergence at ITERns=',itermax
           write(*,*)'@@ another round'
           iconv=2
           goto 100
        else
           do 440 i=1,maxroot
              d(i)=eigval(i)
440        continue
           goto 100
        endif
c       ITERATIONs end
c
500     continue
c
c       CHECK quality
        isuspect=0
        rewind 99
        read(99) stf
        write(*   ,*)'@@ checking quality of eigenvectors'
        write(ilog,*)'@@ checking quality of eigenvectors'
        write(*,*)'@@ SUSPECTS:'
        do 580 i=1,maxroot
           eig=eigval(i)+zlam
           write(*,*)'@@ EIGENv = ',eig,i
           call bandAxd(stf, r(1,i) ,wk ,neq,iband)
           vnorm=0.0
           do 590 n=1,neq
              vnorm=vnorm+wk(n)*wk(n)
590        continue
           call bandAxd(mass, r(1,i) , load ,neq,ibandm)
           wnorm=0.0
           do 600 n=1,neq
              wk(n)=wk(n)-eig*load(n)
              wnorm=wnorm+wk(n)*wk(n)
600        continue
           vnorm=sqrt(abs(vnorm))
           wnorm=sqrt(abs(wnorm))
           if (vnorm .lt. 10e-12) vnorm=10e-12
           enorm=wnorm/vnorm
           dj(i)=0
           icluster(i)=i
           if (enorm .gt. .01) then
              isuspect=isuspect+1
              icluster(i)=0
              dj(i)=100
              write(*,'(a,i4,a\)') ' [',i,']'
           endif
           write(ilog,*)'@@ NORM error: ',i,enorm,icluster(i)
580     continue
        write(*,*)' '
        write(ilog,*)'@@ SUSPECT eigenvalues ',isuspect
        write(*   ,*)'@@ SUSPECT eigenvalues ',isuspect
c
c       FLAG spurious eigens and count others
        ibound=0
        do i=1,maxroot
           ic=icluster(i)
           if (ic .gt. 0) then
              ibound=ibound+1
              icluster(i)=ibound
           endif
        enddo
c
c       STURM SEQUENCE for total number
c       Decompose effective stiffness matrix
        eigup=1.01*eigval(modemax)
c
        inc=0
399     inc=inc+1
```

```
      eignex=0.99*eigval(modemax+inc)
      if (eigup .ge. eignex) goto 399
      write(*,*)'@@ SHIFT total: ',eignex
      maxeigs=icluster(modemax+inc-1)
c
      call shiftip( stf, mass, neq,iband,ibandm,eignex )
      ier1=0
      call udu(stf,neq,iband,ier1)
      if (ier1.eq.0) then
         write(*,*)'@@ ZERO diagonal term AGAIN: give up'
         return
      endif
      call countd(stf,neq,iband,itotal)
      write(ilog,*)'@@ should be',maxeigs,' actually',itotal
      write(*   ,*)'@@ should be',maxeigs,' actually',itotal
      if (itotal .gt. maxeigs) then
         write(*,*)'@@ MISSing eigenvalues'
      elseif (itotal .eq. maxeigs) then
         write(*,*)'@@ OK eigenvalues'
      elseif (itotal .lt. maxeigs) then
         write(*,*)'@@ SPURious eigenvalues'
      endif
c
      write(*,*)'@@ EIGENvalues & vectors in <<MODDYN.SNP>>'
      rewind(isnp)
      mdx=0
      do 510 i=1,modemax
         if (dj(i) .gt. 1.0) goto 510
         mdx=mdx+1
         ev=(eigval(i)+zlam)*sumk/summ
         sqrm=sqrt(summ)
         write(isnp) ev,(r(j,i)/sqrm, j=1,neq)
510   continue
      modemax=mdx
c
      return
      end
c
      SUBROUTINE SHIFTIP(stf,mass,neq,iband,ibandm,zlam)
c     SHIFT the stiffness matrix in place by zlam*mass
      implicit real*8 (a-h,o-z)
      real*8 stf(neq,iband), mass(neq,ibandm )
c
c     Shift matrix by eigenvalue estimate
      if (ibandm.eq.1) then
         do 20 i = 1, neq
            stf(i,1) = stf(i,1) - zlam*mass(i,1)
20       continue
      else
         do 24 i = 1, neq
            do 26 j=1,iband
               stf(i,j) = stf(i,j) - zlam*mass(i,j)
26          continue
24       continue
      endif
c
      return
      end
c
      SUBROUTINE ZERODIAG(stf, mass,  neq,iband,ibandm,zlam,
     >                    dzz,ilog,ier2)
      implicit real*8 (a-h,o-z)
      real*8 stf(neq ,iband ), mass(neq ,ibandm )
c
      write(*,*)'INPUT:  shift    [recommend > ',dzz,']'
      write(*,'(a\)')') --> '
      read(*,*) dzz
      write(ilog,*) dzz ,' ::shift '
      zlam=zlam-dzz
      call shiftip( stf, mass,  neq,iband,ibandm,zlam)
c     Decompose effective stiffness matrix
      ier2=0
      call udu(stf,neq,iband,ier2)
      if (ier2.eq.0) then
         write(*,*)'@@ ZERO diagonal term AGAIN: give up'
         return
```

```
      endif
      return
      end
c
      SUBROUTINE RANVEC(wk,neq,iseed)
      real*8 wk(neq)
c
cUNiX    use     ranv( )
      call msseed(iseed)
      sum=0.0
      do 36 i=1,neq
         call random(ranval)
         wk(i)=ranval-0.5
         sum=sum+wk(i)*wk(i)
36    continue
      enorm=sqrt(sum)
      do 37 i=1,neq
         wk(i)=wk(i)/enorm
37    continue
      return
      end
c
      SUBROUTINE COUNTD(a,neq,iband,iless)
c     Count negatives on UDU diagonal
      real*8 a(neq,iband)
      iless=0
      do 20 i=1,neq
         if (a(i,1) .lt. 0.0) iless=iless+1
20    continue
      return
      end
c
      SUBROUTINE BANDAXD( matrix, vecin, vecout,neq,iband )
c     Multiplies   [ banded ]{vector] = {vector}
      implicit real*8 (a-h,o-z)
      real*8 vecout(neq), matrix(neq,iband), vecin(neq)
c
      do 10 i= 1, neq
         sum=0.0
         do 20 j= max(1,i-iband+1),i-1
            ii = j
            jj = i-j+1
            sum = sum + matrix(ii,jj)*vecin(j)
20       continue
         do 30 j= i, min( i+iband-1, neq)
            ii = i
            jj = j-i+1
            sum = sum + matrix(ii,jj)*vecin(j)
30       continue
         vecout(i)=sum
10    continue
      return
      end
c
      subroutine msseed(iseed)
      integer*2 i2
      i2=iseed
      call seed(i2)
      return
      end
cccc
```

# References

[1] Allen, D.H., and Haisler, W.E., 1985, *Introduction to Aerospace Structural Analysis*, Wiley, New York.

[2] Argyris, J.H. and Kelsey, S, 1960, *Energy Theorems and Structural Analysis*, Butterworths, London.

[3] Balfour, J.A.D., 1986, *Computer Analysis of Structural Frameworks*, Nichols, New York.

[4] Bathe, K-J. and Ramaswamy, S., 1980, "An Accelerated Subspace Iteration Method," *Journal Computer Methods in Applied Mechanics and Engineering*, Vol 23, pp. 313-331.

[5] Bathe, K-J., 1982, *Finite Element Procedures in Engineering Analysis*, Prentice-Hall, New Jersey.

[6] Bishop,R.E.D., and Johnson D.C, 1960, *Mechanic of Vibrations* , Cambridge, London.

[7] Bishop,R.E.D., Gladwell, G.M.L., and Michaelson, S., 1965, *The Matrix Analysis of Vibrations* , Cambridge, London.

[8] Bland, D.R., 1960, *The Theory of Linear Viscoelasticity*, Pergamon Press, London.

[9] Brigham, E.O., 1973, *The Fast Fourier Transform*, Prentice-Hall, Englewood Cliffs, New Jersey.

[10] Chandrupatla, T.R. and Belegundu, A.D., 1991, *Introduction to Finite Elements in Engineering* Prentice–Hall, Englewood Cliffs, NJ.

[11] Clough, R.W. and Penzien, J., 1975, *Dynamics of Structures*, McGraw-Hill, New York.

[12] Cook, R.D., 1981, *Concepts and Applications of Finite Element Analysis*, Wiley, New York.

[13] Crandall, S.H., Dahl, N.C., and Lardner, T.J., 1972, *An Introduction to the Mechanics of Solids*, Wiley, New York.

[14] Doyle, J.F., 1989, *Wave Propagation in Structures*, Springer-Verlag, New York.

[15] Doyle, J.F., and Farris, T.N., 1991, "Analysis of Wave Motion in 3-D Structures using Spectral Methods," *Computational Aspects of Contact, Impact and Penetration*, Eds: R.F. Kulak and L.E. Schwer, Elmepress International, Switzerland. pp. 197-219.

[16] Eisley, J.G., 1989, *Mechanics of Elastic Structures*, Prentice-Hall, New Jersey.

[17] Ewins, D.J., 1984, *Modal Analysis: Theory and Practice*, Wiley, New York.

[18] Goldstine, H.H., Murray, F.J., and Von Neumann, J., 1959, "The Jacobi Method for Real Symmetric Matrices," *Journal of the Association for Computing Machinery*, Vol. 6, pp. 59-96.

[19] Hilderbrand, F.B., 1965, *Methods of Applied Mathematics*, Prentice-Hall, New Jersey.

[20] Hoff, N.J., 1956, *The Analysis of Structures*, Wiley, New York.

[21] Hughes, T.J.R., 1987, *The Finite Element Method*, Prentice-Hall, New Jersey.

[22] Hurty, W.C. and Rubinstein, M.F., 1964, *Dynamics of Structures*, Prentice-Hall, New Jersey.

[23] Humar, J.L., 1990, *Dynamics of Structures*, Prentice-Hall, New Jersey.

[24] Ibrahim, S.R., 1985, "Modal Identification Techniques: Assessment and Comparison," *Sound and Vibration*, August, pp. 10-15.

[25] James, M.L., Smith, G.M., Wolford, J.C. and Whaley, P.W., 1989, *Vibration of Mechanical and Structural Systems*, Harper and Row, New York.

[26] Kerr, A.D. and Accorsi, M.L., 1985, "Generalization of the Equations for Frame-Type Structures; a Variational Approach," *Acta Mechanica* Vol. 56, pp. 55-73.

[27] Langhaar, H.L., 1962, *Energy Methods in Applied Mechanics*, Wiley, New York.

[28] Lalanne, M., Berthier, P. and Der Hagopian, J., 1983, *Mechanical Vibrations for Engineers*, Wiley, New York.

[29] Lanczos, C., 1966, *The Variational Principles of Mechanics*, University of Toronto Press, Toronto.

[30] Martin, H.C., 1966, *Introduction to Matrix Methods of Structural Analysis*, McGraw-Hill, New York.

[31] Meirovitch, L., 1986, *Elements of Vibration Analysis*, McGraw-Hill.

[32] Melosh, R.J., 1990, *Structural Engineering Analysis by Finite Elements* , Prentice-Hall, New Jersey.

[33] Oden, J.T., 1967, *Mechanics of Elastic Structures*, McGraw-Hill, New York.

[34] Press, W.H., Flannery, B.P., Teukolsky, S.A., and Vetterling, W.T., 1986, *Numerical Recipes*, Cambridge University Press, Cambridge.

[35] Przemieniecki, J.S., 1986, *Theory of Matrix Structural Analysis*, Dover, New York.

[36] Richards, T.H. and Leung, Y.T., 1977, "An Accurate Method in Structural Vibration Analysis," *Journal of Sound & Vibration*, Vol. 55, pp. 363-376.

[37] Reiger, N.F., 1986, "The Relationship Between Finite Element Analysis and Modal Analysis," *Sound and Vibration* January, pp. 18-31.

[38] Sehmi, N.S., 1989, *Large Order Structural Eigenanalysis Techniques*, Ellis Horwood Limited Publisher, Chichester.

[39] Stasa, F.L., 1985, *Applied Finite Element Analysis for Engineers*, Holt, Rinehart and Winston, New York.

[40] Thomson, W.T., 1981, *Theory of Vibrations with Applications*, Prentice-Hall, New Jersey.

[41] Timoshenko, S.P. and Gere, J.M., 1988, *Theory of Elastic Stability*, McGraw-Hill.

[42] Tuma, J.J. and Cheng, F.Y., 1983, *Theory and Problems of Dynamic Structural Analysis*, Schaum's Outline Series, McGraw-Hill, New York.

[43] Warburton, G.B., 1976, *The Dynamical Behavior of Structures*, Pergamon Press, New York.

[44] Wells, D.A. 1967, *Theory and Problems of Lagrangian Dynamics*, Schaum's Outline Series, McGraw-Hill, New York

[45] Weaver, W. and Gere, J.M., 1980, *Matrix Analysis of Framed Structures*, Van Nostrand, New York.

[46] Weaver, W. and Johnston, P.R., 1984, *Finite Elements for Structural Analysis*, Prentice-Hall, New Jersey.

[47] Williams, F.W. and Wittrick, W.H., 1984, "Exact Buckling and Frequency Calculations Surveyed," *J. of Structural Engineering*, Vol. 110, pp. 169-187.

[48] Wilkinson, J.H., 1962, "Note on Quadratic Convergence of the Cyclic Jacobi Process," *Numerische Mathematik*, Vol. 4, pp. 296-300.

[49] Yang, T.Y., 1986, *Finite Element Structural Analysis*, Prentice-Hall.

[50] Zaveri, K., 1985, *Modal Analysis of Large Structures — Multiple Exciter Systems*, Bruel & Kjaer, Denmark.

[51] Ziegler, H., 1968, *Principles of Structural Stability*, Ginn and Company, Massachusetts.

# Index

"Those who have meditated on the beauty and utility of the general method of Lagrange — who have felt the power and dignity of that central dynamical theorem which he deduced from a combination of the principle of virtual velocities with the principle of D'Alembert — and who have appreciated the simplicity and harmony which he introduced by the idea of the variation of parameters, must feel the unfolding of a central idea.

Lagrange has perhaps done more than any other analyst to give extent and harmony to such deductive researches, by showing that the most varied consequences may be derived from one radical formula; the beauty of the method so suiting the dignity of the results, as to make of his great work a kind of scientific poem."

*W. R. HAMILTON*

# Mechanics

From 1990, books on the subject of *mechanics* will be published under two series.

## *SOLID* MECHANICS AND ITS APPLICATIONS

*Series Editor:* G.M.L. Gladwell

*Aims and Scope of the Series*

The fundamental questions arising in mechanics are: *Why?, How?,* and *How much?* The aim of this series is to provide lucid accounts written by authoritative researchers giving vision and insight in answering these questions on the subject of mechanics as it relates to solids. The scope of the series covers the entire spectrum of solid mechanics. Thus it includes the foundation of mechanics; variational formulations; computational mechanics; statics, kinematics and dynamics of rigid and elastic bodies; vibrations of solids and structures; dynamical systems and chaos; the theories of elasticity, plasticity and viscoelasticity; composite materials; rods, beams, shells and membranes; structural control and stability; soils, rocks and geomechanics; fracture; tribology; experimental mechanics; biomechanics and machine design.

1.  R.T. Haftka, Z. Gürdal and M.P. Kamat: *Elements of Structural Optimization*. 2nd rev.ed., 1990                                                                        ISBN 0-7923-0608-2
2.  J.J. Kalker: *Three-Dimensional Elastic Bodies in Rolling Contact*. 1990
                                                                            ISBN 0-7923-0712-7
3.  P. Karasudhi: *Foundations of Solid Mechanics*. 1991             ISBN 0-7923-0772-0

## *FLUID* MECHANICS AND ITS APPLICATIONS

*Series Editor:* R. Moreau

*Aims and Scope of the Series*

The purpose of this series is to focus on subjects in which fluid mechanics plays a fundamental role. As well as the more traditional applications of aeronautics, hydraulics, heat and mass transfer etc., books will be published dealing with topics which are currently in a state of rapid development, such as turbulence, suspensions and multiphase fluids, super and hypersonic flows and numerical modelling techniques. It is a widely held view that it is the interdisciplinary subjects that will receive intense scientific attention, bringing them to the forefront of technological advancement. Fluids have the ability to transport matter and its properties as well as transmit force, therefore fluid mechanics is a subject that is particularly open to cross fertilisation with other sciences and disciplines of engineering. The subject of fluid mechanics will be highly relevant in domains such as chemical, metallurgical, biological and ecological engineering. This series is particularly open to such new multidisciplinary domains.

1.  M. Lesieur: *Turbulence in Fluids*. 2nd rev. ed., 1990           ISBN 0-7923-0645-7
2.  O. Métais and M. Lesieur (eds.): *Turbulence and Coherent Structures*. 1991
                                                                            ISBN 0-7923-0646-5
3.  R. Moreau: *Magnetohydrodynamics*. 1990                          ISBN 0-7923-0937-5
4.  E. Coustols (ed.): *Turbulence Control by Passive Means*. 1990    ISBN 0-7923-1020-9

Kluwer Academic Publishers – Dordrecht / Boston / London

# Mechanics

From 1990, books on the subject of *mechanics* will be published under two series:
**FLUID MECHANICS AND ITS APPLICATIONS**
*Series Editor:* R.J. Moreau
**SOLID MECHANICS AND ITS APPLICATIONS**
*Series Editor:* G.M.L. Gladwell

Prior to 1990, the books listed below were published in the respective series indicated below.

## MECHANICS: DYNAMICAL SYSTEMS
Editors: L. Meirovitch and G.Æ. Oravas

## MECHANICS OF STRUCTURAL SYSTEMS
Editors: J.S. Przemieniecki and G.Æ. Oravas

# Mechanics

3. E.B. Magrab: *Vibrations of Elastic Structural Members.* 1979    ISBN 90-286-0207-0
4. R.T. Haftka and M.P. Kamat: *Elements of Structural Optimization.* 1985
   *Revised and enlarged edition see under* Solid Mechanics and Its Applications, Volume 1
5. J.R. Vinson and R.L. Sierakowski: *The Behavior of Structures Composed of Composite Materials.* 1986    ISBN Hb 90-247-3125-9; Pb 90-247-3578-5
6. B.E. Gatewood: *Virtual Principles in Aircraft Structures.* Volume 1: Analysis. 1989
   ISBN 90-247-3754-0
7. B.E. Gatewood: *Virtual Principles in Aircraft Structures.* Volume 2: Design, Plates, Finite Elements. 1989    ISBN 90-247-3755-9
   Set (Gatewood 1 + 2) ISBN 90-247-3753-2

## MECHANICS OF ELASTIC AND INELASTIC SOLIDS

Editors: S. Nemat-Nasser and G.Æ. Oravas

1. G.M.L. Gladwell: *Contact Problems in the Classical Theory of Elasticity.* 1980
   ISBN Hb 90-286-0440-5; Pb 90-286-0760-9
2. G. Wempner: *Mechanics of Solids with Applications to Thin Bodies.* 1981
   ISBN 90-286-0880-X
3. T. Mura: *Micromechanics of Defects in Solids.* 2nd revised edition, 1987
   ISBN 90-247-3343-X
4. R.G. Payton: *Elastic Wave Propagation in Transversely Isotropic Media.* 1983
   ISBN 90-247-2843-6
5. S. Nemat-Nasser, H. Abé and S. Hirakawa (eds.): *Hydraulic Fracturing and Geothermal Energy.* 1983    ISBN 90-247-2855-X
6. S. Nemat-Nasser, R.J. Asaro and G.A. Hegemier (eds.): *Theoretical Foundation for Large-scale Computations of Nonlinear Material Behavior.* 1984  ISBN 90-247-3092-9
7. N. Cristescu: *Rock Rheology.* 1988    ISBN 90-247-3660-9
8. G.I.N. Rozvany: *Structural Design via Optimality Criteria.* The Prager Approach to Structural Optimization. 1989    ISBN 90-247-3613-7

## MECHANICS OF SURFACE STRUCTURES

Editors: W.A. Nash and G.Æ. Oravas

1. P. Seide: *Small Elastic Deformations of Thin Shells.* 1975    ISBN 90-286-0064-7
2. V. Panc: *Theories of Elastic Plates.* 1975    ISBN 90-286-0104-X
3. J.L. Nowinski: *Theory of Thermoelasticity with Applications.* 1978
   ISBN 90-286-0457-X
4. S. Łukasiewicz: *Local Loads in Plates and Shells.* 1979    ISBN 90-286-0047-7
5. C. Firt: *Statics, Formfinding and Dynamics of Air-supported Membrane Structures.* 1983    ISBN 90-247-2672-7
6. Y. Kai-yuan (ed.): *Progress in Applied Mechanics.* The Chien Wei-zang Anniversary Volume. 1987    ISBN 90-247-3249-2
7. R. Negruţiu: *Elastic Analysis of Slab Structures.* 1987    ISBN 90-247-3367-7
8. J.R. Vinson: *The Behavior of Thin Walled Structures.* Beams, Plates, and Shells. 1988
   ISBN Hb 90-247-3663-3; Pb 90-247-3664-1

# Mechanics

## MECHANICS OF FLUIDS AND TRANSPORT PROCESSES
Editors: R.J. Moreau and G.Æ. Oravas

1. J. Happel and H. Brenner: *Low Reynolds Number Hydrodynamics*. With Special Applications to Particular Media. 1983     ISBN Hb 90-01-37115-9; Pb 90-247-2877-0
2. S. Zahorski: *Mechanics of Viscoelastic Fluids*. 1982     ISBN 90-247-2687-5
3. J.A. Sparenberg: *Elements of Hydrodynamics Propulsion*. 1984   ISBN 90-247-2871-1
4. B.K. Shivamoggi: *Theoretical Fluid Dynamics*. 1984     ISBN 90-247-2999-8
5. R. Timman, A.J. Hermans and G.C. Hsiao: *Water Waves and Ship Hydrodynamics*. An Introduction. 1985     ISBN 90-247-3218-2
6. M. Lesieur: *Turbulence in Fluids*. Stochastic and Numerical Modelling. 1987
     ISBN 90-247-3470-3
7. L.A. Lliboutry: *Very Slow Flows of Solids*. Basics of Modeling in Geodynamics and Glaciology. 1987     ISBN 90-247-3482-7
8. B.K. Shivamoggi: *Introduction to Nonlinear Fluid-Plasma Waves*. 1988
     ISBN 90-247-3662-5
9. V. Bojarevičs, Ya. Freibergs, E.I. Shilova and E.V. Shcherbinin: *Electrically Induced Vortical Flows*. 1989     ISBN 90-247-3712-5
10. J. Lielpeteris and R. Moreau (eds.): *Liquid Metal Magnetohydrodynamics*. 1989
     ISBN 0-7923-0344-X

## MECHANICS OF ELASTIC STABILITY
Editors: H. Leipholz and G.Æ. Oravas

1. H. Leipholz: *Theory of Elasticity*. 1974     ISBN 90-286-0193-7
2. L. Librescu: *Elastostatics and Kinetics of Aniosotropic and Heterogeneous Shell-type Structures*. 1975     ISBN 90-286-0035-3
3. C.L. Dym: *Stability Theory and Its Applications to Structural Mechanics*. 1974
     ISBN 90-286-0094-9
4. K. Huseyin: *Nonlinear Theory of Elastic Stability*. 1975     ISBN 90-286-0344-1
5. H. Leipholz: *Direct Variational Methods and Eigenvalue Problems in Engineering*. 1977     ISBN 90-286-0106-6
6. K. Huseyin: *Vibrations and Stability of Multiple Parameter Systems*. 1978
     ISBN 90-286-0136-8
7. H. Leipholz: *Stability of Elastic Systems*. 1980     ISBN 90-286-0050-7
8. V.V. Bolotin: *Random Vibrations of Elastic Systems*. 1984     ISBN 90-247-2981-5
9. D. Bushnell: *Computerized Buckling Analysis of Shells*. 1985     ISBN 90-247-3099-6
10. L.M. Kachanov: *Introduction to Continuum Damage Mechanics*. 1986
     ISBN 90-247-3319-7
11. H.H.E. Leipholz and M. Abdel-Rohman: *Control of Structures*. 1986
     ISBN 90-247-3321-9
12. H.E. Lindberg and A.L. Florence: *Dynamic Pulse Buckling*. Theory and Experiment. 1987     ISBN 90-247-3566-1
13. A. Gajewski and M. Zyczkowski: *Optimal Structural Design under Stability Constraints*. 1988     ISBN 90-247-3612-9

# Mechanics

# Mechanics

Kluwer Academic Publishers - Dordrecht / Boston / London